Cartridges and Firearm Identification

Cartridges and Firearm Identification

Robert E. Walker

CRC Press
Taylor & Francis Group
Boca Raton London New York

CRC Press is an imprint of the
Taylor & Francis Group, an **informa** business

CRC Press
Taylor & Francis Group
6000 Broken Sound Parkway NW, Suite 300
Boca Raton, FL 33487-2742

First issued in paperback 2021

© 2013 by Taylor & Francis Group, LLC
CRC Press is an imprint of Taylor & Francis Group, an Informa business

No claim to original U.S. Government works

Version Date: 2012907

ISBN 13: 978-0-367-77830-9 (pbk)
ISBN 13: 978-1-4665-0206-2 (hbk)

Library of Congress Cataloging-in-Publication Data

Walker, Robert E., 1973-
  Cartridges and firearm identification / Robert E. Walker.
  pages cm. -- (Advances in materials science and engineering)
  Includes bibliographical references and index.
  ISBN 978-1-4665-0206-2 (hardback)
  1. Firearms--Identification. 2. Cartridges--Identification. 3. Evidence, Expert. I. Title.

TS534.W34 2013
683.4--dc23   2012032950

Visit the Taylor & Francis Web site at
http://www.taylorandfrancis.com

and the CRC Press Web site at
http://www.crcpress.com

To my lovely wife Jayne and our adorable children Sam, Ava, and Max.

Contents

Foreword   xv
Preface   xvii
Acknowledgments   xix
About the Author   xxi

1 The World of Firearms and Ammunition   1

Introduction   1
Balancing Firearm Safety with Evidence Preservation   1
Crimes Involving Firearms and Ammunition   5
Firearm "Expert"? A Complex Question   8
    Exculpatory Evidence   10
    The *Frye* and *Daubert* Standards   11
    Gunsmiths versus Armorers   12
    Firearm Examiner   14
    Firearm Subject-Matter Expert   14
Basics: Understanding What Makes a Firearm Operate   15
    Trigger Travel   16
    Single-Action Firearms   17
    Double-Action Firearms   18
    Sequence of Operation   19
Methods of Operation   20
    Gas Impingement   20
    Gas-Piston Operation   21
    Recoil Operation   23
    Blowback Operation   23
    Recoil-Operated versus Blowback Firearms   25
    Striker-Fired Firearms, a Contemporary Trend 27
    Emergence of the Polymer-Framed Pistol   28
Manually Operated Firearms   30
    Breech-Loaded Firearms   30
    Lever-Action Firearms   31
    Bolt-Action Firearms   31
    Slide- or Pump-Action Firearms   32

2 Ammunition Cartridges   33

Ammunition: A Backdrop   33
Cartridge Identification by Dimensions, Names, and Identifiers   38
    Caliber   38

Hyphenations, Names, and Other Identifiers   40
Metric System   42
Cartridges Having Caliber and Metric Designations   43
Other Identifiers   43
Shot Shells   44
Components of Fixed Ammunition   46
  Cartridge Casing   46
  Materials Used in Construction of Cartridge Casings   49
    Brass   52
    Aluminum   53
    Steel   54
    Nickel Plating   54
    Shot Shells   54
    Paper   55
    Plastic   56
    Polymer-Cased Cartridges   56
    Caseless Ammunition   58
  Ignition Systems and Propellants   60
    Center-Fire Cartridges   61
    Berden Priming   61
    Boxer Priming   62
    Rim-Fire Cartridges   62
    Black Powder   62
    Smokeless Powder   63
    Lesmoke Powder   65
    Action Time   65
    Electric Priming   65
  Projectiles   66
    Commonly Used Abbreviations to Identify Projectile Styles   66
    Cannelures   67
    Ball Projectiles and Variations   67
    Controlled-Expansion Projectiles   72
    Open-Tip Match   75
    Wad Cutter and Semi-Wad Cutter   78
    Armor Piercing   78
    Tres Haute Vitesse (THV)   79
    Van Bruaene Rik (VBR)   79
    Frangible Projectiles   80
    Blended-Metal Projectiles   82
    Glaser Safety Slug   82
    Tracer Ammunition   83
    Explosive Projectiles   84
    Dummy, Drilled, and Inert   86
    Blank Cartridges   86
    Subsonic Cartridges   88
    Sabots   89
    Gyro Jet   90

        Shot Pellets   90
        Slugs   91
        Duplex Shot Shells   92
        Other Types of Shot Shell Loads   93
        Less-Lethal Munitions   93
        Training Ammunition   96
        Marking Cartridges   96
        Short-Range Training Ammunition   97
      Projectile Jacketing   98
        Metal Jacketing   99
        Metal Washes and Plating   99
        Nyclad   100
        Other Composite Jacketing Materials   100
        Nonjacketed   101
    Projectile Mass   101
    Projectile Material Composition   101
      Materials Selection   101
      Lead   102
      Solid-Metal Projectiles   102
      Shot Pellets   103
      Wooden Projectiles   103
    Hand Loads, Reloads, and Wildcats   104
    Caliber Interchangeability   106
    Identification Markings   110

3   Cartridge Head Stamps   115

  Cartridge Head Stamp Overview   115
    Head Stamp Elements   115
    Counterfeit Head Stamps   116
    Rim-Fire Head Stamps   117
    NATO Identification Marking   117
    U.S. Government Arsenal Markings   118
  Process of the Identification of a Cartridge, Cartridge Casing, or Projectile   118
  Head Stamp and Manufacturing Practices for Selected Nations   125
    Albania   125
    Argentina   125
    Australia   126
    Austria   127
    Bangladesh   127
    Belgium   127
    Bolivia   128
    Brazil   128
    Burkina Faso   128
    Burma   129
    Cambodia   129
    Cameroon   129

Canada   129
Chile   130
China   130
Colombia   131
Cuba   131
Denmark   132
Dominican Republic   132
Egypt   132
Ethiopia   133
Finland   133
France   133
Germany (Prewar, West Germany, Reunified Germany)   134
Greece   135
Guatemala   136
India   136
Indonesia   136
Iran   137
Ireland   137
Israel   137
Italy   138
Japan   138
Kenya   139
Lithuania   139
Malaysia   139
Mexico   140
Morocco   140
Netherlands   140
New Zealand   140
Nigeria   141
North Korea   141
Norway   141
Pakistan   141
Peru   142
The Philippines   142
Portugal   142
Saudi Arabia   143
Scotland   143
Singapore   143
South Africa   143
South Korea   144
Spain   144
Sudan   145
Sweden   145
Switzerland   145
Taiwan   146
Tanzania   146
Thailand   146

Turkey   146
Uganda   146
United Arab Emirates   147
United Kingdom   147
Venezuela   148
Zimbabwe   148
The Russian Federation, Soviet Union, Warsaw Pact, and Eastern Europe   148
Soviet Factory Codes   150
The Russian Federation   151
Federal State Enterprise Production's Amursk Cartridge Plant "Vympel"   152
Joint Stock Company's Barnaul Machine Tool Plant   152
Joint Stock Company Tula Cartridge Works   153
Klimovsk Specialized Ammunition Plant   154
Novosibirsk Low Voltage Equipment Plant   154
Ulyanovsk Machinery Plant   155
Wolf Performance Ammunition   155
Other Eastern European Sources   156
Bulgaria   156
Czechoslovakia (Czech Republic and the Slovak Republic)   157
East Germany   158
Hungary   159
Poland   159
Romania   160
Ukraine   160
Yugoslavia   161

4   Firearms   163

Introduction   163
United States Domestic Firearm Production for 2011   165
Firearm Import Trends   166
Contemporary Trends in the Federal Prosecution of Firearms Offenses   166
Firearm Classification Types   167
Handguns   167
Revolvers   168
Semiautomatic Pistols   178
Single-Shot Handguns   187
General Safety Issues   188
Hang Fires   188
Squib Loads   188
Slam Fire   188
Long Guns   189
Long-Gun Safety Systems   189
Rifles   189
Short-Barreled Rifles   196
Weapon Made from a Rifle   199
Shotguns   199

Drillings   203
Short-Barreled Shotguns   203
Weapon Made from a Shotgun   204
Destructive Devices   205
Suppressors, Silencers, Moderators, Mufflers, and Cans   206
Machine Guns   215
Improvised Firearms   221
Any Other Weapons   222
Antique Firearms   223
Air Guns   224
Ammunition-Feeding Devices   227
Magazines   227
Shared Magazine Architecture   229
Drum Magazines   229
Magazine Components   230
The Clip   231
Ammunition Belts   233
Internal Magazines   233
Firearm Alterations and Modifications   233
Finish, Refinishing, Colors   234
Firearm Chassis   235
Mechanical and Functional Modifications   236
Machine Gun Conversions   237
Field-Testing Procedure for Automatic Weapons   239
Closed Bolt-Action Field-Test Procedure   240
Open-Bolt-Action Procedure   240
Case Studies of Commonly Converted Firearms   240
MAC-Pattern Firearms   240
Heckler & Koch   242
AR-Pattern Firearms   244
Replacing Semiautomatic Parts with Machine Gun Parts   245
Drop-In Auto Sear   246
Lightning Link   247
Open-Bolt Modification   248
Indicia of a Converted AR Firearm   248
Receiver Markings   248
Intratec Pistols   250
Glock Pistols   251
SKS (Simonov Carbine Self-Loading)   252
AK-Pattern Firearms   253
M1 Carbine   254
Assorted Other Conversions   256
Colt 1911 and 1911A1   257
Browning Hi Power   257
Marlin/Glenfield Model 60   258
STEN Gun   258
Devices Designed to Simulate Fully Automatic Firing   259

Firearm Receiver   260
DEWATS, Unserviceable Firearms, and Inert Firearms   262
Manufacturers   262
Manufacture by Multiple Firms   265
Historical Footnotes on Older, Common Nameplates   265
    Iver Johnson   265
    Marlin Firearms   266
    Harrington & Richardson   266
    High Standard   266
    Savage Arms   267
    Firearms Imports & Exports   267
    RG Industries   268
    Hermann Weihrauch   268
Department Store Firearms   269
    Sears & Roebuck   269
    Montgomery Ward   271
    Western Auto Supply Company   271
    Kmart   271
    J.C. Penney   271
Firearm Designs Manufactured by Multiple Manufacturers   274
    Walther PP and PPK   274
    Model 1911 and 1911A1   276
    Beretta 92   278
    M14, M1A, and Variants   278
    Armalite Rifle   280
    Kalashnikov-Pattern Firearm   285
    Heckler & Koch G.3   291
    FN FAL and the FN FNC   293
Modular Weapon Systems Concept   294
Firearms Identification Methodology   299
    Identification of a Firearm from Photographic or Video Images   300

5   Firearm Markings   303

Serial Numbers   303
    Serial Number Structures and Practices   304
    Firearms Absent Serial Numbers   307
    Firearms Bearing Multiple, Different Serial Numbers   308
    Renumbered Firearms   308
    BATFE Serial Number Traces and E-Trace   309
    Methods Used to Apply Serial Numbers   310
    Obliteration, Alteration, or Defacing of Serial Numbers   311
    Methods Observed Used to Alter, Obliterate, or Deface Serial Numbers   312
        Scratching   313
        Gouging   313
        Grinding, Sanding, or Other Abrasives   314
        Drilling   314

        Punch   314
        Peening   314
        Filling   314
        Chemical   315
    Mechanical Serial Number Restoration Techniques   315
    Chemical Serial Number Restoration Techniques   316
    Nondestructive Serial Number Restoration Techniques   317
        Digital Imagery   317
        X-Ray   318
        Ultrasonic Cavitation   318
        Magnetic Particle Inspection Process or Magnetic Flux   318
        Variations of Temperature as a Restorative Technique   319
    Best Practices for Serial Number Restoration   319
        Photographic Considerations   320
        Frozen, Fixed, and Environmentally Exposed Firearms   320
    Model Designations   321
    Import Markings   323
    Other Markings   325
        National Crests   326
        Arsenal Markings   327
    Patent Dates, Legends, and Other Information   329
    Proof Marks   329
        U.S. Commercial Proof Marks   330
        Proof Marks Used around the World   330
        United States Military   331

References   337
Index   343

Foreword

At a time when crime scene television shows are all the rage among the civilian population, knowledge of forensics—especially firearms—is of paramount importance to the crime scene analyst, police detectives, and attorneys for both the prosecution and the defense alike. *Cartridges and Firearm Identification* encapsulates everything the firearms practitioner needs to know, whether he or she is an analyst, law enforcement officer, or attorney.

As a former prosecutor for both the state and federal governments, I can attest to the fact that knowledge of firearms and their inner workings is essential to any investigation of a firearms offense, as well as any litigation that follows. As a former special assistant with the U.S. Attorney's Office for the Middle District of Florida, I prosecuted many firearms offenses under President George W. Bush's Project Safe Neighborhoods initiative. As a prosecutor for the state of Florida, I prosecuted hundreds of firearms cases under Florida's 10-20-Life law, which set forth some of the country's most severe criminal penalties for the commission of a gun crime. Because firearms offenses have become more and more prevalent, it is essential that both prosecutors and defense attorneys know as much as possible about how firearms function to properly try a firearms case.

Cartridges and Firearm Identification not only provides anyone who reads it (whether he or she has little knowledge or extensive knowledge of firearms) with an easy-to-read, in-depth analysis of how firearms operate, but also the essentials of an often-overlooked area of information—ammunition.

Ammunition is important because, more often than not, shell casings are the main evidence at a firearms crime scene. More importantly, even though the intricacies of cartridge identification can be tedious, *Cartridges and Firearm Identification* simplifies this particular area enough so that even someone without any knowledge of firearms and ammunition can easily understand how to identify a cartridge and its origin. For a prosecutor and a detective, this can become another area of investigation. For a defense attorney, this can be an area of attack on the prosecution's case when the detective does not investigate such things. *Cartridges and Firearm Identification* will also assist both the analyst as well as the detective in preparing for their testimony at trial. If one knows all of the avenues of investigation, one would know all of the avenues of attack by defense counsel.

As firearms offenses become more and more prevalent, and the penalties for such offenses become more severe, juries will expect more from the analyst, the detectives in charge of the case, prosecutors seeking convictions, and defense attorneys seeking to place a reasonable doubt in the minds of those jurors. Therefore, *Cartridges and Firearm Identification* is an absolute must-have for any of the aforementioned criminal justice participants.

Jean-Paul Galasso, Esq.
Criminal trial lawyer
Fort Myers, Florida

Preface

Of the innumerable texts written about ammunition and firearms, few titles sought to sell themselves as a reference to meet the unique needs of the legal and law enforcement professional, criminalistics student, apprenticing firearm examiner, analyst, or regulatory professional. Traditionally, the publications have focused more toward collectors, historians, readers of general firearm interest, and especially those interested in specific firearms, i.e., Colt Single Action Army revolvers, Axis pistols, and the like. Ammunition texts have focused on loading and reloading datum for those who make their own ammunition, or have focused on cartridges from a historical perspective. Firearm- and shooting-centric periodicals center more on editorializing products, a mix of factual data blended with the writer's opinion and comparison/contrast with comparable products. Is an attorney or criminal investigator seeking clarity in a judicial controversy going to find answers in a book that covers machine guns used by the armed forces of the world? My own background and experience—as a firearms instructor, collector, researcher, shooter, and student—brings me a unique perspective to the topics of ammunition and firearms. Firearms evidence does not differ from any other form of physical evidence: It can exonerate as well as implicate; it proves or disproves that the event occurred; and it corroborates or disputes statements that have been made.

This text brings together a unique, multidisciplined approach to questions that arise regarding ammunition and firearms within the context of investigation. The sheer volume of laws and regulations pertaining specifically to firearms calls for the need for knowledgeable professionals to do the technical heavy lifting. The text does not address to significant depth the microscopic comparisons of the unique, individual striae of cartridge casings, projectiles, and candidate firearms. Unquestionably, these are critical analyses that are crucial to investigations involving the discharge of firearms, but the topics at hand go beyond the lab setting. Beyond the lab and the microscope are the questionable areas that arise within the context of an investigation involving firearms: obliterated serial numbers, unlawful modifications and alterations, contraband articles, identifying firearms and ammunition and the particulars about them, and understanding what evidence can be gleaned from firearms and ammunition within the perspective of that particular crime scene. Those whose profession it is to evaluate crime scenes have heard that "the crime scene is trying to tell you something; are you listening?" It can be equally said that "the firearm is trying to talk to you; are you listening?" This text is intended to be the translator. A knowledgeable investigator can extrapolate all sorts of valuable investigative intelligence and evidence strictly from the evidence present at the scene, long before the paperwork is even drawn up for transmittal to a lab. Such information may provide a substantial investigative lead or even a breakthrough.

The readers should not look to this text to guide them in making an informed firearm purchase. There is no "this is the best gun"; "this is the most effective ammunition ever made"; or "that particular brand is no good" commentary made. There are certainly plenty of sources out there that would eagerly opine in response to these statements. Certain brand names, and certain firearms and ammunition are referenced here, but do not look for specific recommendations.

This text goes beyond skin deep into a firearm. The reader will see that mere appearances can be relatively meaningless, and can often be misleading. This text introduces the reader to the standards of defining ammunition dimensions and the subtleties to these definitions. The reader is immersed with the various types of projectiles in existence, and the text challenges some of the commonly held dogmas. Perhaps most importantly, extensive research has gone into providing the reader with a global perspective on ammunition, as well as a cross section of manufacturers around the world and what their manufacturing and respective marking practices are.

The new entrants into these topics who aspire to achieve a level of expertise will greatly benefit from this text as one of the first they acquire on their journey forward. Any professionals in the administration of law and justice in society are offered a treasure trove of knowledge in this text; lawyers, judges, and investigators of all sorts will learn much here. I attempt in this book to bring you ever closer to the firearm without actually putting it in your hands.

Above all, go out and find your prey. Aspiring botanists will learn nothing by sitting in their study hoping that, by some good fortune, a rare species of plant will be carried in by the breeze. Students of ammunition and firearms go out and seek to acquire and expand their knowledge by finding specimens to study. Start off by looking at the overt features and characteristics: the ergonomics, the physical shape and size, the layout and location of the controls, the finish, and the materials that are used in the construction. Look at the study subject closer; pay attention to the markings and the locations of those markings. Note the practices used to affix the markings and which are unique to that example and those that are common with like firearms. Copious note taking should accompany any studies. When you are ready, field strip or disassemble some firearms. Schematics and diagrams are especially helpful references, telling you what a particular part is, what it looks like, and its function. The same applies to ammunition; take the time to just look at it whenever you get the chance. Pay attention to what the manufacturers are up to, follow the industry trends and latest developments. Using this text as your guide, a great deal of information can be learned and understood by simply handling a firearm or ammunition. The single greatest impediment to getting a more diverse understanding of firearms is that certain types, machine guns and such in particular, are heavily restricted and regulated, so the frequency of familiarizing oneself with such examples may be uncommon, and every opportunity to do so should be taken advantage of.

Finally, in your journey, be mindful of what you hear and from whom. Contemplate where their knowledge came from; not all cops are gun experts just because they carry one. There was a time when police officers, stereotypically speaking, all were gun experts by virtue of their occupation. Today, modern law enforcement officers are required to possess so many skills to do their job that technical firearms knowledge is not one of them; it suffices to simply train them in the use of their own duty arms. There are many self-professed "experts" out there ready to express their opinion, so don't buy into the hyperbole. Enjoy the journey. And no, not every Luger ever made was carried by a German officer, especially those chromed or plated "presentation examples."

Acknowledgments

Learning is a perpetual, lifelong pursuit. No one can possibly know everything as long as there is progress. For as long as there is progress, there will be something more to learn. I have been very fortunate to have known and learned from individuals who all had a common interest, that being ammunition and firearms, but who came into the conversation from different perspectives—tactical, historical, and technical—but all focused on the central theme of acquiring factual, credible, firsthand knowledge. Sadly, some of these individuals are no longer among us, and whatever knowledge they possessed that was not passed along is now lost and must be rediscovered. As long as I am engaged in this subject, I will consider myself a student, and I gratefully acknowledge these individuals from whose mentorship I have gained so much: the late Lieutenant Vincent L. Zarifian, the late John K. Little, Todd Salmon, Special Agent (retired) Randolph D. Eubanks, the U.S. Army CID, and the late Major Terry Branscome.

Furthermore, I owe a great debt of gratitude to those who gave me opportunities to learn and expand my horizons, which ultimately brought me to penning this work: David Lounsbury, PhD; Charley Mesloh, PhD, Florida Gulf Coast University; Ross Wolf, EdD, University of Central Florida; Senior Special Agent (retired) Daniel O'Kelly, Bureau of Alcohol, Tobacco, Firearms and Explosives; Senior Special Agent (retired) Ann Myers, Bureau of Alcohol, Tobacco, Firearms and Explosives; Harvey Kane, Brandon Pannone, L. Frank Thompson, King Brown, and Dawn Watkins.

Special thanks to Jarod Ford, Paul Shipley, AAI Corporation, and Christian Sahlberg for assistance with several of the photographs that appear in this text.

About the Author

Robert E. Walker began working in the field of crime scene investigation in 1995. In 1996, he entered into an apprenticeship under the tutelage of a noted firearm expert and first qualified to render expert testimony on a firearm case in 1999. Robert continues to work in crime scene investigation and remains engaged in firearm examination.

Robert earned his certification as a crime scene investigator from the International Association for Identification in 2010. He has been qualified as an expert in firearms, among other subjects, in the state of Florida's Twentieth Judicial Circuit and in the Federal Middle District of Florida. During his tenure, Robert has taught courses such as crime scene investigation and forensic photography at various colleges. He is an active police academy instructor, teaching investigative subject matter and also serving as a range officer. Robert routinely presents lectures related to firearms and shooting investigations to trade organizations, law enforcement agencies, and in the university setting.

Robert is a graduate of Barry University in Miami Shores, Florida, receiving a Bachelor of Public Administration. Robert earned his general instructor certification and firearm instructor certification from the Florida Criminal Justice Standards and Training Commission in 2003. During his career, he has completed a diverse series of courses in topics related to forensics and crime scene investigations. Robert has completed numerous armory courses and attended other firearms courses presented by various federal entities.

The World of Firearms and Ammunition

1

Introduction

Ammunition and firearms are broad, diverse, and dynamic topics. New products are regularly introduced to the marketplace; new companies appear on the landscape; and there is constant change. In addition, the uncommon, rarely seen older example comes out of the woodwork, occasionally challenging even the most knowledgeable person in the field so that, in effect, the students of these subjects must look in both directions to where the industry is at the moment and what has already transpired to round off their knowledge. Inevitably, the examiners who claim to have seen it all are taken to task when that which they did not know suddenly appears after they already thought they knew it all.

Ammunition and firearms are academic subjects unto themselves and can be viewed from any number of different viewpoints:

- Historians seeking to understand the implications and uses of firearms and how they affected past events
- Collectors of arms and ammunition seeking to expand their knowledge base about their particular interests and ascertain the value of the artifacts they hold
- Designers, engineers, and gunsmiths seeking to innovate and develop new products or improve an existing one
- Legislative and regulatory entities contemplating legislation or regulation with respect to firearms and ammunition
- Special-interest groups seeking to forward their respective agendas
- Law enforcement and investigative agencies focusing on the criminal aspects of the use of firearms and ammunition
- Marketers and users evaluating the suitability of firearms and ammunition for particular uses or markets: tactical, sporting, or otherwise

Balancing Firearm Safety with Evidence Preservation

The paramount consideration when handling firearms in any situation is safety. The devotion to safety overrides evidentiary and investigative concerns. Of course, the intent is always to avoid disturbing potential evidence, but a certain amount of handling is inevitable to ensure that an accident will not occur. In the case of firearms, the procedures needed to render them safe are, unfortunately, generally in the reverse order of those applied in preparation for safe use. Familiarity with the potential hazards can establish a comfort level that will allow the evidence collector to minimize the impact of safety considerations on the evidentiary value of the firearm.

In general, the rules for handling firearms are as follows:

1. Always treat a firearm as if it were loaded.
2. Always keep the muzzle of the firearm pointed in a safer direction.*
3. Never press or allow anything to contact the trigger unless you are ready to and intent on discharging the firearm.
4. Never point a firearm at anything that you do not intend to destroy or damage.
5. Be sure of what your target is and what lies beyond.
6. Understand the mechanical and operational characteristics of the firearm you are handling.
7. Always ensure that the proper ammunition is used.
8. Ensure that the barrel is clear of obstructions before firing.
9. Never rely on the mechanical safeties of the firearm to prevent it from discharging.
10. Always handle the firearm in a diligent, conscientious manner.

For the law enforcement/investigative professional or firearm examiner, there are additional rules to consider given the context of evidentiary concerns, agency policies, and the particular circumstances under which the contact with the firearm has occurred. When encountering a firearm in an investigative context, the following procedures should be observed, keeping the basic rules of firearms safety in mind:

1. Document the condition of the firearm *in situ*† by way of photographs and field notes.
   a. Does the firearm appear loaded or unloaded?
   b. Note the position of a manually selectable safety device, if the firearm is so equipped.
   c. Note the position of an external hammer, if the firearm is so equipped.
   d. Note the presence of a malfunction.
      Stovepipe: There is a spent casing stuck in the ejection port of a self-loading firearm. A stovepipe can have several root causes. From the perspective of a firearms instructor, the stovepipe indicates that the shooter is not maintaining a proper amount of grip pressure on the firearm to allow it to cycle, such as if the grip is relaxed while shooting or an insufficient amount of grip and wrist strength were applied during shooting, a condition often referred to as "limp wristing." From the perspective of an investigator, the stovepipe condition may be observed in cases of suspected self-inflicted gunshot wounds because the person's grip relaxed when the individual succumbed to the gunshot injury. It is important to note that the presence or absence of a stovepipe condition in such a scenario is not singularly indicative that the injury was self-inflicted, but rather it can be used in consider-

---

* Fired projectiles can travel through walls and ricochet unexpectedly. An absolutely safe direction may not exist, but a *safer* direction will minimize the risk of injury or damage should an accidental discharge occur.

† A Latin term meaning *literally*, or *in place*. Contextually, the term is used to describe when something is unaltered from its original condition. The term has gained acceptable use in other fields, such as crime scene photography, where the first series of scene photographs can be described as *in situ*.

ation of the totality of circumstances surrounding death or injury when an apparent suicide is being investigated.

Double feed: A combination of live cartridge and spent casing in the chamber that jammed the firearm. As in the case of a stovepipe, the double feed has several root causes, and can be a shooter-induced error, a mechanical malfunction on the part of the firearm, a lack of preventive firearm maintenance, a magazine-related failure, or the result of ammunition with a powder charge below specification.

Misfeed: A live cartridge did not properly load, causing the action to jam. Misfeeds are generally attributed to a firearm-related failure, such as a mechanical malfunction or maintenance issue; however, magazine-related issues cannot be ruled out, particularly a weak magazine spring. Misfeeds may also be the result of improperly sized ammunition being loaded.

Misfire: A live cartridge failed to detonate in the chamber. A misfire can be caused by mechanical failure in the firearm, such as the firing pin or striker failing to contact the primer of the loaded cartridge, or the strike being weak and thus having insufficient pressure to cause detonation. A slight impression in the primer of the loaded cartridge will reveal a weak firing-pin strike. It may also be possible that the firing-pin surface is sufficiently worn to be incapable of operating as designed. Ammunition is the other possible cause, such as a bad primer, a primer that is harder than normal specification, a dud cartridge (no powder charge), or ammunition that has somehow been rendered completely inert, such as by exposure to the elements.

   e. If the firearm has a detachable ammunition magazine, is a magazine inserted?

   f. Note the presence of blood or other biological material that presents a biohazard to the handler.

2. When adequate documentation is made, pick up the firearm; keep away from the trigger and note additional observations as necessary.

   a. When handling handguns, it is often easiest to grasp the right and left sides of the grips.

   b. With long guns (rifles and shotguns), handling is often easiest by the edge of the stock and by the fore grip or hand guard forward of the trigger.

   c. If the action can be opened, do so slowly and carefully so as not to eject a chambered cartridge. If there is a cartridge in the chamber, take additional photographs with the action opened slightly to depict the presence of the loaded cartridge. When documentation is done, slowly close the action.

3. Identify and isolate the ammunition source. Determine where on the firearm the ammunition source is located. When the ammunition source has been identified, remove or unload the ammunition source.

   a. Remove the detachable magazine using the magazine release button or lever.

   b. Open the action of a breech-loaded firearm using the release lever, button, or knob.

   c. Open the cylinder of a revolver using cylinder-release mechanism, or remove the cylinder.

   d. Remove the magazine tube assembly from a tubular magazine and remove cartridges.

e. Open the action of a belt-fed weapon and remove the belt.
f. Slowly pull the slide back on a slide- or pump-action firearm. A loaded cartridge will reveal itself from the chamber. This should be done gradually, and the working of the action should be short, or not completely open. This process may require manipulation of an action bar release knob or lever to allow the action to open. If the firearm does not have separate loading and ejection ports, it may only be possible to slowly work the action to eject any loaded ammunition from the magazine.
g. If the firearm is a bolt action, open or remove the floor plate from the internal magazine; caution should be exercised because the loaded cartridges will spill out. Otherwise, once the ammunition source has been identified and isolated from the action, open the bolt.

4. When the ammunition feed source has been isolated and removed, unload the chamber and leave the action open. Inspect the chamber visually and by feel, using a flashlight if necessary.
5. It may be required that the person clearing the firearm then band the weapon with a wire tie to visually confirm that the firearm is unloaded.

It is recommended, if possible, that two persons participate in rendering a firearm safe, even if one or both are fluent with firearms. One person handles the firearm while the second person can aid by having a camera handy to photograph the presence of the cartridge or casing in the chamber, which can be instrumental in certain types of investigations and recoveries. Figure 1.1 depicts an example of such an image; the action was slowly opened to reveal that a live cartridge was chambered, although no magazine was inserted in the weapon at the time it was inspected. In Figure 1.2, there is a ring of burnt powder surrounding the charge hole in the 12 o'clock position; however, the other charge holes do not indicate this same appearance, suggesting that only a single round had been discharged. Investigators should refrain from excessive handling or "dry" (unloaded) operations to the firearm (pressing the trigger, cycling the action, etc.) during the recovery phase. Consideration must also be given for the presence of latent fingerprints or potential DNA sampling, should the circumstances of the case call for it. The investigator is reminded that the recovery and documentation process should be treated as a

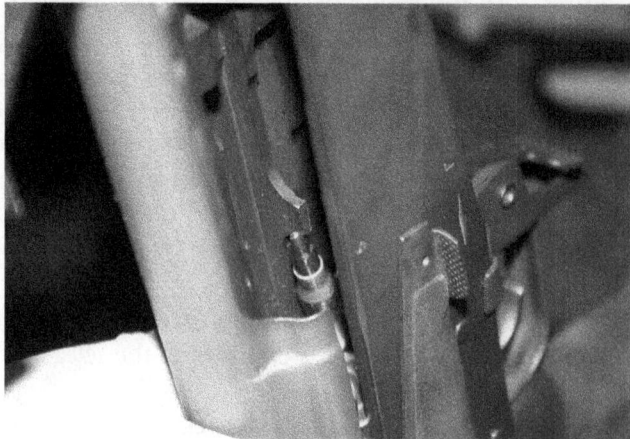

Figure 1.1 (See color insert.) A semiautomatic rifle action is opened, revealing the presence of a live cartridge in the chamber. Such a find can be of great investigative interest. (Image from author's collection.)

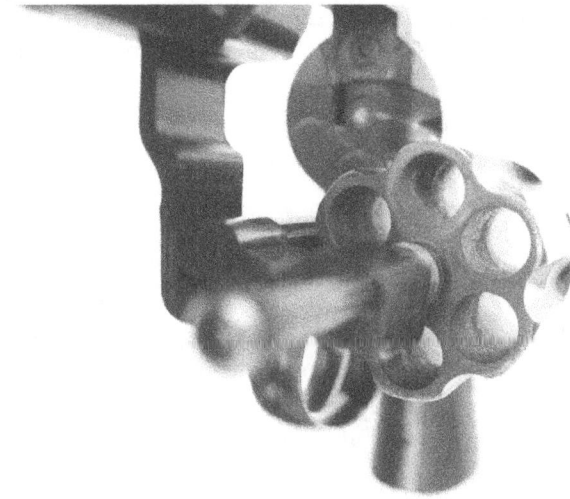

Figure 1.2 A close-up photograph of the front side of a revolver cylinder. Note the burnt powder ring around the charge hole at the twelve o'clock position. (Image from author's collection.)

"one time shot," and evidence not collected at the time may not be available later if the need presents itself. It is inevitable that the evidence may be compromised to a certain extent due to the necessities of collecting and rendering the arm safe, and this must be factored into the equation as the cost of doing business. Investigators making the collection under more controlled circumstances can exercise more diligent handling. The first responding officers who must take control of a loose weapon in the interest of their own safety are not generally afforded this luxury.

Crimes Involving Firearms and Ammunition

The value and accuracy of capturing and assessing accurate firearm and cartridge evidence early in the investigation cannot be understated. This statement must be balanced with other investigative, safety, and evidentiary concerns, as the condition of a firearm *in situ* is subject to change very quickly in the course of a criminal inquiry, so as many details as possible must be captured. In the criminal context, firearm-related crimes do not necessarily include the actual discharge of a firearm or that anyone was injured. Broadly, firearm-related crimes can be broken down into several categories, each with a subtly different evidentiary concern that supports the element of the crime:

1. Possession of a firearm or ammunition by a prohibited person:
   a. Unlicensed possession in venues were licensing or permits are required
   b. Possession by convicted felons whose civil rights have not been restored
   c. Persons adjudicated mentally incompetent
   d. Possession by a person who is an unlawful user of narcotics or who habitually uses narcotics
   e. Possession by illegal aliens
   f. Possession by underage persons
   g. Possession by a fugitive from justice

   h. Possession by a person who has renounced U.S. citizenship
   i. Possession by a person discharged from the armed forces under conditions
       other than honorable
   j. Possession by persons otherwise prohibited by statutory objection
   k. Possession in sterile areas such as airports, public buildings, and other similar
       places
2. Possession of an unlawful or contraband firearm:
   a. A firearm deemed unlawful by its present configuration
   b. Unregistered automatic weapons either as manufactured, converted, or reman-
       ufactured into such
   c. Unregistered firearm suppressors or silencers
   d. Unregistered improvised weapons
   e. A firearm with an obliterated, defaced, or altered serial number
3. Use of a firearm as an instrument in crime:
   a. Homicide
   b. Assault and battery
   c. Robbery
   d. Poaching
   e. Reckless or careless handling or display
   f. Storing of a firearm in a manner readily accessible by minors
   g. Situations involving culpable negligence
4. Ammunition offenses:
   a. Possession by prohibited persons
   b. Contraband ammunition
   c. Possession of ammunition in a controlled or sterile environment
   d. Possession of ammunition in a criminal context, such as in connection with
       drug trafficking

In each instance, evidentiary concerns are subtly different and must be considered on
a case-by-case basis. The focus of the investigation from the standpoint of the technical
examination is either to satisfy the elements of the crime under investigation and assist in
establishing the requisite burden of proof, or to contradict investigative presumptions. In
an investigation where a prohibited person is in possession of a firearm and/or ammunition, the focus of the evidence lies in merely demonstrating that the possessed article is, in
fact, a firearm or ammunition by the applicable legal definition of such. The firearm examiner or subject-matter expert must be granted access to the evidence and render an opinion
of whether the article meets the applicable definition or not.

If a firearm is suspected of being a contraband firearm, the examiner must be able
to ascertain that the exhibit is in a configuration that does define it as being unlawful or
contraband. If, in the opinion of the examiner, the exhibit in question is one that is contraband, the examiner should be prepared to fully explain the rationale for this opinion as
well as the technical reasons that justify the opinion. These observations should specifically
outline the observed modifications or alterations. Modifications can take several forms to
include modification of the firearm itself, replacing factory specification components with
nonstandard ones, or the modification of existing components. Hence, the examiner is
able to recognize modifications, replacements, or configurations and call them out specifically to support the assertions. The examiner also understands what effects were desired

by the modifications or alterations by understanding the purpose and function of the components in question, whether the desired results were realized or not. Configuration questions ordinarily focus on such items as the overall length or the barrel length of a long gun; if the article is a machine gun by manufacture, conversion, or remanufacture; if the article has some other form of alteration that renders it contraband; or if a device is defined as a firearm suppressor. Above all, examiners must be fair, honest, and ethical in their rendering of the opinion, even if the opinion does not coincide with the lay interpretation of the evidence. This may put the examiner at odds with others, but it must be understood that the opinions of others, regardless of their credentials, are not applicable in the court of law; the only opinion that matters is that which the examiner renders and goes on the record to support.

A number of years ago, an attorney in private criminal defense practice sought out the author in a case. His client had been arrested for various charges, including possession of a short-barreled rifle. In the venue where the crime was alleged to have occurred, as well as under U.S. Code, the minimum barrel length for a rifle is 16 inches. The seized firearm was stored in evidence by the investigating law enforcement agency, which permitted the article to be viewed by the defense as part of the discovery process. When the author viewed the exhibit, the firearm was identified as having been manufactured by Universal Firearms, a firearm manufacturer then based in Hialeah, Florida. The exhibit was the Enforcer model, caliber .30 M1 carbine, equipped with an 11-inch barrel, short wooden pistol grip, no buttstock, and perforated metal upper hand guard. The Enforcer model was a postwar commercially manufactured copy of the World War II–era U.S. M1 carbine. What differed about the Enforcer was that it was manufactured as a handgun, not as a rifle; thus neither the minimum barrel length requirement of 16 inches nor the overall minimum length of 26 inches applied. When seized, the supposition by officers was that the firearm was a cut-down M1 carbine—not an unreasonable assertion to make. Given the author's opinion, the firearm charge was dropped, as no violation of law had in fact occurred with respect to the configuration of that particular firearm. In another instance, the author was asked to examine a shotgun with an obliterated serial number. The shotgun had been seized during a criminal investigation and was being held as evidence. The author obtained the shotgun for review and determined that the serial number had not been obliterated from the shotgun. The shotgun in question, a department store brand dating back to the 1950s, never had a factory serial number affixed, as it predated the serial number requirement enacted in the Gun Control Act* of 1968. As such, the criminal case was dismissed, as, once again, no violation of law had taken place.

In cases of murder, assault, battery, robbery, and other violent crimes where a firearm is brandished or used as an instrument, the evidence lies not only in that the article is a firearm, but it opens the dimension of the comparative forensics aspect. The comparative ballistics aspect seeks to establish the presence or lack of a relationship between candidate firearm/s and cartridge casings and ammunition that may have cycled through or been fired by the candidate firearm/s. Ballistic comparisons focus on the markings that occur because of the interaction of the firearm with the ammunition. Cartridge casings may bear a firing-pin impression, breech face markings, scratches from the magazine-fed lip if cycled from a magazine, feed-ramp marks, and ejector or extraction marks. The fired

---

* Title 18, Chapter 44, United States Code.

projectiles may have sufficient transfer of the barrel profile impressed onto them to effect a comparison. Even without a firearm, casings and projectiles suspected of being fired from the same firearm can be compared to one another to form a nexus between the events should a match be established by examination and formation of opinion by a credentialed examiner. Ballistic comparisons are separate from firearm examinations in that not every firearm case requires that comparisons of fired projectiles, spent casings, and suspect firearms be made, rather just that the submitted exhibit be evaluated and defined.

In crimes where gunfire has occurred, it may be of benefit to bring the firearm subject-matter expert into the scene investigation early on to assist in assimilating the available firearm evidence into the overall investigation. Since the behavior of firearms is often in question within the scope of certain investigations, the firearm subject-matter expert may be able to interpret the scene and make certain determinations that might otherwise be overlooked. If possible, it is always advisable to conduct such inquiry at the scene itself, with minimal scene disturbance, as reconstructive efforts *ex post facto* may result in incomplete analysis due to gaps in documentation or a loss of context when removed from the original scene. Shooting-scene reconstruction is a discipline all unto itself, but this reconstruction can only be enhanced by bringing in a firearm subject-matter expert who can explain firearm "behaviorisms" within the context of the broader investigation.

As with firearms cases, criminal cases where the ammunition is the evidence can be prosecuted. Persons prohibited from possessing firearms are likely prohibited from possessing ammunition as well. In court, a subject-matter expert would be consulted to determine if the suspected exhibits are ammunition, as well as any particulars about the ammunition. From the investigative standpoint, the expert should be consulted about identification of ammunition that may serve as good intelligence during the investigative process. Certain examples of ammunition or the firearms chambered for that particular cartridge are rare or unique, and may prove to be valuable investigative intelligence to possess.

Firearm "Expert"? A Complex Question

What defines a firearm expert? The topic of firearms is so broad that the term is somewhat elusive. Persons with in-depth knowledge of the subject have tended to follow firearms with some degree of specificity, including interest in specific makes and models and firearms of a particular era. The person may have a specific interest in sport and competition shooting, law enforcement and military hardware, hunting, or purely as a collector. From a pure forensic standpoint, the examiner is familiar with all varieties and types of firearms. There will be voids in any examiner's knowledge by virtue of exposure, and there are certain firearms that seldom, if ever, are encountered within the criminal context. This is not to mean that the credentials of the examiner should be impeached on this point. Defined, the *expert* becomes recognized as such based on having gained comprehension of the topic from five sources: knowledge, skill, experience, training, and education. Moreover, the expert's comprehension of the subject matter is greater than that of the layperson. Experts are able to theorize about controversies that are presented to them and engage in hypothetical discussions, avenues not open to the fact witness whose testimony is limited to direct firsthand knowledge. Article VII, Rule 702 of the Federal Rules of Evidence state that a "witness who is qualified as an expert by knowledge, skill, experience, training, or education may testify in the form of an opinion."

Individuals seeking recognition as an expert collectively draw upon these five sources to support their claim, and such individuals should expect such claims to be thoroughly vetted. In the courtroom, the prospective expert should reasonably expect to undergo the legal process of voir dire, that is, to present and clarify the witness's credentials to the court and jury, as well as permitting opposing council the chance to challenge the credibility of the witness under cross examination.

- *Knowledge*: Knowledge is in direct proportion to the proficiency one has with information obtained from the other four sources. In other words, how much information does a person have related to the subject in question? This information can be obtained by research, independent thought and observations, as well as through practice, training, and education relevant to the field. In general, knowledge can be taken from both formal and informal settings, but it must be credible. Where did the shared information come from in the first place, and how was this information obtained?
- *Skill*: Skill is the adeptness that one has for successfully undertaking and completing a task. Anyone, regardless of their competency, is capable of making a mistake; however, the apparent skill level of a person can be used as a general indication of that person's competency in the field. Skill is acquired over time through prolonged and repetitious exposure. Some practitioners can learn very quickly, and the practice of the craft becomes second nature, whereas others require more exposure. Skills may diminish and become obsolete over time if the practitioner is not actively engaged in the field.
- *Experience*: How long has the person been involved with the subject? Experience can be somewhat misleading, and it must be balanced between several considerations to establish what an accurate relative experience level is. Questions to ponder are how long a person has been involved in the field versus what that person's exposure to the field has been during that tenure. Another factor to consider in experience is how productive the person has been in the field, such as caseload, success rate on cases, complexity of cases, as well as outstanding achievements. Prospective witnesses should be prepared to answer not only how long they have been in their profession, but also be prepared to discuss their workload. In an ideal setting, practical experience bolsters learning through education or training, and they become complementary.
- *Training*: Training is generally associated with improving or expanding job skills, such as changes in work processes or completion of work tasks. Training can take the form of an apprenticeship, topic-specific courses (armory course given by a particular manufacturer about a particular product or the product line), technical seminars, trade conferences, and the like. Training may have been obtained from the law enforcement community and/or military service. The advantage of training is that it tends to be concurrent with the state of the art, more relevant to the consumer, and typically can be applied immediately. Reference materials that accompany training sessions tend to be technically focused and are designed to be workbench aids. Although these materials may become obsolete, they never go completely out of date; you never know when an old source will come in handy.
- *Education*: Education involves the presentation of programs for the student to engage in long-term learning goals. The line between training and education can indeed be thin; however, education can most generally be applied to formal

education in the setting of an academic environment, especially the university or college setting. The gateway into career paths in modern forensics is a degree, and this is true even with firearm examiners, although it is unlikely to find firearms taught as a specific course of study in any university or college catalog. Nonetheless, a degree rounds off the resume or curriculum vitae of persons seeking to become subject-matter experts, and this is no less true with firearms.

Rule 702 also stipulates that "the expert's scientific, technical, or other specialized knowledge will help the trier of fact to understand the evidence or to determine a fact at issue." In furtherance of this, "The testimony is based on sufficient facts or data; the testimony is the product of reliable principles and methods; and the expert had reliably applied the principles and methods to the facts of the case." In addition, Rule 703 states,

An expert may base an opinion on facts or data in the case that the expert has been made aware of or personally observed. If experts in the particular field would reasonably rely on those kinds of facts or data in forming an opinion on the subject, they need not be admissible for the opinion to be admitted. But if the facts or data would otherwise be inadmissible, the proponent of the opinion may disclose them to the jury only if their probative value in helping the jury evaluate the opinion substantially outweighs their prejudicial effect.

The expert must therefore prepare to answer the following questions:

- What underlying facts and information were used to form the opinion?
- Was relevant case information reviewed or not reviewed when the expert contemplated the question?
- What was the process of reasoning, and what theories and methods were used to form the opinion?

Recognition as an expert does not necessarily give the person carte blanche within the field; the subject matter must be appropriate to the expertise demonstrated by the prospective expert witness. As an example, a recognized, renowned expert in the field of the Austrian Rast-Gasser revolvers and Steyr-Hahn pistols may not necessarily meet the definition of *expert* if the judicial controversy at hand concerns defining an article as a firearms suppressor by virtue of design and construction—unless, of course, this same person is prepared to offer evidence supporting expertise in that subject as well.

Exculpatory Evidence

Exculpatory evidence is any evidence that tends to prove, or at least suggest, the innocence of a party against whom an allegation has been levied. When exculpatory evidence is uncovered, it must be disclosed to the defense, even in the absence of formal requests for discovery of the state's evidence. If the examiner is working for the defense, it is obviously in the client's best interest to make the prosecution aware of this evidence in the timeliest manner. The expert cannot simply choose to ignore exculpatory evidence when confronted with it. A common error on the part of inexperienced examiners is the failure to research the question put before them. The examiner is reminded that, as part of due diligence, a reasonable amount of research must be invested before formalizing an opinion that will serve as the basis of the examiner's expert conclusion. To frame this in the context of the firearm examiner, as a hypothetical example: Was the examiner aware that the firearm in question was subject to a manufacturer's recall due to a faulty component that could have caused an accidental discharge to take place? If the examiner was aware of this recall and had evaluated the significance of such a recall with respect to the behavior exhibited by the firearm at the time of the event, then the examiner can still form an opinion but must be prepared to address the possibility that the errant behavior that prompted the recall did or did not come into play within the circumstances of the case. Nothing could hurt the overall credibility of the expert or the soundness of the case more than being blindsided by the startling development that counsel for a defendant produces such evidence that the examiner was completely unaware of.

Experts should remember that they are objective purveyors of fact, and although most firearm examiners are employed by law enforcement organizations, this should not serve as a justification to cause them to present their testimony in such a way that attempts to sway the jury toward a guilty verdict. Simply put—the facts are the facts. This should in no way dissuade the expert from rendering a well-founded opinion and the basis for it; however it should be presented in such a way that is firm without being prejudicial or inflammatory.

The *Frye* and *Daubert* Standards

Legally, the recognition and the definition of experts has fallen upon the two legal standards: *Frye* and *Daubert*. Both standards are used in the United States, but which standard is applied varies from state to state. The basic premise of the *Frye* Standard is that the underlying method or procedure utilized is accepted in the relevant scientific community. The *Frye* Test—that is to say, the question that must be answered to qualify under the standard—is that the *proponent* (the proposed expert) must first identify the pertinent scientific circle, and then the theory, instrument, circle, or test is accepted—not the conclusion. The *Frye* Standard does not apply to purely scientific opinion but, rather, asks if the expert testimony is reliable. An opinion that is based upon training and experience versus the application of scientific methods, principles, and testing is not subject to the *Frye* Standard. Recall that the expert relies not only upon training and experience, but also education, skill, and knowledge. There is a very fine yet distinctive line separating the two.

As an example, a law enforcement officer encounters a particular firearm that is suspected of being a machine gun. The officer forms the opinion because of military experience and training, having become familiar with this particular type of weapon in the course of military service, and having received specific training in weapons that would include the particular type and model of firearm in question. The officer acts on the basis of training and experience as substantial reason to seize the weapon and further the investigation. In this illustration, the officer did not apply scientific methods, principles, or testing to arrive at a conclusion that the article in question was, in fact, a machine gun. At face value, the actions taken by the officer would not be scrutinized under the *Frye* Standard, as the officer did not make any conclusive determinations concerning the exhibit. When the exhibit is later in the hands of a firearm examiner, who conducts examinations and draws conclusions based upon an examination, the conduct and conclusion of the examiner becomes subject to the *Frye* Standard as to the methods and techniques used and the principles that were applied.

The U.S. federal courts and certain states observe the *Daubert* Standard. *Daubert* is both similar and dissimilar to *Frye*. There are parallels between them, but distinct differences as well. Under *Daubert*, the presiding judge acts as a form of gatekeeper in determining the reliability, the relevance, and hence the admissibility of the expert testimony that is being offered. The role of gatekeeper permits the judge some flexibility, subject to some basic criteria. The judge must decide if the offered scientific method, procedure, or technique has been subjected to testing or could be tested, and considers whether it has widespread acceptance within the relevant scientific circle. The judge may also clarify whether the method, procedure, or technique has been tested; what the margin of error is; what controls or standards were in place when the work was conducted; and whether the testimony is scientifically oriented or not. To properly credentialed subject-matter experts acting within their field of knowledge, *Frye* and *Daubert* are practically transparent.

It is not the position of the examiner to attempt to interpret a statute and apply it against contemporary case law per se; rather, it is the function of the examiner to conduct reviews and report findings. This is separate and distinct from the contemporary legal interpretations of any particular law as it relates to a firearm or ammunition, which are solely the responsibility of the attorneys arguing the controversy before the court. The results of an objective, scientific examination of the evidence may yield a result that is not desirous for the prosecution. However, examiners must bear in mind that they are, in effect, a witness of the court that is called upon to render objective, unbiased testimony in the legal question or controversy at hand. The expert cannot opine on questions of law, the guilt or innocence of the accused, or the credibility of other experts. The expert, however, can challenge the methodology applied by another expert or comment on the deficiencies in the opposing expert's methodology, short of personal attacks and making libelous or slanderous claims and statements. Experts should view themselves as educators to the jury and seek to facilitate a level of understanding of the subject matter within the parameters of the testimony they are providing.

The firearm examiner or subject-matter expert often plays a more intrinsic role beyond just the typical investigation and prosecution of an alleged crime. The expert may be called in to opine when an organization is in the process of selecting potential candidate firearms. The expert may evaluate a firearm for its uniqueness, rarity, technical properties, or historical significance that may call for the preservation of the firearm for future study. Even if the organization in possession of the firearm does not have the desire or facilities to retain it, countless museums, laboratories, and reference collections may be eager to take possession of the firearm. When firearms of an unusual nature or having unusual characteristics are encountered, the examiner conducts a thorough inquiry and publishes the findings for fellow examiners and other concerned parties in the form of intelligence bulletins or other mediums of communication. All examiners benefit from the network of information sharing and cooperation.

Gunsmiths versus Armorers

Gunsmithing is the oldest profession acquainted with the firearm. Before mass production, guns were handmade creations that were the work of a knowledgeable artisan. Every component was fabricated from raw material and hand fitted together into a workable product, as much a work of industrial art and artisanship as a tool. With mass production, the role of manufacturer transitioned from the gunsmith to the machinist, who fabricated components for production, and engineers, who fine-tuned the ideas of the inventors and turned them into working products.

The role of the gunsmith did not disappear with the industrial revolution, as there were always firearms needing repair or owners who wanted some type of custom work or modification to their firearms. The gunsmith's trade was to address the functional and aesthetic issues of a firearm. It may have been a modification, restoration, repair, or replacement of the stock. It could have been the refinishing of the firearm, but frequently it concerned repair of a broken firearm. The gunsmith had to diagnose the symptoms of the malfunction and then begin deducing what the issue was: a broken part, a slight modification required to a certain component, replacing a spring, and even fabricating a replacement part that might not be available through ordinary channels. Their knowledge was gained through such avenues as apprenticeships, trade schools, experience, and learning from new projects that came through the door. It was a matter of routine for a gunsmith to fabricate a firearm from a prefabricated receiver (the frame of a firearm) as well as modifying an existing firearm to suit the desires and needs of the client. As such, gunsmiths became experts in their own right, having an understanding of any number of firearms they handled during the course of their careers. Gunsmiths worked from reference guides and schematics in addition to using their own eyes, hands, and experience to guide them. Modifying an action was not an exact science; gunsmiths relied on their competence and best judgment to achieve the desired result.

The armorer possesses a similar, yet distinctly different skill set from the gunsmith; a principle distinction being that gunsmiths are involved in the fabrication of arms and parts, whereas armorers are not. The armorer is not trained in the same skills and, in fact, while performing similar labor, is going about the task differently. The armorer still diagnoses the functional issues of a firearm, but the philosophy of the armorer is affecting repair through replacement of parts, and not necessarily modifying the firearm or tweaking a firearm action through machine work and hand fitting. This is especially true in law enforcement and military circles. There is no need to try to rebuild a part; simply replace it through spares kept on hand.

Firearm design has evolved a great deal in a relatively short time. For the most part, firearms contain fewer internal components and offer a degree of inherent flexibility that in previous generations of arms would have required the services of the competent gunsmith to successfully modify or repair. This approach has strong merits: It allows customization to suit the user while retaining predictable behaviors and a margin of safety in keeping with the manufacturer specifications. In contrast, the modification of parts creates the potential for undesirable firearm behavior that can result in adverse actions, such as increased potential for accidental discharges. Liability issues permeate firearm design, and manufacturer specifications and parts keep the behavior of the firearm well within a predictable and acceptable tolerance, something that simply cannot be guaranteed otherwise.

Most major firearm manufacturers present courses that cover their specific products. Third-party instructors often fill the gap for weapon systems when there is no factory support or when no single manufacturer is available to offer the training. Armorer training is often restricted to official entities, licensed firearm dealers, and recognized gunsmiths; thus such training is generally not available to the general public. Armory courses are a great way for novice examiners to get started, regardless of what direction they wish to take their career. Lasting one to five days, these courses are very much hands on, requiring that the student assemble and disassemble firearms, identify components, and diagnose various mechanical failures. The courses often end with a written and practical exam, ensuring that the student demonstrates a level of proficiency and competency with the product. Once an armory course is successfully completed, the manufacturer will generally permit the student to order parts and perform work that will be sanctioned by the factory.

Firearm Examiner

The firearm examiner works with firearms on a different level altogether from gunsmiths and armorers. Examiners may receive armorer training as a matter of course in their career; this allows them to see the concepts, principles, and theories behind a particular firearm or weapon system. However, examiners typically do not apply their knowledge to repair firearms. Examiners apply their accumulated expertise to conduct a review of a submitted firearm as a matter of investigation. The examiner is able to make functional determinations and obtain whatever data is required from the firearm—identifying unknown firearms, restoring obliterated serial numbers, or obtaining other pieces of information that the circumstances may call for.

In the modern paradigms, most firearm examiners are far removed from the scene of the crime, and they typically do not play any role in the investigation other than evaluating the firearm submitted to the crime lab. This approach has both pros and cons. This approach is beneficial in that it allows the examiner a measure of independence from the investigation, eliminating the suggestion of improprieties or conflicts of interest, or that the examiner's opinion will somehow be tainted by having direct contact or knowledge of the particular circumstances of the case. The downside is that because of a general lack of context given the examiner, there is the possibility that the results returned from the examination will be incomplete in the context of the investigation. This distance may also create a communication barrier between the examiner and the investigator; thus pertinent information that could be shared in either direction may not ever be exchanged.

Firearm Subject-Matter Expert

Within the community of firearm subject-matter experts are those that choose to specialize in one or more specific fields: ammunition, firearms, and comparative ballistics. To draw an analogy, neurologists and cardiologists are both physicians; however, their fields of practice and expertise are different, although both practice their respective specialty under the broad field of medicine. Firearm examiners, in particular, may be engaged in comparison work as well as other firearm-related matters, but there are those who work with firearms to an extent that does not include comparative examinations. Such examiners are engaged in the practice of identification, classification, and answering other questions related to firearms, short of doing comparative ballistic work. Such examiners require a degree of competence commensurate with the work they conduct. A simple function test can be completed by anyone familiar enough with a firearm to safely use it. However, a simple function test may not rise to the level of expertise as defined by the *Frye* or *Daubert* Standards, as it can involve testimony that is based on firsthand knowledge as opposed to being based on expert opinion.

Not every case presented is simple. Consider the question of a submitted exhibit that appears to be a firearm but does not function. Does inoperability disqualify the exhibit as a firearm under applicable legal language? For example, Florida State Statute 790.001(6) states:

"Firearm" means any weapon (including a starter gun) which will, is designed to, or may be readily converted to expel a projectile by the action of an explosive; the frame or receiver of any such weapon; any firearm muffler or firearm silencer; any destructive device; or any machine gun. The term "firearm" does not include an antique firearm unless the antique firearm is used in the commission of a crime.

The term *operable* should not be confused with the term *functional*. In an instance where the exhibit is inoperable, can the examiner explain why this could be the case? What defect is present that prevents the exhibit from operating as designed? If the examiner were to remedy the problem so that the exhibit could then function as a legally defined firearm, how could this discrepancy be explained? An operable firearm cannot be deemed nonfunctional purely on the occurrence that it did not fire when tested. In such a scenario, this is where the expert opinion comes into play. The examiner must be able to diagnose the probable cause(s) of the inoperability, and then perhaps seek to experiment with the different possibilities to ascertain the actual cause(s). The opinion therein lies in the articulation of a device that "is designed to, or may be readily be converted to expel a projectile." Based on the legal definition, it is also possible that the mere presence of a firearm frame or receiver would be sufficient to meet the definition of a firearm. The examiner must be able to identify the exhibit as a frame or receiver. Once again, as a matter of opinion, is the frame or receiver present, and could it be readily restored to function/operation?

For example, most pistol receivers can be made into fully functioning firearms in a matter of minutes if the balance of parts are at hand. Thus a frame or receiver that is disassembled could be articulated to be a firearm by definition, despite the fact that, in its current state, it is inoperable yet functional. Consider further, within the definition of a firearm, all of the following terms apply: *antique*, *machine gun*, *muffler* or *silencer*, and *destructive device*. Examiners must be able to articulate each one in relation to an exhibit before them if it is to be classified as a firearm. If an exhibit is a homemade or improvised device, does it meet any portion of the definition? In the opinion of the assessor, is the device or apparatus designed to act as a firearm? That is, is it designed, or can it be converted, to expel a projectile by the action of an explosive?

Basics: Understanding What Makes a Firearm Operate

For all the technological advancements that have occurred in firearm design, the majority of these advancements have affected everything else about a firearm except the basic tenets of function, which is how a firearm operates. The firearm has a sole purpose, but the means by which this purpose is achieved are, from a mechanical perspective, quite varied. The majority of operating systems and principles are old, well-established ideas that have changed very little.

It is ironic that one of the most popular handgun designs in the world, the Colt-Browning Model 1911, remains virtually unchanged since first conceived. Functionally, the 1911 design is still a 1911, regardless of which manufacturer produced the article and what aesthetic qualities may have separated one version from another. There have been subtle improvements to the basic design during its century-long existence, but the basic premise remains unchanged. The Mauser 98 design went into service in 1898 and was directly descended from the Mauser Gewehr Model 1871. Today, it remains a top choice for bolt-action rifle builders to copy, as there apparently is not much room for improvement to the basic mechanisms of the action. Before the introduction of the Soviet AK pattern firearms, the Mauser 98 was the most prolific battle rifle the world had ever seen, in terms of both quantity manufactured and adaptation by nations on every continent. Nations from Argentina to Israel and from Iraq to the Orange Free State used a Mauser at some time during the first half of the twentieth century and even later. Ordinarily, these rifles can readily be identified by the appearance of national crests, a manufacturer's mark, and a date on the receiver ring. Numerous firms manufactured the 98, including firms not located in Germany, such as Fabrique Nationale in Herstal, Belgium, and the Radom arsenal in Poland. Between 1934 and 1945 alone, a combination of German and foreign firms turned out over 12 million rifles, roughly 1 million per year. A Model 98 rifle can still turn up anywhere.

A firearm is basically a machine. A machine fulfills a specific purpose and accomplishes the work by a process. The purpose of a firearm is to expel a projectile under the force of an explosive. To accomplish this work or operation, the ammunition cartridge is first introduced into the chamber or breech of the firearm. This introduction, called *loading* or *charging*, can come by hand insertion to the breech or chamber, by loading from a clip into an internal magazine, by inserting a detachable magazine, by charge holes in a cylinder, or by being fed from an ammunition belt. Once ammunition is loaded into the chamber and the action is closed, the firearm is ready to fire. A breech-loaded firearm would be closed by hand, whereas a firearm that that is self-loading would be closed by working the action manually or by the manipulation of a release in preparation for firing.

Once loaded, the firearm is fired by the press of a trigger. The trigger press is subject to resistance that can be measured in pounds or other metrics of force that must be overcome by the pressure exerted by the shooter. There is not a universal standard that dictates how much resistance a trigger should have, but it is noted as a matter of safety that the trigger press should not be excessively light. An exception to this would be weapons dedicated to sporting competition, where the trigger press is greatly reduced to increase the speed of the shooter.

Firearms can be capable of two different firing conditions—double and single action—each having a separate trigger-resistance weight. Many firearms are capable of either firing condition, whereas other firearms are capable of acting under one to the exclusion of the other.

Trigger Travel

The *trigger travel* should not be confused with the mechanical force required to overcome the trigger resistance. The trigger travel is the physical distance the trigger must move to the rear to cause the firearm to discharge. When the trigger travels a given distance, the firing train or mechanical linkage that connects the trigger to the hammer or firing mechanism disengages, releasing the trigger from the firing mechanism and causing the firearm to discharge. This point is often called the *trigger break*. Once a complete pull of the trigger is made, the trigger must be allowed to return forward. The trigger *reset* is the distance that the trigger must travel forward to reengage the firing mechanism and enable the firearm to be capable of firing again, assuming there is ammunition chambered and ready to fire. Experienced shooters train their trigger finger to know the exact point of trigger return where the reset takes place, allowing for minimal trigger press, thus firing more quickly and likely more accurately as well. Trigger *overtravel* is the physical distance that the trigger can continue to travel rearward after the discharge has occurred. Certain firearms have adjustable triggers that permit the amount of trigger overtravel to be tailored to shooter preference.

Figure 1.3 A modern replica of the single-action-only 1847 Colt Walker manufactured by Uberti. (Image courtesy of A. Uberti/Benelli USA.)

Single-Action Firearms

Single-action firearms require that the hammer be cocked, either manually or by the action of the firearm, such as working the action by using the charging handle, pulling back on the slide, or by pulling the hammer back if the firearm has an exposed hammer. Single-action automatic firearms automatically reset the hammer during the cycling action. The slide or bolt travels to the rear and cocks the hammer automatically.

There are semiautomatic firearms that have exposed hammers, found at the rear of the slide above the grip. However, there are also semiautomatic firearms that have internal hammers whose actions and firing condition are not visible to the shooter until the firearm is disassembled. Firearms with internal hammers are often called *hammerless designs*, but this may not be technically correct. The hammer may not be exposed to view from the exterior, but there is a part that transfers force by impact onto the firing pin. This is frequently the case with long guns, where there is a hammer, in the purest technical sense of the word; however, it is not visible, as it is contained within the receiver. There are true hammerless designs that completely omit the hammer, instead relying on the trigger to directly interact with the other fire-train components to cause the loaded cartridge to discharge.

Single-action-only revolvers represented the first major step toward modern firearms and were among the first to employ modern self-contained cartridges. Figure 1.3 shows a modern replica of the Colt Walker model, an atypical example of a single-action-only revolver. The single-action-only mechanism exists in all varieties of firearms. Specification data for such firearms will abbreviate the action as SA (single action) or SAO (single action only). The typical range for a single-action trigger press is generally set between 3 and 6 pounds; however, it can be significantly less if modifications or adjustments have been made to the trigger mechanism. Certain firearms are capable of user-adjustable trigger-resistance weights or are equipped with a set trigger. A set trigger is a second trigger, located within the trigger guard, which is pressed not to discharge the firearm, but to preset the amount of trigger resistance on the firing trigger. The Swiss Vetterli 1871 carbine* is an example of a firearm equipped with a set trigger. The set trigger is located behind the firing trigger. A set trigger must not be confused with double triggers used to fire individual chambers on multiple-barreled arms such as double-barrel shotguns and drilling-type firearms.

---

* A carbine is defined as a short rifle. Carbines initially were developed by reducing the barrel length of a full-sized main battle rifle, primarily intended for use by cavalry or other specialized types of infantry. Carbines were simply shorter versions of existing firearm platforms. Later, unique pattern firearms expressly designed as carbines were introduced, and were not necessarily reconfigured from an existing platform.

Double-Action Firearms

A *double-action* firearm requires that the press of the trigger must first cock the firing mechanism and then cause it to release. The shooter presses the trigger, which, through mechanical linkage, first cocks the hammer and then releases it once there is sufficient trigger travel. Suffice it to say, a double-action trigger press is generally much greater than a single-action press, both in the physical distance the trigger must travel and the amount of effort that must be exerted by the shooter to cause the discharge to occur.

Many double-action firearms are also capable of single-action operation by simply cocking the hammer or working the action in preparation to fire. A firearm capable of both double and single action will behave exactly like a single-action firearm in this condition. Many firearms are double action only, often abbreviated as DAO; likewise, single-action-only firearms are abbreviated as SAO; and firearms capable of either firing mode are abbreviated SA/DA. The typical range for a double-action trigger press ranges between 5 and 15 pounds, although there is no set standard that must be observed. Like single-action-only firearms, double-action trigger resistance may be altered by modifications made to the trigger and the other components in the firing train.

The revolver in Figure 1.4 is capable of both single- and double-action operation. The first visual cue is the spurred hammer, which allows for thumb cocking to a single-action condition. Modern revolvers of this design are not single action only. Figures 1.5 and 1.6 demonstrate the movement of the revolver's mainspring and how it relates to the behavior of the hammer. There are variations in mainspring design and exact placement within the grip, which varies by designer and manufacturer; however, they all serve the same purpose.

There are two distinct types of double-action trigger. The first type has to be cocked, such as the Glock or Springfield XD. Some view this type of trigger as having a distinct disadvantage. In the event of an ammunition failure, the shooter must perform a clearance drill to strip the dud cartridge from the chamber, reset the action, and prepare again to fire. However, not all surveyors of this fact deem it to be a disadvantage from a tactical perspective. The other type of double-action trigger is one that offers a second strike or restrike capability. In such a system, the function of the trigger cocks and releases the action, such as in the case of a double-action revolver or a semiautomatic pistol such as the Smith & Wesson Sigma, the Beretta 92, the Beretta 96, or the H&K USP series. In the event of a cartridge failure, the shooter simply presses the trigger again.

Figure 1.4 Most modern revolvers are capable of single- and double-action firing, as is the case with this Taurus Model 856, chambered in .38 Special. (Image courtesy of Taurus International Manufacturing.)

Figure 1.5 A Smith & Wesson Model 19 in double action firing condition, the hammer is "at rest", note also the position of the mainspring, the flat metal piece revealed by removal of the grip. (Image from author's collection.)

Figure 1.6 A Smith & Wesson Model 19 in single action firing condition, note the movement of the mainspring as compared with its position in Figure 1.5. The side plate has also been removed, revealing the hammer block safety and the linkage between the hammer and trigger. (Image from author's collection.)

Sequence of Operation

Regardless of whether the firearm is double or single action, when the hammer releases its fall, or when travel causes it to impact directly (or through intermediate parts) onto the firing pin, the force of the impact is passed on to the primer or rim of the loaded cartridge. This is the percussion detonation that is used in modern, self-contained cartridges, as opposed to archaic ignition systems such as the wheel lock, matchlock, or flintlock.

In some firearms, the firing pin is integral to the hammer or is attached to the bolt face; in other designs, the firing pin or striker is a separate piece from the hammer or bolt. Various safety designs exist that may act as intermediaries that prevent direct hammer/firing-pin connection, instead passing the impact pressure through this intermediary, such as the case with a transfer-bar-type safety on a revolver.

Upon impact, the primer or priming compound detonates, which causes an intense super-hot fire to pass into the main cartridge chamber through small holes called *flash holes*, which detonate the contained powder mixture. Once ignited, the contained powder charge starts to deflagrate, creating gas as a by-product. Once sufficient gas pressure has been achieved within the casing, the gas forces its way through the path of least resistance: the seated projectile. The casing is sitting snuggly within the walls of the firearm receiver and is thus well reinforced. When this is not the case—if the chamber is eroded or if sub-sized ammunition has been chambered into the firearm—a ruptured casing may occur as a result of this gas pressure. This may jam or even damage the firearm. Firearms have been known to come apart when an overpressure cartridge detonates within the chamber, damaging the host weapon and possibly resulting in death or injury of the shooter or others in proximity. Once the loaded cartridge is fired, the cycle of operation must be successfully repeated to allow fresh ammunition to be loaded, either by hand, by manual manipulation of a control (as in the case of a bolt-action or lever-action firearm), or automatically by independent operation of the firearm without input by the operator short of pressing the trigger.

Methods of Operation

Automatic* or self-loading firearms function by capturing the gas pressure created by the discharge of the ammunition cartridge that pushes the projectile out of the barrel. This energy can be captured, redirected, and utilized to cause the firearm to continue to operate by automatically ejecting and extracting the spent casing, then loading a fresh cartridge into the chamber while resetting the action to permit an immediate follow-up shot, assuming there is ammunition available. This principle is called *gas operation*. There are numerous methods and subvariations of gas operation, and although all gas-operated weapons do work by the same principle—by way of this gas pressure—mechanically they work differently. The first cartridge must be loaded by interaction of the operator, typically by inserting a magazine or loading cartridges into an internal magazine, and the action must be manually cocked. The automatic sequence only occurs when discharge has taken place.

Gas Impingement

A firearm that operates on gas impingement, which is also called direct gas operation, uses gas pressure returned to the receiver from the muzzle that automatically cycles the firearm's action. The gas pressure returns to the receiver through a gas tube or channel, where such gas pressure strikes, collides, or impinges directly on the firearm action. The Swedish Ljungmann AG42, introduced in 1942, was such a self-loading rifle that used a gas impingement system, and it became the inspiration for a series of early gas-operated rifles used around the world. The best-known application of gas impingement is the AR-15, a derivative of a prior rifle, the AR-10, and developed in the 1950s. The AR-15's gas-impingement system has been subjected to criticism since the rifle was adopted in 1962 by the U.S. Air Force and in 1964 by the U.S. Army. A fact that is often overlooked historically is that

---

* The term *automatic* should not be inferred in this text to exclusively define a firearm capable of sustained automatic fire with a single press of the trigger, called a machine gun. An automatic firearm is also meant to include a semiautomatic firearm, which still operates automatically when cycling, yet fires once per press of the trigger.

Eugene Stoner, the designer of the AR-15, did in fact develop the AR-10 as a gas piston-driven system. However, as history would have it, the gas-impingement system won out.

Gas-Piston Operation

Gas-piston operation represents an alternative to the direct gas action and may be referred to as an indirect gas-action system because the gas pressure does not directly work with the firearm action; instead, the piston acts as the intermediary between the two. A gas-piston-operated firearm uses the gas pressure generated by the expanding volume of gas in the fired cartridge to move a piston. The stroke of the piston may be long, or short, relatively speaking. As is the case with all types of firearm actions, there is no singular method by which the gas piston system is employed; there are numerous variations of how the gas-piston action is set up.

John Garand's M1 rifle was a piston-driven semiautomatic, although it is often just called a gas-operated rifle. The Garand's piston pushes an operating rod that interacts with the bolt, allowing the automatic cycle of extracting, ejecting, loading, cocking, and chambering to occur. A counteracting spring returns the operating rod and piston to close the action in preparation for repeat firing. The components in the Garand's gas piston system are located underneath the barrel. In contrast, during World War II, German forces fielded a gas-operated semiautomatic rifle, the G.43 (*gewehr* meaning rifle), which also used a gas-piston system, albeit more complicated than the Garand. The G.43 used a fixed gas piston that was acted upon by a moving gas cylinder. Gas entered the system from ports in the barrel and propelled the cylinder, not the piston. The cylinder linked to a connecting rod that linked to an actuating rod that worked the action. At the end of the train was a spring that countered the actuator rod, closing the gas system in preparation for repeated fire. When viewed, it looks like an excessively complicated system; it contained more pieces than the Garand, and in contrast to the Garand, these parts were located on top of the barrel. The G.43 was not an innovation; there was likely influence from the Soviet SVT 38, also a gas-piston-operated rifle.

The SVT 38 and its simplified successor, the SVT 40, used a short-stroke gas piston that, like the previous examples, received gas pressure from gas ports cut in the barrel. The piston pushes an operating rod, which pushes the bolt. The operating rod is spring loaded to cycle the action. A forward-looking and novel feature of the SVT gas system was an adjustable gas-regulating valve. The presence of an adjustable gas valve is almost a universal feature on gas-piston-operated firearms since 1945.

The Austrian Steyr AUG (Armee Universal Gewehr) and copies of it are examples of contemporary rifles operated by the gas piston method. Introduced in 1977, the AUG is still a rather futuristic looking weapon that made extensive use of polymer construction and was a compact, "bull pup"* design. An interesting feature of the AUG is that its gas piston serves double duty, as it also acts as one of the two bolt guide rods in the receiver, a design touch that reduces the number of parts and contributes to the compactness of the design.

Mikhail Kalashnikov, the designer of the AK47, AKM, and AK74 series of rifles, took the gas piston concept in a slightly different direction. He further simplified the method by combining the piston and the bolt carrier.

---

* A bull pup is a long-gun design concept that moves the receiver to the rear of the firearm, often incorporating it into the shoulder stock, allowing for a shorter overall dimension while maintaining a full-length barrel.

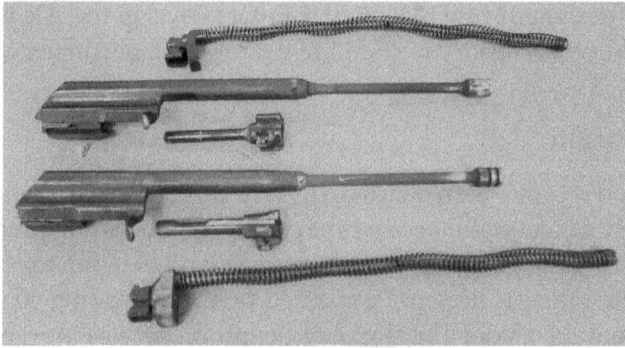

Figure 1.7 Two styles of bolt carrier, bolt, and recoil spring for an AK-pattern weapon. The piston threads into the bolt carrier to form one piece. (Image from author's collection.)

The piston threads into the bolt carrier and essentially makes the two one piece, eliminating the need for additional operating or actuating rods and other linkage that is found in other designs. This approach to the gas-piston system is widely acclaimed for its reliability and general lack of required maintenance to ensure reliable operation. Figure 1.7 shows two different designs of the Kalashnikov combination piston, operating rod, and bolt carrier. The bolts are shown, as are the recoil springs and recoil spring guides. The recoil spring attaches to the rifle by two lugs at the back of the receiver, and the front part of the recoil spring sits within the piston/operating rod. The bolt carrier rotates within a channel on the bottom of the unit.

Gas-piston systems have been widely touted as the replacement for gas impingement in the AR-15/M16-pattern weapons. Recent entrants to the market, such as the Heckler & Koch 416 and 417 rifles,* the FN SCAR 16 and 17, Sig 556, and the Remington/Bushmaster ACR (Adaptive Combat Rifle), are all gas-piston-driven designs. Two potential disadvantages of piston-driven systems versus gas impingement are the added weight of the additional pieces required in a piston system as opposed to a gas tube and a lower cyclic rate of fire if the firearm in question is a machine gun. Both of these potential disadvantages can be abated by engineering adjustments that make the components lighter to reduce weapon weight and increase the rate of fire.

Gas-operated semiautomatic pistols are not typical, but they do exist. There is nothing technically or inherently wrong with a pistol that functions using gas, if one can get past the complication and additional engineering of such a pistol, relative to the simplicity of the recoil-operated or blowback designs. Akin to gas-operated rifles, a gas-operated handgun must have a means or method to capture and channel the gas generated by the discharged cartridge, which ordinarily entails a gas trap and the associated plumbing to return the gas to the action. The Israeli Weapons Industries manufactures the Desert Eagle line of handguns, one of the few examples of gas-operated pistols. The Desert Eagle, marketed by the American subsidiary Magnum Research, features a fixed-barrel design in a configuration somewhat reminiscent of a blowback pistol, but looks can be deceiving; the pistol uses gas bled from the bore that drives a piston that acts upon the slide, pushing it open. The design features a rotating bolt, like the AR-15/M16. The smallest pistol of the

---

* The H&K 416 and 417 are Heckler & Koch's interpretation of the M16 rifle, but feature the gas-piston system of operation as well as some ergonomic and other improvements over the basic M16 rifle in its current configuration.

line, the Micro, also uses gas-assisted blowback. The Desert Eagle is best known for its big bore caliber, .50 Action Express, rivaling the .500 Smith & Wesson Magnum, .460 Smith & Wesson Magnum, and others for the largest handgun cartridges available. The Desert Eagle is physically not a small pistol, and it features the ability to change calibers quite easily. Pistolized versions of rifles, such as the AR or AK pattern, retain fidelity to their original operating principle of gas operation.

Recoil Operation

Semiautomatic handguns, also called pistols, are almost universally recoil operated save for those few gas-operated examples. Recoil operation allows pistols to be simplified in parts content and keeps them reasonably compact. Recoil or blowback pistols rely on a recoil spring (or springs) that serves to compress against the rearward motion of the slide as the projectile travels down the barrel, generating the "opposite and equal reaction" caused by the detonation of the cartridge. The rearward travel of the slide ejects the spent casing and resets the hammer. The recoil spring then rebounds, forcing the slide forward, loading a fresh cartridge, locks the action, and prepares the firearm for a subsequent shot. This impulse occurs within fractions of a second.

The recoil spring can be installed on a recoil spring guide, essentially a metal or plastic shaft, and is installed beneath the barrel. Most designs in modern times have centered on recoil-operated pistols that employ locking lugs or cams at the bottom of the barrel that interact with a locking block in the frame during the recoil sequence of firing. Some designs have a locking block that is integral to the frame; others use a locking block that is a separate component and is installed in the frame. There are also dual recoil spring designs, as is the case with the Walther P.38 and its descendents such as the P-1. The Walther P.38 featured dual recoil springs on steel guide rods on either side of the frame above the grips and beneath the barrel. Another interesting feature of the Walther P.38 is its locking block, which is a separate piece from the barrel that pushes down at an angle into a recess in the frame.

Single-recoil springs on a recoil spring guide that are fixed underneath the barrel are most prevalent on modern pistols. Other pistols, such as the Colt Woodsman, Colt Huntsman, Browning Buck Mark, and the Smith & Wesson 22 series, have a barrel fixed to the frame, but the slide does not fully enclose it, nor does it rest upon the entire length of the frame. In these instances, the slide is behind the barrel and reciprocates, extending forward only to the breech.

Blowback Operation

The blowback method of operation is the simplest means to make an automatically repeating firearm function. A blowback-operated firearm functions by the pressure generated in the breech when a loaded cartridge is expelled, causing the entire bolt assembly or slide to reciprocate when cycling. When cartridge detonation occurs, it pushes the reciprocating part rearward, which then returns to battery by the counteracting pressure of a recoil spring. Blowback firearms have fixed barrels that are integral to the receiver or locked in place by other means. Blowback firearms generally are extremely reliable because, with so few pieces, there is little that can go wrong. They can be lightweight and compact because they do not require the added parts and plumbing of other gas systems; everything is contained within the receiver.

Due to the simplicity of the blowback principle, it has been a regular choice for semi-automatic handgun designs. The Israeli UZI submachine gun, one the most easily recognized examples of a blowback firearm, is well known for its simplicity, ease of manufacture, low maintenance, and utter reliability. Gordon Ingram certainly looked to the UZI both in form and in function when designing the MAC pattern firearms. The AR-15 platform firearms chambered in 9×19mm and in .22 operate on the blowback principle, since these cartridges develop insufficient gas pressure to work effectively using gas impingement; this includes the Colt Model 635 9×19mm submachine gun. The 635 is often called the CAR-9, a name apparently meant to indicate that the firearm was a carbine chambered in 9mm. Pistols such as the Mauser HSc, Walther PP and PPK, and the SIG Sauer P230 series are classic blowback pistols, having fixed barrels and a recoil spring that shrouds the barrel.

As with all operating systems, there are variations to the basic blowback principle. The delayed blowback pistol is a subtle variation of straight blowback. In a delayed blowback system, there is an intermediate step that prevents the immediate retraction of the slide or bolt under the force of the recoil. The Fabrique Nationale (FN) Five-seveN pistol is such a design, owing to the chamber pressure developed by the 5.7×28mm cartridge. Figure 1.8 shows the Five-seveN pistol field stripped.

The simplicity of the blowback system meant that it was ideal to develop a more complex system that was, in effect, blowback but with some twist or addition. The toggle action of the Luger pistol is one such example. The slide recoiled very little; however, the toggle action absorbed the recoil energy and returned the gun to battery. The roller-lock bolt system is another example of delayed blowback, and is one of the few successful examples where blowback worked with full-power military cartridge applications. The roller-lock bolt system was developed in Germany in the late stages of World War II and found its way back to Germany by way of Spain, where the system was more fully developed. The firm Heckler & Koch is best known for employing the

Figure 1.8 The FN Five-seveN is a delayed blowback pistol. Although there is no externally visible hammer, this is not a hammerless design. The hammer is seen at the rear of the frame, next to the ejector, which has a hooked appearance. (Image from author's collection.)

system in its entire long-gun line, from precision rifles to general purpose machine guns. In an ironic twist, the readily recognized Thompson 1928 submachine gun was originally designed to use a form of delayed blowback and ended up as a straightforward blowback when the design was simplified for mass production as the M1/M1A. The simplification did not seem to hamper the Thompson's reputation as a reliable and effective submachine gun, and it probably went unnoticed except by the technical aficionados.

The Israeli Galil, which borrowed heavily from the Kalashnikov system, operates on a gas-piston system that also provides for a delay where gas pressure is reduced. Rather than accomplish this delay by mechanical means, it is done by having loose tolerances in the bolt carrier parts, including a notched gas piston, which allows some gas pressure to bleed off but reserves a sufficient quantity to cycle the weapon. The Beretta Model 92 operates by delayed blowback. This is accomplished by a locking block that locks the barrel and slide until chamber pressure has abated. The barrel maintains a flat plane during the course of cycling.

Other methods of delaying blowback include using a lever to moderate the movement of the bolt; using gas pressure to retard the rearward movement of the bolt, in essence a reverse of gas operation; or the recessed-ring delayed blowback. The recessed ring is unique to the Seecamp LW32 and LWS380 pistols, which are identified as *retarded blowback* by the manufacturer, although the company website references a recessed ring in the chamber "into which the case expands on firing, making the weapon a retarded blowback" (L.W. Seecamp Co. n.d.). Unsuccessful attempts at blowback include the screw-delayed and the Pederson hesitation lock, both of which are unlikely to be encountered or appear again.

Recoil-Operated versus Blowback Firearms

What differentiates a recoil-operated firearm from a blowback firearm is that the recoil-operated firearm has a measure of barrel travel. There are two approaches to recoil operation: short and long recoil. As the names imply, either recoil operation is based purely upon the physical distance that recoil takes place. The classic example of short recoil, the Colt 1911, uses a swinging link. The barrel has a hinge that is connected to the frame by the takedown pin, and there is a locking cam on top of the barrel to fix it to the slide when the action is closed (see Figure 1.9). Interestingly, Browning abandoned the swinging link approach when he designed the heir apparent to the Model 1911, the Hi Power, which uses the now familiar locking lug located underneath the barrel, and which locked during recoil into a steel bar in the pistol frame. The continuation of this popular design has a barrel equipped with locking lugs that connect to a locking block as the firearm is fired and the barrel is pushed slightly rearwards because of the recoil. Nearly every handgun manufacturer uses or has used this method at one time or another, and this system is still defined as a short recoil or modified Browning action, as the barrel and slide travel rearward together for only a short distance before the barrel is locked into place by the locking lugs. Figure 1.10 shows the modified Browning action as applied in the H&K USP Tactical 45 pistol. The barrel has locking lugs on the bottom side, and the recoil spring guide locks to the barrel and to the guide on the frame. Another design variation to recoil operation does not require that the barrel necessarily be captured by locking lugs or by any other linkage, but instead the barrel travels along a rail or guides that are part of the receiver, and the barrel does not tilt as it travels.

Figure 1.9 The classic Browning short-recoil action. Note the swinging link under the barrel. Although this is a Colt 1991A1, not much has changed from the original 1911. (Image from author's collection.)

Figure 1.10 The Heckler & Koch USP Tactical pistol chambered in .45 ACP. The USP is recoil operated, using a modified Browning action. Note the differences between this recoil action, swinging link, and blowback designs compared with Figure 1.9. (Image from author's collection.)

Long-recoil-operated arms are primarily long guns, as handguns generally are physically too small to accommodate a long recoil action. A long-recoil-operated firearm operates in much the same way as a short-recoil firearm, save for the fact that in a long-recoil arm, the action and barrel remain locked together in the recoil sequence and are counteracted by recoil springs, whereupon the barrel returns in advance of the bolt, which recoils after the barrel has returned forward. This delay allows the bolt to complete the sequence of ejecting and loading a fresh ammunition cartridge and chambering same when return-ing forward. An unusual form of recoil operation is the *blow-forward* action. As the name implies, it works by recoil but in reverse of a blowback design. A blow-forward action pushes the barrel forward by the force of the recoil.

Striker-Fired Firearms, a Contemporary Trend

The recent trend in pistols is a polymer frame using a striker-fire design. The striker is not a new idea, having appeared quite early in the twentieth century at the dawn of the automatic pistol era. The Colt Model 1908 Vest Pocket is an example of an early striker-fired firearm and was quite novel for the time. The striker design eliminates a hammer as part of the fire-control train. The striker instead can rely upon a direct action of the trigger sear releasing a spring-loaded firing pin. In a single action, striker-fired design, the firing-pin spring is placed under tension, and therefore the pistol would have to be charged by pulling back the slide, which would in turn cock the firing-pin spring. In the case of the Colt 1908 Vest Pocket, two safety devices were built into the design. The first was a grip safety, consisting of a pressure plate located on the back of the grip. The pressure plate disengaged the trig-ger from the firing train and thus disabled the firearm until it was depressed, presumably by gripping the pistol. Later Colt further enhanced the safety of the model by providing a manual safety operated by a lever on the left side of the frame, but it was really redundant given the other positive safety features. The 1908 Vest Pocket was in stark contrast to the Colt Model 1905. Although they shared the grip safety, the 1905 used an internal hammer.

The firearms designed by Bruce Jennings that emerged in the early 1970s were blow-back striker-fired pistols. The design used a fixed barrel on the cast-metal alloy frame. In 1971, Raven Arms entered the market, followed by Jennings Firearms in 1978, Bryco Arms, and later CalWestCo. Phoenix Arms succeeded Raven Arms in 1992. A Raven MP-25, the .25 ACP pistol, is shown in Figure 1.11. The name of the company that made

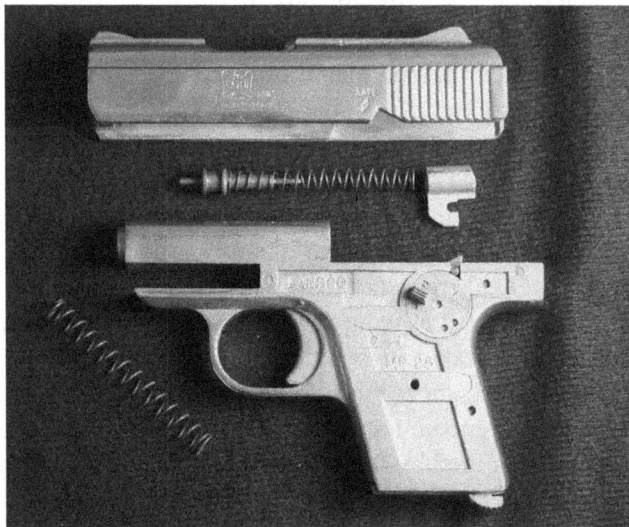

Figure 1.11 The Raven Arms MP-25, chambered in .25 ACP, is a striker-fired blowback pistol. The striker spring is cocked when the slide is retracted. The other spring is the recoil spring, which sits in the shroud surrounding the barrel. (Image from author's collection.)

the casting, Lansco, as well as the MP-25 is visible with the grips removed. The simplicity of the design is readily apparent from this perspective. Jimenez Arms has continued manufacture of the basic Jennings designs, albeit with a slightly more modern appearance. Cobra Enterprises also mimicked the basic Bryco design, with subtle cosmetic differences. These firearms were available in several calibers during their production run: .22 Long Rifle, .25 ACP, .32 ACP, .380 ACP, and even 9×19mm in a larger frame. Like the Colt 1908 Vest Pocket, these firearms were single action. The pistol must first be loaded by inserting a magazine and then the slide worked to charge the weapon. The firing train was a simple trigger bar married to a cam that locks and releases the sear. These designs suffered from two major flaws: The firing pins have a tendency to break due to substandard material used to construct them, and the feed ramps can be of irregular quality. It is not uncommon to encounter misfeeds due to ammunition being unable to chamber due to a rough feed ramp. This is especially true with the .22 firearms, particularly when the same ammunition has been cycled through the firearm a number of times by loading and unloading.

The Hi Point family of pistols, to include Haskell, Iberia Firearms, and Bee Miller, are also striker-fired, blow-back designs using a fixed barrel. These firearms feature a cast frame with a distinctly large slide, relative to the size of the frame. They are available chambered in .380 ACP, 9×19mm, .40 S&W, and .45 ACP.

Emergence of the Polymer-Framed Pistol

The polymer-framed, double-action, striker-fired pistol appeared in the form of the 9×19mm H&K VP-70, introduced in 1970 and produced until 1989. There were two variations of this pistol: The VP-70Z was semiautomatic. The second version, the VP-70, was capable of operating as a submachine gun by attaching a shoulder stock. The stock contained the selector switch to permit full-auto firing, as attaching the stock interfaced with the sear within the pistol. The VP-70 had a cross-bar-type safety that was manipulated by a push button located by the trigger, similar to many rifles and shotguns. H&K marketed the VP-70 in the late 1970s, but neither version achieved significant commercial success. However, they set the stage for the polymer-framed firearm that featured high-capacity double-stacked magazines, hammerless double-action-only operation, and a matte parkerized finish.

The appearance of the Glock 17 ushered in the new era of handguns. The tremendous success of Glock in the commercial, law enforcement, and military markets has prompted nearly every handgun manufacturer in the world to follow suit. The Glock's patented "Safe Action" is a variation on the striker system that incorporates three internal safety devices: a trigger safety, a drop safety, and a firing-pin safety. None of the Glock safety features is set manually, but each is defeated in series by a complete press of the trigger, and is reset automatically when the trigger is released. The trigger is designed to have a "slack" travel, whereupon resistance is encountered only after the first ½ inch of trigger travel. The amount of resistance is determined by the installed trigger spring and trigger connector. To the shooter, the physical feel of the trigger is more akin to single action, not double action; however, the Glock is truly a double-action firearm. The press of the trigger, in addition to defeating the safeties in succession, causes the trigger bar to move, which causes the firing pin to cock against the firing-pin spring until the trigger bar releases the firing pin when a complete trigger press is made. Since the trigger first cocks the action and then

Figure 1.12 The Smith & Wesson Sigma SW40E and magazine. The Sigma series were the first polymer-framed pistols from Smith & Wesson. The trigger press had a more traditional double-action feel: long and smooth. Smith & Wesson chose to use steel-bodied magazines instead of polymer. (Image from author's collection.)

releases it, this is a double action in the true sense of the definition, even if it does not physically or tactilely conform to the expectation of the shooter in that sense.

The overwhelming commercial success of Glock has prompted many other manufacturers worldwide to adopt a similar product. In 1994, Smith & Wesson responded to Glock with their Sigma series handguns featuring a polymer frame for a striker-fired, double-action-only pistol. The Smith & Wesson SW40E, chambered in .40 S&W, is depicted in Figure 1.12. The appearance of the Sigma prompted litigation between Glock and Smith & Wesson, later settled with Smith & Wesson paying an undisclosed sum in damages and making a design alteration to the Sigma. The movement toward the polymer-framed, striker-fired pistol did not end with the Sigma series; Smith & Wesson released their M&P series pistols that featured polymer frames and were double action and striker fired. The M&P series also includes AR pattern rifles and a series of revolvers. The success of pistols in this configuration has prompted Taurus International, Springfield Armory, Ruger, Kel-Tec, and others to develop their own similar pistols. Even manufacturers that have resisted the movement toward polymer have accepted polymer or composite frames as the new norm, such as Beretta; the H&K USP pistols; Fabrique Nationale's FNP, FNX, and Five-seveN pistols; Walther's P99, PPS, and P22; and numerous others, especially in the very compact concealed-carry pistol market. Brazilian manufacturer Taurus produces an extensive line of hammerless automatic handguns that are both single and double action. This slight technical variation is given away only when the trigger is pressed. There is a lighter press if the slide is retracted and released. Due to the fact that these pistols are hammerless, there is no visual indication of the firing condition, prompting many manufactures to install some form of indicator that reflects a loaded chamber condition.

The double-action-only striker-fired firearm is deceiving. Many have difficulty in grasping that such a design is truly double action only, because the trigger feel does not behave in a manner consistent with that expected of the traditional double-action trigger press as a shooter rooted in the traditional single/double action would expect. In the traditional sense, the single-action trigger press was very short, crisp, and light compared to a double-action press, which had a physically longer travel and heavier resistance; however, modern double-action-only firearms have engineered this out of the design.

Manually Operated Firearms

Figure 1.13 shows examples of different types of manually operated rifles.

Breech-Loaded Firearms

Breech-loaded firearms are single shot. The breech is opened, and a single cartridge is hand loaded. The breech is opened by a release lever, and the receiver hinges open to allow loading or unloading. Typically, breech-loaded firearms have spring loaded, automatic ejectors built in that eject the fired casings from the breech when opened. These firearms operate on single action and require the exposed hammers be cocked in preparation to fire. Double-barrel and combination firearms are breech loaded. Double-barrel shotguns in particular have two separate triggers, one for each chamber (see Figure 1.14). Breech-loaded firearms are also called break action, break tops, or hinged action. Gilbert Harrington, cofounder of Harrington & Richardson, patented the automatic shell ejection system and first applied it to a revolver-type firearm (Harrington 1871). The Harrington & Richardson firm later expanded this concept to include breech-loaded shotguns.

Figure 1.13 Examples of different types of rifles. Top: lever action. Bottom: bolt action. (Image courtesy of BATFE.)

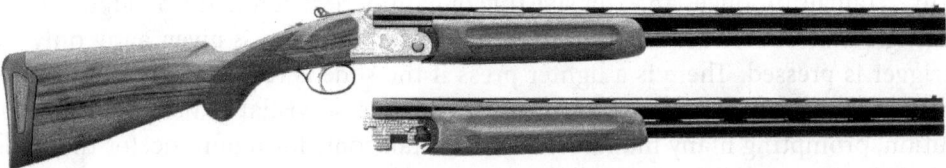

Figure 1.14 The Benelli Renaissance Classic, an over/under configured shotgun. The 20- and 28-gauge barrels are shown. (Image courtesy of Benelli USA.)

Lever-Action Firearms

Lever-action firearms were the first practical repeating arms to appear as the era of the breech-loaded firearm was being eclipsed. The lever action operates by a lever, usually comprising the trigger guard with a loop to permit the insertion of fingers through it. The lever is flipped, which ejects the spent casing, loads a fresh cartridge, and cocks the hammer in preparation to fire. The first lever-action rifle, the Spencer, appeared at the start the American Civil War. Unlike later lever actions, the Spencer did not automatically cock the hammer when the lever was worked; instead, the Spencer's lever only unloaded and loaded the firearm. Later designs combined these two separate operations into a single operation by working the lever. American firearm manufacturers are well known for long and distinguished lines of repeating rifles from such companies as Winchester, Marlin, Stevens, and Remington. Lever-action firearms are fed by tube magazines located underneath the barrel. Repeating rifles remain popular with hunters and cowboy action shooters and are basically unchanged.

There are few examples of lever-action shotguns. The Winchester Model 1887 is probably the only significant example, and it has been copied by other companies subsequent to Winchester discontinuing production. The Model 1895 by Marlin is likely the only lever-action shotgun currently in production.

The popularity of the lever-action rifle has kept it in production for over a century. The Winchester Model 1894 practically set the standard for lever-action rifles. The Model 94 was the product of continuous improvement on the part of Winchester and was continuously manufactured until 2006.

Bolt-Action Firearms

Bolt-action rifles such as the Lee Enfield, Mauser 98, Mosin Nagant, Mannlicher, and the Carcano rely on the operator to manually operate the bolt per shot fired. On such weapons, there is a handle attached to the bolt. Bolt-action rifles can either be a straight-pull bolt or a turning bolt. In the case of a turning bolt, the bolt must be opened by lifting the handle up to unlock the action and eject the spent casing. Sliding the bolt forward and turning the bolt handle downward then loads a fresh cartridge and locks the action in preparation to fire. Straight-pull bolts work differently than turndown bolts. The Steyr Mannlicher M95 is an example of a straight-pull bolt. In such a system, the bolt is pulled straight back from the receiver, ejecting the spent casing; when the bolt is returned forward, the action is cocked, and a live cartridge is then chambered. The bolt action was quite an advancement in firearm technology. The first bolt-action rifles began to appear in the early 1870s, and they remained in front-line military service until the end of World War II, when they were superseded by the appearance of automatic rifles, submachine guns, and the introduction of intermediate-caliber firearms. Despite this, bolt-action rifles remained in service with smaller nations and were scattered all over the world during the late nineteenth and the first half of the twentieth centuries.

Bolt-action rifles can draw their ammunition supply from an internal magazine, from a detachable magazine, or they can be single-shot firearms. Bolt-action arms are equipped with an extractor on the bolt that automatically ejects the loaded casing when the action is opened. Despite their age, bolt-action rifles are still very much revered for hunting, match target, and precision-shooting operations. U.S. Marine Corps Gunnery Sergeant Carlos Hathcock used a Winchester Model 70, a bolt-action rifle, as a sniper during his service in the Vietnam War. The most recent addition to the arsenal of rifles available the U.S. Army is the XM-2010, a heavily modified Remington 700. It may appear to be a totally different firearm, but its action is still that of the 700. It is chambered in .300 WIN-MAG (Winchester Magnum).

Bolt-action shotguns mimic their rifle counterparts in operation and may be single shot, contain an integral magazine, or have a detachable magazine. Mechanically, the bolt-action firearm is extremely reliable and not generally subject to malfunction. Manual safety devices are generally flip-type levers that lock out the trigger from the firing pin, or crossbar-type safeties that consist of a push button on the trigger guard.

Slide- or Pump-Action Firearms

Slide- or pump-action rifles and shotguns are manually operated by the action of a slide or pump that ejects and reloads the firearm each time the trigger is pressed, which unlocks the action and permits the slide to be moved to the rear and then pushed forward. The slide action operates by means of an action bar that connects the slide to the bolt. Some designs use two action bars; other designs use a single action bar. When the action is cycled, it locks the action closed and can only be released with a press of the trigger or by using an action-bar lock-release lever.

The majority of slide-action firearms have internal hammers and are fired from a single-action firing condition. Retracting the slide cocks the hammer, pushing the slide forward, which loads the chamber and locks the action. Typical slide-action malfunctions involve a "short stroke," where the slide is not pulled rearward with enough force or the travel of the slide is incomplete, hanging up the action. The result is typically a double feed, where the casing from the chamber cannot be ejected because a cartridge has been partially pushed into position in preparation to load.

The Winchester Model 97 shotgun, perhaps one of the most famous shotguns ever produced, has an external hammer. This allowed the Model 97 to be cocked or carried loaded but with the hammer uncocked. The hammer could be cocked by action of the slide or by hand. Another unique feature of this shotgun was its ability to "slam fire"; as long as the trigger was held down, it would continue firing each time the slide was worked because the design omitted a trigger disconnector. The Chinese company NORINCO later copied the Model 97 and sold several versions of it in the United States, although redesigned with a disconnector so that it would not slam fire. The Winchester Model 12 is also capable of slam fire.

Slide-action firearms can be fed through internal tube magazines located underneath the barrel or by a detachable magazine. There are slide-action firearms that do not have separate loading and ejection ports. One such example is the Remington Model 10 shotgun; it has a single loading/ejection port located on the bottom of the receiver. Slide-action shotguns (see Figure 1.15) are practically universal for use with law enforcement and military organizations. Slide-action rifles were never as popular as shotguns, probably because the availability of auto-loading and lever-action rifles negated market interest.

Figure 1.15 Benelli Super Nova 12-gauge slide-action shotgun. The tactical configuration model is depicted. (Image courtesy of Benelli USA.)

Ammunition Cartridges

2

Ammunition: A Backdrop

The ammunition cartridge, also known as the *integrated cartridge, self-contained cartridge,* or *fixed ammunition*, is the product of a long and persistent process of evolution, refinement, and improvement in technology. Development of the ammunition cartridge has been a combination of chemistry, metallurgy, and industrial technology. It was the advent of the self-contained cartridge that permitted a quantum leap forward in firearms technology—from arms that were muzzle loaded to those that were breech loaded; from weapons capable of single shots to manually repeating arms and onward to self-loading firearms and machine guns.

It can be said that cartridge technology developed in concert with small arms. As cartridges became more powerful, firearms technology advanced to take advantage of them. The reverse is also true: With progressive advancements in firearms, cartridges were developed, often specifically for adaptation into this new technology. Like all human endeavor, there have been many innovations, some proving to be enduring, almost timeless; yet others fell by the wayside without so much as a footnote in history to mark their passing. Concepts, ideas, and theories come and go in the pursuit of the cartridge *par excellence*. If ever there was a perfect cartridge, no one has ever agreed to what it was. To this end, cartridge development continues to be an ongoing affair, with new ones emerging all the time. These developments take the form of new projectile designs; new materials, revisions, or reworking of existing cartridges; or entirely new cartridges in dimensions not previously seen.

There are so many different types of ammunition because of the broad range of consumers and uses. Ammunition has been the subject of at least as much hyperbole as any given firearm. This exaggeration has gone in both directions, with the designers making wildly overstated claims to sell more products, while those who oppose firearms present equally outlandish and factually incorrect claims to suit their agendas.

The end of the Second World War saw a geopolitical shift occur, and two broad standards in small arms[*] ammunition practices emerged, guided by the two diametrically opposed sociopolitical ideologies that represented either side: the Western standard that continued the use of ordnance-grade brass, and the practices of the Soviet Union, where the use of brass was discarded in favor of less expensive but practical materials. The formation of the North Atlantic Treaty Organization (NATO) and the Warsaw Pact further solidified the political lines that divided the world. Nations that were influenced by either of the superpowers tended to mimic the practices of whomever they were receiving technical assistance and aid from, whether to establish an indigenous munitions production capability or

---

[*] A small arm is typically defined as any firearm capable of being handled by an individual and having a caliber less than one-half inch.

to rebuild an existing industry. The net result was vast stockpiles of ammunition that were amassed throughout the world, a mixture of cartridges ranging from postwar developments to those whose use dated back to the turn of the twentieth century.

With the end of the Cold War and the subsequent collapse of communism in the Eastern Bloc, production practices have again shifted. Producers once located behind the iron curtain have continued to manufacture cartridges as they had during the Cold War, but many have expanded to produce cartridges to the Western standard. Economics can be considered the driving engine behind this, and there is the desire to reach out to a worldwide market as well as a movement of nations into NATO membership.

With the turn of the twenty-first century has come a series of conflicts across the globe, and worldwide demand for ammunition has increased. In an April 2, 2009, press release, Alliant Techsystems (ATK) reported it had increased production at the U.S. Army's Lake City Ammunition Plant "to more than 1.4 billion rounds annually," a marked increase from "350 million rounds annually" when ATK had assumed control of plant operations in April 2000. The cartridges produced there include 5.56×45mm, 7.62×51mm, .50 BMG (Browning machine gun), and 20 mm (Alliant Techsystems 2009). Even with this output, Winchester Ammunition announced in June 2010 that it had been awarded a contract valued at US$43.4 million to manufacture .50 BMG SLAP (Saboted Light Armor Piercing) cartridges for the U.S. military (Winchester Ammunition 2010). ATK announced on October 28, 2010, that it had received more than $200 million in orders to supply 5.56×45mm, 7.62×51mm, and .50 caliber ammunition from the U.S. Army Contracting Command. This ammunition was manufactured at the Lake City Army Ammunition Plant, where ATK claims to have manufactured more than 10 billion rounds since 2000 (Alliant Techsystems 2010). In August 2005, General Dynamics Ordnance and Tactical Systems was awarded a five-year contract by the U.S. Army to deliver 5.56, 7.62, and .50 caliber ammunition. In a press release dated March 15, 2011, the company claimed to have delivered 1 billion rounds under that contract (General Dynamics Ordnance and Tactical Systems 2011).

Munitions producers seeking to capture a share of the worldwide ammunition market retooled to meet the ammunition standards of potential customers, which generally shifted to the Western standards in materials and calibers of ammunition. Producers in Eastern Europe, the Middle East, and the Far East have shifted their ammunition production to meet NATO requirements and produce NATO-caliber cartridges, even if they had not been previously tooled up to do so. Demonstrative of this point, in 2003 the Russian munitions producer Barnaul started producing NATO-standard caliber small-arms cartridges, and other Russian makers soon followed suit. Currently, all munitions manufacturers in the Russian Federation manufacture NATO-caliber cartridges.

As defined by the United Nations, "arms, ammunition; and parts & accessories thereof" imported into the United States rose steadily over the period between 2006 and 2009, as seen in Table 2.1. In 2006, the value of these commodities imported into the United States totaled $1,867,306,951. By 2009, the value had risen to $2,855,837,614, an increase of some $988,530,633, an average of $329,510,211 annually. Great Britain was the largest trading partner, totaling some $429,266,163. Germany, Italy, Austria, and Brazil accounted for the rest of the top five, with imports totaling $274,790,890, $204,758,317, $189,795,857, and $184,772,488, respectively. The remaining balance of $1,599,957,644 was shared by Norway, Canada, the Russian Federation, China, Israel, and Spain, in that order (United Nations Comtrade n.d.).

Table 2.1   Value of Imported Arms, Ammunition, Parts, and Accessories into the United States

Millions (US dollars)

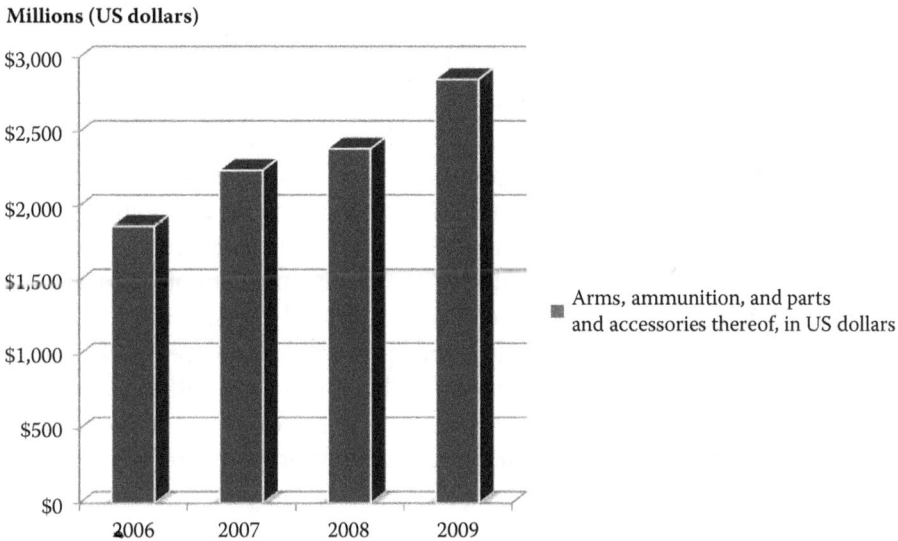

Source: United Nations Comtrade (2009).

According to the U.S. Census Bureau's 2007 Economic Census, there were 107 establishments involved in small-arms ammunition manufacturing operating within the United States. In 2007, domestic production of small-arms ammunition was valued at $2,338,684,000. The census data does not specify what percentage of the product went to domestic commercial sales, export, or military consumers (U.S. Census Bureau 2007). Perhaps some context can be referenced by the U.S. Army's own numbers. The army's demand for small-caliber ammunition soared from 426 million rounds in 2001 to 1.5 billion rounds in 2006, according to the Joint Munitions Command at the Rock Island Arsenal in Illinois. "'The government spent $688 million on ammunition last year (2006), up from $242 million in 2001,' said Gail Smith, a Joint Munitions Command spokeswoman. 'The most common rounds ordered are 5.56 mm, 7.62 mm and .50 caliber,' she said" (Wilson 2007). These numbers correspond to demands placed upon available ammunition reserves and production by U.S. military commitments in Iraq and Afghanistan, in addition to its basic needs outside of the context of the combat theater.

A June 15, 2006, briefing note published by Oxfam International* stated, "At least 76 states are known to industrially manufacture small arms ammunition." This capacity suggests an estimate of the annual global output of ammunition "to be in the region of 10–14 billion rounds, or between 27–38 million rounds per day" (Anders 2006). The primary contributors to this volume as reported were "thirty-nine per cent located in the Americas," "thirty-six per cent" in Europe and the Commonwealth States, and the balance, approximately 25%, originating from various smaller nations in Asia, the Pacific region, Africa,

---

* According to information published on their website, Oxfam "is an international confederation of 14 organizations working together in 99 countries and with partners and allies around the world to find lasting solutions to poverty and injustice."

and the Middle East, including such states as China, India, South Korea, the Philippines, Egypt, Sudan, Uganda, the United Arab Emirates, and states within the Russian Federation.

Ammunition preferences in the criminal context are entirely dependent upon a number of variables. The nuances of guns and crime vary greatly, ranging from matters of possession under unlawful circumstances, poaching and other sporting violations, robbery with a firearm, and assaults and homicides involving gunshot as the manner of injury or death. There is no reliable statistical data that can be drawn upon to truly determine what the most common firearms, and hence ammunition preferences, used in crime are, and the argument that certain types of firearms are "crime conducive" is an illogical statement at best. At present, the best method to determine the most common firearms used in crime relies on firearm trace data that is generated by the Bureau of Alcohol, Tobacco, Firearms, and Explosives (BATFE), although that data has admitted flaws. Firearms, as a general rule, are specifically chambered to handle a single cartridge, and are legally identified by that specific cartridge, even if said firearm is suitable for use with other cartridges. There are exceptions to this statement, because firearms can be reconfigured to use a different cartridge, and many are purpose-built specifically to support using different-sized ammunition cartridges.

Another variable in addressing the prevalence of certain ammunition cartridges over others may be dependent upon geography. In venues where hunting game is common, there may be a greater proportion of firearms, and hence ammunition, within the sporting calibers than in other venues where such activities are less common and the population of firearms in that region is different. As there is little demand for hunting arms in urban environments, the prevalence of firearms in such a hypothetical scenario suggests that other types of firearms may be more directly attributed to criminal activity, such as small-caliber handguns that would have greater demand on the secondary gun market due to ease of concealment and market preferences. Neither of these statements takes into account the possibility that any firearm can appear anywhere under a criminal context. This argument is made moot by the presence of more general-purpose arms such as shotguns, which are abundant in all areas.

A 2002 revision of a 2001 study by the Bureau of Justice Statistics surveyed prison inmates about where they obtained the firearms in their possession at the time of the current offense that caused their incarceration. It was reported that 13.9% of those surveyed obtained the firearm by purchase or trade from a retail gun outlet (pawn shop, gun shop, etc.). A high percentage of those surveyed, 39.6%, obtained the firearm from a friend or family member in the form of a purchase, trade, rental, or loan. A nearly equal percentage, 39.2%, obtained the firearm via illegal sources such as theft, black market, or from secondary markets such as street transactions. Other, unidentified means accounted for the remaining 7% (Wolf Harlow 2001). Since a firearm is of little or no value without ammunition, the sources of acquiring ammunition for illicit purposes must also be sought. According to a 2004 RAND Corporation survey of 2,031 persons who purchased ammunition in the Los Angeles area, 52 persons had felony convictions or were otherwise legally prohibited from making an ammunition purchase (Tita, Pierce, and Braga 2006). This relatively small percentage suggests that the majority of prohibited persons who acquired ammunition must have obtained it from other sources, perhaps coupled with their obtaining the firearm, including theft of the ammunition with the firearm, ammunition supplied from street sources, or ammunition received from the friend or family member from whom the arm was obtained.

Ammunition, regardless of the intended market or user, contributes greatly to the success of a weapon that uses it. Even the most efficient and practical weapons platform can be compromised by ineffective ammunition. Effectiveness or ineffectiveness can be subjective, strictly subject to the surveyor. The cartridge can be thought of as a platform, a base where a projectile is expected to behave within a certain set of parameters based upon its size, mass, and powder charge. But within the platform are a multitude of options to redesign or reconfigure some design element to enhance its performance, or to explore a different application of the platform. The most popular cartridges can be configured into a wide variety of applications, particularly if the platform is carried in the inventories of military forces and different operational needs have to be met. The .45 ACP (Automatic Colt Pistol) is one of the most widely used cartridges for personal defense, military and law enforcement applications, as well as sport and target shooting. Evolutions in cartridge design take the form of different gunpowder formulations to achieve specific design goals, such as increased velocity, reduced muzzle flash report, environmental and health concerns, etc. The projectile performance can be enhanced by engineering and design changes as well, making it lighter or heavier by using different materials, and by tuning its physical shape and structure. Projectiles are purpose-designed for specific applications, though any projectile, regardless of design intent, can be capable of inflicting serious or lethal injuries. Quite frequently, an existing cartridge platform serves as the inspiration for a new cartridge altogether.

There are many hundreds, if not thousands, of different cartridges, the majority now being obsolete or passing into obsolescence. More cartridges are extinct today than are commonly encountered, having passed into history with improvements in technology, either in some element of the ammunition or the guns that used them. The classic full-power, full-sized rifle cartridges that served the world's armies for nearly 50 years for the most part gave way to changes in firearms technology as they transitioned from bolt action rifles for the infantryman to automatic weapons. What keeps many of these cartridges alive is the huge number of arms that still exist and shooter interest in them for recreational shooting, collecting, and hunting. Archaic guns can still be found used in conflicts around the world, for sustenance, or protective purposes, but clearly their days are numbered as they wear out or are replaced by more modern equipment. In spite of technological advances, many cartridges are practically timeless, remaining popular and viable platforms to use. Some are so prolific that it is unlikely that they will disappear from the landscape any time soon. Any given cartridge can be the subject of specific and in-depth study. This is particularly true of military cartridges because numerous variations exist; there simply are no such things as the "normal" or "regular" kinds of ammunition cartridges. Ammunition is designed from the outset with an intended purpose. The three principle purposes of ammunition are:

- *Lethal*: Intended to cause injury that is serious, permanent, incapacitating, or deadly
- *Training*: Intended to improve shooting skills, with cost effectiveness and safety in mind
- *Less lethal*: Intended to obtain compliance from an otherwise noncompliant subject using pain and discomfort

Lethal ammunition is designed with the express intent to cause deadly or at least incapacitating injuries to the target. This is not to say that there is any guarantee that injuries caused by such cartridges will cause death; there are many variables beyond the control of the cartridge design that ultimately determine this. Factors such as shot placement within the target body, total amount of traumatic injuries delivered, accessibility to qualified medical care, and physiological factors on the part of the target will play a significant role in the net result realized by application of lethal munitions. That being said, it must always be presumed that the application of lethal munitions will cause death. Lethal ammunition is within the purview of military, law enforcement, personal defense, and hunting purposes.

Training ammunition is designed with the intention of facilitating a further understanding and development of marksmanship skills. Training ammunition can take on many forms, but dedicated training ammunition cannot be looked upon as being incapable of causing death or serious injury. Training ammunition is a broad term; it could mean wholesale low-grade ammunition sold at discount, or specialized, premium, match-grade cartridges that are practically handmade for specialized competition use.

Less-lethal ammunition is designed to cause pain, discomfort, and perhaps temporary incapacitation to gain the acquiescence of an otherwise noncompliant individual. As is the case with training ammunition, less-lethal ammunition cannot be assumed to be incapable of causing death or serious injury. Less-lethal ammunition may cause either due to misapplication of the ammunition or by unintended, yet unfortunate, consequences of the application of less-lethal munitions. Less-lethal munitions largely fall within the purview of law enforcement users.

Ammunition is referred to in any number of different terms: *cartridge, self-contained cartridge, bullets, shots,* or *rounds,* to name a few. Technically speaking, these terms are not all interchangeable, as they do not truly represent the same article. In less technical speech, terms such as *bullets* are often applied but can lead to some confusion, particularly since the term itself does not specify whether it is fired or unfired bullets. It is imperative that the investigator use proper nomenclature and terminology and maintain consistency in its use throughout the investigation. Defining an artifact as a "bullet" and then later referring to the same artifact as a "round" or "cartridge" can be misleading to the users of that material down the line, as well as potentially creating issues for that particular artifact. For best practices, it is recommended that whatever term is deemed best to use, that it be described as either being spent or live, i.e., spent projectile, spent casing, fired casing, etc. The term *bullet,* most properly applied, refers to the projected portion of a cartridge, or simply the projectile. A properly applied adjective—spent or live—best accompanies whatever term the user is most comfortable using. Other variables in ammunition are described in the text and may be best applied when describing a particular artifact when it is encountered.

Cartridge Identification by Dimensions, Names, and Identifiers

Caliber

The use of the imperial system, using fractions of an inch and designated as *caliber* to indicate the interior bore diameter of the arm in question, originated with smooth-bore arms. Since projectiles were bare lead, they were often undersized relative to the interior bore diameter of the host firearm, allowing for a projectile to have a linen or paper patch that was used to encase the projectile, a rudimentary form of jacketing, prior to ramming the projectile down the barrel with the intention of reducing the fouling of the bore. With the introduction of self-contained cartridges, the definition of *caliber* was amended slightly to accommodate this change in technology. However, there has never been an established, standardized protocol or convention for determination of caliber. The advent of the rifled

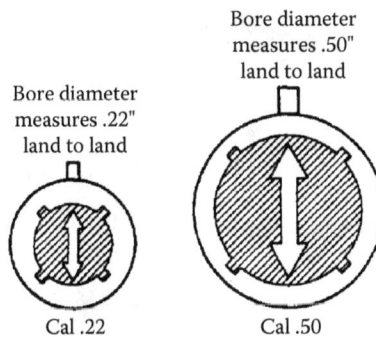

Figure 2.1 Interior bore diameter measurement to ascertain caliber. (Image by U.S. Army.)

bore only complicated the situation, because now the caliber could be determined either by measuring land to land or groove to groove. The land is the raised portion of the rifle profile of a bore. The recessed portion is the "groove" (see Figure 2.1). Traditionally, the European practice has been to use a deeper groove in a rifled barrel, mating it with a slightly undersized diameter projectile, relative to the groove diameter. American preference traditionally has been to use a projectile that equals the groove diameter. The term *caliber* remains somewhat elastic and can be defined by several different meanings:

- The interior diameter of the firearm bore, measured from land to land
- The interior diameter of the firearm bore, measured from groove to groove
- The interior diameter of the cartridge casing neck opening
- The diameter of the projectile at its widest circumference
- A value as assigned by the designer or underwriting firm

In its simplest form, the caliber is an expression of the decimal representation of a fraction of an inch, using the number ten as a base. The caliber is generally expressed in terms such as, for example, .25, .44, or .68. In this context, it is not necessary to follow the numerical expression with the term *caliber*, as it would be redundant. In speech, the term would be stated as "25 caliber"; however, verbal expression of the term has included "point 25 caliber" (stating the period as "point"). How the speaker wishes to express the term is entirely based on one's preference, as long as the term is used consistently in the course of the dialogue.

A cartridge or projectile that is expressed as .25 would suggest a literal caliber of 25/100″ or ¼″; however, this numerical value may be somewhat misleading. Many cartridges are of dissimilar dimensions, as represented by their names. Many of the .44-class projectiles have a true diameter of .429 instead of the advertised .44. Other examples include the .380 ACP (actual projectile diameter .355), the .38 Special (actual projectile diameter .357), and the .357 Sig (actual projectile diameter .355). To further the point, the diameter of the .38 Super projectile is actually .355, just 1/250″ in diameter narrower than the .38 Special, which in turn is the same diameter as the more powerful .357 Magnum cartridge.* The .30-30 Winchester, .30-06, .300 Winchester Magnum, and the .300 Holland & Holland Magnum all share a common projectile diameter of .308″.

---

* With the exception of the length of the casing, the .357 Magnum is exactly the same as the .38 Special. The term *Magnum* in firearms parlance indicates an extended or elongated cartridge casing, allowing for a greater amount of powder to be filled.

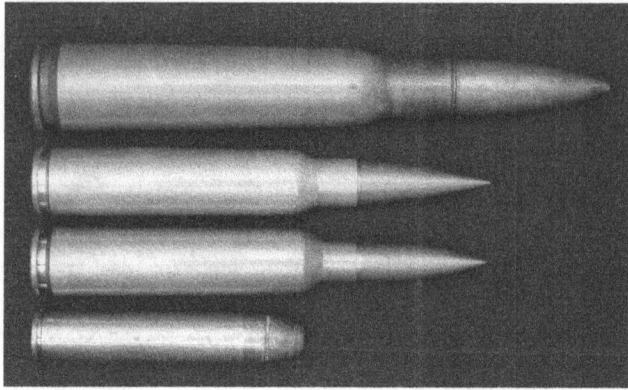

Figure 2.2 The term *caliber* can be deceiving, as evidenced by this cartridge comparison, from top to bottom: .50 Browning Machine Gun (12.7×99mm), .408 Chey Tac, .338 Extreme, and .444 Marlin. (Images from author's collection.)

When loading data is consulted, any specific "caliber" can be found to have a projectile diameter range that will seat in a particular casing. The range expresses a minimum and maximum diameter, and the differential is measured in literally hundredths or thousandths of an inch. Caliber, at best, can be used as an estimation to determine the diameter "class" of a particular projectile. The .40 S&W is a truly .40-class projectile, developed by the gun maker Smith & Wesson. The .357 Sig, also known as .357 Sig Auto or simply .357 Auto, was developed jointly by gun maker Sig Sauer and Federal Cartridge. Not to be confused with the .357 Magnum cartridge, the Sig cartridge was designed for use in autoloading pistols, while the .357 Magnum was designed for use in revolver-type handguns. Taking the two calibers strictly at caliber designation, there is no apparent difference unless they are viewed, whereupon the differences are immediately noticeable. The .357 Magnum has a longer overall length, and has a cylindrically shaped casing with a full rim so that it will fit in the step of a revolver cylinder, allowing for proper seating and ejection from the cylinder. The .357 Sig cartridge is shorter and has a bottleneck-type cartridge casing, appearing to be more of a small rifle cartridge than that expected of a handgun.

The most significant problem with caliber is that it fails to indicate the overall length of the projectile or cartridge casing. In the early years of the self-contained cartridge, this typically was not much of an issue, given the fact that there were relatively few cartridges on the market. However, this changed so rapidly that the market filled up with synonymous sounding ammunition. It would surprise many to discover that the .357 Magnum, .30 Carbine, and the .41 Magnum cartridge casings can be the same maximum length, 1.290″, whereas the .44 Magnum and the .45 Colt are ever so slightly shorter at 1.285″. It is a common assumption that a larger caliber, say .44 Magnum, would be dimensionally larger than a .41 Magnum, although they both are Magnum cartridges. Figure 2.2 depicts various calibers, although the comparison of caliber values can vary greatly from the physical dimensions.

Hyphenations, Names, and Other Identifiers

Self-contained metallic cartridges were first identified using a hyphenated system. As there was no convention or norm, this was voluntary, and not every manufacturer chose to follow it, so there are exceptions to the general guidelines. Cartridges were hyphenated in two or three sets, such as the .50-70, .45-70, and .30-40. The first number designated the projectile caliber, followed by the propellant charge defined in grains of black powder. In three-set designations such as the .45-70-350, the third number indicated the projectile weight, expressed in grains. The practice of using hyphenations continued until the turn of the twentieth century. The switch from black powder to smokeless powder rendered the need to indicate the black powder charge redundant.

During this transitional period, it was not uncommon for cartridges to became known by two different names, such as the .38 WCF (Winchester Centerfire), also called the .38-40. There were still cartridges identified by hyphenations even well into the era of smokeless powder, such as the .30-30 Winchester, which was never commercially loaded using black powder, but Winchester stayed with the hyphenated nomenclature, which was still in use when the .30-30 was introduced, and remains so to this day. When the U.S. Army switched their standard cartridge from the .30-40 Krag, the new cartridge went by a hyphenated name, .30-03. The .30-03 derived its name from its .30-class projectile and the year of its introduction into service, 1903; thus the last two digits of the year now replaced other information, i.e., the weight of the contained powder charge. The .30-03 was superseded in three years by the .30-06, which promised superior ballistic performance. The .30-06 drew its name in the same manner: .30 caliber-class projectile introduced in 1906. The .30-06 would later become universally known as .30-06 Springfield, for the U.S. Arsenal at Springfield, Massachusetts (not to be confused with the commercial gun maker Springfield Armory, located in Genesco, Illinois).

During World War II, the U.S. military brought into service another .30 caliber-class cartridge; however, this cartridge had nothing in common with the .30-06. The .30 Carbine defied the convention of hyphenated nomenclature and instead replaced it with a descriptor, in this case indicating the firearm, the M1 carbine. The cartridge was developed concurrently with the firearm and specifically for such.

Cartridges developed after the passing of the black powder era may still use hyphenated names that may indicate other information that varies greatly, depending on the intention of whoever named the cartridge. The Savage .250-3000 is a cartridge that was developed by the Savage Arms company, using a .250-class projectile with a reported velocity of 3,000 feet per second. This cartridge is not to be confused with the .250/3000 Improved, a different cartridge altogether. Another example is the .338/50 Talbot, which uses a .50 BMG casing that has neck diameter reduced to .338. Obviously, the intention was to use the large capacity offered by the .50 BMG casing to propel a relatively small projectile a long range at an extremely high velocity. Such identifications are most commonly reported in cartridges that fall under the categories of "Wildcat" or "Proprietary," meaning that they are not standard cartridges that could be expected to be encountered outside of customized, special order, or limited-production guns or in specific circumstances. A slash may be used in substitution for hyphens. The .38/200 cartridge is also known as the .38 British Service. This cartridge replaced the .455 Webley as the standard sidearm caliber in British and Commonwealth Armed Forces. The projectile was .38 class but weighed 200 grains, a heavier projectile than used in other .38-class cartridges of the day, such as .38 Smith & Wesson or .38 Super. In fact, the .38/200 was derived from the .38 Smith & Wesson cartridge, but with the heavier projectile to satisfy the British preference.

It has been customary for cartridges to be named or to include an adjective as part of the cartridge identification. Quite often it was the case that the firearm manufacturer also was the sole producer of the cartridge for their firearm, hence the need to identify exactly what cartridge was suitable for what firearm. This became particularly important, and more and more cartridges were added, many of them centering on certain popular calibers, such as the

.45 class. Ammunition was often named for the designer, sponsoring firm, the manufacturer, the intended user, or some other adjective that described its purpose or provided a catchy name. It is not uncommon for a cartridge to be known by several names, even if it is not technically correct. In the case of the .30-40 Krag, the cartridge was named using the old hyphenation system despite being designed for use with smokeless powder instead of black powder. The cartridge featured a .30 projectile, a 40-grain smokeless powder charge, and Krag was the name of the rifle it was chambered for, which in turn was named for the designers, Johannes Krag and Erik Jorgenson. This same cartridge was later called the .30 Army; however, this term never really gained as much acceptance as simply calling it the "Krag."

Similarly, there are innumerable cartridges that are called the "Remington," "Marlin," "Winchester," "Smith & Wesson," "Colt," "Norma," etc., so named for the firms that brought them to market. These names may also provide a clue in providing the examiner with more detailed information, such as "Long Colt," "Short Colt," "Browning Long," "ACP" (Automatic Colt Pistol), "Bergman," "Bayard," "Largo," or "Browning Short." The name may describe the intended user, such as "Police," "Government," "NATO," or "Army." The name may be an adjective describing something about the cartridge itself, even as a marketing ploy or something catchy: "Fireball," "Express," "Magnum," or "Nitro." In cases of national calibers, terms such as *Spanish*, *Turkish*, *Chilean*, and so forth designate a particular cartridge, even in a class of similarly dimensioned cartridges.

Metric System

The metric system was signed into accord by 17 nations in 1875 (International Bureau of Weights and Measures n.d.). For the most part, signatories of the convention immediately went to metrics and discarded the caliber. The United States and the United Kingdom were loath to adopt the system and chose to remain on the Imperial system for the purposes of small-arms ammunition. Ironically, during the same period that the metric system was taking hold around the world, the state of the art of ammunition was changing dramatically. As a result, cartridges originating from nonmetric nations continued the practice of caliber, whereas in nations that went to metrics, their cartridges were identified by their metric measurement. The continental European military cartridges of the day were largely centered on the 11 mm class, nearly one-half inch in diameter. By the early 1880s, this had started to change, in keeping with the practice of the United States and Britain, who were switching standard military cartridges from the larger calibers to those around .30″, roughly 7 to 8 mm in diameter. These remained unchanged until after World War II and the start of the Cold War. In Europe, ammunition was firmly based upon national preference. Each nation developed cartridges for their own use; some exported it out to less industrialized nations for use or armed their remote colonies with it as well. Unlike the caliber cartridges, metric cartridge measurements provide not only the stated or indicated projectile diameter, but also the cartridge casing length (not the overall length of the loaded cartridge that includes the projectile, but only the casing). A metrically measured cartridge such as the 7.92×57mm would indicate a projectile diameter of 7.92 mm with a casing length of 57 mm.

The prevalent conventions for cartridge measurement did not prevent companies and inventors from naming a particular cartridge after themselves; using a descriptive term or adjective in conjunction with the cartridge dimensions; or defining the application for the particular cartridge, such as 9mm Parabellum (the term *Parabellum* being Latin meaning "prepare for war"). This cartridge is known by many names, such as 9mm Para, 9mm Luger (named for Georg Luger), 9×19mm (the projectile diameter and casing length), and 9mm NATO (when it was officially adopted as a standard NATO cartridge). In this particular instance, it is important to note that a cartridge designated as 9mm NATO is precisely the same dimensionally as any other 9×19mm cartridge; however, the 9mm NATO is loaded to a considerably higher pressure and may not be suitable for use in all firearms chambered for the 9×19mm cartridge. The 9×19mm L7A1, manufactured by the Austrian firm Hirtenberger, is the classic example. Although it is a 9×19mm cartridge by dimensions, the pressure load is so high that it is deemed unsafe for use in anything other than submachine guns or other long guns chambered for 9 mm. Another high-pressure 9×19mm cartridge was made by Israeli Military Industries and sold under the UZI brand ammunition line. The cartridges were identified by their black tip, but were of the ball variety. The cartridge head stamp read "IMI 9mm" and "CARB," presumably for carbine.

Cartridges Having Caliber and Metric Designations

There are cartridges measured in both caliber and metric measurement. The practice of dual dimension is not new, starting in the late 1800s. There was concurrent firearm and cartridge development in the United States and Europe, and these products were widely exported. An example of dually identified cartridges includes the .380 ACP, also called the 9×17mm, but it is also seen as the 9mm kurz (*kurz* is German for "short"). The .25 ACP is the 6.35mm, and the .32 ACP is the 7.65mm. Some manufacturers have labeled their .22 ammunition as 5.5 mm (but is actually closer to 5.4 mm); this is uncommonly used, but is so identified in cartridges originating from China, the Russian Federation, and some other nations. The .416 Barrett, a relative newcomer to the ammunition world, is metrically expressed as 10.5×83mm and the use of the descriptor "Barrett" for the developing firm, Barrett Firearms Manufacturing. Similarly, NATO standard cartridges must be identified by a metric measurement, as the majority of the NATO partners use metrics. Therefore, any cartridge that has any hope of becoming a NATO standard or recognized cartridge must be measured metrically, even if originally designated by caliber. The NATO standard cartridges include the 9×19mm, 5.56×45mm, 7.62×51mm, and the 12.7×99mm.

Other Identifiers

In addition to hyphenations, adjectives, or other names, some cartridges have a suffix that presents additional information. The presence of an *R* indicates that the cartridge casing is a rimmed type. A 7×57mm is a rimless design, whereas the 7×57R is a rimmed casing. The Soviet 7.62×54mmR indicates that the cartridge has a full rim; however, the British .303 cartridge, while also having a full rim, does not carry such indication. A caveat to this guideline is the .500 Smith & Wesson Magnum cartridge, which may be designated with an *R* suffix. In this case, the R indicates that a rifle-type primer is used to prime the cartridge, as opposed to the cartridge having a full rim. The term *SR* indicates a cartridge casing that is semi-rimmed. The .25 ACP, .32 ACP, and the .38 Super are calibers used in semiautomatic pistols and all are all examples of semi-rimmed cartridges although the term "SR" or semi-rimmed rarely appears in cartridge identifiers. The term *JS*, as in the case of the 8×57mmJS cartridge, reflects a rather sordid history. The *J* has no meaning and was applied due to confusion in the translation; however, the *S* indicates a "spitzer"-type projectile. Despite this minor historical inconvenience, the term has stuck, probably as a practical matter. Why create confusion when it was already confused?

When a cartridge is taken into military service, regardless of the commercial caliber designation, it will receive a military model designation. These systems are entirely dependent upon the practices of the force utilizing the cartridge, and formal nomenclature has been known to change from time to time. For any given cartridge in use by a military force, there may be variations of the same cartridge, and each would receive particular designations to differentiate one from another. Such variations include tracers, armor piercing, training, blank, and match grade, among others. The same cartridge adopted by different nations would receive different identifying nomenclature that would be unique to each nation. Such is the case of the 5.56×45mm NATO cartridge loaded with the SS109 projectile, which is known by different nomenclatures by different nations: in the United States as the M855, in Australia as the F1, in Canada as the C77, in the United Kingdom as the L2A1, and in Germany as the DM11.

Shot Shells

Shot shells, as well as shotguns, generally are measured by *gauge*. The gauge is not a caliber as defined using metrics or imperial measurements. The gauge of a shotgun or the diameter of a shot shell, that is to say the internal bore diameter of the shotgun barrel, is determined by the number of lead balls of that diameter required to equal one pound. In other words, a 12-gauge bore means that 12 lead balls of that diameter (approximately 72/100″ or .72 caliber) would weigh one pound. In European circles, the term *bore* is often used in lieu of the term *gauge*; however, gauge has more universal acceptance. The terms are interchangeable for all practical purposes. The only common modern exception to the use of the term *gauge* in shotguns is the .410, which is the caliber, and may only be properly referred to as the caliber, and not the gauge.

From the definition of gauge, it is surmised that the smaller the numbers, i.e., 10 gauge versus 12 gauge, the larger the bore. This definition should not be taken to mean that the projectile(s) fired from a shotgun will be of that exact size of the bore itself; indeed, quite the contrary is true. Conversely, the larger the number, 16 gauge versus 12 gauge, the smaller is the bore diameter. Shotguns have been manufactured in gauges ranging from 4 up to 32 gauge. Gauges other than 12, 16, 20, and 28 are uncommonly encountered. Figure 2.3 compares various gauges. The largest gauge firearms are themselves highly specialized and rare weapons indeed. As a subtle variation, certain high-end safari-style arms are smooth bore and may be referenced either by gauge or caliber. Shotguns have been in existence for so long, produced in so many different parts of the world, and for so many different applications that esoteric and oddball sizes are to be expected. Shotgun gauges are generally not associated with caliber other than to provide a form of context for understanding the actual bore diameter of a shotgun, using inches as the guide. Table 2.2 shows the caliber equivalent (measured in inches) for different shotgun gauges.

In addition to the gauge, shot shells are measured by the overall length of the shot shell itself. Shells measuring 2¾″ (70 mm) in length are the modern standard; 3″ shells are referred to as *Magnums*, like other elongated cartridges; and 3½″ shot shells are *Super Magnums*. Shot shells shorter than 2¾″ length exist, as do shotguns that chamber them, but they are largely collector's items and oddities now, although the Mexican ammunition manufacturer Aguila manufactures 12-gauge 1¾″ "mini shells." These mini shells are available in a variety of loads from slugs, light birdshot, and duplex. Unloaded shotgun shells, either spent or virgin, are referred to as *hulls*. It is possible to inadvertently interchange smaller

Figure 2.3 (See color insert.) A cross-section of various size shot shells, from left to right: 12-gauge 3½" Super Magnum, 12-gauge 3" Magnum, 12-gauge 2¾", 16-gauge 2¾", 20-gauge 2¾", Aguila 12-gauge 1½", and .410 3" shell. (Images from author's collection.)

Table 2.2   Shotgun Gauge to Caliber Equivalency Table

Gauge	Caliber Equivalent
8	.83″
10	.77″
12	.73″
14	.69″
16	.66″
18	.63″
20	.61″
24	.58″
28	.55″
32	.52″
36	.410″

shot shells into larger gauge shotguns, such as loading a 20-gauge shell into a 16-gauge shotgun, or even a 20 gauge into a 12 gauge. Such an interchange creates a hazardous situation that, at a minimum, could jam the weapon, and in a worst-case scenario could cause serious injury or even death. Generally speaking, it is safe to use a shell shorter than the chamber, such as the case of shooting 2¾″ shells from a 3″ chambered shotgun.

Aside from the bore and shell length, the shot shell is identified by the size, weight, and type of shot that is loaded into the shell. When examining the shot shell, aside from the gauge and the shell length, there are three additional numbers to note. These three numbers are marked in a sequence, each separated by a space, hyphen, or slash. The first number of the sequence is what is called the *dram equivalent*[*]; the second number is the shot

---

[*] A *dram* is defined as being equivalent to 1/8 ounce or 60 grains in the apothecary system of weights; it is also defined as 1/16 ounce, 27.34 grains, or 1.77 grams in the avoirdupois system of weights.

charge; and the third number represents the shot size number, for example, the numerical series 1 1/8-8. The dram equivalent is an archaic method, but still in use, that indicates the equivalent amount of black powder that would be needed to produce the same projectile velocity as a smokeless powder charge. The shot charge indicates the amount of shot loaded into the shell by weight, given in ounces. A higher weight equates to a larger load; however, this should not be taken to indicate that there are more pellets. Larger shot is not expressed by weight but, rather, by the number of pellets loaded. The third number is an expression of the shot size loaded in a particular shell. Internationally, there are slight variations in shot size standards, as well as some variations of shot identification, depending upon the material from which the shot itself is constructed.

Components of Fixed Ammunition

The diversity of legal venues defines ammunition in subtlely different manners. In some venues, the term *ammunition* defines any singular component (i.e., primer, projectile, powder, or casing) that could be used in the construction of a functional unit of ammunition, called a *cartridge* or *fixed ammunition*. Fixed ammunition would constitute the presence of all necessary components assembled in a manner that renders it ready to operate when inserted into a suitable firearm. Nonfixed ammunition refers to ammunition loaded into weapons individually, such as the case of an artillery piece where the projectile is independently loaded from the propellant charge, which is separately contained. The definition of *cartridge* or *fixed ammunition* may further be construed to include combinations of any of the components; yet other venues define *ammunition* or *cartridges* as only when all the necessary components are present and only when they are assembled into the functional product. The legal lines may further be blurred by various exceptions, such as age, recognized or articulable collector value and rarity, availability through routine commercial channels, size, and design intent and purpose (such as whether it is deemed for sporting purposes, armor piercing, etc.). Furthermore, many venues prohibit certain persons from having in their possession ammunition or even ammunition components. Such persons routinely prohibited from possession of ammunition could include those convicted of felony crimes, the mentally incompetent, those present in areas where specific laws broadly prevent possession, and those prohibited by action of a court having jurisdiction over them.

The self-contained cartridge is comprised of four components:

1. A casing or hull constructed of metallic or nonmetallic material
2. A primer or internally contained ignition composition that detonates when acted upon by an external force
3. A contained propellant charge that provides the energy for expulsion of the projectile(s)
4. Projectile(s) either recessed within the casing or pressed into the opening at the casing mouth

Cartridge Casing

The casing or hull is the container that holds all of the other components together as a single unit. The ignition source, propellant charge, and projectile(s) are all contained within

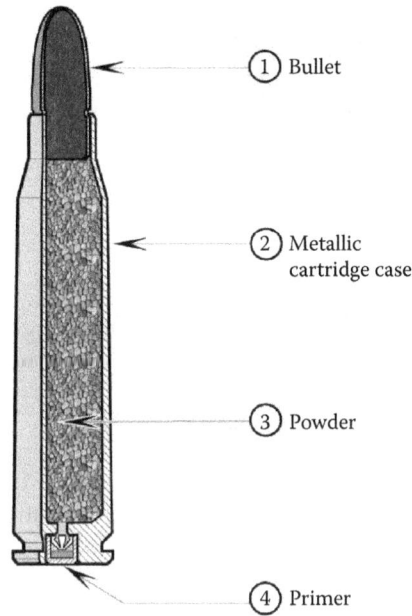

Figure 2.4 A cross-section of a generic self-contained cartridge showing components. (Image by Bureau of Alcohol, Tobacco, Firearms, and Explosives.)

the casing. Figure 2.4 is a generic anatomy of a typical center-fire cartridge. In addition to providing the container that brings the sum of the parts together, the casing serves the additional purpose of acting as a gas plug. The casing itself must be strong enough to withstand pressure, however instantaneously, when the contained powder is detonated and the gas created as a result of the detonation begins to expand. As a plug, it must seal against the breech of the firearm, thereby preventing gas leakage back into the firearm, which could not only create a hazardous condition, but may also deprive the projectile of performance by diverting energy away.

One of the first observed characteristics of ammunition, in particular the casing, is the shape. There are obvious differences in the shapes of casings: straight, tapered, cylindrical, and the bottleneck (see Figure 2.5). The basic cylindrical-style casing can be of any proportion relative to its length versus diameter. Many cylindrical-shaped casings have a slight forward taper; however, this may be barely discernible to the eye, whereas there are casings that are radically tapered, as is the case with the .22 Remington Jet cartridge. The design of cartridge casings changed quickly, and a bottleneck shape took hold to replace the straight casings of full-sized, full-powered rifle and, later, machine-gun cartridges. Bottleneck-style casings, in contrast to cylindrical casings, start with a wide base and taper down into a smaller diameter through a neck that resembles a soda bottle. The tapering is referred to as the shoulders. The exact dimensions of where the cartridge begins to taper, called the shoulder, and how radical the taper is vary greatly. There are casings with very long necks as well as cartridges that have comparatively short necks in proportion to the overall cartridge casing length.

From a design standpoint, the bottleneck cartridge casing is ideal. As the trend in rifle cartridges changed from larger bore diameters in the .50″ class and transitioned into smaller bores, centered in the .30″ class, the bottleneck casing allowed more space for a larger powder charge pushing the smaller projectile. Bottleneck-shaped cartridges are not

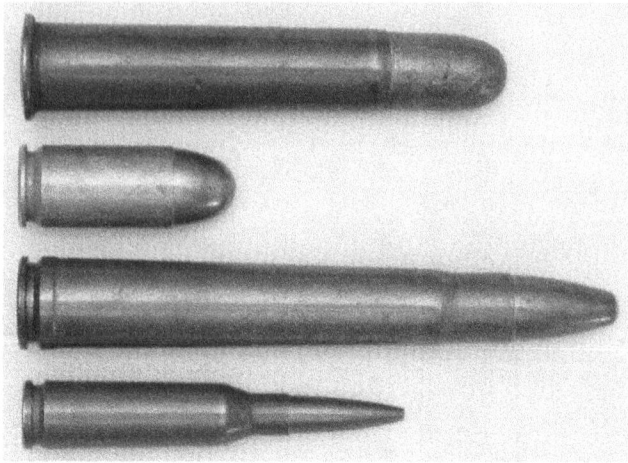

Figure 2.5 Casing styles (from top to bottom): .45-70 Government, .45 ACP, .375 Holland & Holland, and 5.45×39mm Soviet M74. In addition to the bottleneck design, note the tapering of the .375 Holland & Holland casing from base to shoulder. (Images from author's collection.)

limited exclusively to rifles; bottleneck cartridges developed for handguns were developed as well. The 7.63×25mm Mauser (introduced in 1896), the 7.65mm Luger cartridge (introduced in 1898), and the Soviet 7.62×25mm Tokarev (introduced in 1930) are all early examples of the bottlenecked handgun cartridges, but not the only examples. It is not a coincidence that the 7.62×25mm Tokarev is dimensionally similar to the Mauser cartridge, the Russians having drawn their inspiration from it, with the caveat that the Tokarev cartridge is slightly more powerful and, therefore, firearms chambered for the Mauser cartridge would not be suitable for use with the Tokarev cartridge. As with rifle cartridges, the bottleneck handgun cartridges allowed for larger powder charges propelling smaller projectiles, truly a potent combination. The 7.65×25mm Mauser was a preferred Magnum cartridge of its day and was ideally suited for the gun originally chambered for it, the Mauser C/96, affectionately called the "Broom handle." The C/96 had the ability to accept a shoulder stock, turning it into a compact rifle, and in this configuration, the power of the ammunition was not a hindrance.

The bottleneck handgun cartridges came and went, with the exception of the Tokarev, its continued use guaranteed by the Soviet Union and then later by nations receiving military assistance from them, as well as numerous copies made in various countries. It is still a common cartridge, given the millions of examples of handguns and submachine guns that were chambered for it. The bottlenecked handgun cartridge was reintroduced to the world in the early 1990s, in the form of the .357 Sig Auto. The Belgian firm Fabrique Nationale and the German firm Heckler & Koch have both brought bottleneck handgun cartridges to market: the FN 5.7×28mm and the H&K 4.6×30mm.

A second identifiable feature of a cartridge casing is the rim design. The rim is located at the base, or bottom, of the casing in the area identified as the cartridge head. There are six basic rim designs: grooveless belted, semi-rimmed, rimmed, rimless, rebated rim, and rimless grooveless.

The belted type has a pronounced "belt" of additional casing material running around the circumference of the cartridge base above the groove. A belted casing is customarily attributed to large full-powered cartridges used for hunting or long-range precision rifles and is the strongest casing type, capable of handling very high pressure loads. An example of a belted casing is the .300 Winchester Magnum (normally known by the abbreviated .300 Win Mag).

Semi-rimmed casings have a rim that projects slightly wider than the casing diameter. Rimmed casings have a rim that is markedly larger than the diameter of the casing. Rimmed casings are most easily equated to cartridges intended for use with revolvers, such as the .38 Special, .44 Magnum, .38 Smith & Wesson, .32 H&R Magnum, and so forth. Rimless casings, despite the name, are not literally without a rim; the rim is simply the same diameter as the casing. Rimless casings are the norm for autoloading firearms, commonly the .40 S&W, 10mm Norma, and the .32 ACP.

A rebated rim is a rim that is undersized relative to the casing diameter. Rebated rims are an uncommon variety; recent iterations of rebated rim cartridges are the .458 SOCOM and .50 Beowolf cartridges. The likely rationale behind the use of the rebated rim was to simplify adaptation of these cartridges into the AR-15/M16 weapons platform. Rimless, grooveless cartridges are archaic, having seen only limited use in the late 1800s and then quickly discontinued. A rimless, grooveless cartridge literally lacks any form of rim and groove, essentially giving the projectile a clean, cylindrical appearance.

The basic function of the rim and groove is to provide for the seating and subsequent expulsion of the casing from the host firearm. The small recessed area around the circumference of the casing between the casing body and the rim is called the groove. The groove provides the channel for an extractor claw to hold the casing and then subsequently expel it, as is the case with an automatic firearm. Automatic firearms ordinarily chamber rimless or rebated rim cartridges. The most notable exceptions were the .303 British and the Russian 7.62×54mmR cartridges. The .303 cartridge was used in the Enfield series of rifles, the BREN light machine gun, and a version of the Colt-Browning M1919 machine gun. The Russian 7.62×54mmR (the R indicating a full rim) has remained the staple full-size, full-powered cartridge for Russian weapons since its inception in 1898. Arms chambered for the cartridge include the Mosin Nagant rifles and carbines, the Russian DP and its variants. The DP's later successors, the PK and its variants, continue to use the venerable 7.62×54mmR, as does the infamous Dragunov sniper rifle. Rimmed cartridges, such as the two aforementioned, do not have grooves, instead relying on the rim flange for the extractor to grasp onto. Figure 2.6 shows a representative sample of rim designs.

Materials Used in Construction of Cartridge Casings

Since the conceptualization of self-contained cartridges, various materials have been used to manufacture them. The first self-contained cartridges were constructed of linen or paper. These early cartridges contained the black powder charge and the projectile, but were ignited from an external source, having no contained means of ignition; the shooter still had to rely on an external primer such as a percussion cap. Paper cartridges were very flimsy and likely to break apart if not handled with utmost care. Paper was supplemented by combustible linen, which was less fragile but was essentially the same for all practical purposes. To ensure complete consumption and reduce the incidence of fouling residues, the linen was often cured or impregnated with an inflammable substance.

The advent of the percussion cap in the early nineteenth century opened the door to the development of the self-contained cartridge. By 1836, Prussian designer Johann Nicolaus Von Dreyse had developed a self-contained cartridge using paper as the casing material, as

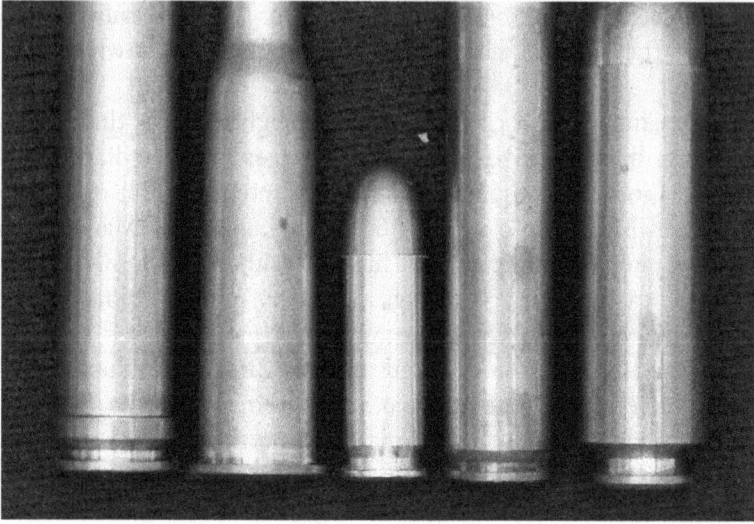

Figure 2.6 Examples of the cartridge rim styles (from left to right): .300 Winchester Magnum (belted), 7.6×54mmR Russian (full rim), .38 Super (semirim), .308 Winchester (rimless), .50 Beowulf (rebated rim). (Images from author's collection.)

well as and the firearm to use it. Called the *needle gun*, the firearm did not literally fire a needle; instead the term was coined because of the unique firing mechanism, which used a needle to puncture the paper cartridge, detonating the percussion cap and expelling the egg shaped projectile under the force of an explosive. It is worth noting that the Von Dreyse needle system operates in a method that is exactly the opposite of modern, self-contained cartridges, but the net effect is the same. In the Von Dreyse system, the percussion cap was not seated at the base of the casing; rather, it was inserted well into the casing body, and perhaps the majority of the powder charge was physically behind the cap. The needle was required to pierce the casing and travel through the powder charge to initiate the reaction. Despite some technical shortcomings, the idea was certainly very novel, and it stands to reason that the Dreyse held tremendous advantages over its muzzle-loaded contemporaries. Using a self-contained cartridge could only allow a greater rate of fire, even from a single-shot weapon. Contemporarily, other concepts were developing, including completely self-contained cartridges constructed of metallic and nonmetallic (envelope) materials, both containing the means of ignition at the base of the casing.

The various systems that were developed in the early to middle nineteenth century were quite often specific to the gun makers of the day and intended expressly for their particular firearms. Ammunition was developed by the same interest that developed the firearm system, and it was a logical business decision to try to maintain proprietary ammunition sources and capitalize on that side of the market as well. According to Blair (1962),

On the 10th of December, 1847, an English patent was granted to one Stephen Taylor for a bullet containing the propellant charge in its base together with a lever-action repeating rifle with which it was to be used. Taylor was merely the English agent for two Americans, Walter Hall and Lewis Jennings, of whom the former was granted an American patent for the cartridge on the 10th August, 1848.

Horton Smith and Daniel Wesson received a U.S. patent for their "Improvement In Cartridges" on August 8, 1854. Smith & Wesson specified, as part of the invention, that the void or chamber between the impressed projectile and the powder charge within the casing be filled with tallow, essentially a form of animal fat, which likely served two purposes. The first would have been as a weather sealant and the second as a form of projectile lubrication or bearing surface against the firearm bore. The primary claim to the invention was

> the employment, in the cartridge, of the metallic or indurated disk or seat plate, so that it shall rest directly on the powder, in combination with arranging the priming or percussion powder in the rear of said disk, or on that side of it opposite to that which rests against the powder, our said arrangement of the disk and priming affording an excellent opportunity for applying the force of the blow by which the priming is inflamed, such force being applied in the line of the axis of the cartridge. (Smith & Wesson 1854)

On December 15, 1863, U.S. Patent 40,988 was granted to Thomas J. Rodman and Silas Crispin. This patent concerned an "Improvement In Metallic Cartridges" and described it as a "metallic case formed of thin wrapped sheet metal" (Rodman and Crispin 1863). This patent references brass, as well as other metals, as a suitable substance for the manufacture of small-arms cartridges. Note, however, that the process of manufacture was not drawn metal, but rather thin strips of metal that were wrapped to form the desired cartridge shape. The result was a cartridge with a somewhat coarse outward appearance. These are but two examples of a plurality of cartridge patents that were filed and granted concerning the introduction of self-contained cartridges and the various "improvements" that different persons put forth on that principle.

A composite or hybrid option was proposed by I. M. Milbank, who in 1872 claimed as his invention "a cartridge-case made with a metallic cup-shaped base and paper cylinder, connected together and rendered waterproof by soluble glass or silicate of soda" (Milbank 1872). There can be little doubt that his hybrid cartridge casing served to inspire later inventors when the question of material choice for casings came back around and alternatives to metallic casings were sought.

In 1866, the United States officially adopted its first metallic center-fire cartridge, the .45-70 Government. The British followed suit shortly thereafter in 1867 when Great Britain officially adopted its first metallic cartridge. With the general acceptance of the self-contained metallic cartridge, millions of percussion-fired weapons were effectively rendered obsolete. Designers went about developing methods of converting these existing stockpiles of arms to chamber the new type of ammunition. Such was the case of American Jacob Snider. His ideas were rejected by the U.S. Ordnance Bureau, but his work caught the attention of the British government. The British pattern rifle was converted and became known as the Enfield/Snider, chambered in .577″. Initially the cartridges of this caliber, designated the Mark I, were fabricated from paper, but they had a metallic case head, consistent with American Hiram Berden's cartridge design, which had been patented in 1866. The later Mark II cartridges were formed from rolled brass, giving them a distinctive rough-hewn and coarse appearance (Hamilton 1916). "Cartridge cases made from drawn brass were not introduced into England, except for machine gun use, until after the Egyptian campaign of 1885" (p. 4).

During the 1870s and into the 1880s, most European nations had discarded the musket for military purposes and had transitioned to single-shot or repeater-type breech-loaded firearms. The transition of commercial and sporting arms from the muzzle loaders to repeater arms occurred simultaneously. The decision on the cartridge size to be used was largely a national one. The first self-contained cartridges did maintain the musket-based thinking of large bores, which led to the first self-contained cartridges such as the black powder loaded 11×60mm Mauser, which came into service in 1871 in the Mauser Gewehr 71. The big-bore concept was relatively short-lived, and by the 1880s, cartridge and firearm thinking had shifted to cartridges in the 6–8-mm bore-diameter class. Cartridge preferences continued to be one of national identity, so any basic rifle platform, whether a Mauser, Mannlicher, Schmitt Ruben, Enfield, or Krag, could be found chambered for any given cartridge, depending on the demands of the customer.

Brass remained the material of choice in cartridge production until the Second World War. During the war, munitions producers in the United States moved away from using brass exclusively and manufactured steel casings for small-arms caliber cartridges, most notably in M1911 .45 ACP cartridges, as well as some production of the .30-06 cartridge. Germany and the Axis allies used brass during the rearmament period in the years preceding the Second World War and into the early part of the war. By midwar, however, substitutes for brass started to appear, and casings made of steel, zinc, and other lesser metals that put less strain on strategic materials and industrial commitment were used. The United Kingdom and the Commonwealth nations produced ammunition predominantly using brass casings. The Japanese ammunition and small-arms output suffered as the tide of battle turned against them. As the end of the war approached, Japan had resorted to soft iron for cartridges and even projectiles because of the widespread scarcity of strategic materials. The Soviet Union had utilized brass but moved away from it and started using steel in the mid-1930s and had completely switched over by the late 1940s, again as a means of conserving more valuable materials. With respect to modern ammunition—rim fire or center fire—brass and steel continue to be the materials of choice. Figure 2.7 shows examples of cartridges using different casing materials.

Brass

The term *brass* is often used by the layman to refer to fired or spent casings, even if they are not actually made of brass. Brass is a mainstay material used in the construction of cartridge casings. The use of brass has distinct advantages: It is strong yet elastic, so that it is easy to form and work with in mass production; it is durable enough to ensure that the cartridge will function in various environmental conditions and exposures; and it creates little friction and thus is also not likely to cause sparks in a volatile environment (such as a munitions plant). This low friction also reduces the chances of binding up the mechanical action of a firearm and therefore reduces the possibility of a malfunction. Since brass is slightly malleable, it is ideal with respect to its ability to flex and bend when sealing the firearm breech against gas leakage back into the chamber, forcing the gas forward down the barrel, called *obturation*. Leakage or loss of gas pressure into the chamber not only results in reduced ballistic performance, but it could potentially create a hazardous condition should the pressure rise above the ability of the action to sustain it.

Brass is not the perfect material and it does have drawbacks, notably weight and cost. Weight is typically not a factor for the general consumer; however it is a factor for military personnel, who must shoulder sufficient quantities of ammunition in the course of a military operation. It is also an expensive material when compared to other potential candidates. Brass will likely remain the material of choice for the foreseeable future, but research

Figure 2.7 Examples of .223/5.56×45mm cartridges using different casing materials. From left to right: commercial brass, green steel, gray steel, U.S. military brass, and nickel-plated brass. The variety of projectiles is also worth noting. (Images from author's collection.)

into alternatives is ongoing, and it is reasonable to presume that alternative materials will at least supplement the use of brass. It has been long established in the Western standard to use brass; however, brass is not a universal material worldwide.

Brass used in munitions manufacture is often called cartridge-grade brass and generally follows an alloyed composition of 70% copper and 30% zinc. New brass has a bright and shiny appearance, but it tarnishes when exposed to the elements or simply by exposure to oils present in fingerprints or other solvents and oils associated with firearms handling and maintenance. In some instances, brass cartridge casings may have a distinct burnished look that is the result of the casing being heat treated (annealed). It is almost universal that brass cartridge casings are annealed as a measure of protection against degradation due to environmental exposure. This is the norm for ammunition manufactured to U.S. military as well as NATO specifications. Commercially sold ammunition cartridges made of brass may have also been heat treated, but typically the discoloration is buffed away in the interest of aesthetics. Investigators searching for spent casings made of brass should be mindful that the brass will dull and tarnish over time; however, it will remain intact for a very long period of time.

Brass is readily and easily located by contemporary metal detectors when investigators search for spent casings or live ammunition. When using metal detectors that discriminate between metals, the users should be aware that they may not only be seeking brass, but other potential metals as well, which depend entirely on the material used to construct the casing. When searching for fired casings, the metal detector may indicate the presence of copper, steel, or nickel.

Aluminum

Aluminum-cased cartridges are manufactured by Blazer, a division of ATK Commercial Products. These aluminum-cased cartridges are marketed primarily for general target shooting and training applications. These cartridges were formerly marketed by CCI

(Cascade Cartridge International) and later as CCI/Speer, so it is possible to encounter these casings marked in either manner. This ammunition also uses a lead-free priming compound, and the primer is marked LF (lead free). In addition to aluminum-cased cartridges, Blazer produces a separate line of cartridges using brass. Blazer's sibling, CCI, currently manufactures Rimfire cartridges in the .22 class and in .17 HMR (Hornady Magnum Rimfire) as well as handgun-caliber shot shells. The shot shells use an aluminum casing that is topped with a plastic cap. The M35 shot shell entered service with the U.S. Navy in the early 1950s. Sufficient quantities remained in stock such that this vintage ammunition has been offered for sale commercially in case quantities. The M35 was used for competition, training, and was also intended for use in shotguns that were standard equipment in emergency survival kits. The M35 was loaded with No. 6 shot.

Steel

Steel casings are the most commonly seen alternative to cartridge-grade brass casings. Steel casings primarily originate from the Russian Federation, China, North Korea, and the Eastern European nations, as well as Second and Third World nations with domestic munitions production capability that were influenced by these nations. The steel casing comes in two varieties, including a base-steel casing and a bimetal casing, consisting of base steel that is coated with a secondary metal such as zinc, nickel, brass, or copper. Such casings have an appearance of being constructed from the coating material. Base-steel casings have an appearance ranging from dark brown to galvanized gray, and various shades of green. In 2010, Hornady Manufacturing, a U.S.-based ammunition producer, announced they were introducing steel-cased cartridges in the Russian 5.45×39mm and 7.62×39mm to their product line. This steel-cased ammunition is lacquer coated as a weatherproofing agent. This move by a major American manufacturer indicates the apparent popularity of these cartridges with the American shooter. In general, the use of steel casings in certain firearms not specifically designed for them is discouraged, as the casings tend to prematurely wear ejectors and extractors in autoloading arms. Steel-cased ammunition is not recommended for use in firearms that have a fluted chamber.

Nickel Plating

The major ammunition manufacturers such as Speer, Hornady, Winchester, Magtech, and others offer nickel-plated casings. The nickel-plated casings often serve as a way of differentiating a particular line of ammunition when the manufacturer produces several different lines of cartridges, particularly a premium line. Nickel is said to have certain performance advantages over brass in that a nickel-plated casing will feed more smoothly and reliably in a self-loading firearm. Nickel casings have a bright, silvery appearance at first but can dull or tarnish over time. Chemically blackened nickel is also used, providing a matte tone to these casings, again to distinguish that particular product line and perhaps as a measure to subdue the ammunition for tactical purposes. Inert dummy cartridges may be completely blackened and overtly resemble live ammunition; thus careful inspection must be made to ensure that they are properly identified as either inert or live to avoid a hazardous situation.

Shot Shells

Three major materials have been used to manufacture hulls for shot shells: brass, paper, and plastic (see Figure 2.8). Brass was the first material to be used, and the practice continues today. It is possible to purchase newly minted brass shot shells either as loaded

Figure 2.8 Examples of .410 shot shells constructed of brass and paper. The paper shot shells are vintage Winchester Super X; the brass shells were produced by the Russian firm Barnaul. (Images from author's collection.)

cartridges or empty hulls ready for custom loading. Brass shot shells are more expensive to purchase than plastic- or paper-hulled shot shells, but brass shells afford the shooter the economy and convenience of reuse and reloading to suit individual needs. Brass shot shells are uncommon compared with paper or plastic hulled shells, but their use has persisted over the years. The life span of a brass shot shell can be well in excess of 150 firings so long as the casing is not ruptured or otherwise damaged. The most likely cause of failure in a brass shot shell would be (a) repetitive use in a self-loading or slide-action shotgun, where the rim would be exposed to repeated extraction and ejection, which would cause the rim to wear and bur and possibly lead to malfunctions, or (b) overcharging the cartridge by using too much powder. Brass shells could be expected to have the longest life span in a breach-loaded, single-shot shotgun. Brass shot shells are manufactured by the Brazilian munitions manufacturer Companhia Brasileira de Cartuchos and the Russian based firm Barnaul. Barnaul also manufactures steel shot shell hulls that can be either brass plated or left in natural base-steel appearance (Barnaul Cartridge Plant n.d.).

Paper
Paper-hulled shot shells appeared quite early as an alternative to brass shot shells. The paper was almost universally a laminated type or pasteboard that had some weather-resistant characteristics about it, although when exposed to moisture they would inevitably swell and become unusable. Paper shot shells were cheaper than their solid brass counterparts, but the use of paper shot shells fell out of favor with the market for quite some time, most likely because of the fragile nature of paper shells versus brass or plastic. However, paper shells have not totally disappeared from the landscape. They can be found on the second-hand market quite regularly; often the boxes are more valuable than the shells as collectible pieces of firearms ephemera of the bygone era. The shells themselves may or may not be functional, depending entirely on the environment they were stored in. New production paper shot shells are manufactured by Federal Cartridge, Fiocchi, and Sellier & Bellot. Chinese-sourced paper-hulled shot shells were in circulation in the United States as

of the mid 1980s and can still be expected to be encountered from time to time. Paper shot shells may be especially prevalent in venues where ownership of firearms or ammunition is heavily restricted; the use of paper shells may deter stockpiling and certainly does prevent fired hulls from being reloaded.

Plastic

The material of choice in modern hulls for shot shells is plastic. By 1960, plastic-hulled shot shells came on the market, rapidly took hold, and have become the standard for all gauges and loads with all manufacturers. The colors used are varied, and red, yellow, orange, black, green, and even clear plastic will be encountered. Some companies mint plastic hulls in certain colors to complement a particular line of shell. Remington Arms has manufactured its law enforcement–oriented shot shells with green hulls, while Speer has manufactured them in blue and red. Whatever the color, manufacturers use deliberately bright colors for high visibility, helpful to hunters who inadvertently drops their shells in the hurry of trying to load while watching the game slip away. Hence, no particular emphasis can be placed upon mere observation of a particular hull coloring.

Regardless of the material used to manufacture the hull, the case head, which is the lower portion of the shell that contains the primer, is made of metal, typically zinc, brass, or nickel. A distinguishing characteristic of shot shells is the length of the case head. Referred to as "high" brass or "low" brass (even if the case head is not literally made of brass), the length of the head is readily visible by observation. High-powered shot shells use high brass bases, while low brass shells are for less powerful loads. The mistake is often made that the Magnum shot shells are all high brass; this is not the case, as there are low-pressure versions of Magnum shot shells manufactured. The top of the plastic shot shell is typically sealed with a crimp. This crimp is the folding of the hull material into itself, giving it the appearance of triangles folding into a center point. An alternative to a crimped top can be a circular insert that is pressed into the top and seals the contents. In the case of slugs, the leading portion of the slug is visible.

Polymer-Cased Cartridges

The use of polymers in the manufacture of cartridge casings other than shot shells has come and gone numerous times since the development of modern plastics. Plastic casings have appeared at various times in modern history in an attempt to develop a viable and cost-effective alternative to metallic casings. There have been suggestions that plastic-cased cartridges could have been manufactured in France as early as the late 1800s. The German munitions firm Dynamit Nobel (now part of the RUAG Ammotec Group) manufactured polymer-cased cartridges in 7.65mm Browning, 9×19mm, 5.56×45mm, .30 M1 (carbine), .30 M1 (.30-06), 7.62×51mm NATO, .303 Enfield, 7.92×57mm, and .50 BMG (12.7×99mm). The company produced two different types of plastic ammunition: the PM (Plastic Man oeuvre), a blank; and the PT (Plastic Training), a live cartridge that expelled a projectile when detonated. The PM cartridge would not cycle an automatic weapon without the use of a BFA (blank fire adapter). The PT-type cartridges also required a special adapter to properly function in an automatic firearm. The PT cartridge was designed for short-range training and was produced primarily in blue casings; however, white casings were also produced. The PM cartridges can be distinguished from PT by a series of serrations or cuts along the nose of the cartridge. The PT cartridges have been observed colored red, black, or olive drab green; however, other colors likely were produced as well. The

Norwegian munitions firm Nammo AS manufactures special training cartridges designated as PSRTA (plastic short-range training ammunition) in NATO-centric small-arms calibers (Nammo 2012).

Polymer-cased ammunition manufactured by the Washington State-based United States Ammunition Company appeared in the mid-1980s. The cartridge had a white polymer casing with a metallic base. The production was apparently short-lived, and very few examples probably exist today. Production appeared to be limited to handgun caliber cartridges: .38 Special, 9×19mm, .44, and possibly .45 ACP.

Another attempt to bring the polymer-cased cartridge to market came in the form of the PCA (polymer-cased ammunition) Spectrum, produced by a company called Natec, formally Amtech. Inspection of period ammunition known to have been produced by this firm shows that they manufactured .223 ammunition loaded with either 55- or 62-grain projectiles. Once again, the obvious technical problem with polymer casings (reliable feeding and extraction without tearing the back of the casing off, with resultant malfunction) was met by making the case head of brass, and then using polymer for the balance of the casing body. These cartridges have been observed in five different colors: blue, white, gray, black, and tan. The head-stamp marking consisted of "223 REM PCA" and "03" or "04," indicating that manufacture took place between 2003 and 2004. Preceding the production of .223 cartridges, Amtech manufactured a polymer-cased .38 Special cartridge. The casing itself is black polymer that is seated on an aluminum case head. The loaded projectile is a truncated nose type. These cartridges apparently were produced starting in 1992, with a production run lasting approximately one year.

In 2003, the Polytech Ammunition Company (no apparent relation to the Chinese firearms manufacturing consortium) produced .38 Special cartridges that outwardly resembled other polymer-cased cartridges. Polytech cartridges used a black cylindrical polymer sleeve that met a metallic case head. These cartridges were marketed under the trade name Spitfire. Once again, it does not appear that these cartridges were produced in significant numbers or for a great length of time, and the remaining examples are more at home in collections than in use, but it is always possible that some new old stock may appear. The Spitfire name should not be confused with the 5.7mm Spitfire cartridge developed in the early 1960s.

A recent development in the polymer-cased cartridge concept has appeared in the form of the cased-telescoped cartridge. This technology is currently being developed by AAI/Textron under the U.S. Army's LSAT (Lightweight Small Arms Technology) Program. The cased-telescoped cartridge in its current form is an all-plastic cartridge with a projectile that is recessed into the casing. The cartridge, by design, seeks to eliminate the immediate concerns of plastic-based cartridge casings, i.e., cyclic reliability* and thermal resistance. The cased-telescoped cartridge is a cylindrical shape that lacks the traditional extractor groove located at the rear of the casing; hence, the need for a metallic case head is eliminated. Test platform weapons using cased-telescoped cartridges have a separate barrel and chamber to eliminate "cook off" (unintentional firing of a chambered cartridge occurring as a result of heat transfer from the firearm that

* The reliability issue is twofold. The first is that polymer casings can tear when subjected to the force of a metallic extractor interacting with the polymer. The second is the unsupported portion of the casing that may flex or distort, causing a weapon failure. Both issues are compounded by heat accumulation within the firearm that may soften polymers.

causes the powder to ignite) and heat-induced casing deformity. The cartridge is fed from a feed tray into the chamber in a straight-through motion. The chamber itself fully supports the cartridge, which is to say that there is no part of the casing body that is not contained within the chamber when the cartridge is loaded. After firing, the spent casing is pushed out of the chamber by the incoming cartridge as it moves into firing position. This movement further works to disperse heat, complementing the separate barrel and chamber design (Shipley 2010).

At the 2009 National Small Arms Conference in Las Vegas, Nevada, a partnership between Colt Defense and BML Tool & Manufacturing presented two alternatives to brass cartridge casings in military applications. The two approaches were based upon the size of the cartridge in question. The concept proposes that larger caliber (above .50 BMG) casings could be fabricated from a hybrid steel and polymer in a "modular casing" approach. This modular casing, as proposed, is an inert skeleton. The powder charge comes in the form of a "charge vessel" that is stored separately and installed into the skeletonized casing on demand. In this concept, it is possible to stockpile components while not stockpiling live ammunition. There is also a second variant that takes on a more conventional approach to assembly. This approach to small-arms cartridge cases is a "spiral casing" molded from polymer with a separately molded casing base. The presented prototype has a distinctively spiral appearance that is claimed to add "approximately 50% more perimeter for bonding and joint strength." The spiral casing also claims a 40–47% weight reduction as compared to brass casings (Brown and Battaglia 2009).

In 2011, PCP (Lightweight Polymer Cased Ammunition), a company based in Orlando, Florida, announced it was producing polymer-cased cartridges. Images available for viewing on the company's website indicate a casing that has a brass rim and groove coupled to a white casing body and a black upper portion that comprises the upper one-third of the casing, including the neck.

Caseless Ammunition

Caseless ammunition eliminates the traditional cartridge casing as a container, making the powder charge itself the casing. There is no container per se, as the name *caseless* implies. The concept has been subject to extensive research and has proven to be a viable and attractive alternative to traditional cased ammunition. A January 1965 report published by the U.S. Army Ordnance Corps indicated that the concept of caseless ammunition was well underway using a design by the Olin Mathieson Chemical Corporation. The cartridge was described as a .30″ primed projectile configuration for a caseless liquid-propellant/dual-projectile gun system (Scanlon, Quilan, and Vanartsdalen 1965). The report recommended that further research be continued because the concept demonstrated merit. From a technical standpoint, a potential benefit of using caseless ammunition is that the firearms using such ammunition could potentially have a higher rate of fire because the post-ignition processes of ejection, extraction, and reloading in a traditional action would be eliminated if there were no residue (i.e., fired casing) to expel, as the container had been consumed. In addition, caseless ammunition has the potential to reduce the encumbrance of the ammunition bearer, as the cartridge casing alone can account for roughly 40% of total cartridge weight. The principle disadvantage is that the price per cartridge unit has not been able to match the cost of traditional cased cartridges, as well as having to bear the significant investment required in issuing the new cartridges and training personnel on the weapons systems that would employ such a cartridge.

Figure 2.9 (See color insert.) Caseless-telescoped cartridges. (Image courtesy of Paul Shipley, AAI Corporation.)

Heckler & Koch, in partnership with Dynamit Nobel, developed a caseless ammunition designed exclusively for the H&K G-11 rifle. The rifle appeared in the NATO rifle and ammunition test trials that took place between 1976 and 1979. While no agreement could be reached concerning a standardized NATO rifle during those trials, the project continued to be developed until the early 1990s, when the reunification of Germany prompted the partnership to abandon the project, and it passed into the hands of the U.S. Army. Research and development continues in caseless ammunition, now designated the *caseless-telescoped cartridge*. Testing and evaluation of the caseless-telescoped cartridge is conducted under the auspices of the U.S. Army LSAT (Lightweight Small Arms Technology) program. The cartridge itself has undergone numerous technical changes. The current iteration of this cartridge has a projectile that is recessed into the cartridge body. This projectile is enclosed in a plastic cup that is somewhat synonymous to a shot cup in a shot shell. Upon detonation, the projectile punches a hole in the cup as it expels from the firearm. The cup itself is not consumed but instead is broken up, and its remnants can be found on the ground where the firing occurred, having also been expelled with the projectile. The caseless-telescoped cartridge is shown in Figure 2.9 (Shipley 2010).

The Austrian firm Voere is the single entity that produces a commercially available rifle chambered for caseless ammunition. According to the information published by the manufacturer, the cartridge is composed of a nitrocellulose-based compound that is detonated by a booster ignition charge in lieu of a conventional percussion primer. The high cost of the ammunition and rifle contributes to the relative rarity of these firearms.

Daisy, better known for manufacturing BB and pellet air guns, briefly manufactured a rifle that fired a caseless .22 cartridge called the Daisy V/L (see Figure 2.10). According to the company, it was invented by Jules Van Langenhoven, whom the V/L is named after. The cartridge itself consisted of a .22 projectile coupled to a propellant compound that was ignited by superheated air. This product was announced to the public on August 20, 1962. Production was short-lived, however, as the BATFE determined that the V/L constituted a firearm, and Daisy was not a licensed manufacturer of firearms. New old-stock .22 V/L and the Daisy rifles regularly appear on the second-hand market as collector's items. The .22 V/L cartridges have a yellowish appearance, with a bare lead projectile seated on top. The cartridges came supplied in clear plastic tubes.

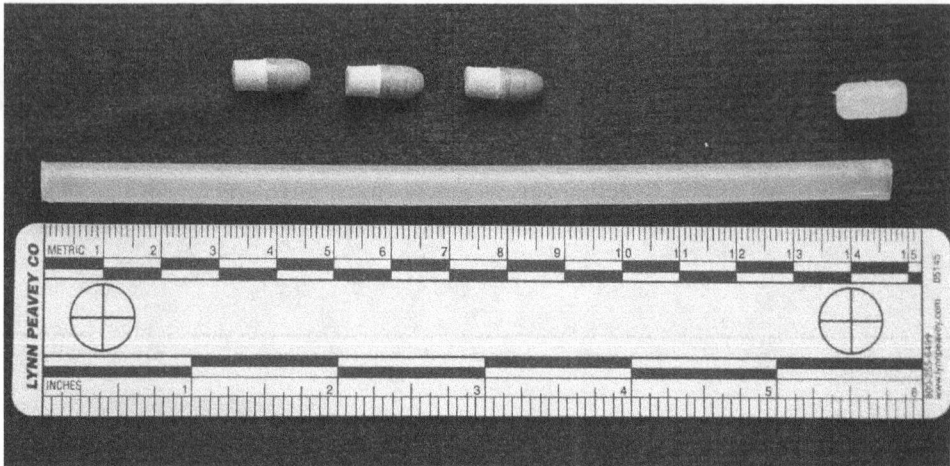

Figure 2.10 **(See color insert.)** Daisy V/L caseless cartridges. The bare lead projectile is seated against the yellowish, granular propellant charge. The plastic tube is the container the cartridges were shipped in. (Image from author's collection.)

Ignition Systems and Propellants

The ignition system is the means or method of initiating the discharge of the propellant charge. The earliest firearms used various means to get a flame or sparks into the firearm chamber where the propellant charge was contained. These means evolved rather quickly. The simplest form was an aperture where flame could be directly applied to touch off the propellant. Another approach was to use fuses or wicks that were lit using an open flame such as a match or kindling, the fuse or wick acting as the conduit to get the fire to the charge: the so-called matchlock ignition system.

With the furthering of mechanical developments, mechanisms such as the wheel lock and the flintlock superseded the earlier systems. These systems permitted a mechanical striking action, using a shock-sensitive material such as flint, to create sparks that would ignite the propellant charge instead of the operator being required to ignite the system manually, which meant having a ready source of flame at hand.

Alexander John Forsyth, a Scottish clergymen, is credited with the invention of *fulminate*, achieved by subjecting metals to the action of fulmanic acid. The preferred metal was mercury, and thus was born mercury fulminate and what was to become the first primer, as modern terminology would describe it. The residuals of the fulminate reaction are an explosive material susceptible to detonation by impact or friction. Mercury fulminate primers have long been discontinued, as the residues of the explosion are highly corrosive to metals unless quickly and thoroughly cleaned. By the late nineteenth century into the turn of the twentieth century, potassium chloride and antimony sulfide priming compounds had replaced fulminate mercury; however, these compositions were also corrosive, which led to chamber and barrel erosion and pitting if the weapon was not promptly cleaned. Lead styphnate–based primers had replaced these corrosive primers by the mid-twentieth century. However, corrosive primed cartridges, especially older military surplus cartridges from around the world, can still be had that are primed using corrosive priming compounds. Almost concurrently, a number of noncorrosive primer compounds were used as well.

Center-Fire Cartridges

The center-fire cartridge uses a small explosive device called a *primer* that is seated in a pocket at the center of the cartridge casing's base. The primer is struck by force acting upon it by way of a firing pin or striker in the firearm that detonates the volatile priming compound. When the primer is detonated, the flame created by the detonation passes through one or more "flash holes" in the interior of the casing and reaches the contained powder charge, causing it to deflagrate (burn rapidly). There are several sizes of primer based on the application, which is often dictated by the size of the cartridge: shot shell, Magnum shot shell, small rifle, large rifle, small pistol, and large pistol are the classifications used. The industry has been shifting priming chemistry toward formulations that have reduced or completely eliminated heavy elements such as lead, antimony, and barium. This change has been prompted by increased environmental concerns and health issues surrounding repeated or prolonged lead exposure. These issues have coupled with an increased use of indoor gun ranges, where leaded ammunition may be prohibited altogether. Such primers tend to be conspicuously marked by abbreviations that indicate a lead-free or clean-burning primer compound.

Berden Priming

There are two center-fire primer designs currently in use: the Berden primer and the Boxer primer. The Berden primer was the product of American Hiram Berden. On March 20, 1866, Berden was granted U.S. Patent 53,388, "Improvement in Priming Metallic Cartridges" (Berden 1866). Berden indicated in his patent that he was interested in the safe transportation of cartridges and the reuse of metallic cartridge casings. It is also interesting to note that Berden advocated the use of brass for cartridge casings. Furthermore, Berden was concerned that the fulminating priming compounds used during that period had a relatively short shelf life and would render the cartridge useless if stored for an excessive period of time. A short shelf life was not advisable for military ammunition because it would prove impractical, if not impossible, to accumulate and maintain a strategic stockpile in the event of conflict. Ammunition would have to be immediately manufactured on an almost impossibly large scale in such a scenario, thereby creating an issue of national security. Although this problem did not affect civilian consumers, a short shelf life for commercial cartridges did present its own problems.

Physically, the Berden primer consists of a cup containing the priming compound. When seated, the primer is met inside the casing by the anvil, a small appendage that crosses the opening between the casing body and the primer pocket. In this design, the anvil is not part of the primer, but rather part of the cartridge casing. On either side of the anvil can be one, two, or three flash holes, through which the flame created by an ignited primer will travel to the loaded powder charge. Berden's design was an improvement over other center-fire designs of the day such as the Benet or Bar Anvil, which did not use a separate primer component but instead relied upon a priming "pellet," which was seated within an abscess at the casing base.

Despite Berden's claim that his system was ideal for recycling fired casings, it is quite the contrary. Berden primed cartridges are quite common, and many munitions producers around the world utilize this design, as it tends to discourage collection of spent casings with the intention of reloading them. Another reason that Berden priming is still in widespread use is the cost; it is significantly cheaper to produce Berden primed cartridges in comparison to Boxer primed cartridges (see the following section). Although the cost differential may be measured in just fractions of a cent per unit, the macroeconomics of cartridge production net a tremendous savings between the two designs. This is especially true in military cartridges, where it would seem impractical to detail soldiers to scrounge the battlefield for spent casings for the purposes of reloading them. That being said, this was a major consideration of the U.S. Ordnance Corps for peacetime operations, allowing costs to be contained by collecting and reusing brass. Thus it made sense to not use the Berden system, instead opting for the Boxer system, which provides easier reloading of fired casings. Even today, the Boxer system remains the preference in the United States.

Boxer Priming

Edward M. Boxer devised the ignition system that carries his name while working for the Royal Arsenal in Woolrich, England. Boxer accomplished the same goal as Berden while going at the same question from the opposite approach. The Boxer design, like the Berden design, contained an anvil; however, the two are completely opposite. In the Boxer primer design the anvil is a separate piece from the casing, whereas in the Berden primer design, the anvil is part of the casing. Another difference is that rather than having one and up to three flash holes, the Boxer primer uses a single, larger diameter flash hole. For all practical purposes, whichever system a particular cartridge uses, the firearm does not know the difference, and they are interchangeable: Both are still identified as center-fire cartridges. American manufacturers tend to favor Boxer priming, which is somewhat ironic, given the fact that the Boxer is of English origin. It is customary for cartridge manufacturers to indicate on the specification sheet whether a particular cartridge is Boxer or Berden primed, and this, in and of itself, can be used as a class characteristic for the purposes of identification.

Rim-Fire Cartridges

Rim-fire cartridges do not have primers. Instead of a primer seated in the center of the cartridge casing, the base of a rim-fire cartridge is a pressure-sensitive cap where the ignition compound is spread around the periphery of the rim of the casing. Impact on any portion of the surface will compress the charge and cause the ignition to occur. Modern rim-fire cartridges tend to be smaller caliber, almost exclusively .17- or .22-class cartridges; however, there are exceptions, such as the seldom-seen .22 Remington Jet and the .22 Hornet. Historically speaking, rim-fire cartridges for military purposes appeared around the time of the American Civil War, but their use was short-lived as self-contained cartridge technology rapidly advanced. The Swiss Vetterli, chambered in .41 rim fire, was probably the last military rifle using rim-fire cartridges when it was removed from service in the early 1890s.

Black Powder

An effective propellant compound must be able to deflagrate rapidly without the introduction of outside oxygen, generate a tremendous amount of gas relative to the quantity of powder, and create heat. The earliest propellant used in firearms, called *black powder*, is a simple chemical combination of sulfur, charcoal, and saltpeter (potassium nitrate); however, there are innumerable other recipes. The proportions of the reagents were subject to variation.

The true history surrounding the advent of powder is a subject of some dispute. Some sources attribute black powder as a development of ancient China that was subsequently brought to the West by traders over the centuries. Hamilton (1916, 10) wrote that "as far as it can be ascertained, it was first produced in England in the thirteenth century." Around 1242, the formula may have been deduced or reformulated, perhaps surreptitiously, by Franciscan friar Roger Bacon (1214–1294 CE). Bacon is known to have dabbled in alchemy, and the attempt to either formulate or break down the existing formula may have been part of his work. Hamilton (1916, 10) also credits a German monk named Berthold Schwarz with being an original inventor of black powder, although it may have been a simultaneous yet independent invention between the two men. Other claims include the involvement of Archimedes (287–212 BCE), who is also alleged to have developed black powder, perhaps independently of the Far East. It is entirely possible that the development occurred concurrently in various parts of the world, with variations in the formulas.

While universally known as black powder, the substance can range in color from dark black, brown, or gray, to metallic silver. Black powder is considered a low explosive in that it does not explode but rather deflagrates. Physically, black powder grains vary in appearance and can vary in size from a very fine powder to relatively large chunks. Black powder can often be discriminated from smokeless powder because black powder is generally irregular in shape but of uniform size. The size of the powder flakes directly affects the burn rate, with smaller flakes burning faster than larger ones. When exposed to moisture, black powder will not burn; however, if it is allowed to dry, it will become usable again. Hence, the concept of having an encased powder charge that was reasonably resistant to the elements was a logical progression in firearms technology.

Black powder firearms predated self-contained cartridge-based firearms. The introduction of self-contained cartridges that included powder and a projectile did not render black powder obsolete. These first self-contained cartridges used black powder as the propellant. This permitted many firearms of the era to be retrofitted to fire self-contained cartridges; however, due to the differential in energetic potential between black powder and smokeless powder, these retrofits tended to be relatively short-lived and were rapidly replaced by newer armaments designed specifically to handle self-contained smokeless powder–based munitions. Despite this advancement, black powder–based cartridges remained sufficiently popular to be commercially viable to manufacture until the onset of World War II. Firearms produced in the era between the introduction of smokeless powder and until the 1940s will often be marked "smokeless powder" to indicate that the firearm was suitable for use with smokeless powder–loaded cartridges.

Black powder does not have nearly the energetic potential of smokeless powder, but it remains a dangerous substance. The relative availability and seemingly innocuous nature of black powder makes it a prime candidate for use in improvised explosive devices such as pipe bombs. It is often erroneously thought that powder used in pyrotechnics is a form of gunpowder, but quite the contrary is true. Standard pyrotechnic powder is generically called *flash powder*, although numerous formulas are mixed and used. Brown powder is akin to black powder, save for a slower rate of deflagration, and is archaic.

Smokeless Powder

Smokeless powder, like black powder, is also a low explosive; thus it deflagrates rather than detonates. Smokeless powder is also called *gunpowder* and is considerably more powerful than black powder. Deflagrating black powder expands to some 250–300 times its original volume; smokeless powder can expand 900 to 1,000 times its original volume and generate pressures in excess of 44,000 pounds per square inch.

Modern gunpowder compounds are single-, double-, or triple-based compositions; however, triple-based powder formulas are not seen in small-arms applications. A single-base composition contains nitrocellulose as the main ingredient, which is dissolved in alcohol to form a sticky byproduct often called *colloid*. The colloid is cut to the desired size, and the alcohol is then allowed to evaporate, leaving a hard granular powder. Double-base compositions are made by dissolving *gun cotton** in nitroglycerin. Coatings and other nonexplosive constituents may be added to the basic powder formula to control the powder burn rate, and other additives may be mixed to achieve a desired effect such as flash reduction. Smokeless powder has a clearly defined shape and is of uniform size in a particular application, although the size and shape are part of the engineering process when a cartridge is developed.

Smokeless powder has been the subject of continuous refinement and experimentation since its inception and into modern times. The actual invention of smokeless gunpowder has been attributed by Hamilton (1916) to a German chemist named Christian Friedrich Schöenbein, and is said to have taken place in 1845–1846 when he discovered gun cotton. In correspondence dated August 25, 1846, to his colleague Michael Faraday, an English chemist, Schöenbein described his "explosive cotton" and was apparently already cognizant of its potential, characterizing it as a "dangerous rival to gunpowder" (Kahlbaum and Darbishire 1899). From the context of the letter, it is apparent that he was not debuting the development of gun cotton, as it had already been subjected to significant testing by the time the correspondence was written. Gun cotton was not usable in that form because of its erratic behavior, and it had to be processed. The discovery of gun cotton was not the only significant achievement in Schöenbein's life. He also discovered ozone (Schöenbein spelled it as "ozon") and invented what is now called the *fuel cell*.

Since the introductions of smokeless powder and gun cotton, they both have been subjected to a great deal of refinement and reengineering. Nitroglycerine followed soon thereafter, developed by Ascanio Sobrero, an Italian chemist. Noted explosives industrialist Alfred Nobel used nitroglycerine to produce dynamite, but he also mixed nitroglycerine with nitrocellulose to produce his proprietary smokeless propellent called *ballistite*, which was patented in 1887.

Smokeless powder is not truly smokeless, as it does produce visible smoke when detonated; however, it is smokeless in the sense that it produces considerably less smoke than black powder as a by-product. Aside from the increased power of smokeless powder in contrast to black powder, it is exactly like black powder in many respects. It comes in various shapes and sizes, generally flake, ball, or cylindrical. Also like black powder, there are various recipes for making smokeless powder, with most recipes focusing on the burn rate. As smokeless powder is considerably more powerful than black powder, confusing the two will certainly result in an unfortunate accident.

In the last 20 years, there has been serious effort made to reformulate smokeless powder to reduce or eliminate the use of lead and other heavy elements, and many manufacturers have discarded the traditional powder formulations in favor of these newer formulas. Newer gunpowder formulations have coincided with new primer compounds that reduce or completely eliminate the presence of heavy elements such as lead. In England, smokeless powder

---

* Gun cotton is made by treating cotton, wood pulp, plant fibers, or other cellulosic materials with sulfuric and nitric acids; it is also called *nitrocellulose*.

was called *cordite*, presumably because the powder was rolled into strings resembling cords and cut to the desired length. Cordite was one of the earliest smokeless powder formulations.

Lesmoke Powder

All varieties of powders have always been subject to proprietary blends manufactured by powder makers, and they have gone under innumerable trade names. The DuPont Company, well known today as a manufacturer of a wide variety of goods from countertops to communications, was founded as a powder manufacturer in 1802. "In 1857 Lammont du Pont developed a new method of black powder manufacture which substituted South American sodium nitrate for the more expensive, British controlled potassium nitrate" (DuPont 2011). This product, called *B blasting powder*, "was the first notable change in black powder composition in over 600 years" (DuPont 2011). DuPont also produced a now-archaic hybrid powder called *Lesmoke*. This powder formulation was a combination of gun cotton and black powder, and likely contained some other ingredients to separate it from other contemporary products by other firms and avoid the liabilities of infringing on intellectual property. Lesmoke is reported to have had the appearance of fine gray sand. References dating back to the early twentieth century state that self-contained cartridges were loaded commercially with Lesmoke powder.

Action Time

The *action time* is the time interval between the ignition of the primer, propellant burn, and the projectile exiting the barrel. In modern self-contained cartridge systems, the action time is practically instantaneous. Delayed action times are called *hang fires* and have several possible causations, including defective primer; ammunition that has been exposed to solvents, oils, and the environment; and irregularities in the powder charge. When the trigger is pressed and there is no gunshot on a loaded firearm, the recommended practice is to wait at least 30–60 seconds before attempting to clear the chamber to allow the time for a potentially slow-burning cartridge to complete the reaction. Muzzle discipline is critical when a hang fire occurs because of the possibility that the firearm could discharge at any moment without warning. Archaic ignition systems have a delayed action time relative to self-contained cartridge arms.

Electric Priming

Electric priming is an alternative to traditional percussion-detonated cartridges, but it has never achieved widespread use outside of military circles and a handful of commercial oddities. Electrically primed cartridges use a high-intensity electrical arc, rather than percussion by mechanical means, to detonate the primer. Electrically primed cartridges are used in high-rate-of-fire weapons such as the 7.62×51mm M-134 "mini-gun" and M-61 20mm Vulcan cannon.

It is thought that Smith & Wesson experimented with electric priming in the late 1960s as an alternative technology to advance their Model 76 submachine gun; however, the idea apparently never got beyond testing. Unfortunately for Smith & Wesson, the Model 76 was not as commercially successful as had been hoped. Although the design was sound enough, borrowing heavily from the Swedish K—known by many as the Karl Gustav—Smith & Wesson was unable to find a market in the world's militaries, and it appears that most of the 76s went to American law enforcement agencies, while others passed into private hands. In the American machine gun market, the Model 76 is one of the more common submachine guns that can be found, ironically many of them bearing the property markings of their former owners—law enforcement agencies.

Voere, an Austrian gun maker, markets a 5.7mm caseless cartridge. The cartridge is not only unique in that it is a caseless* cartridge, but it also is electrically primed, having no percussion primer but, instead, a semiconductor that is attached to a booster charge. The rifle chambered for this cartridge, the Model VEC91, uses two 15-volt batteries instead of having a traditional mechanical firing apparatus. The manufacturer reports that the 3.6-gram (approximately 55 grain) 5.7mm projectile travels at approximately 930 m/s (3051 ft/s). Given these figures, such performance is on par with other comparable, traditionally constructed cartridges. The significant difference is the high cost of the rifle and ammunition in contrast to conventional arms and ammunition.

In 2000, Remington Arms started marketing a family of electrically primed cartridges called EtronX. Remington selected the Model 700 bolt-action rifle to serve as the platform for the new technology, and the 700 was the only model manufactured to use EtronX. The only significant differences between a conventional 700 rifle and the EtronX version was a battery compartment and an LED that indicated the weapon's condition and acted as a form of safety. Various calibers were offered, including .22-250, .243, and .220 Swift. The concept was not successful, and Remington discontinued the product within a few years. EtronX ammunition is still available.

Projectiles

Most cartridge platforms are flexible and are readily reengineered to suit specific market needs and purposes. In the design and manufacturing process, the selection of materials, design characteristics, and powder charge are combined to produce a cartridge to maximize results for the intended use. This is not to say that any given projectile will meet the performance expectations placed upon it. Over the course of modern cartridge development, various materials have been explored to either accomplish a specific objective or to improve some measure of projectile and, hence, cartridge performance. Projectiles can be described as having three effects when they become terminal, i.e., when they impact the target: nonexpansion, controlled expansion, and fragmentation. Any of these effects can be engineered into the projectile design. However, design intent does not necessarily imply that the desired result will be realized. The pursuit of better projectiles has led to the diversity of designs and materials that exist today.

Commonly Used Abbreviations to Identify Projectile Styles

For the purposes of packaging, advertising, and identification, ammunition manufacturers have adopted universally recognized abbreviations to sort, at least by class, the type of projectile loaded in a particular cartridge. Most of the abbreviations are relatively broad in scope, but others are proprietary to the projectile design that they represent. There are numerous styles or shapes of projectiles. The abbreviations that are used to identify the projectile shape are largely universal, save for proprietary projectiles that are developed by a company. Table 2.3 shows the common abbreviations for ammunition.

---

* A caseless cartridge is one wherein the casing itself comprises the powder charge.

Table 2.3   Common Ammunition Cartridge Abbreviations

A-BOND	Accubond	MATCH	Match Grade cartridge
AP	Armor Piercing	OTM	Open Tip Match
BT	Boat Tail	Pb	Lead
BND	Bonded	PART	Partitioned projectile
BDD	Banded	PLF	Plated Lead Free
BPA	Blank Plastic Bullet	PROOF	Proof loaded cartridge
BS	Ballistic Silver Tip	RN	Round Nose
B-TIP	Ballistic Tip	RNFP	Round Nose Flat Base
BTHP	Boat Tail Hollow Point	RNSWC	Round Nose Semi Wad Cutter
BTSP	Boat Tail Soft Point	RPPB	Round Point Plastic Bullet
CP	Cone Point	SBT	Spitzer Boat Tail
DEWC	Double End Wad Cutter	SCHP	Solid Copper Hollow Point
FB	Flat Base	SL	Soft Lead
FMC	Full Metal Case	SLD	Solid
FMJ	Full Metal Jacket	SJHP	Semi Jacketed Hollow Point
FMJBT	Full Metal Jacket Boat Tail	SJ	Semi Jacketed
FMJSWC	Full Metal Jacket Semi-Wad Cutter	SMP	Semi Point
FN	Flat Nose	SP	Soft Point or Spire Point
FP	Flat Point	SPHJ	Soft Point Heavy Jacket Tail
GDHP	Gold Dot Hollow Point	SS	Semi Spitzer
HP	Hollow Point	SSP	Single Shot Pistol
HPBT	Hollow Point Boat Tail	STHP	Silver Tip Hollow Point
HSP	Hollow Soft Point	SPT	Spitzer
JHP	Jacketed Hollow Point	SWC	Semi Wad Cutter
JRN	Jacketed Round Nose	TC	Truncated Cone
JSP	Jacketed Soft Point	TMJ	Total Metal Jacket
JSWC	Jacketed Semi Wad Cutter	TNHP	Truncated Nose Hollow Point
L	Lead	WC	Wad Cutter
LRN	Lead Round Nose	WTP	Wide Taper Point
LWC	Lead Wad Cutter		

Cannelures

A feature of many, but not all, projectiles is the cannelure, which is the groove that runs the circumference of a projectile. Some projectiles may have two cannelures. The cannelure can be knurled or smooth in appearance. The cannelure can serve two purposes, the first being to provide a place to set a lubricant. Cannelures are often referred to as "grease grooves" just for this reason. The second reason for a cannelure is to provide the location on the projectile where it will be seated in the casing neck. The cannelure runs the circumference of the projectile and may be visible, partially visible at the casing neck, or completely invisible if the projectile is seated so that the cannelure is within the casing.

Ball Projectiles and Variations

The ball or solid projectiles are simply projectiles constructed of a solid mass of material that have been cast or formed into a particular shape, such as that depicted in Figure 2.11. The term *ball* does not indicate that the projectile is of a round or ball shape. Ball or solid ammunition may or may not be jacketed or coated with a gilding material. While all of

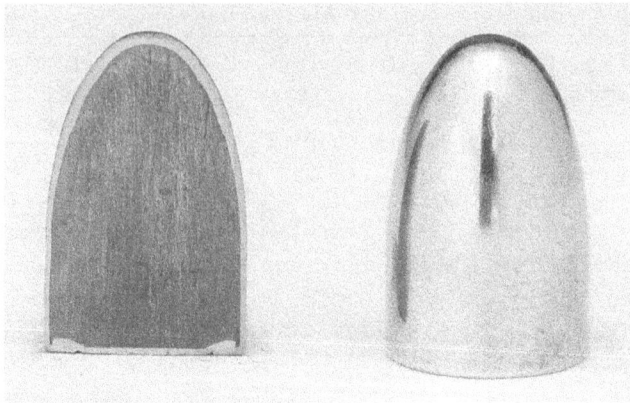

Figure 2.11 A 9mm 115-grain copper-jacketed ball projectile. (Image courtesy of Hornady Manufacturing.)

these are considered to be ball ammunition because they have the same class characteristics, there are variations that have design features intended to enhance their performance. All ball ammunition, whether by design or not, has the possibility of fragmentation when striking an object. The disintegration of the projectile may be partial or total, depending upon the composition of the target, impact velocity, and other variables that are situation specific. If jacketed, portions of the jacket may fragment and scatter about randomly, or the jacket may completely separate from the main projectile mass, either as one piece or in smaller pieces. Upon impact, ball projectiles may deform while not shedding projectile material or jacketing; such deformation may be minimal or to such an extent that it may render the projectile completely useless for comparison studies.

Ball ammunition is designed to be nonexpansive in the context of terminal ballistics, regardless of how the projectile behaves upon impact. One variation of ball ammunition with a performance enhancement built into the design is the *penetrator*, which is essentially a projectile contained within a projectile. The penetrators are made of various materials and are encapsulated by the lead mass and jacketing material. In most cases, the penetrators are denser than the lead and are designed to expel themselves out of the projectile upon impact with an object. The penetrator continues along the trajectory, perhaps with some deviation from the last indicated trajectory that the round was fired upon. NATO specification SS109 projectiles (5.56×45mm) are one such example. This projectile contains a penetrator that is designed (a) to penetrate intermediate barriers such as cinder blocks, glass, and light armor plating and (b) to enhance the injury potential in "soft" targets. The SS109 projectile is formally identified as "heavy ball," weighs 62 grains, and may be identified by a green tip. However, the green tip should not be solely relied upon to identify a 5.56×45mm cartridge as a heavy ball, as some cartridges of this specification are not so identified. British M855 ammunition produced by the Royal Ordnance Factory at Radway Green lacks the color coding, as does M855 ammunition produced by Israeli Military Industries, and even U.S. commercial market ammunition manufactured to M855 specification, so the presence or absence of the green tip cannot be relied upon to be definitive for identification purposes. M855 cartridges are produced around the world by many manufacturers. The M855 was designated *heavy ball* because it replaced the M193 cartridge, loaded with the 56-grain ball projectile. See Figure 2.12.

Figure 2.12 (See color insert.) Example of 5.56×45mm M855 "Green Tip" cartridges on stripper clips. Note the annealed casing neck. (U.S. Marine Corps image; photographer Cpl. Lydia M. Davey, USMC.)

Figure 2.13 Chinese 7.62×39mm projectile and cylindrically shaped mild-steel penetrator. (Image from author's collection.)

Certain examples of Chinese-produced 7.62×39mm and 7.62×54mmR ammunition contain a cylindrically shaped penetrator constructed of mild steel. The use of mild-steel cores is more likely to be attributed to economic rather than traumatic purposes. A Chinese-sourced 7.62×39mm mild-steel core projectile is shown in Figure 2.13. The author's experience suggests that this penetrator-type projectile behaves marginally against building materials and sheet metal.

The M855A1 Enhanced Performance Round (EPR) is the intended replacement for the M855 heavy ball cartridge. The M855A1 is reported to have been systematically engineered to meet a variety of specifications intended to create an improved cartridge over the current M855 cartridge. It is claimed that the cartridge is even more effective than the 7.62×51mm M80 cartridge. The projectile itself is reported to be more accurate and to offer improved performance against both "hard" and "soft" targets, as compared to the M855, while offering a more environmentally sensitive product in terms of production and in residuals left

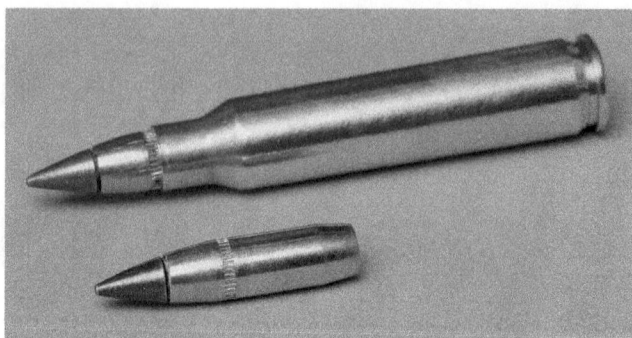

Figure 2.14 (See color insert.) The 5.56×45mm M855A1 EPR. Note the gap between the tip and projectile body and the cannelure on the projectile itself. (U.S. Army image; photographer Todd Mozes.)

after use. The M855A1 is a completely lead-free design, and while the M855A1 retains the penetrator design, it has been redesigned to resemble an arrowhead instead of a cylindrical shape. The cartridge features a different propellant formula to reduce muzzle flash and increase the projectile velocity. The cartridge is identified by its bronze-colored tip and may feature a subtly visible gap between the tip and the projectile (see Figure 2.14). The U.S. Army began general issue of the M855A1 in June 2010. ATK reported that it received an order from the U.S. Army for "nearly 300 million rounds of the new M855A1 EPR" after completing an initial order of "20 million rounds of M855A1, which were delivered to the troops in Afghanistan earlier this year" (Alliant Techsystems 2010b). The technology embodied in the M855A1 looks to be infused into the new 5.56×45mm lead-free tracer cartridge, tentatively identified as the M856A1, as well as the M995 armor-piercing cartridge. If the M855A1 proves to be a viable cartridge platform in application, it will likely serve as the inspiration for a new generation of cartridges that draw upon the technology pioneered in it. The 7.62×51mm M80 is reported to be in the process of a similar redesign, also incorporating lead-free components. A seemingly likely scenario is that it will mimic the M855A1 in all respects except in dimensions.

Ball projectiles can be of any shape—round nose, pointed, truncated nose, tapered, etc. For example, the cartridges depicted in Figure 2.15 are U.S. military M41 .38 Special ball cartridges, which have a round nose. A significant improvement in ball projectile design, known as the *spitzer*, was first developed in France and used in the 8mm Lebel cartridge around the turn of the twentieth century. Germany took off on the design shortly thereafter, calling their version of it the "spitzer," and the name stuck. Nearly every nation followed, abandoning the round-nose projectiles then in use in favor of the pointed, slender spitzer projectile. The round-nose projectile was considered more appropriate for then-contemporary rifle designs, many of which used a tubular magazine where the projectile tip bumped into the primer of the cartridge loaded ahead of it. Round-nose projectiles were considered to be less likely to cause an inadvertent detonation when compared to pointed-nose projectiles. As rifle designs evolved and ammunition was fed from internal or detachable box-style magazines, the pointed-nose projectile was no longer considered an impediment. Ironically, many full-sized, full-powered rifle cartridges developed at the turn of the nineteenth to the twentieth century that are still in use have retained the basic round-nose design.

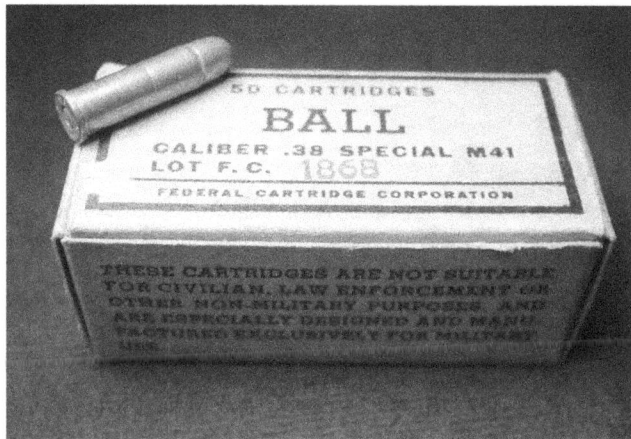

Figure 2.15 Vintage U.S. military M41 .38 Special cartridges. The warning on the box indicates that this ammunition is strictly for military purposes; however it is nothing more than ball .38 ammunition. (Image from author's collection.)

The spitzer projectile is identified by its pointed nose, tapered shape, and defined waistline. The taper expands from the pointed nose to the waist. In some cases, a spitzer will have a flat base, whereas other spitzer projectiles taper down to the base. The actual length of the waist can be long or short, relative to the projectile length. Spitzer projectiles with the tapered base may also be called a *boat tail,* a reference to the classic speedboat hull design. The boat-tail projectile is significantly longer than flat-based varieties. All modern military rifle cartridges are of spitzer or boat-tail design, and a great many civilian rifle cartridges are as well.

The British interpretation of the spitzer projectile for their standard .303 rifle cartridge was designated the Mark VII. Previous .303 cartridges, each designated by the Mark nomenclature, were round nose. The Mark VII was not a boat-tail-shaped spitzer but, rather, had a flat base. A significant difference between the British Mark VII and other contemporary spitzer-type projectiles was that it had a hollowed cavity behind the nose. To avoid the appearance of using a hollow-point projectile, which is prohibited in armed conflict by the Hague Convention, the projectile had a lightweight filler material contained within the cavity. There was also no abscess at the tip communicating the exterior of the projectile to the cavity. The purpose was to shift the center of gravity toward the back of the projectile, which weighed more than the front, as the bulk of the mass was behind the center of the projectile. When the projectile impacted a target, it would cause the projectile to yaw, thereby, in theory, increasing the Mark VII's wounding potential. British munitions producers marked the cartridge casing base with Arabic or Roman numerals to indicate what projectile was loaded; hence, VII or 7 would indicate that a Mark VII projectile was seated. Some of these casings may also be marked with a *z,* indicating the use of smokeless powder, so that they would not be confused with old-stock .303 cartridges loaded with black powder.

Apparently, there was enough faith placed in this design that the Soviets copied it when they developed the 5.45×39mm M74. This cartridge was likely inspired by analyzing the performance of the U.S. 5.56×45mm cartridge used during the Vietnam War. The 5.45×39mm 5N7 cartridge is a boat-tail projectile having a full-metal gilded jacket. Like the Mark VII, it has a hollow-cavity nose. A thin layer of lead covers an unhardened steel

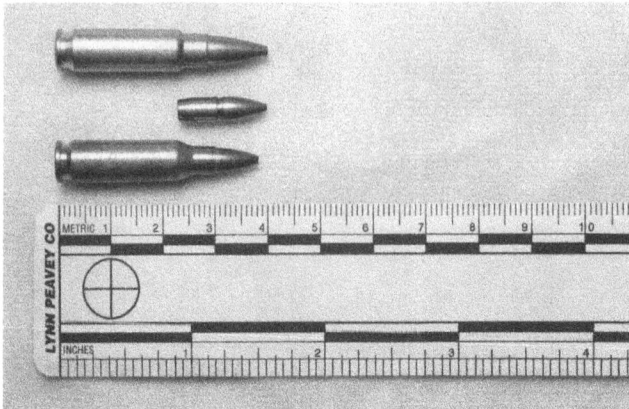

Figure 2.16 The FN 5.7×28mm (top), and the H&K 4.6×30mm (bottom) are very small, high-velocity cartridges intended for use with handguns and submachine guns. (Image from author's collection.)

core, and there is a small lead plug crimped in place at the base of the projectile. This bias shifts the center of gravity to the back, making the 5N7 susceptible to yaw and deformity when it impacts a target. The 5.45×39mm M74 is a very long, slender projectile, with quite a bit of the projectile projecting from the casing. Although the Soviet M43 7.62×39mm is represented as the same casing length, the M74 is actually just a millimeter longer, as the M74 is 39.37 mm in length over the M43's 38.60 mm. The M855A1 is reported to not be yaw dependent, "causing the same effects when striking its target, regardless of the angle of yaw" (Woods 2010).

By and large, handgun-caliber cartridges do not employ spitzer-type projectiles, but this should not be interpreted to mean that they are not used. The FN 5.7×28mm and the H&K 4.6×30mm both can use spitzer projectiles, and there are other loadings that also can use a spitzer projectile (see Figure 2.16). Older model semiautomatic handguns in particular have tended to be very susceptible to jamming when a cartridge loaded with anything other than a round-nose projectile was used. This is a direct function of the feed-ramp design of the firearm in question. Historically, semiautomatic handguns were designed to cycle ball cartridges, and older designs can be found with failure-to-feed malfunctions when truncated-nose or other style projectiles are used, as they tend to hang up at the feed ramp or the breech. Newer handgun designs, as a rule, are not susceptible to feed-ramp issues, as they have been engineered to cycle various projectile varieties, especially hollow-point-type projectiles that tend to have truncated noses. The projectile design is of no consequence to the revolver, insomuch as the cartridge being used is suitable for use with the revolver in question.

Controlled-Expansion Projectiles

The term *hollow point* is a colloquial term used to describe a projectile that is more formally defined as a *controlled-expansion projectile*. These projectiles get their name from their most distinctive characteristic: a cavity or abscess in the projectile nose that is readily apparent. This particular class of projectile is engineered to expand from its original diameter to perhaps twice or even three times the original diameter. Hollow points are marketed toward law enforcement, security providers, and private citizens seeking personal defense or hunting ammunition. In accordance with the 1899 Hague Convention, military forces that abide by the protocol are prohibited from using hollow-point-type projectiles in combat.

Figure 2.17 Magtech First Defense controlled-expansion projectiles. The projectile is constructed of solid copper, thus eliminating the possibility of the jacket separating from the projectile. (Image courtesy of Magtech Ammunition.)

The hollow point is still sometimes erroneously called a *dum dum*, an archaic term that dates back to the nineteenth century. A false premise of hollow-point projectiles is that they are often thought of as exploding bullets. This premise probably originates from the fragmentation effect that could occur in some projectiles after striking a target, even if the fragmentation effect was not intended or engineered into the projectile design. Hollow-point projectiles are not the only projectile variety susceptible to fragmentation when impacting a target; ball-type projectiles are also susceptible to this behavior. The designed purpose of the hollow point is to allow the projectile to expand upon impacting a target. This expansion is caused by the outer edge of the projectile nose around the ogive* folding open and to the rear when meeting the resistance of the target medium. Figure 2.17 demonstrates an ideal fully controlled expansion. Figure 2.18 depicts the generic anatomy of a controlled-expansion (hollow point) projectile. Figure 2.19, although depicting a different brand than those in Figure 2.17, demonstrates that the hollow point is a class characteristic of like-designed projectiles and basically is the same regardless of brand or manufacturer.

When a hollow point expands, it has two effects: The first is to increase the size of the temporal and permanent wound cavities in the target, thus increasing the incapacitating or lethal potential of the projectile. The second effect is that because the projectile impact surface has been increased, often more than doubled from the original diameter, there is a greater surface area to fully transfer the kinetic energy from the projectile into the target, thus reducing the risk of overpenetration and keeping the projectile within the target. The appearance of an ideally expanded hollow point is that of a star. Most major munitions manufacturers produce a hollow-point-type projectile and market their particular products under different trade names, such as Speer Gold Dot, Winchester Silver Tip and the Supreme Elite PDX1, Federal Hyrda-Shok, Hornady TAP, Nosler Accubond, and the Barnes VOR-TX line (which can be loaded with the Barnes TSX, Tipped TSX, and TSX FN projectiles). Capitalizing on the zombie phenomenon, Hornady Manufacturing announced in October 2011 a new line of ammunition specifically designed to kills zombies (see Figure 2.20). In spite of the tongue in cheek spirit of the packaging and marketing, Zombie Max projectiles are of the controlled expansion variety, that are at least as likely to be equally effective on the living as the undead. Nevertheless, all of these are still hollow points by definition and class characteristics.

---

* The ogive is the rounded, tapered, leading end of the projectile, sometimes called the *nose*.

Figure 2.18 (See color insert.) A cutaway image of the Hornady Critical Defense cartridge. (Image courtesy of Hornady Manufacturing.)

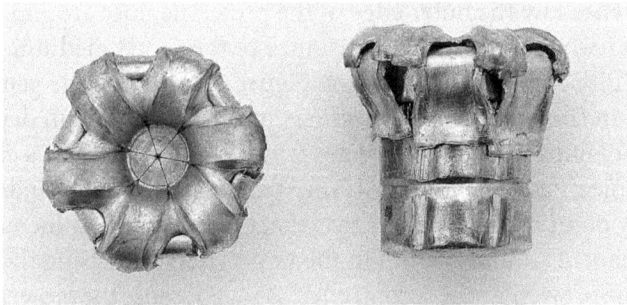

Figure 2.19 Hornady 9×19mm Critical Duty projectile that has passed through heavy clothing. The cartridge was a +P load. (Image courtesy of Hornady Manufacturing.)

The major handicap of hollow-point-type projectiles has been their ability to properly expand and remain as a unified, coherent projectile while the expansion is occurring. Projectile fragmentation, failure to expand, inconsistent expansion, and jacket separation have been performance issues that affect the final result achieved when using hollow-point projectiles. A variation of the hollow-point projectile is one that is partitioned or segmented. This variety of hollow points has one or more additional hollowed partitions within the projectile. Demonstrated examples of partitioned hollow points suggest that controlled expansion can nearly flatten the projectile. The Nosler Partition projectile is an example of such a design.

A recent entrant into the market from Federal Cartridge is the Guard Dog line. The Guard Dog cartridge, currently available in 9×19mm, .40 S&W, and .45 ACP blurs the line

Figure 2.20 (See color insert.) The pop culture phenomenon of a world taken over by zombies prompted Hornady to release a line of "zombie killing" ammunition, a rather jocular bit of marketing. (Image courtesy of Hornady Manufacturing.)

between ball and controlled-expansion projectiles. Overtly, the projectile appears as a ball type, lacking the abscessed nose that is traditionally associated with the controlled-expansion-class projectiles. Federal Cartridge states on its website that the Guard Dog is not a hollow point, although it behaves as one, expanding upon target impact. A distinctive feature of the Guard Dog is a series of blue fillers in the projectile tip. These fillers are not visible to the eye before the projectile is expended, but they are revealed after the projectile has expanded. It would appear that there are cuts or serrations within the projectile that cause it to petal open, much like a traditional hollow point, yet eliminate the need for an opening at the tip. This cartridge could prove to be especially popular in venues that prohibit controlled-expansion hollow-point projectiles, as this does not appear to meet the requisite design features of such. A second unique feature of this line is that the three currently represented calibers use projectiles that are of reduced mass by comparison to other projectiles of those caliber classes. The 9×19mm projectile weighs a mere 105 grains; the .40 weighs 135 grains, and the .45 ACP weighs 165 grains (Federal Premium Ammunition 2011).

Open-Tip Match

The Open-Tip Match (OTM), is a class of projectile that has been subject to debate as to whether or not it can be qualified or defined as a hollow-point-type projectile. The OTM differs from a traditional hollow point in appearance. The OTM projectile has a very narrow, shallow opening at the projectile tip. There are no serrations, or petal cuts, as found in typical hollow-point projectiles. OTM projectiles are associated with traditional rifle calibers such as 5.56×45mm or .308 (7.62×51mm Winchester), among others. The apparent design purpose of the OTM-type projectile is to enhance long-range accuracy through certain characteristics specifically engineered into the design, not to provide for a capacity to expand in a terminal ballistic situation.

A memorandum dated October 12, 1990, written by U.S. Marine Corps Col. W. Hayes Parks, chief of the Judge Advocate General's International Law Branch, addressed to the commander of the U.S. Army Special Operations Command, discussed the legalities of using OTM ammunition in the conduct of war. The memorandum first clarified the significance of the open tip.

The purpose of the small, shallow aperture in the MatchKing* is to provide a bullet design offering maximum accuracy at very long ranges, rolling the jacket of the bullet around its core from base to tip; standard military bullets and other match bullets roll the jacket around its core from tip to base, leaving an exposed lead core at its base. Design purpose of the MatchKing was not to produce a bullet that would expand or flatten easily on impact with the human body, or otherwise cause wounds greater than those caused by standard military small arms ammunition. (Parks 1990)

The author opined on the question of military application of such a projectile:

The purpose of the 7.62mm "open-tip" MatchKing bullet is to provide maximum accuracy at very long range. Like most 5.56mm and 7.62mm military ball bullets, it may fragment upon striking its target, although the probability of its fragmentation is not as great as some military ball bullets currently in use by some nations. Bullet fragmentation is not a design characteristic, however, nor a purpose for use of the MatchKing by United States Army snipers. Wounds caused by MatchKing ammunition are similar to those caused by a fully jacketed military ball bullet, which is legal under the law of war, when compared at the same ranges and under the same conditions. The military necessity for its use—its ability to offer maximum accuracy at very long ranges—is complemented by the high degree of discriminate fire it offers in the hands of a trained sniper. It not only meets, but exceeds, the law of war obligations of the United States for use in combat. (Parks 1990)

In September 2006, ATK was awarded a development contract for enhanced cartridges in 7.62×51mm and 5.56×45mm, resulting in the creation of the MK 318 MOD 0 (5.56×45mm) (Figure 2.21) and the MK 319 MOD 0 (7.62×51mm) (Figure 2.22). Both projectile types are open-tip match. The MK 318 projectile has a mass of 62 grains, and the MK 319 projectile has a mass of 130 grains.

The MK 318 MOD 0 and MK 319 MOD 0 cartridge designs are a variation of the Federal Trophy Bonded Bear Claw round developed by Federal Cartridge for large game hunting. The Trophy Bonded Bear is a commercial item that can be found in a number of commercial publications such as Cabelas. The design for these rounds have been modified to meet all law and military requirements as required while meeting all performance Key Performance Parameters (KPPs). (Department of the Navy Crane Division 2009)

The front of these projectiles is designed to defeat intermediate barriers, while the base, which is solid copper, is designed to act as a penetrator. This style of projectile has been termed to be *barrier blind* because of the guiding design philosophy, which is to effectively engage a target through intermediate barriers such as automotive glass, doors, and so forth (Marsh, Stoll, and Leis 2009). The U.S. Marine Corps is reported to have adopted the MK 318 and discontinued the use of the standard M855 cartridge (Lamothe 2010b). The Marine Corps is also reported to be testing the M855A1 cartridge, but remains skeptical in contrast to the MK 318 (Lamothe 2010a). "The MK 318 and MK 319 were developed in conjunction with the Special Operation Forces Combat Assault Rifle (SCAR)" (Department of the Navy Crane Division 2009). Federal Cartridge had already developed the MK 316

---

* The MatchKing is a specific projectile manufactured by Sierra Bullets. The cartridge subject of the memorandum is the Gold Medal Match, manufactured by Federal Cartridge using the MatchKing projectile.

Figure 2.21 The 5.56×45mm Ball, Carbine, Barrier MK 318 MOD 0 cartridge. Note cannelures and classic boat tail or spitzer shape. (U.S. Navy image.)

Figure 2.22 The 7.62×51mm Ball, Carbine, Barrier MK 319 MOD 0 cartridge. Note the flat projectile base. Compare the differences between this cartridge and Figure 2.21. (U.S. Navy image.)

(Special Ball, Long Range) for the U.S. Navy. The MK 316 is a 7.62×51mm cartridge loaded with the 175-grain Sierra MatchKing OTM projectile.

The MK 262 MOD 0 is another open-tip match projectile developed at the request of the U.S. military. Officially, the cartridge was designated the Special Ball, Long Range, MK 262 MOD 0. The projectile is significantly heavier than that used in the M855, weighing 77 grains versus 62 grains. The cartridge was sole-sourced to Black Hills Ammunition for manufacture; however, the projectile is the Sierra MatchKing OTM. The MK 262 MOD 1 is the same in all respects, save for the addition of a cannelure in the projectile. Although the cartridge will function in any firearm chambered for 5.56×45mm, it was specifically intended for use in the Mark 12 Special Purpose Rifle (Buxton and Marsh 2003). The Mark 12 SPR is a derivative of the basic M16 rifle, with various updates and modifications.

Wad Cutter and Semi-Wad Cutter

A wad cutter is a cylindrically shaped projectile that has a flat nose. The appearance of a wad cutter is similar to that of a segmented cylinder, and most wad cutters have several cannelures. Wad cutters can be jacketed or unjacketed. The purpose of a wad cutter is to cut clean, round holes in paper for target shooting, making the target easier to count. Wad cutters are generally bare lead, but jacketed versions have been made as well. Outwardly, a cartridge loaded with a wad-cutter projectile appears to have no projectile seated at all, much like a blank, as the projectile is seated such that the front of the projectile sits flush with the opening of the casing neck. Although intended for target use, wad cutters can certainly cause lethal injury.

A semi-wad cutter is similar in appearance and description to the wad cutter with the exception of a truncated nose. Like wad cutters, their primary application is for shooting paper targets and creating a clean hole in the paper for scoring. The majority of wad cutters that were ever manufactured were produced in calibers popular with competition shooters, primarily in the .38 and .45 class. To see a wad cutter or semi-wad cutter in other calibers would be unusual but cannot be ruled out. Wad-cutter hollow points are hybrids that combine the physical shape of a wad cutter but have the abscessed-nose characteristic of a hollow-point-type projectile.

Armor Piercing

Armor-piercing ammunition is designed to perform exactly as the name states: It is designed to overcome armor of a given specification. Most venues qualify a cartridge as being armor piercing by the material that constitutes the projectile. Materials such as tungsten carbide, hardened steel, iron, brass, bronze, beryllium, copper, and even depleted uranium have been or are currently used to construct armor-piercing projectiles.

Dedicated armor-piercing (AP) cartridges should not be confused with non-AP cartridges that are, in fact, capable of defeating certain types of armor by virtue of other factors such as projectile mass coupled with velocity, or simply by the velocity of the projectile alone. Hollow-point projectiles are typically unable to penetrate armor because their engineered expansion occurs upon impact with the hardened surface, whereas ball-type projectiles may be more likely to pierce or penetrate armor, relative to other types of projectiles.

In addition to the composition of the projectile mass itself, armor-piercing capability may be complemented by the jacketing, either by the material used or by the mass of the jacketing relative to the overall projectile mass. The Swedish M39B 9×19mm cartridge is one such example. This particular cartridge has a disproportionately thick jacket relative

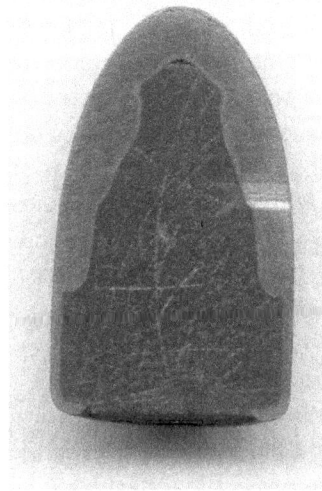

Figure 2.23 A cross-section of the Swedish M39B 9×19mm projectile. Of particular interest is the disproportionately thick jacket, relative the overall projectile mass. (Image courtesy of Christian Sahlberg.)

to the overall projectile size. The proportions become particularly evident when compared to other 9×19mm cartridges. A cross-section of the M39B is shown in Figure 2.23. The German munitions firm MEN manufactures a 9×19mm cartridge specified for military applications, designated the DM91. The DM91 overtly resembles an ordinary 9×19mm cartridge; however, it contains a steel penetrator core encapsulated by the jacketing. Surrounding the penetrator is a layer of lead, which makes the DM91 a sabot-type projectile. Sabots are a common approach to armor-piercing projectiles.

Tres Haute Vitesse (THV)

The Tres Haute Vitesse (THV) was developed by Société Française Munitions (SFM) and appeared in the United States in the early 1980s. The THV was a radical approach in projectile design and was built on existing caliber platforms such as .357 Magnum, .38 Special, 9×19mm, .32 ACP (7.65×17mm or 7.65 Browning), and .45 ACP. The apparent intention was to create hypervelocity projectiles on cartridge platforms that were already in use by law enforcement agencies, permitting the use of the same firearms but with a projectile whose performance was well in excess of any other cartridge in the given calibers at the time. The projectiles were very distinct in appearance, and there were several designs, including a cylindrical body with a conical tip and a similar version that differed in that the tip of the cone had a blunted, rounded nose. The .357 Magnum THV projectile mass was 45 grains (2.9 grams) compared to other .357-class projectiles, which range from 125 to 180 grains. The reported velocity of the .357 Magnum THV was approximately 2,625 ft/s. In contrast, a typical 158-grain .357 Magnum hollow point achieved a velocity at the muzzle of some 1,250 ft/s. THV was very expensive for the time, costing roughly $1 per cartridge in 1984. The high cost was no doubt due to the projectile composition: They were made of solid brass and machined to shape.

Van Bruaene Rik (VBR)

Van Bruaene Rik (VBR) is a Belgian-based ammunition firm. The company specializes in the cartridge that it developed, the 7.92×24mm VBR, but it also manufactures cartridges in other calibers using its proprietary projectiles. The company manufactures several varieties of projectiles designed around specific applications ranging from general purpose ball projectiles to armor-piercing penetrators, and fragmentation-causing types. The company markets barrel reamers that allow an existing barrel to be resized to chamber the 7.92×24mm. Custom replacement barrels are also available to retrofit an existing firearm. The VBR-PDW is purpose-built to chamber the 7.92×24mm, but can be rebarreled to accommodate other calibers. In addition to the 7.92×24mm, VBR cartridges are available in calibers 4.6×30mm, 5.7×28mm, 9×19mm, 9×18mm, .357 SIG AUTO, .38 Special, .357 Magnum, .40 S&W, 10mm Auto, .44 Magnum, .45 ACP, and .45 GAP (F.S.D.I.P. and VBR-Belgium n.d.).

Frangible Projectiles

A frangible projectile is one that is designed to disintegrate into inert dust upon impacting materials such as steel targets and backstops at gun ranges. There is still a risk of backsplash, where larger chunks of the projectile reflect back in the direction of the shooter or bystanders on the firing line, so proper eye protection is a must. The concept of a frangible projectile is not a new one, but it is one that has seen a tremendous resurgence of interest in recent years.

During the Second World War, frangible projectiles using a lead/plastic blend were used by U.S. forces for aircraft gunnery training purposes. During the war, the German armaments industry developed (and German forces fielded) sintered-iron projectiles as a wartime expedient measure to reduce the use of strategic materials. Although not a frangible projectile in the truest sense as defined today, it would appear to be the first practical application of compressing a powdered metal under high temperature to produce a solid projectile, an approach that would emerge years later as a means of producing a purpose-designed frangible projectile.

The concept of a frangible projectile was explored at the Oak Ridge National Laboratory by the Surface Processing and Mechanics Group. Powder metallurgy was determined to be the best route; options such as ceramics were discarded. Their research concluded that blending higher density, harder materials such as tungsten coupled with softer materials such as tin or zinc produced a projectile that was practically similar to a traditional lead projectile (Lowden and Vaughn 2009). Figure 2.24 shows fired frangible projectiles that have broken apart.

In general, a frangible projectile is manufactured from a blend of powdered metals such as copper, bismuth, tungsten, or tin. These projectiles are formed in their requisite shape and held together through a bonding agent such as epoxy, nylon, or another matrix-type bonding additive, or they are sintered (adhered by heat). Frangible projectiles may or may not be jacketed. One approach is to produce a projectile that is bimetal in nature, using an inert powdered metal for the core and another metal to form the outer edge. A second approach is to blend the powdered-metal constituents and combine them with the bonding agent.

Frangible projectiles have proven themselves to be a viable alternative to traditional lead-based ammunition in training settings, and are particularly well suited for use in indoor gun range environments. Frangible ammunition has been widely touted as the "green" alternative to traditional lead-based ammunition and is almost universally primed using a clean or "green" primer and powder compounds devoid of the heavy elements, particularly lead.

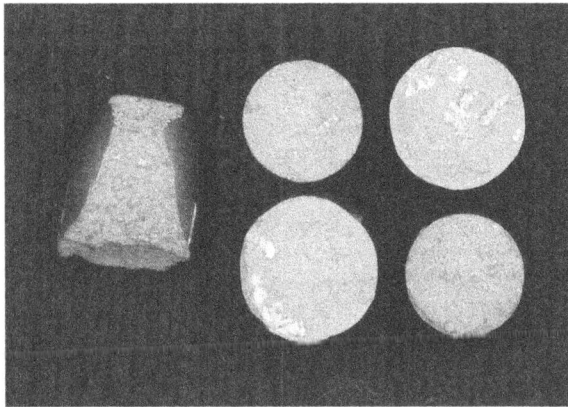

Figure 2.24 Fired frangible projectiles that have broken apart. Note the consistency and structure of the interior of the projectile. (Image from author's collection.)

General Dynamics Ordnance and Tactical Systems of Canada manufacture two versions of frangible ammunition: Short Stop and Greenshield, both of which are labeled as nontoxic. RWS, a part of RUAG Ammotec USA, manufactures frangible training ammunition using a copper matrix. Prvi Partizan, a munitions manufacturer based in Serbia, markets dedicated training frangible ammunition as its PPU Green Line GL. Speer's frangible ammunition is called Clean Fire and has dedicated law enforcement and general use product lines. Remington Arms markets two varieties of frangible ammunition: Disintegrator CTF, which uses a jacketless copper/tin lead-free projectile, and Disintegrator Plated Frangible, which has a metal-particle projectile encased in an electroplated copper jacket. Winchester offers frangible ammunition under its Ranger line in popular handgun and rifle calibers as well as shot shells. International Cartridge Corporation and Kilgore Ammunition Products also manufacture frangible ammunition.

Frangible ammunition is not solely for training applications. Cartridges loaded with frangible projectiles have also been developed for lethal applications. International Cartridge Corp. manufactures frangible ammunition lines for training, hunting, and lethal force (law enforcement, military, and personal defense) in the popular handgun and long-gun calibers. Serbian munitions manufacturer Prvi Partizan manufactures a sintered frangible projectile cartridge for lethal use. The Duty, or lethal, frangible cartridges are of the hollow-point class (Prvi Partizan 2009). Corbon MPG (Multi-Purpose Green) is advertised on the Dakota Ammo website to be dual application: "a perfect load for target shooting and lead free ranges. It uses a gilding metal jacket and a compressed copper core. This low penetrating round can also be used as a self-defense round" (Dakota Ammo/Glaser 2012).

Another application for frangible-projectile technology has been in shot shells used for door- or barricade-breaching operations. A breaching cartridge permits the explosive defeat of doors by destroying hinges, locks, and handles. Breaching rounds can also be used to defeat other barriers and even windows. Frangible projectiles are ideal for breaching applications in that they reduce the risk of injury to both the operator of the breaching weapon and anyone standing in the periphery of the point of breach. The 12-gauge shotgun can serve as a platform for breaching tools. Like other specialty munitions, breaching rounds are conspicuously marked to alert the operator.

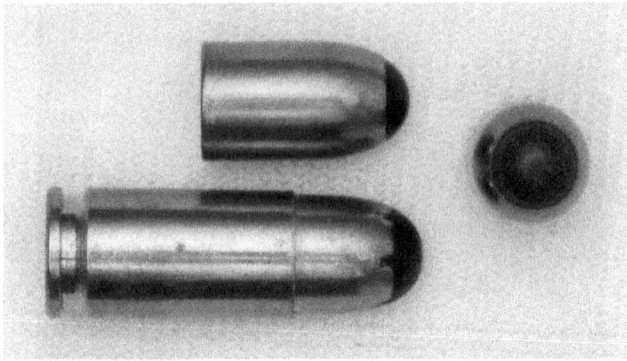

Figure 2.25 Enhanced Penetration Round (EPR) manufactured by ExtremeShock. The pictured 9×19mm projectiles have a black composite tip insert, which should not be confused with painted color markings. Also pictured is a loaded ExtremeShock 9×19mm cartridge. (Image from author's collection.)

Blended-Metal Projectiles

A blended-metal projectile has been developed by Dynamic Research Technologies (DRT). According to information supplied by DRT, the projectile is not sintered or bonded. The projectiles are claimed to be 100% lead free and completely devoid of impurities. The projectile is encased in a jacket, and the projectile itself is composed of a very hard, fine metallic powder. The projectile is designed to completely disintegrate when encountering materials that are hardened or denser than the projectile itself. DRT ammunition is strictly dedicated for lethal-intended applications for hunting, personal defense, military, and law enforcement (DRT 2011).

ExtremeShock is another company that currently markets lethal-application frangible ammunition. Like other lethal-use frangibles, it is designed exclusively for use against soft targets. The "projectile" is composed of tungsten powder encapsulated by a traditional copper jacket. The powdered medium itself does not comprise a solid-mass projectile, per se, but it is tightly packed and contained within the jacket, which acts as the container. Like all frangible projectiles, these projectiles return to powder when impacting an object. ExtremeShock cartridges are available in all modern popular calibers and some other unusual calibers as well. Figure 2.25 shows a loaded ExtremeShock 9×19mm cartridge as well as the projectile and the black cap.

Glaser Safety Slug

The Glaser Safety Slug was a milestone in the development of small-arms ammunition, as it represented the first practical effort to develop a cartridge with lethal potential yet designed to minimize, if not completely eliminate, the risk of overpenetration; that is, the risk of injury or damage beyond the intended target. Born of the era when the airline passenger could smoke a cigarette while enjoying a complimentary cocktail, this cartridge was developed after the airliner became a popular target for hijacking. The Glaser Safety Slug was designed to deliver lethal force or at least incapacitating injury to the soft target while preventing the penetration of a hard target, such as an interior wall or an aircraft fuselage. It is presumed that the first generation of the Safety Slug was .38 Special caliber, as most of the preferred concealed-carry revolvers of the day were so chambered. The composition of the original Glaser Safety Slug resembled that of a small shot shell, a hollowed lead projectile completely filled with small-diameter shot and topped with a polymer or Teflon tip. Undated literature produced by Glaser Safety Slug describes the projectile as a "pre-fragmented projectile" that, upon "impact with tissue simulants causes immediate and complete fragmentation, release[ing] the core particles in a cone shaped pattern of over 330 sub projectiles" (Glaser Safety Slug n.d.).

Although it has passed through several advancements in the last 30 years, the concept remains the same. Currently, the Glaser Safety Slug is manufactured by Corbon. According to the literature released by the manufacturer, the Glaser Safety Slug is made in two configurations: one loaded with #12 shot and capped with a blue ball; the second loaded with #6 shot and topped with a silver ball. Silver-ball cartridges are reported to have increased penetration over the blue version. Glaser Safety Shot should not be confused with shot shells produced by CCI designed for rodent and pest control.

Tracer Ammunition

Tracer ammunition contains a pyrotechnic component within the projectile that burns upon detonation and continues to burn for a given period of time as it travels downrange from the weapon. Tracer ammunition is viewed as a streak of light that travels along the general trajectory that a projectile would take from the weapon, giving shooters a visual indication about their point of aim. It is common in military applications to insert one tracer per five rounds loaded on ammunition belts. Figure 2.26 shows belted tracer cartridges (circled).

There are variations of tracers: a traditional long-burning type that is visible over a great distance, reduced illumination that burns with reduced light, and those whose emissions are visible only to operators using night-vision equipment. The "headlight" tracer round has an illuminating effect downrange, as though the viewer were observing using vehicle headlights. Military tracer munitions are conspicuously marked in transportation vessels as well as on individual cartridges. Depending on the source of origin and era of the tracer, the cartridge will be color-coded on the projectile tip, cartridge body, and cartridge casing base. The color of the illumination varies from yellow to red to green, depending on the ammunition.

Figure 2.26 Belted 7.62×51mm cartridges. Note the orange-painted tip, circled, indicating that the cartridge is a tracer. (U.S. Air Force image; photographer Technical Sgt. Jeremy Lock, USAF.)

Tracer ammunition has an incendiary effect, and the use of tracers can result in fires in vegetation or flame-sensitive structures and can ignite fuel tanks and other volatile substances. More than one fire has been inadvertently started when tracer ammunition was fired into dry brush or when it ignited inflammables. Armor-piercing and incendiary-type projectiles often have a tracer component to them, these combinations referenced as an API (armor-piercing incendiary) or APTI (armor-piercing tracer incendiary) type projectile. Tracer shot shells have been manufactured but are uncommonly encountered.

Explosive Projectiles

The fragmenting action of projectiles impacting a target may often be described or erroneously associated with an "explosive" occurrence. This association may be drawn particularly from controlled-expansion projectiles (hollow points), which may have a tendency to break apart upon impact. Explosive small-arms ordnance was manufactured by every nation and on a limited commercial basis for most of the twentieth century. For example, during World War II, a variant of the Japanese Type 92 7.7mm semi-rimmed cartridge was loaded with a lead projectile surrounding a core of PETN (pentaerythritol tetranitrate). The jacketing material was cupronickel. The projectile weighed 162 grains and was intended for use exclusively with machine guns and likely in an antiaircraft or antivehicle role. This particular cartridge was color-coded with a purple band painted where the projectile meets the casing (Departments of the Army and the Air Force 1953). A schematic of the cartridge is depicted in Figure 2.27. The Argentinean-manufactured 7.65×53.5mm Type R, an observation cartridge, is a similar explosive cartridge. There are innumerable other examples historically of explosive cartridges, even in the context of small-arms ammunition; such cartridges were manufactured wholesale across the world in all eras for any possible military force as the consumer.

In the late 1970s, self-described "exploding projectiles" started to appear on the market and were available in the popular calibers such as the .22 Long Rifle, .380 ACP, .38 Special, .357 Magnum, 9×19mm Luger, .45 ACP, .44 Special, and .44 Magnum. Note the absence of the .40 S&W, an exploding projectile that would not be developed for nearly 10 more years. A study published in 1980 that sought to confirm or refute the claims made of the Exploder brand cartridge as being a truly "explosive projectile" did not report any form of explosion; instead, the projectile was observed to fragment when fired into ballistic gelatin. It was reported that the 9mm Exploder cartridges overtly resembled standard 9mm jacketed hollow points, having a cavity in the nose, although the Exploder had two cavities. A more detailed inspection of the projectile revealed that "approximately ¼ grain of a non-perforated disk type smokeless powder is placed on the lower cavity and a small pistol type primer (.175″) seated in the upper cavity" (Thompson and Amble 1979).

Another such cartridge was the Devastator cartridge. These were marketed through Bingham Ltd. of Atlanta, Georgia. The most notable appearance of the Devastator cartridge was during the 1981 assassination attempt on the late former President Reagan by John Hinckley. The results of the six fired projectiles are self-evident: No explosive occurrence was documented or recorded, and precautionary measures taken at the hospital to shield the attending staff from an unexploded "bomb" inside of George Washington University Hospital—and more particularly lodged in the sitting president of the United States—have been attributed to the angst and misinformation during that most confusing time. News coverage of the event continued to reference explosive bullets, and even in more contemporary works that revisit the event, the Devastator is still defined as an exploding bullet.

Aluminum

Lead
Steel

Lead
Copper

Tracer
composition

Armor piercing   Tracer

Cupro-nickel

P.E.T.N.
Copper sleeve
Brass
White phosphorus
Lead

Ball   Incendiary   Explosive

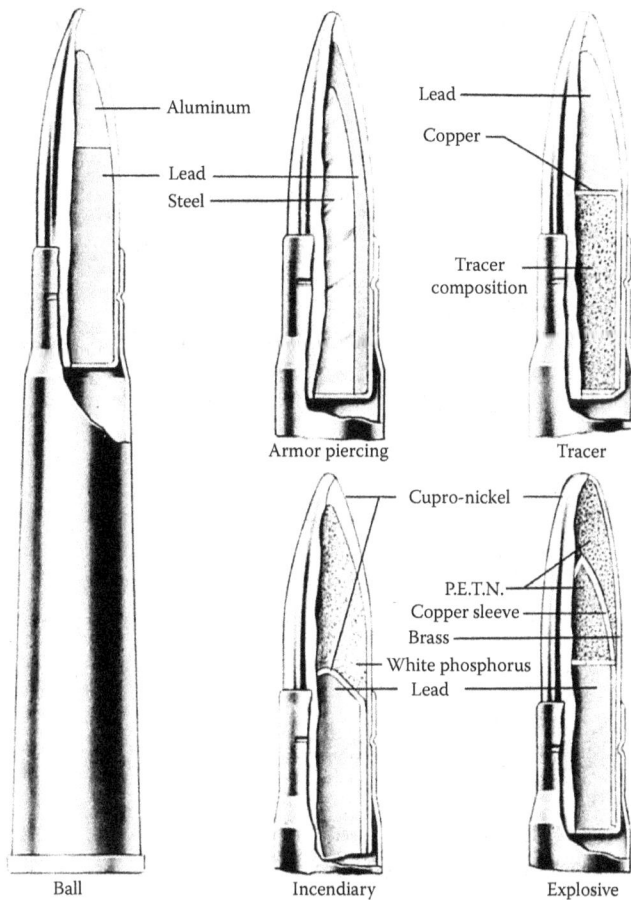

Figure 2.27 Wartime Japanese ammunition. (From U.S. Army Technical Manual TM 9-1985-5, 1953.)

Hinckley's decision to use a .22 in his assassination attempt may be considered as a contributing factor to the underperformance of the Devastator, due to its small size. However, it seems more objective to discard the broader claims of such a projectile as being "explosive" in a context that is more readily accepted: that such an amount of energetic material could not possibly be loaded into a pistol projectile.

The Velex and the Velet cartridges were of similar construction to the Exploder and Devastator cartridges. In 1978, the Velet cartridge, manufactured by the Velet Cartridge Co. in Spokane, Washington, had been subjected to study as a possible explosive projectile. The Velet was claimed to offer significantly greater "stopping power" than similar-caliber hollow-point designs. The results of the testing found that when the Velet ammunition was fired into ballistic gelatin "to observe the 'explosion' with the result that the bullet makes a very dirty wound track as it appears that the nose is filled with 'black powder'" (Lutz 1978). According to Gerns in a 1984 report, the Velex/Velet projectile resembled a hollow-point-type projectile with the cavity containing "black powder, #4 lead shot, and a red percussion cap. More recently produced rounds have the black powder replaced with Pyrodex,* and the percussion cap with a pistol primer" (Gerns 1984). These designs were not generally

* Pyrodex is a trademarked black-powder propellant produced by Hodgdon.

accepted as having any greater wounding effect than other hollow-point designs of the day and gradually disappeared with the passage of time.

Historically, it would appear that the intention of these cartridges was to overcome engineering challenges of the day that caused controlled-expansion projectiles to perform unsatisfactorily. Perhaps the thought was that by inserting a trace amount of black powder and a percussion primer as a booster of sorts would create a small "explosion" that would almost guarantee that the projectile would expand or fragment, causing an enhanced wounding potential. In the case of the Velex/Velet projectiles, it stands to reason that the addition of the #4 shot would have only enhanced the wounding potential of the projectile by introducing more projectiles in the form of shot into the equation.

Dummy, Drilled, and Inert

Dummy, drilled, and inert cartridges refer to the same object: a cartridge analog that is incapable of operation by design and intent. These cartridges are for nonfiring training, demonstration, and instructional purposes such as the manual of arms, the proper loading and unloading of firearms, cartridge nomenclature, ammunition handling practices and procedures, and other similar purposes. There are numerous variations of inert dummies ranging from pure polymer cartridges molded into the shape of a particular cartridge to metallic casings with plastic inserts.

Great lengths are taken to ensure that dummy or inert cartridges are conspicuously marked "INERT," color coded, or otherwise immediately identifiable in such a way as to alert the handler and other observers that such rounds are not live and are therefore suitable for "dry" or nonfiring operations. The entire cartridge may be molded in a high-visibility color such as blaze orange, which has become an industry standard safety color. Metal-cased dummies may have plastic projectiles, again molded in high-visibility colors for identification, as well as having either no primer seated or a high-visibility colored insert in the primer pocket at the base of the casing. Metallic versions can be manufactured from high-grade steel and intended for repetitive use, especially for armory work where a weapon function is inspected and approved for service. These cartridges themselves may also be marked and used as a tool, allowing for headspace and throat erosion to be checked. Another variant of the steel dummy has a corrugated casing, or lines that run longitudinally down the casing. These all-steel variants are of one-piece construction and cannot be disassembled. For inert dummies that are produced using actual casings and with a projectile seated, the primer pocket will be empty or filled in with a high-visibility plug. The casing itself will be drilled through in several places, preventing powder from being stored inside the casing, as well as providing a visual clue that the object is an inert article. Figure 2.28 shows examples of various dummy, drilled, and inert cartridges.

Blank Cartridges

Blank ammunition is manufactured in a wide variety of calibers (see Figure 2.29). Blank ammunition may overtly appear to be the same as projectile-loaded ammunition. An obvious difference might be the appearance of a casing mouth that is closed by crimping or by having a small disc covering the opening of the casing. Blanks, when fired, produce the characteristically loud report of a gunshot but without expelling a projectile. Blanks are used as a way of simulating gunfire without the obvious danger of expelled projectiles in situations such as signaling, drill and ceremony, theatrical productions, training, starter pistols, and even in industrial tools such as nail guns that use gas pressure.

Figure 2.28 (See color insert.) Examples of various dummy, drilled, and inert cartridges (from left to right): safety orange marked 9×19mm, drilled casing 9×19mm, solid black plastic 5.56×45mm, corrugated casing 5.56×45mm, and a 12-gauge shell marked DUMMY. (Image from author's collection.)

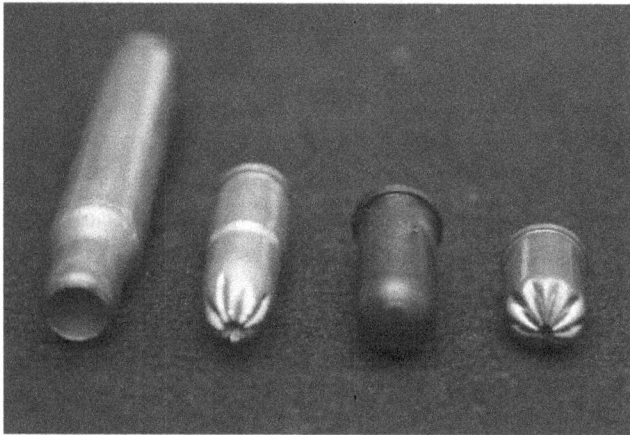

Figure 2.29 An assortment of blank cartridge styles. Some have "rosette" or crimped necks, while others use a cap to cover the interior. (Image from author's collection.)

This is not to imply that blanks are not dangerous; blanks are extremely dangerous, and there have been many fatal accidents involving blanks. One misconception about blanks is that, as there is no projectile, there is no projection of any sort from the fired weapon, but quite the contrary is true. Although no projectile is expelled per se, there is the potential expulsion of unburned powder flakes, high-pressure hot gases, and other matter that could cause serious, if not fatal, injury. It is often the case that blank firing causes serious injuries because blank firings often occur in close quarters.

Live ammunition is frequently remanufactured into blanks by first removing the projectile and removing the powder charge. In such instances, the projectile may or may not be reseated into the casing. Regardless, a tremendous hazard still exists, as there is sufficient pressure in the detonation of the primer alone to cause the expulsion of the projectile.

If no projectile has been reseated, there is still a danger from debris exiting the casing from the detonation of the primer. The author is aware of a training scenario where a law enforcement officer was wounded when he was shot in the stomach by a handgun loaded with commercially produced blanks. The death of actor Brandon Lee on a movie set on March 31, 1993, was attributed to the young actor being shot by a firearm used as a prop to film a scene. Gas operated automatic firearms may require the use of a blank-firing adapter to reliably cycle the action of the weapon. Such devices attach to the end of the muzzle and plug the barrel, allowing for sufficient gas pressure to be contained within the firearm to allow it to automatically cycle. Ordinarily, simple blowback automatic weapons may not require such an adapter due to the simplicity of their system of operation, assuming that enough gas pressure is produced by the blank to cycle the action.

Another application for blanks is for use in the firing of rifle grenades. Blank cartridges are loaded to facilitate the firing of a grenade from a muzzle-mounted grenade launcher. Such delivery systems are not used in contemporary military firearms; however, older weapon systems may employ a grenade-launching attachment. For the most part, rifle grenade launchers were muzzle-attached devices that were removed when not in use. The Yugoslavian version of the Russian SKS semiautomatic rifle, the M59/66, had a permanently attached rifle grenade launcher that also acted as a muzzle brake. The rifle grenade launcher is often called a *spigot*. In certain less-lethal munitions-delivery systems, there are similar devices used to provide the energy sufficient for propulsion. U.S. military rifle grenade-launching cartridges include the M64 (7.62×51mm), the M3 and M6 (.30-06), and the M195 (5.56×45mm).

Subsonic Cartridges

Many cartridges are capable of propelling the projectile above the speed of sound. For situations where this may not be desirable, subsonic loads have been manufactured. A subsonic load contains a reduced powder charge, sometimes in conjunction with a heavier projectile, to reduce the projectile's velocity below the speed of the sound, and thus a suppressor or silencer need only abate the report of the discharging of the cartridge and subsequent muzzle blast. In this situation, it is counterproductive to then fire a projectile that creates a crack as it breaks the sound barrier when the suppressor is attempting to muffle noise.

Certain firearms with integral suppressors are often specifically designed for use with such special loads. Subsonic .22 cartridges are manufactured by Aguila (see Figure 2.30), CCI, Federal Cartridge, Remington Arms, RWS (Dynamit Nobel), Winchester, and Eley. Fiocchi, Israeli Military Industries, Remington Arms, and others manufacture 9×19mm subsonic cartridges. ExtremeShock produces subsonic cartridges in .308, .223, .40 S&W, and .45 ACP. Corbon manufactures subsonic loads in .223, .308, .338 Lapua, and 6.8mm SPC. Especially in .22 subsonic cartridges, some manufacturers opt not to load a propellant charge, so that the projectile is expelled by the force of the priming compound alone. While not specifically marketed as subsonic cartridges per se, reduced-propellant-load cartridges may result in subsonic performance in cartridges that otherwise would achieve supersonic velocity.

The .300 Whisper by SSK was designed from the outset to be a full-sized, full-powered cartridge intended to travel subsonically to reduce noise signature. To achieve optimal results, a suppressed weapon would be used. The .300 Whisper name is trademarked, so similar cartridges may be found marketed under other names. In addition to the manufacturers listed, subsonic ammunition is also produced by private reloaders or by smaller firms that produce specialty ammunition.

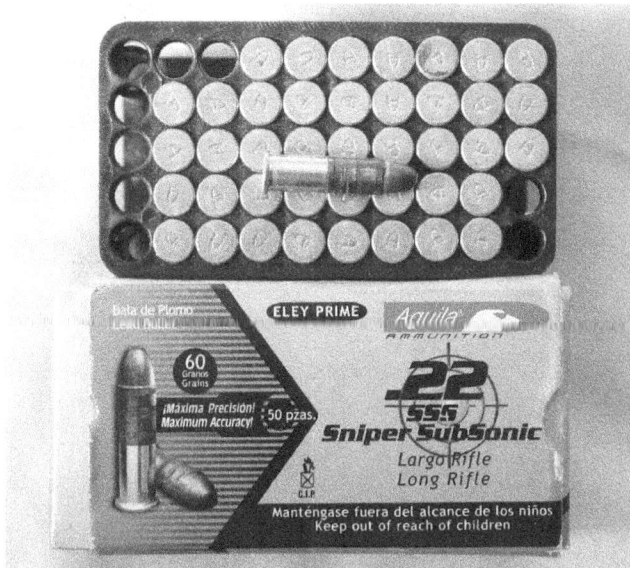

Figure 2.30 Aguila brand .22 subsonic cartridges. The casing is more consistent in size with a .22 Short, but these cartridges are for a .22 Long Rifle. Note that the projectile is of a disproportionate length relative to other .22-class ammunition. (Image from author's collection.)

Sabots

Sabots are undersized projectiles, relative to the nominal projectile diameter, and are essentially a projectile within a projectile. The sabot provides the balance of the diameter that differentiates the undersized projectile to the host weapon's interior bore diameter (caliber). Modern sabots are typically plastic, but metal has also been used. Many have a pedaled or claw appearance, allowing the projectile to be firmly seated within it. Sabot-type cartridges are not uncommon and are found not only in center-fire cartridges, but also in shot shells and even in muzzle-loading weapons. Functionally, sabots allow for a lighter projectile to be propelled by a larger powder charge due to the oversized casing. This is a win/win situation that benefits the two major variables affecting projectile performance: the projectile mass and the amount of energy that can be put onto it.

The M903, a .50 SLAP (Saboted Light Armor Piercing) cartridge, is utilized by the U.S. military. The M903 contains a .30 projectile encapsulated in a .50 sabot, itself loaded into a .50 BMG casing. The projectile weighs 355 grains and is constructed of tungsten. This is in contrast to the M33 ball projectile, which weighs 706 grains. The reported velocity of M903 is approximately 4,000 ft/s. The M962 is the counterpart of the .50 SLAP, the difference being that it contains a tracer component. The Remington Accelerator was a center-fire sabot. The Accelerator used a .22-class projectile loaded into .30-30, .308, and .30-06 cartridges. A plastic shot cup partially encapsulated the projectile and was separated from the projectile in flight. The Accelerator was only manufactured for a brief period of time in the late 1970s.

Figure 2.31 depicts a sabot-type projectile with a plastic shot cup. A sabot should not be confused with a subcaliber insert that allows for undersized cartridges to be fired from large-caliber firearms. Inserts are not ammunition cartridges, as they contain none of the requisite components to properly identify them as ammunition. Rather, they are better defined as components to a firearm. The use of inserts in certain situations can raise

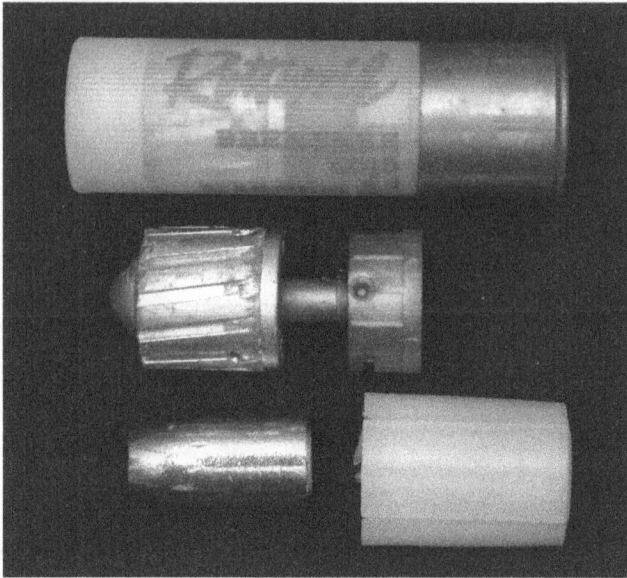

Figure 2.31 Rottweil Brenneke slug (top) and an expanding-style sabot slug complete with plastic shot cup. (Images from author's collection.)

questions of legality, such as using an insert in an apparatus not deemed a firearm to permit the firing of ammunition.

Gyro Jet

On November 29, 1962, a company called M.B. Associates located in San Ramon, California, filed for a patent to the U.S. Patent Office for a handgun designed to fire miniature ballistic rockets. Patent 3,212,402 was granted October 19, 1965. The system has colloquially become known as the gyro jet. On November 26, 1968, an improvement to the original design was patented under U.S. Patent 3,412,641. The gyro jet cartridge, when ignited, lit the contained propellant and accelerated the projectile out of the barrel. Unlike conventional projectiles, the rocket continued to accelerate even as it traveled after leaving the barrel. There were several variants of the original handgun, as well as rifles, that were marketed and sold commercially, but they were not sold in great numbers. Today, gyro jet firearms are only rarely seen. M.B. was the only company to bring the concept to market, and the design did not go forward. A major hurdle faced by the gyro jet cartridge was that the projectile diameter was 13mm, thus giving it a bore diameter in excess of .50″ and causing it to be classified as a destructive device in the United States. Even with a redesigned gyro jet cartridge of 12.7mm (.50″) in diameter, the concept did not have sustained marketability. BATFE has since reclassified the gyro jets as Curios and Relics. Examples of gyro jet guns and their unique cartridge survive today in private hands and museum collections. The gyro jet ammunition itself is quite rare, more so than the firearms.

Shot Pellets

The construction of a shot shell differs slightly from other center-fire cartridges. In addition to the four basic components of a typical cartridge, the shot shell adds a fifth internal component, the wadding. A wadding itself may be divided into three separate parts: a powder wadding, cushion, and shot cup. Between the powder charge and the contained

Figure 2.32 Speer Lawman 12-gauge shot shell loaded with 00 buckshot. This particular load uses eight copper-coated pellets instead of the more traditional nine-pellet load that has been preferred by the industry. (Image from author's collection.)

projectiles is the powder wadding, which acts as a seal. The cushion is between the powder wadding and the shot cup. The cushion literally cushions the force of the powder detonation against the projectile load. When the round is fired, the cushion also acts as a piston to propel the projectiles down the barrel. The shot cup lines the walls of the shell interior where the projectiles are located. Upon detonation, the projectiles exit the hull and travel down the barrel, with the shot cup acting as a liner between the projectile and interior of the bore. In most shot shells, these parts may be made completely of clear or off-white colored plastic, or a combination of a felt or similar material with a plastic cushion and a shot cup. Figure 2.32 shows a 12-gauge shot shell loaded with 00 buckshot. Table 2.4 lists U.S. lead shot sizes.

Slugs

The shotgun slug is a single projectile contained within the shot shell. A slug can be viewed as a reiteration of the musket ball, but fired from a shotgun loaded with a self-contained cartridge instead of a breech-loaded musket. Slugs are used in hunting more hardy game that may not be taken down by shot, or simply as a matter of shooter preference. Military, law enforcement, and self-defense shooters employ slugs in lieu of shells loaded with shot. In certain conditions, the use of shotgun slugs may offer tactical advantages over rifles and offer a new dimension to deployment of shotguns into scenarios with the potential for the application of lethal force. The slug offers a greater range than shot while eliminating the scattering of pellets that may errantly strike surrounding objects or persons unintentionally.

There are two basic slug designs. The *Brenneke*-style slug is of solid construction. In contrast, the *Foster*-style slug has a hollow base. Both feature distinct circular cuts on the exterior that often lead either variation to be called *rifled slugs*, although functionally these cuts do not act as rifling in the true sense of the word. Traditionally, European slugs tend

Table 2.4 United States Lead Shot Sizes

Shot Size	Diameter (inches)	Diameter (millimeters)
000 Buckshot	.36	9.14
00 Buckshot	.34	8.64
0 Buckshot	.32	8.38
#1 Buckshot	.30	7.62
#2 Buckshot	.27	6.90
#3 Buckshot	.25	6.83
#4 Buckshot	.24	6.10
F	.22	5.59
T	.20	5.16
BBB	.19	4.83
BB	.177	4.50
1	.16	4.00
2	.15	3.81
3	.14	3.40
4	.13	3.25
5	.12	3.00
6	.11	2.70
7	.10	2.50
7½	.095	2.41
8	.09	2.30
9	.08	2.03
10	.07	1.78
11	.059	1.50

to be of Brenneke design, while American preference is more toward the Foster. It is not uncommon to see European shells marked to identify which variety of slug is loaded.

A variation of the slug is the *sabot*, as the slug itself is encapsulated in a sleeve and, like other sabot projectiles, is subcaliber relative the bore diameter it is intended to be fired from (see Figure 2.31). This sleeve prevents physical contact between the bore and the slug as the slug travels down the barrel, reducing barrel wear and maintaining the continuity of the slug. Sabots may also act as a capsule for other subcaliber projectiles, allowing for a smaller projectile to be fired from a larger diameter bore safely without having the projectile "skip" off the bore, possibly damaging the barrel and degrading accuracy.

Duplex Shot Shells

The duplex cartridge contains multiple-sized projectiles within a single cartridge casing. The duplex is not to be confused with the standard shot shell, which contains a plurality of pellets that are all the same size. Duplex and triplex cartridges were the subject of research and development in military circles. The rationale behind such loads was that, with the average soldier carrying a machine gun, the marksmanship usually was a little lacking between trigger-control discipline and the high rate of fire and muzzle rise inherent in automatic weapons. A duplex or triplex cartridge permitted the primary projectile to travel along to the point of aim as set by the shooter using the sights. The secondary projectile, which was seated behind the primary projectile, would follow and strike a point slightly randomly off from the point of aim, potentially enhancing hit probability. Duplex cartridges are oddities at this point, likely to be seen only in collections.

There are also duplex shot shells. Duplex shot shells are loaded with two or more different sizes of shot; normal shot shells that contain a single type of shot are not considered duplex. Currently, Remington Arms manufactures duplex shot shells on their Premier line for turkey hunting. These 12-gauge shells come in either 2¾″ or 3″ lengths, containing #4 and #6 shot. Previously, Remington Peters produced a duplex shot shell that was similar. The only observed examples of this were 12 gauge. They had a blackened case head, an olive drab green plastic hull, and were marked 4×6. An Italian line of shot shells marketed under the Centurion name offers a duplex shot shell in 12 gauge, containing a single .65 slug coupled with six #1 buckshot-size pellets. Winchester law enforcement ammunition offers a combination slug/buckshot-loaded shot shell.

Other Types of Shot Shell Loads

The fact that a shot shell of any gauge is simply a wide-mouthed cylinder lends itself to great versatility. Less-lethal munitions, chemical munitions, and other esoteric lethal loads such as flechettes, bolo rounds, tracers, and even projectiles from nonshot shell cartridges can be fired from shot shells. A bolo round consists of two steel balls joined by a length of steel wire. Flechettes resemble small arrows or crossbow bolts, having a rodlike body pointed on the end. Flechettes have been used not only in shot shells, but also in field artillery–sized munitions and in rocket warheads. The U.S. military actively used flechettes during the Vietnam conflict, and they appeared to be effective in the jungle environment. The standard U.S. military shot shell contained 20 flechettes weighing 8 grains apiece. These shells had typical plastic hulls with a brass case head. Surplus or homebuilt flechette cartridges are encountered, and the flechettes themselves are available commercially, sourced from demilled ordnance and sold for scrap value.

Homemade shot shells have been found loaded with everything from small coins to nails and even rock salt. Said to date from the 1920s as a poor man's expedient, *cut shot* is a reconfigured shot shell that behaves more like a slug than shot. Cut shot is fashioned from a shot shell by making an offset circular incision around the circumference of the shot shell, approximately one-half to two-thirds down from the top of the hull. The ends of the incision overlap slightly but do not meet, as it is not the intention to cut the shell completely open but to create a weak point in the hull. The incision penetrates the hull body but ideally does not penetrate into the wadding or shot cup contained inside the shell. Cut shot has the effect of keeping clustered groups of shot together when expelled from the shotgun, in essence a form of rudimentary slug. It is an interesting thought to ponder if the cut shot concept was used as a basis for later prefragmented and other frangible-type projectiles.

Less-Lethal Munitions

The term *less lethal* has replaced the former terminology of *less than lethal*. This subtle shift in language came after the recognition that force options that could be deployed that were not intended to cause serious permanent injury or death could, in fact, result in same, as there are risk factors when force is applied that are beyond anyone's practical control. There is a myriad assortment of less-lethal-force options, ranging from chemical munitions, electronic control devices, and impact projectiles. This topic has become a field of study unto itself. For the purposes of this text, impact-projectile munitions will be covered on the basis that they are, in fact, employed around the world by civilian law enforcement,

security providers, and even military forces. The delivery systems for kinetic-impact projectiles tend to be firearms-based platforms that have been dedicated to less-lethal-force
delivery, as well as platforms that are specifically designed around less-lethal-force delivery.

Examples of converted firearms platforms include 12-gauge shotguns and 40mm grenade launchers for delivering less-lethal munitions. The 37mm grenade launchers were
developed specifically for law enforcement applications and never intended to deliver lethal
munitions; the bore is too small to accommodate the lethal 40mm munitions. This has not
prevented clandestine manufacture of 37mm munitions that are lethal. Other handguns
and rifles have also been used as delivery platforms for less-lethal-force options, primarily to fire impact projectiles. The general term used by media outlets and others in talking
about impact projectiles is *rubber bullets*, although the projectile need not necessarily be
made of rubber, nor do they necessarily constitute "bullets," even as the term is casually
applied. Ordinarily, lethal firearms that have been dedicated to less-lethal-force options
are conspicuously marked by way of changing the color of the stock to orange, blue, or
another attention-getting color and even marking the weapon for rapid identification as a
less-lethal-force option. Orange has been adopted as the industry standard for less-lethal
delivery platforms, phasing the colors blue and red out.

Less-lethal munitions fired from shotguns include kinetic-impact projectile options
such as bean bags and rubber projectiles, including shot and stabilized projectiles. Bean
bags are literally small fabric bags filled with dried beans, lead birdshot, or other media
that give the bag some mass (see Figure 2.33). These bags are generally not round in shape:
They can have a teardrop shape, take on the form of a square, or have drag fins to provide
a form of stability. Bag-type projectiles often have ultraviolet dyes that will transfer onto a
subject's skin or clothing that may remain for weeks on end. In the event that the subject
eludes apprehension immediately but is later encountered, an ultraviolet light could be
used to confirm that the person had been exposed to the marking dye, perhaps placing
that individual at the previous encounter. Bag-type projectiles may also contain a chemical
agent that deploys upon impact.

Rubber-ball shells are simply shotgun shells where the lethal metallic shot load has been
replaced with a rubber shot load, generally the size of #1 or 0 buckshot. Another option

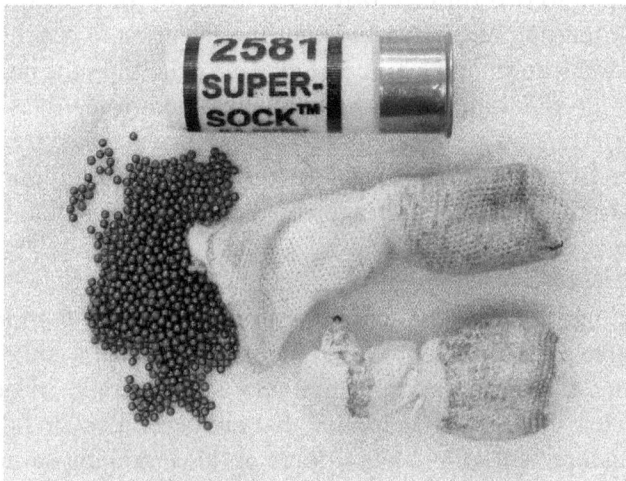

Figure 2.33 A 12-gauge "Super Sock" less-lethal munition. The projectile is a small fabric bag
loaded with lead shot. (Image from author's collection.)

is single rubber-finned projectiles, loaded one per shell. These resemble small bombs or mortar rounds in their outward appearance. Foam, rubber, or wooden baton-type kinetic-impact projectiles are cylindrically shaped, typically having flat edges to avoid piercing injuries, and are intended to deliver the force by direct, blunt impact onto the intended target. Such baton-type projectiles were often called *knee knockers*, as they were intended to be skipped off pavement in front the intended target and then deflected into the lower legs. Shot shells can also be loaded with any number of chemical agents ranging from OC (oleoresin capsicum) to the tear gases CS and CN in powder or liquid forms. Although they are not less-lethal munitions, shot-shell-based distraction loads are available. These rounds are a smaller version of the "flash bang" devices commonly used by law enforcement and military forces to disorient persons preceding an assault. The report of such a device is a bright flash of light accompanied with a loud explosion. All less-lethal munitions are con-spicuously marked as such to avoid confusion with lethal munitions and to clearly indicate what load is contained.

The FN 303 is a dedicated chemical agent delivery system. Developed by Fabrique Nationale, the system is available as either a stand-alone delivery system or as an ancillary device that can be attached to another weapons platform or system and that can be used as an alternative to the lethal-force option also available to the operator. The 303 operates by compressed air, delivering 8.5-g, .68 caliber projectiles, which have a circular fin pattern running the circumference of the projectile. The principle intention of the 303 system is to deliver a kinetic-impact projectile to induce pain compliance of the intended target. Secondary effects can be caused by using a projectile loaded with a chemical munition. Several variations of munitions are available to authorized entities, including washable and indelible paints, a PAVA/OC (pelargonic acid vanillylamide/oleoresin capsicum) chemical munition, a basic impact projectile, and a dedicated training variant. The 303 munitions have a body constructed of polystyrene and are fin stabilized. The front of the projectile is filled with bismuth powder; the rear of the body is loaded with a secondary payload (FNH USA n.d.).

The purpose of less-lethal-force options is to obtain lawful compliance while reducing the chances for serious injury or death. A less-lethal-force option relies on the ability to cause temporary incapacitation and/or pain compliance. This incapacitation or voluntary compliance is intended to allow a person to be taken into custody without further resis-tance, as they would acquiesce to the control of authorities. The presence and application of less-lethal-force options, regardless of the nature of the particular instrument in ques-tion and technological enhancements, is here to stay as a viable option for law enforcement operations. The tools and technology have evolved, but the basic premise that makes less-lethal options work has not changed.

From an investigative standpoint, the application of less-lethal force presents no less a challenge than the application of lethal force, particularly if the application of the less-lethal force results in serious injury or death. In such a scenario, the investigator must carefully document the delivery system used as well as the presence of residues of force application, such as spent munitions casings, projectiles, dye markings, and physical dam-age to structures that occurred during the scenario. The presence of stimuli that precipi-tated the application of the force should also be identified and documented. Examples of the stimuli include the presence of readily identified weapons (firearms and edged objects), potential weapons (broken glass objects, sticks, and the like), and the actions of persons over the course of the scenario (constructing barricades, acts of vandalism, presence of narcotic substances and paraphernalia, notes and recordings stating intentions, physical indicators of ideological extremism, etc.). Other more traditional forms of physical evidence cannot be overlooked within the context of the investigation. In the case of a misidentification of munitions, hypothetically speaking, such as the intent to use less-lethal kinetic-impact munitions and the inadvertent loading and use of lethal-force munitions fired from a weapon, the investigator must be diligent in documenting the expended munitions, the other munitions loaded into the arm, and other munitions carried on the person of the operator.

Training Ammunition

Dedicated training ammunition was developed shortly after the adoption of the self-contained cartridge. It was developed as a less expensive and safer alternative to full-power cartridges intended for duty or combat use. Training ammunition is generally most recognized by the use of alternative projectiles and is clearly marked as such to prevent accidental issue and use in combat situations.

The first training cartridges typically had projectiles made of wood, paper, or metal with reduced propellant charges, sometimes replacing smokeless powder with black powder as an added margin of safety. Such training cartridges were not particularly well suited for marksmanship training due to their different ballistic behavior from standard ammunition, but they may have fit this role on reduced-distance ranges, where targets were proportionally reduced to simulate greater ranges.

In lieu of dedicated training cartridges, many nations had arms produced that resembled issued weapons but fired .22 ammunition, or they issued kits that allowed the use of subcaliber cartridges by replacing parts of the action to accommodate the smaller cartridge. Companies such as Sig Sauer and aftermarket, third-party sources manufacture .22 conversion kits. Offshoots of training ammunition include reduced-power cartridges designed for use as "gallery" or "guard" applications. These cartridges featured reduced power for safety purposes, as in a shooting-gallery setting or when issued to prison guards as an alternative to full-power cartridges, presumably with the intent to be used in a less-lethal manner as opposed to causing lethal injury. Such cartridges may be color coded to identify them as such, or the packaging material would be marked accordingly. As previously discussed, frangible ammunition and plastic-projectile cartridges are manufactured as dedicated training ammunition.

Marking Cartridges

FX marking cartridges and CQT (Close Quarters Training) ammunition are marketed under the name Simunition by General Dynamics Ordnance and Tactical Systems of Canada. Simunition represents five different lines of training ammunition, including training blanks and frangible projectiles. FX cartridges are designed to allow participants in the training environment to shoot one another while reducing the risk of injury. The FX cartridge is a marking cartridge; the plastic projectile contains a colored dye that shows where hits occurred. The FX cartridge has a short brass casing with a plastic insert that completes the length of the cartridge casing. The projectile itself has deep cuts in the nose, allowing the dye to spatter in a starlike pattern when impact occurs. Glock has continuously produced a dedicated version of its model 17, called the *Glock 17 FX*, that is expressly designed to use FX cartridges. Outwardly, the Glock 17 FX has a blue-colored slide to distinguish it from standard Glock handguns. The 17 FX is shown in Figure 2.34, and

Figure 2.34 (See color insert.) The Glock 17T is dedicated exclusively for training. The frame is blue polymer to clearly identify it as a training aid. Also depicted is the magazine with blue floor plate for identification. Note the Simunition cartridge loaded in the magazine. (Image from author's collection.)

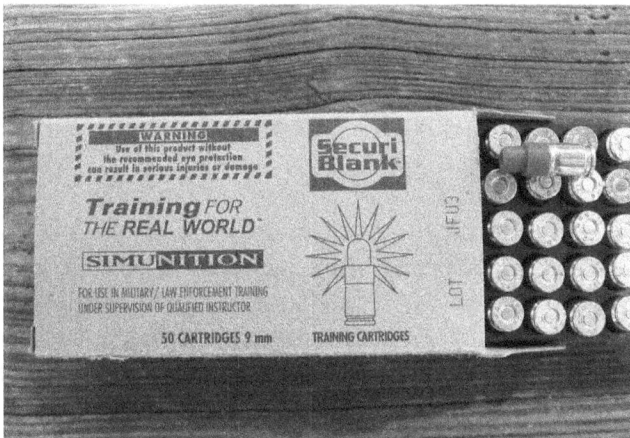

Figure 2.35 A box of 9×19mm Simunition marking cartridges. Note the IVI head stamp, which indicates manufacture by General Dynamics Ordnance & Tactical Systems. (Image from author's collection.)

Simunition marking cartridges are shown in Figure 2.35. Conversion kits exist for other models of handguns and rifles that are popular with law enforcement. Speer LE markets a similar product, called *Force-on-Force*, which also uses a modified firearm in a manner similar to that of Simunition. An aluminum casing is used, and the projectiles also mark impacts using a high-visibility dye.

Short-Range Training Ammunition

SRTA (short-range training ammunition) represents a variation of training ammunition that uses plastic projectiles in lieu of conventional metal projectiles. SRTA is designed to mimic the performance of standard projectiles while enhancing the safety margins by using nonleaded projectiles that reduce environment impact and the risk of projectile ricochet or overpenetration in training environments. SRTA is in use by the U.S. military in 5.56×45mm (designated M862), 7.62×51mm (M973), and 12.7×99mm (M858). In addition to these, there are tracer versions of the 7.62×51mm (M974) and 12.7×99mm (M860), designated SRTA-T (short-range training ammunition, tracer). SRTA is used in conjunction with the Close Combat Mission Capability Kit (CCMCK). The CCMCK is a series of different kits for different weapons: The M1041 is configured for the standard M9 (Beretta 92) or M11 (Sig Sauer P228) 9×19mm handguns; the M1042 is configured for the 5.56×45mm M16 and M4 family of firearms; and the M1071 is configured for the 5.56×45mm M249 Squad Automatic Weapon (SAW). As an added safety margin, installation of these kits prevents the use of standard ammunition. SRTA is a dedicated training cartridge; however, it should be treated as lethal. Like any projectile, it is certainly capable of causing serious injuries or death. CQT (Close Quarters Training), developed and marketed by General Dynamics Ordnance and Tactical Systems of Canada, is the same as SRTA. The Prvi Partizan Ecology Line is a plastic-projectile training cartridge that can be considered a SRTA-type cartridge.

Projectile Jacketing

A projectile jacket acts as a bearing surface between the projectile and the firearm's bore. With the introduction of rifled bores, bare-lead projectiles were found to be unsatisfactory, as the lead from the projectile tended to shave off into the groves in the bore, over time creating a smooth bore due to lead accumulation in the rifling grooves. This fouling resulted in a reduction in accuracy, and thus gilding or jacketing material came into use to encapsulate the lead. Like the musket balls of the muzzle loaders before them, the first center-fire cartridges were often undersized relative to the diameter of the bore to permit the use of paper or cloth patches to coat the projectile, thereby eliminating the fouling effect of bare lead. As projectile design and technology improved, it was also realized that the jacketing material had a secondary effect: The jacketing helped maintain the shape of the projectile as it impacted the target. Bare lead would tend to fragment, deform, and perhaps even disintegrate upon impacting the target surface. Consequently, various projectiles with varying styles of jacketing exist, based upon the designed intent of the projectile.

Hunting cartridges are often semijacketed, leaving a bare tip of lead to facilitate expansion and perhaps the fragmentation effect, offering greater assurance of a single incapacitating shot to game. Other projectiles are fully jacketed so as to encapsulate the projectile core, ensuring a certain measure of projectile adhesion and to prevent any form of lead exposure whatsoever from the projectile. The Blazer brand center-fire cartridges feature a total metal jacket (TMJ) that encapsulates the entire projectile, including the base. This ensures that no bare lead is exposed to the environment, thus addressing environmental and health concerns about lead exposure. In the case of a fully jacketed projectile, the jacket or gilding material completely encompasses the projectile from tip to end, but likely not the projectile base. Any of the variety of projectiles can be fully jacketed. Traditionally, full metal jacketing did not enclose the projectile base; thus, if observed from reverse, the lead-core portion of the projectile would be visible. Full-metal-jacketed projectiles are practically requisite for use in self-loading firearms, as bare-lead tips have a tendency to deform when traveling from the magazine to the chamber on the feed ramp, with a resultant jam or misfeed. This is especially true with older semiautomatic firearms that were intended to be used with full-metal-jacketed projectiles. All military projectiles are fully jacketed.

Semijacketed projectiles are partially clad with the jacketing material but leave the tip or nose exposed. How much nose is left exposed as bare lead is subject to the individual design characteristics of that particular projectile. The edge of the jacketing material can be serrated in appearance, or it can be a smooth edge. Quite often, semijacketed projectiles are of the hollow-point variety, and thus the design intent is to leave some lead exposed to encourage the mushrooming effect of hollow points when terminal ballistics occur. Frequently, dedicated hunting loads are semijacketed, as are some revolver-based caliber cartridges, as they do not present feed issues that would be present in semiautomatic arms.

Jacket separation is a significant issue with projectiles. All varieties of projectiles may suffer from jacket separation. Given the various materials that projectiles may impact and subsequently penetrate—various types of glass, sheet metal, soft tissue, construction materials, etc.—the behavior of projectile jackets is hard to reliably and consistently predict because it is based on innumerable variables. Jacket separation directly affects the performance of the projectile, especially with controlled-expansion types. These projectiles rely heavily upon retaining their jacketing to fully effect expansion and remain intact for maximum performance, as a displacing or fragmenting jacket will reduce the effectiveness of any projectile. The author is familiar with a situation where a large semijacketed projectile penetrated automotive glass, causing the projectile and jacket to completely separate from one another. The two individual pieces continued, although the trajectory of the jacket shifted slightly, where upholstery was perforated before striking an individual. The two pieces, now acting as independent projectiles, impacted one slightly above the other, leaving two distinct gunshot wounds.

Metal Jacketing
Copper has overwhelmingly remained the material of choice for jacketing material, but it is not the only option. Steel has been used, particularly in military cartridges. Bimetal jacketing uses two layers of metal to coat the projectile mass. The Winchester Silver Tip was not truly jacketed in silver; instead it was a bimetal jacketing that used a nickel outer jacket over a copper jacket. Many Eastern Bloc projectiles are bimetal jacketed, typically copper over steel. Tombac (also spelled Tombak), a malleable form of the brass alloy, is another jacketing material that is used. Like steel, copper alloys such as cupronickel, cuprotin, and cuprozinc have been used in military cartridges, particularly on armor-piercing projectiles, but they were not uncommon on military cartridges in the early twentieth century. Figure 2.36 shows a projectile clad with copper over a steel jacket.

Metal Washes and Plating
Copper or gilding metal washes are used on .22 cartridges. These are not true jackets per se, since the coating material is not a separate metallic component affixed to the projectile. Rather, it is literally "washed" onto the projectile. Plating is the application of a very thin coating of material over the projectile and is not jacketing in the true technical sense. Projectile jackets are a separate component that is bonded or otherwise affixed to the projectile, whereas plating is done by an electrolytic or chemical process that adheres the plating material to the projectile. The process is quite similar to that used in the jewelry industry, where precious metals are applied over a base metal. Plated projectiles are prevalent in the .22 class, especially in .22 Short and .22 Long Rifle. Copper is the material of choice, although another material, called Luballoy, has been seen on a wide variety of calibers, including larger calibers such as .38 Special. Luballoy is an alloy of copper, tin, and zinc.

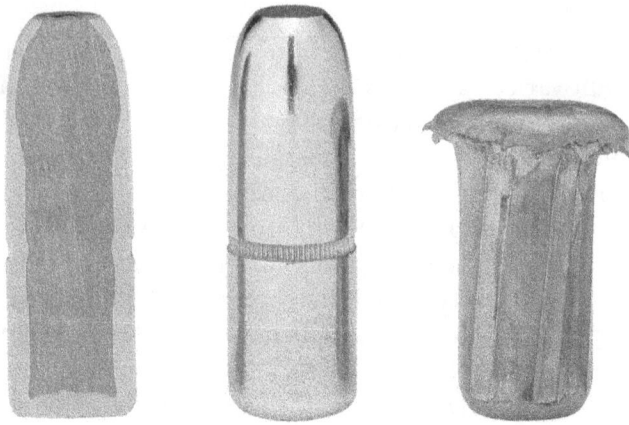

Figure 2.36 The Hornady DGX (Dangerous Game) 500-grain projectile in .45 caliber. The projectile is clad with copper over a steel jacket; also note the flat meplat. (Image courtesy of Hornady Manufacturing.)

Nyclad

The Nyclad projectile was developed by gun maker Smith & Wesson and released into the market in 1982. Almost immediately it drew criticism stemming from allegations that the projectile was purposely designed to defeat soft body armor worn by law enforcement. In fact, the Nyclad was never designed or intended to be an armor-piercing type projectile, although it was branded as such by media outlets and outspoken political figures acting on erroneous information. The use of nylon was simply an attempt to create an alternative to metallic-based bearing surfaces on projectiles. Ultimately the Nyclad passed into the hands of Federal Cartridge Corporation. In 1983, Federal Cartridge announced production of .38 Special cartridges using Nyclad jacketing. Nyclad, manufactured by Federal, is still commercially available.

Other Composite Jacketing Materials

Teflon-jacketed projectiles were apparently only manufactured by the North American Ordnance Corporation on behalf of the KTW Co., and were short-lived. The projectile had a Teflon jacket over a projectile fabricated of solid brass. Teflon, like Nyclad, was another attempt at a nonmetallic projectile bearing surface; however, it suffered the same political backlash and was quickly legislated to contraband status in various venues. Composite jacketing has been revisited in recent years, with CBC reporting that it has developed projectiles using "special resin, graphite, and Teflon" as alternatives to metallic-based jacketing in an effort to reduce barrel wear. Barnes XLC coated projectiles have a distinctive blue color and could be potentially mistaken for polymer projectiles. The coating is sprayed on the projectile and dried. The coating is meant as an alternative to traditional copper jacketing, although the projectile itself is solid copper, and is used in the Barnes X projectiles in hunting calibers and sold to many different firms for loading. Ammunition manufactured by the firm 3-D Police Ammunition used projectiles that were constructed of zinc, thereby eliminating potential lead exposure from the projectile, as no lead was used in the manufacture; however, the project was not commercially successful. Polymer has been used as a bearing surface for bare lead projectiles. Sellier & Bellot uses plastic to coat such projectiles.

Nonjacketed

Nonjacketed projectiles are available as old stock and new manufacture. The use of non-jacketed ammunition is generally not recommended, as the bare lead will foul the bore. Nonetheless, substantial quantities of bare lead projectiles remain in existence and still appear, especially in older calibers. Unjacketed .22 cartridges are still in production, due to economy of manufacture. Most modern bare lead projectiles are manufactured in archaic calibers known as *cowboy calibers*. These cartridges allow shooters with an interest in this type of sport shooting to use ammunition that is historically and technically proper.

Projectile Mass

Any specific cartridge caliber can be viewed as a platform with a diversity of loads and projectiles. These variations are based upon the intended use of that specific loading. Projectile weights are expressed in grains, from the apothecary weight system, and commonly abbreviated *gr*. In the instance of the ubiquitous 9×19mm Luger, the industry has set grain weights of 115, 124, and 147 as "standards" in the sense that the majority of cartridge manufacturers produce the 9×19mm in those weights; however, those values should not be considered the only potential projectile weights to be encountered in that particular caliber class. Projectiles of the class are commercially manufactured in weights between 92.6 grains and 150 grains. Magtech ammunition markets 9×19mm Luger projectiles in 92.6- and 95-grain weights, in addition to the other "standard weights." Sellier & Bellot manufactures 140 grains and 150 grains in 9×19mm in subsonic loads. By increasing mass, the projectile velocity is reduced to a high subsonic figure, which permits reasonable ballistic performance yet also allows a suppressed firearm to abate the report of the muzzle blast and not have to contend with the sonic boom. Obviously, the lighter the projectile, the higher the velocity will be given a standard powder charge.

The most popular calibers have the widest range of weights. Projectile mass, relative to the SAAMI (Sporting Arms and Ammunition Manufacturers Institute) or CIP (Permanent International Commission for Firearms Testing) powder charge, should not be confused with Magnum, +P, Proof, or otherwise heavily charged cartridges, which are a separate issue. When a projectile is recovered intact, the mass can be used as a starting point to ascertain what caliber class the projectile belongs to. The grain weight is strictly an approximation, and very subtle variations will be expected, even if the projectile is wholly intact. When weighing a projectile, it is recommended to use both ounces and grams, as most ordinary scales are not set up to measure grain directly, although scales specifically made for the ammunition reloading market are set up to do so. Both weights can then be converted to grains, although there will not be complete agreement between the two results. Measurements of the projectile mass are taken alongside measurements of the projectile dimensions, yielding suggestions of candidate class cartridges.

Projectile Material Composition

Materials Selection

The diversity of applications for ammunition has led to a broad list of materials that have been used to construct projectiles. For obvious reasons, specialized types of cartridges such as tracers require additional constituents to fulfill their design purpose. As materials science has improved, the list of materials applied to projectile construction has grown.

Lead

Lead—a common, inexpensive, and easily malleable metal—has been the material of choice for projectiles since firearms were developed. Traditionally, lead projectiles were cast by pouring molten lead into molds and allowing it to cool. These molds were in the shape of the final projectile product. Projectiles are commonly called *bullets*; however, this term may create confusion when referring to bullets, either as unfired projectiles, fired projectiles, or loaded ammunition cartridges that are ready for use. Although lead remains the mainstay of projectile materials, other materials have been developed and have gradually phased into the marketplace. Leaded projectiles are not made of pure lead. Lead often contains traces of other elements such as arsenic, antimony, tin, copper, bismuth, and silver, either naturally or by design. Bismuth, for example, is added to increase the hardness of the lead. Other materials have been tried in conjunction with projectile designs to create projectiles that have a dedicated purpose, such as armor-piercing or incendiary characteristics. Armor-piercing projectiles, while perhaps having some lead in the projectile composition, use heavier metals such as tungsten carbide or hardened steel to create a heavier, denser projectile capable of defeating certain levels or varieties of armor, ranging from soft body armor used by persons to lightly armored vehicles. It will likely be a long time before lead phases out, but numerous alternatives to using lead exclusively have appeared on the horizon, and many are already present in the marketplace.

Solid-Metal Projectiles

Solid metallic projectiles are produced that can do away with the use of any lead as well as any form of jacketing. Instead of producing a lead projectile that is then bonded to a jacket, why not just produce a projectile of a metal that has ideal metallurgical properties? Over the course of the twentieth century, projectiles made of metals other than lead were produced. In January 2010, McMillan Group International, based in Phoenix, Arizona, announced that they were putting into regular production .50 BMG projectiles that were machined from solid brass using CNC (computer numerical control). The projectile weighs 746 grains and is represented as match grade. It is apparent that machining individual projectiles is a very time-consuming and expensive proposition; but it must result in a very high quality product.

CBC markets a line of hollow-point solid copper projectiles in .380 ACP, 9mm Luger, .38 Special +P, .357 Magnum, .40 S&W, .44 Magnum, .45 ACP +P, .454 Casull, .45 GAP, and .500 S&W, identified for hunting and defense purposes (shown in Figure 2.17). Component projectiles for loading as well as loaded cartridges are available. Remington Arms features an entire line of solid copper shotgun loads as well as center-fire ammunition. Corbon's DPX line comprises hollow-point-type projectiles constructed of solid copper. According to the company website, when DPX was tested using FBI protocols, the DPX projectile "achieved soft tissue penetration of 12–17" with reliable and consistent expansion. The recovered bullets are 150% to 200% of the original size and 100% weight retention when recovered from the test medium of 10% ballistic gelatin with four layers of 10-oz. denim barrier" (Dakota Ammo 2012). Other firms that manufacture lathe-turned projectiles include Barnes Bullets, Cutting Edge Bullets, Jamison International, and others.

Solid copper projectiles would appear to have inherent advantages over lead projectiles: Solid copper projectiles are less likely to fragment and lose mass; there is no need for a separate jacket or bearing surface; and the environmental issues surrounding lead are addressed. It is expected that solid copper projectiles will expand their presence in the market, much like powdered metal composition or frangible projectiles have.

The U.S. Army experimented with using a projectile core made of a composition containing nylon and tungsten as a lead alternative. This concept was abandoned in 2006. According to the U.S. Army Environmental Command, "190,000 pounds of tungsten, tungsten-nylon core and completed projectiles" were disposed of through auction in September 2008 (U.S. Army Environmental Command 2008).

Shot Pellets

Like other projectiles, shot pellets may be plated with a metal such as copper, nickel, or zinc. Traditionally, shot pellets were made of lead, but alternative pellets made of steel, bismuth, tungsten, tungsten-iron, tungsten-polymer matrix, and extra-hardened lead are readily available. Environmental concerns surrounding the use of lead shot being left behind in the field from the hunt or ending up in game has prompted alternative materials to come to market. In many venues, the use of steel shot for hunting is required, with an outright prohibition on using lead shot at all. The use of tungsten in shot follows the application of this metal in other projectiles. Chilled shot is a softer lead alloy because of the reduced antimony content, relative to standard lead alloys, generally accepted as between 2%–6% antimony content.

Wooden Projectiles

An alternative, if archaic, material that has been used to construct projectiles is wood. Wood, like plastic, was often used in ammunition intended for the purposes of drill, instruction of the manual of arms, short-range training, and perhaps even by guards as a rudimentary form of less-lethal force. Observed examples of wooden cartridges show that they are solid wood and completely inert; they are seldom seen today outside of collector circles. Cartridge casings manufactured of brass or substitute metals that are live cartridges loaded with a wooden projectile do exist. As with modern polymer projectiles, these cartridges were intended to be used for drill, ceremonial, or training purposes. There have been unsubstantiated stories dating back to the Pacific campaign during World War II where marines were reported to have been shot with wooden bullets. The author has never been able to corroborate these stories, and no historical documentation has surfaced to support such accounts. However, if such events were true, it is very likely that such practices came from necessity as the Japanese garrisons on various islands became isolated and were not resupplied as the war entered its latter stages. As ammunition stocks ran low, it could reasonably be assumed that Japanese soldiers may have resorted to shooting anything out of their weapon that would fire, and that may have included training ammunition loaded with wooden projectiles. U.S. Army Technical Manual TM 9-1985-5 (1953) addressed the subject of Japanese munitions during World War II and referenced the Japanese Type 38 6.5mm cartridge loaded with a wooden projectile. Japan is not the sole

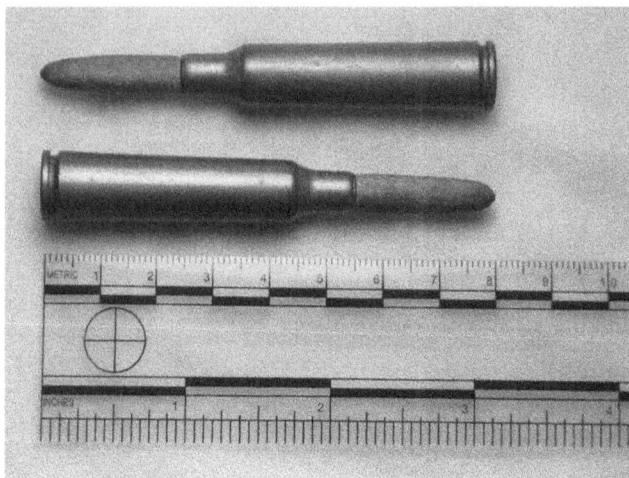

Figure 2.37 (See color insert.) Swedish 6.5mm cartridges loaded with wooden projectiles. (Image from author's collection.)

source for wooden projectile cartridges; they were produced by Sweden, France, Finland, and the Warsaw Pact nations, among others. The primary uses for wooden or paper projectiles were firing rifle grenades, short-range training applications, or for drill, salute, and ceremonial purposes. Figure 2.37 shows cartridges loaded with wooden projectiles.

Hand Loads, Reloads, and Wildcats

Reloading and hand loading of ammunition is a common practice. In areas where it is legal for individuals to manufacture their own ammunition, it is often done for the purposes of economy, as it is often cheaper to purchase the individual components and manufacture or remanufacture ammunition versus buying new factory-loaded ammunition, particularly for the avid shooter. Reloading is also done when the end user has very particular specifications in mind that are not available through ordinary commercial channels, such as bull's-eye and competition shooters. Another reason for reloading is to manufacture archaic or obsolete calibers that are no longer available and remaining stocks have either dried up or transitioned to collector interest. The obvious reason for hand loading or reloading in areas where ammunition is heavily restricted or prohibited is to circumvent such restrictions. Given the state of the art in reloading equipment, it is possible for an individual to produce ammunition on a mass scale.

Many who reload ammunition use salvaged casings collected from gun ranges; however, it is possible to purchase entirely new, unfired casings to load. In fact, many munitions producers sell the individual components required for the hand loader or reloader to either make new ammunition from virgin components or to reuse fired casings when manufacturing serviceable ammunition. When using fired casings, the reloader inspects and prepares the casings; procures powder, primers, and projectiles; and goes about reassembling the components into functional ammunition. Many professional reloaders pride themselves on the shiny, factory-new appearance, and the cleaning process can be a closely guarded secret.

When encountering reloaded ammunition, the head stamps in the sampling provided can be varied, depending entirely on whether the loader used new casings or remanufactured fired casings. Another clue that the cartridges are reloads is that the seated projectile is inconsistent with those used by the original manufacturer. Reloaded ammunition can also exhibit signs that it has been fired more than once, such as burrs, aberrations, or other abnormalities associated with repeated chambering and firing of casings. Plastic cartridge boxes may be used to contain reloads; generic cartridge boxes or even factory boxes with plastic cartridge inserts have also been refilled with reloads.

Factory remanufactured ammunition typically takes military- or government-sourced ammunition that has failed inspection, is a production overrun, or is taken out of service because of age or obsolescence. Such ammunition is purchased and "demilled," a process where the cartridges are disassembled and separated into serviceable and unserviceable components. The serviceable components are then remanufactured back into ammunition. According to a report by the U.S. General Accounting Office, dated June 30, 1999, in response to a congressional inquiry into U.S. civilian acquisition of military-sourced .50 BMG (12.7×99mm) ammunition,

> Talon (the government contractor in question) separates the round and discards the primer. The remaining components can then be (1) sold for scrap, (2) used to manufacture reconditioned ammunition (with a new primer), or (3) sold on the civilian market for customers who reload their own ammunition using the brass casing, projectile, and propellant (gunpowder) components. (Hast 1999)

A common practice within commercial manufacturers who produce finished ammunition is to obtain components from other commercial producers or to outsource ammunition production when demands cannot be met. Consequently, it is not uncommon to encounter ammunition sold under a certain name yet manufactured by some other firm, and this practice extends to projectiles, casings, and primers as well as loaded ammunition cartridges. Certain companies specialize specifically in manufacturing only components and do not manufacture finished cartridges, or they may produce small lots only on special order, preferring instead to sell their product to another manufacturer.

Wildcat ammunition is the term applied to ammunition that is not readily available through routine channels of commerce. In this context, the term *routine commerce* refers to the ability to acquire ammunition from any typical and generally accessible source such as a gun shop, sporting goods store, hardware store, or other physical outlets that sell cartridges. The proliferation of the Internet has stretched the definition somewhat, as the manufacturers of a wildcat load can be contacted with relative ease. However, this should not be taken to mean that such ammunition is readily accessible by means of routine commerce. Wildcat ammunition is typically characterized by its relative unavailability by virtue of there being few, often singular, sources to obtain said cartridges, and it is often produced in very small amounts, frequently by a gunsmith or hand loader. Wildcat ammunition is generally derived from a well-known cartridge, but various parameters of the host cartridge have been adjusted. Many wildcat cartridges draw their influence from existing stock with recognized paternity. Casings are redrawn and redimensioned to change the shoulder geometry or the neck diameter, or they are shortened or lengthened. Wildcat ammunition is often of a completely unknown caliber that is being researched for potential military or commercial application, or simply for novelty or personal research.

When dealing with potential hand-loaded, reloaded, or wildcat ammunition, the investigator is cautioned that the quality of the ammunition is entirely dependent upon the ability of the person who created the load. Shooting such ammunition creates the strong possibility of an accident occurring from an overpressure load or one that is otherwise out of specification. Firearms manufacturers strongly discourage the use of such ammunition, and generally the use of it will void any warranties. If ammunition is encountered that is suspected to originate outside of commercial sources, it should be treated as suspect and not used for testing or investigative purposes until it can be researched sufficiently to allay any safety concerns. However, there may be situations involving archaic firearms where no other alternative exists.

Caliber Interchangeability

A potentially fatal mistake that is often made is the presumption that cartridges of close designations or similar dimensions are interchangeable. This is certainly not the case. There are a great many firearms that will chamber multiple calibers of cartridges; however, the majority of firearms are chambered for a single caliber of cartridge only. There are two guiding principles: chamber size and pressure load. The chamber dimensions of a firearm indicate how wide and how long a chamber is. A cartridge that is either oversized or undersized for the given chamber will result in a hazardous condition, either because (a) an undersized cartridge will result in an insufficient gas check or seal, causing potential casing rupture and damage to the firearm, or (b) in the case of an oversized cartridge, the inability of the action to close entirely, also creating a hazardous situation if the action can be closed enough to permit the firearm to discharge.

Firearms chambered for a particular caliber are designed to handle standardized pressures for that particular round as established by SAAMI and CIP.[*] Both institutions identify suitable cartridge interchangeabilities. Even in those instances where there is reasonable interchangeability between cartridges based on overall dimensions, the pressure created may not be advisable for a particular firearm. This is especially true with older firearms designed for black powder versus smokeless powder. In the transitional era, when both powders were common, firearms designed for use with smokeless or modern gunpowder were clearly marked as such to avoid confusion.

Occasionally, the difference is very subtle. The .223 Remington and the 5.56×45mm demonstrate a dimensional interchangeability that poses two different challenges. The 5.56×45mm is based on the commercial .223 Remington cartridge; 5.56×45mm is the metric dimension and is used in the military designation of that cartridge. The principal difference between the two lies in the casing dimension. The 5.56×45mm cartridge has a slightly thicker casing wall, as it is designed to be fired at a higher pressure due to the different powder formulation per military specification. The differences are slight, but they are present. Firearms that are chambered for the commercial .223 may not be tolerant of the 5.56 cartridge to the extent that it affects the reliable cycling of the firearm. To add

[*] SAAMI is the Sporting Arms and Ammunition Manufacturers Institute, Inc. CIP is the Permanent International Commission for Firearms Testing. SAAMI is based in the United States, and CIP is based in Europe. Both concern themselves with ammunition safety standards, including safe pressure loads for cartridges.

another dimension to this interchangeability issue, commercially chambered arms may not have a barrel rifling profile that is ideally suited for the projectile being fired. Users are cautioned to ensure they are using the proper ammunition that is specifically called for on the firearm being used.

Although firearms are marked to chamber a specific cartridge, it is not uncommon to find that many firearms will fire multiple calibers. The most significant risk comes from similarly dimensioned cartridges loaded to different pressures that may exceed the firearm's ability to safely fire them. A common mistake along this line is attempting to interchange .380 ACP (9×17mm) and 9mm Makarov (9×18mm) cartridges. Although dimensionally similar, there is enough of a dissimilarity to create a hazardous condition. Handguns chambered in .40 S&W can chamber 9×19mm cartridges, which usually results in a functional failure after discharge. The 9mm casing will swell and will likely rupture because the chamber is oversized relative to the casing diameter. This scenario usually entails field stripping the gun to effect the repair. A .357 SIG Auto can be fired in .40 S&W pistols, although the accuracy suffers greatly beyond very short ranges.

Revolvers are more often capable of chambering multiple calibers than auto pistols, simply by design. If the cylinder can accommodate the length of the cartridge and the caliber is such that it will fit snugly into the cylinder, then that revolver likely can safely handle that cartridge. The U.S. Army Model 1917 revolvers manufactured by Smith & Wesson and Colt are one such example. These revolvers were based on commercial designs made by their respective companies during that period. Colt used the New Service revolver, and Smith & Wesson rechambered the N-frame Second Model Hand Ejector revolver. Due to an acute shortage of Colt's Model 1911 to supply the growing demands of the U.S. Army for World War I, both companies were contracted to supply their respective revolvers. Both revolvers were chambered in .45 ACP; however, this necessitated the use of "half-moon clips" due to the rim design of the .45 ACP casing. Since the .45 ACP was designed for use in an automatic handgun, the casing was rimless to facilitate feeding, extraction, and ejection from the firearm. The moon clip, which joined a series of cartridges together, typically three or six, allowed all the spent casings to be ejected from the revolver cylinder simultaneously. Without the use of a moon clip, ejection would not be possible because, without a rim, there is no surface for the ejector in the cylinder to interact with. As an expedient measure, the .45 Auto Rim was developed specifically for use with the U.S. 1917 revolvers. Dimensionally, these cartridges were exactly the same as the .45 ACP, save for the introduction of a full rim to allow the cartridge to seat in the cylinder and properly eject without the use of a moon clip. The .45 Auto Rim has been out of production for many years, but it can be found as collectible ammunition.

Firearms chambered in .38 Special can chamber .380 Center Fire, .38 Short Colt, and .38 Long Colt. A firearm chambered in .44 S&W Special (often just referenced as .44 Special) can accept cartridges in .44 Bull Dog, .44 Russian, .44 S&W American, and .44 Webley. Many firearms are marked to indicate that they can accept multiple cartridges, such as the Winchester Model 9422, the classic Model 94 lever-action rifle chambered in .22. The Model 94 is fed cartridges from an under-barrel tubular magazine. It will fire .22 Long and .22 Long Rifle cartridges, but can also accept the .22 Short cartridge. However, the Model 9422 that is chambered for .22 Magnum is not intended to accept any other .22 cartridge.

Cartridge interchangeability within firearms is dependent upon the dimensions and pressures created by the cartridges in question. Projectiles are arranged by class, that is, a grouping of objects that share the same general characteristics, such as .45-class projectiles, which would include the various .45-class cartridges. Other .45 cartridges that have existed include .45 Auto Rim, .45 Long Colt, .455 Webley, and the .455 Auto Pistol. In 2006, Glock announced that they had developed a new cartridge and released several new pistol models chambered in the round. The new cartridge, designated .45 GAP (Glock Auto Pistol), was designed to mimic the venerable .45 ACP by using the same projectile dimensions—but with a slightly reduced casing length—to produce a cartridge that was somewhat shorter but reportedly had nearly identical ballistic performance. An article that, in part, addressed interchangeabilities of cartridges appeared in the newsletter of the Association of Firearm and Tool Mark Examiners. The article outlined a series of revolvers chambered for the .32 S&W cartridge that would either satisfactorily or loosely fit the .32 A.P. Ctg.[*] Of eight revolvers tested from various manufacturers, all of them were found to chamber a multitude of brands of .32 A.P. cartridges (Smith 1971). There are firearms readily capable of chambering multiple calibers safely.

At the outset of the era of the self-contained cartridge, there was, for a brief period of time, a clear distinction between handgun and long-gun cartridges. For the most part, this distinction was made on the basis of the physical size of the cartridge. The full-sized rifle cartridge could be of such physical size that it would not be practical to produce a handgun of the size required to chamber it. Likewise, the high-pressure load of such a cartridge may have exceeded a smaller-framed firearm's ability to safely fire it. Conversely, physically smaller handgun cartridges might have proven counterproductive in rifles, given their relatively limited performance as compared to rifle cartridges.

In the history of modern cartridges, there does not exist any real exclusivity to identifying a particular cartridge as being dedicated for a handgun versus a long gun. Interchangeability of long-gun and handgun cartridges started as the American Civil War ended with the emergence of the drawn-brass-cased, self-contained cartridge. Many revolvers were chambered for the same cartridges as the rifles of the day, simplifying the logistics for individuals, who could then use a single cartridge in their sidearm and rifle. Examples of such "cowboy calibers" include the .44-40 and the .32-20 Winchester (also known as the .32 WCF Winchester Centerfire).

Rifles chambered in more modern handgun calibers such as the .357 Magnum and the .44 Magnum are quite common. Conversely, there are numerous handguns chambered in modern rifle cartridges such as .223 and 7.62×39mm. It is often the case with handguns chambered in these two calibers to be pistolized versions of the rifle, accomplished by manufacturing the firearm without a shoulder stock or with the provision for attaching one, thereby redefining the firearm as a handgun instead of a long gun.[†] Even this approach bears exception. Thompson Center, a manufacturer of rifles, handguns, and muzzle loaders, produces a line of handguns called the Contender chambered in rifle calibers such as .45-70 Government, .30-30 Winchester, .375 Winchester, .222 Remington, .223 Remington, and .35 Remington, to name a few. These handguns are physically quite large, and the Thompson Contenders enjoy caliber interchangeability by simply removing the barrel from the frame and replacing it with a new barrel of a different caliber.

[*] The author surmises that the .32 A.P. Ctg. referred to in the article is also known as the .32 ACP cartridge, where A.P. would abbreviate the term Automatic Pistol.
[†] The definitions of *handgun* and *long gun* vary due to the various laws and requirements established in different venues.

The .22 class (.22 Short, .22 Long Rifle, .22 Magnum) is probably the most prolific example of this ammunition interchangeability between long guns and handguns. There are innumerable models of all types of handguns and rifles that chamber these cartridges. Quite often, the same model firearm can chamber either .22 Shorts or .22 Long Rifles; however, due to dimensional differences, .22 Magnum generally will not interchange. These technical reasons do not prevent various cartridges from being categorized as handgun or rifle cartridges, but these classifications have no technical basis, instead relying on the traditional application of such cartridges.

A common myth surrounding ammunition is that only ammunition made by a certain manufacturer will fit that same manufacturer's firearms. While it is certainly true that companies such as Remington, Winchester, and Smith & Wesson either did or currently do manufacture firearms and cartridges, it is not to imply that only their specific brand of cartridge would chamber in their particular firearm. For example, Remington manufactures assorted cartridges of all types, but a .45 ACP cartridge they manufacture will work in any firearm chambered for that specific cartridge. When viewing the head stamp, be careful not to confuse the manufacturer abbreviation with a cartridge that is identified by a name. One such example could be head stamped "R-P.38 S&W," meaning that Remington/Peters manufactured this particular cartridge, but it is .38 Smith & Wesson caliber. Truthfully, many cartridges were developed by companies using the prior art of other companies, which resulted in many similar cartridges.

Cartridge interchangeability and the inherent simplification of logistics has become a sales point for companies seeking to sell the simplified logistics of a common caliber between handguns and long guns. The Belgian gun maker Fabrique Nationale (FN) introduced their proprietary 5.7×28mm cartridge for the Five-seveN handgun as well as the PS-90 carbine and P-90 submachine gun. FN has produced an entire line of 5.7 cartridges that suit any application: military, law enforcement, and civilian. In the United States, FN restricts sale of certain 5.7 cartridges to military, law enforcement, and government customers. The SS190 cartridge is a ball-type projectile designed to defeat soft body armor. The projectile has a black tip and its sale is restricted. The L191 cartridge is a tracer, identified by a red tip, and its sale is restricted. The SS192 cartridge was loaded with a hollow-point-type projectile and was not colored coded. The SB193 is a subsonic cartridge, identified by its white-tipped projectile; like the SS192, its sale is restricted. The SS195 cartridge is a lead-free projectile; there is no color marking on the projectile, and it is not subject to restrictions on sale. The SS196 is loaded with a Hornady-supplied V-Max projectile weighing 40 grains and having a colored ballistic tip. The SS197SR is a blue-tipped projectile and is available commercially through Federal Cartridge Company. The SS198 is another lead-free offering; it has a green tipped projectile, and its sale is restricted. A blank cartridge is also available without restriction. It is identified by its red conical projectile.

The German gun maker Heckler & Koch has developed a 4.6×30mm cartridge and markets a submachine gun chambered for it, the MP-7. H&K is reported to have developed a handgun chambered in the 4.6mm; however, it is yet to come to market. Both the 5.7mm and the 4.6mm cartridges represent the concept of the intermediate cartridge taken to a new level (see Figure 2.16). Both cartridges feature very small projectiles propelled to exceptionally high velocities. They both have the overt appearance of scaled-down rifle cartridges: Both are center-fire, bottleneck-shaped casings. The apparent intent to develop these small cartridges was to equip a new generation of firearms called Personal Defense Weapons (PDWs). The PDW is akin to a submachine gun or a carbine and is designed to provide an individual with very compact firepower that is greater than that provided by a handgun yet more manageable than either the traditional carbine or submachine gun.

The same model firearm may be chambered for any number of cartridges. The Remington 870 shotgun has been chambered in gauges 12, 16, 20, and 28, and in .410. Regardless of the gauge, it is still the Remington 870. The Remington 700 has been made in a variety of calibers: .223, .243 Win, .270 Win, 7mm-08, .300 Win Magnum, .30-06, and .308. The Model 1911 and 1911A1 pistols have been made not only in the venerable .45 ACP, but also in .38 Super, .22 Long Rifle, and 10mm Norma, as well as other assorted oddball calibers.

Identification Markings

Cartridges of both military and commercial origin contain markings and other means of identification. These markings are made up of head stamps and color codes. It is not uncommon to encounter surplus military ammunition almost anywhere. Military ammunition or ammunition made to military specification may have a color-coded projectile or a color-coded annulus to aid the user in identifying what the cartridge is. These colored tips should not be confused with polymer or metal inserts in the tips of projectiles that are used by companies such as Nosler or Hornady. These colored polymer tips are inserted into the cavity of hollow point or OTM projectiles to maintain their ballistic performance as well as ensuring smooth and consistent feeding through the firearm, whose feed ramp may have been optimized for feeding pointed-tip projectiles. The world's militaries have long marked projectiles with various colors to identify that particular projectile as being ball, practice, armor piercing, tracer, blank, dummy, and so forth. These color codes were not standardized between nations, and even within nations the codes changed over time. The two prevalent standards, NATO and Warsaw Pact, both sought to standardize markings between members so that ammunition produced by any partner nation could be recognized by other partners. Even in the era of NATO and Warsaw Pact standards, many nations still chose to use internal markings, but this practice has gradually faded away to conform with more internationally recognized markings. See Figures 2.38–2.40 for examples of color-coded projectiles.

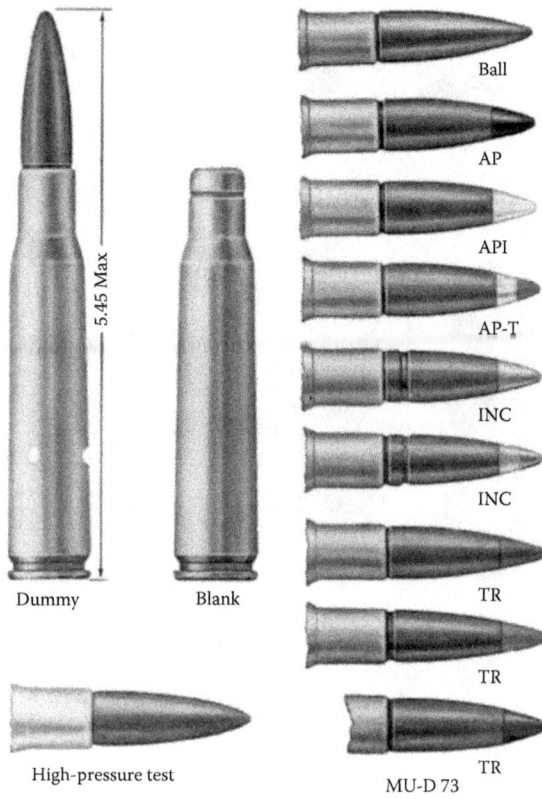

Figure 2.38 (See color insert.) U.S. Military .50 BMG (12.7×99mm) color-coding scheme. (From U.S. Army Technical Manual TM 9-1300-200, 1993.)

Figure 2.39 (See color insert.) U.S. Marine Lance Corporal Richard Mueller mans a turret-mounted M2.50 machine gun. The ammunition load consists of armor-piercing incendiary (gray bands) and armor-piercing tracer (red over gray bands) cartridges. (Image from U.S. Marine Corps; photographer Gunnery Sgt. Scott Dunn, USMC.)

Ball

Blank

High-pressure test (HPT)

Match

Armor-piercing (AP)

Ball, Frangible

Tracer

Dummy, Inert-loaded

Dummy

Duplex

Rifle grenade

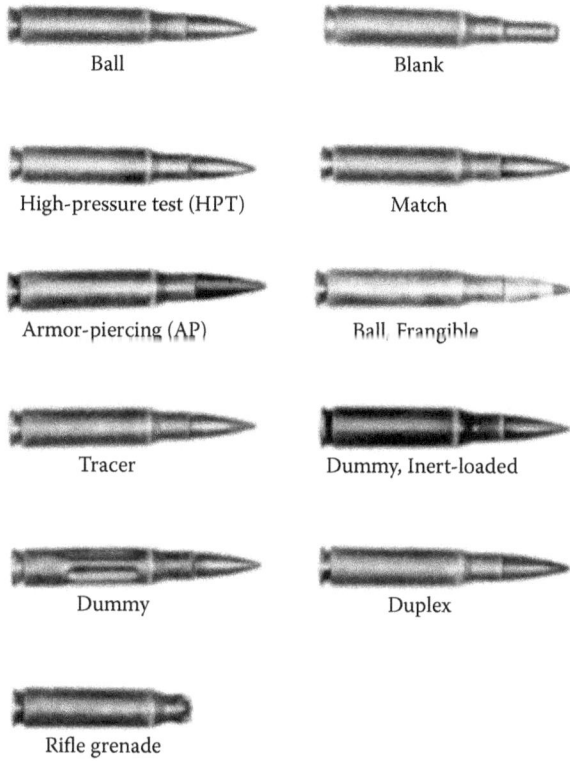

Figure 2.40 (See color insert.) U.S. military 7.62×51mm color-coding scheme. (From U.S. Army Technical Manual TM 9-1300-200, 1993.)

Cartridge Head Stamps

3

Cartridge Head Stamp Overview

A cartridge head stamp is the marking on the base of a cartridge casing that surrounds the primer cup (see Figure 3.1). In the case of a rim-fire cartridge, the head stamp covers the entire casing base. The marking may be pressed into the casing, or it may be a raised marking. There has never been a universal standard or protocol that has been observed with respect to what should be marked on a cartridge; therefore, great variations can be expected. One must remember that ammunition manufacture can be the enterprise of a private company, a state-controlled enterprise, or a governmentally operated arsenal. It must also be considered that commercial manufacturers may produce military ammunition or that government-produced ammunition may be released into the commercial market. There are instances, however rare, of completely anonymous cartridges that bear no markings whatsoever, as none were ever affixed. The marking practices of the manufacturer or nation of origin may reveal any or all of the following information by deciphering the head stamp:

- Manufacturer (by name, initials, or factory code)
- Subcontractor involvement in production
- Casing or projectile material
- Dimensions or chambering (imperial, metric, or gauge)
- Ammunition production lot number
- Date, usually the last one or two digits of the year produced
- National crests or emblems
- Peculiarities of the individual cartridge under inspection, such as match or competition grade, or material supplier subcontractors

Head Stamp Elements

To assist in identification, a head stamp can be broken into elements based on their position on the casing. Due to space considerations, there can only be a finite number of elements and information that can be part of the head stamp. To arrange a casing, orient the casing so that the letters, numbers, or other characters are upright and readable around the circumference of the casing.

There are literally thousands of different head stamps in existence, and deciphering them is truly a subject all unto itself. Given the finite number of letter-and-number combinations, there are duplicate head stamps that have appeared that could be associated with two or more different manufacturers. It is not uncommon for a single manufacturer to use more than one head stamp, changing styles or legends over time.

Head stamps can be used to identify ammunition beyond the mere manufacturer; different type sets, spacing, and placement of information can indicate that a particular cartridge

Figure 3.1 A .45–70 Government cartridge head stamp. The "F 10 87" indicates that this cartridge was manufactured by the Frankford Arsenal in Philadelphia, PA, circa October 1887. (Image from author's collection.)

was manufactured at a particular plant, during a particular period of time, only used on certain caliber cartridges, or some other information that would otherwise be transparent to the reader but would have value internally. A head stamp alone may be somewhat misleading; the use of certain letters, numbers, characters, or other icons and ideograms may not represent the actual manufacturer. Manufacturers often produce cartridges under contract for other companies, who market the ammunition as their own. Ammunition destined for military consumers will often only have a manufacturer stamp and a date or lot code number, whereas ammunition destined for civilian use is typically easier to discern, having relatively complete manufacturer information as well as caliber designation for ease of identification.

There are innumerable head stamps in existence, and proper identification of eccentric or uncommon markings can be difficult. Whenever unfamiliar ammunition is encountered, it is recommended that these exhibits be cataloged and packaging materials also be taken, if available. All crime labs dealing with firearms and ammunition should seek to establish an ammunition reference library that should be actively cultivated.

All forms of alphabet are used to head stamp ammunition: Roman numerals, Hebrew, the Arabic alphabet, Oriental characters, Cyrillic, and others have been used in addition to the English alphabet, obviously presenting some indication of where the ammunition may have been produced or of the intended customer.

Counterfeit Head Stamps

Counterfeit head stamps have appeared, providing completely erroneous information to the investigator. Ordinarily, counterfeit head stamps emerge from sources outside First World manufacturers. There are several instances of counterfeit head stamps used on ammunition originating from China. One instance of this involves 7.62×51mm cartridges produced in China, probably in the late 1960s into the early 1970s and head stamped to mimic British sources (see Figure 3.2). The observed cartridges are marked with the NATO cross, L2A2, and RG for Radway Green. The deception is revealed when the materials used are taken into consideration. The projectile and casing are produced from brass-plated steel, not the ordnance-grade brass used in British-produced, NATO-accepted cartridges. The projectile

Figure 3.2 (See color insert.) A study of two Chinese-sourced cartridge head stamps. On the left is a typical example of 7.62×51mm ammunition of Chinese origin using block characters. The 61 indicates the factory number; 92 indicates that this cartridge was manufactured in 1992. The example on the right, also 7.62×51mm, is a forgery. The head stamp indicates production by Royal Ordnance Radway Green in 1960, complete with NATO standard cross. L2A2 was the British military designation for the 7.62×51mm NATO ball cartridge. Both examples are of similar construction; the casing is copper-plated steel. (Image from author's collection.)

jacketing is also inconsistent with British manufacture. The reasoning behind this forgery is unknown; the most obvious reason put forth is to supply ammunition to a user and obscure the true source. Examples of this ammunition have been observed in Africa, have been sold commercially in the United States, and likely exist elsewhere. Another counterfeit cartridge, also believed to have originated in China, was .30 carbine cartridge marked "LC" for Lake City, and "52" and "53" for production years 1952—53. As is the case with the markings depicted in Figure 3.2, when these are compared with actual examples of U.S. military .30 cartridges, the differences become obvious.

Rim-Fire Head Stamps

Rim-fire cartridges can be particularly elusive in identification, even provided the presence of a head stamp. Generally, rim-fire cartridges provide very limited information, usually some form of character, trademark, a single letter or digit, or an icon. Occasionally an entire name will be marked or a discernible abbreviation such as CCI (Cascade Cartridge International), PMC (Eldorado Cartridge Company, just like center-fire head stamp, however PMC also used a diamond), a stylized W for Western Cartridge, and Rem for Remington. Prior to 1983, Remington Arms used a U to identify their .22 cartridges, then changing to the Rem marking. A delta appears on commercial .22 Long Rifle (5.6mm) ammunition that originated from China. Formerly, it was common for department store brands to be manufactured by another company but marked for retail outlets such as Kmart, Montgomery Ward, or Sears. Sears products were stamped S, and Montgomery Ward with MW, although the ammunition was actually manufactured by another firm. Older .22 ammunition can prove to be especially elusive in definitively identifying the manufacturer, as production was often contracted to other firms and vanity labeled for the retailer.

NATO Identification Marking

Within NATO, there are approved suppliers that furnish the partners with ammunition that meets the requisite standards and thus can be used by anyone within the organization.

The suppliers within the NATO supply structure are immediately recognizable by their head stamp and the bearing of the NATO cross (⊕). These organizations may also produce commercial ammunition and use the same head stamp. The NATO Stock Number, or NSN, will appear on bulk packaging for approved ammunition. The caveat to this, as discussed in Chapter 2, is that NATO-stamped cartridge casings are often reloaded or remanufactured into new cartridges for commercial resale. Thus the mere presence of a NATO marking does not definitively indicate that first-hand military ammunition was used, nor does it imply that military ammunition may have been subject to diversion from stockpiles onto the commercial market, as such ammunition is sold off as surplus, contract production overrun, or cartridges that were rejected by inspectors and sold off as "seconds."

U.S. Government Arsenal Markings

Ammunition produced by ordnance plants owned by the United States from the late 1800s to present times are marked using two- and three-digit letters to denote which facility produced the cartridge, such as the example illustrated in Figure 3.1. Numerous ordnance plants were constructed across the United States, especially during the Second World War, and operated by the government solely for the purpose of meeting military demands. In the years following the war, these plants were gradually phased out, and the U.S. government increasingly outsourced defense munitions needs to private contractors. Many of these plants were involved in producing munitions other than small-arms cartridges, so they are not often encountered. By virtue of the scale of manufacture, and even though much time has passed, these cartridges still appear. The practice observed by these plants called for the stamping of a one-digit year, in addition to the one- or two-digit ordnance plant abbreviation. The Lake City Army Ammunition Plant, currently operated under contract by ATK, is the last remaining U.S. military arsenal still in operation. The appearance of Lake City marked brass casings itself does not indicate that the particular cartridge is of military origin. Casings made at Lake City are used by other companies to manufacture commercial ammunition. Federal Cartridge Corporation, itself a holding of ATK, uses Lake City–marked casings. Surplus or reject brass is sold and reloaded, or new brass is purchased and used to manufacture new ammunition. In addition, former U.S. military–produced brass has been used extensively by reloaders to produce new ammunition.

Figures 3.3–3.6 show head stamps for a wide range of cartridges. Tables 3.1 and 3.2 list head stamps for U.S. plants producing ordnance and ammunition, respectively.

Process of the Identification of a Cartridge, Cartridge Casing, or Projectile

The majority of cartridges that are encountered can be readily identified, but some will prove more elusive than others. When attempting to identify an unknown cartridge, there are several pieces of information that must be gathered from the specimen:

- Physical dimensions
  - Length of cartridge case (inches and millimeters)
  - Overall length of cartridge, to include projectile (inches and millimeters)
  - Diameter of projectile (inches and millimeters)

Figure 3.3 An assortment of 9×19mm cartridge head stamps. Top center: Remington Union Metallic Cartridge; top left: Winchester Cartridge Corp., NATO accepted, 2005 production; top right: Cascade Cartridge International aluminum casing (NR for "not reloadable"); center: Federal Cartridge; lower left: Speer +P cartridge; lower right: Winchester. (Image from author's collection.)

Figure 3.4 (See color insert.) An assortment of 5.56×45mm/.223 cartridge head stamps. Top row from left: Precision Metal Corp.; Poongsan Corp. Korea (October 1981); Federal Cartridge (2005); Hornady Manufacturing; International Cartridge Corp.; Hornady Manufacturing. Bottom row from left: Lake City Ordnance Plant (2002); Lake City Ordnance Plant (nickel plated) (2005); Lake City Ordnance Plant (1975); Winchester Cartridge Corp. (2001); Barnaul Cartridge Plant Russia; Wolf Performance Ammunition. (Image from author's collection.)

Figure 3.5 (See color insert.) An assortment of commonly encountered 7.62×39mm cartridge head stamps. Top row from left: Klimovsk Specialized Ammunition Plant Russia (2000); Barnaul Cartridge Plant Russia (newer logo); Barnaul Cartridge Plant Russia (earlier logo); Barnaul Cartridge Plant Russia (1995); Tulammo Russia; Federal State Enterprise Production Amursk Cartridge Plant Vympel. Middle row from left: Four varieties of Wolf Performance Ammunition head stamps; Igman, Bosnia-Herzegovina (1981). Bottom row from left: Klimovsk Specialized Ammunition Plant (1993); Pretoria Metal Pressing (1988); China State Factory 71 (1991); China State Factory 31 (1971); Winchester Cartridge. (Image from author's collection.)

Figure 3.6 (See color insert.) An assortment of 7.6×51mm/.308 cartridge head stamps. Top row from left: Wolf Performance Ammunition; China State Factory 61; Giraites Ginkluotes Gamykla (2004); Federal Cartridge. Bottom row from left: Gevelot S.A.; Lake City Ordnance Plant; Remington Peters. (Image from author's collection.)

Cartridge Head Stamps 121

Table 3.1 U.S. Ordnance Plant Head Stamps

Alleghany Ordnance Plant	AO
Denver Ordnance Plant	DEN
Des Moines Ordnance Plant	DM
Evansville Ordnance Plant	E
Evansville Ordnance Plant (Chrysler)	EC
Evansville Ordnance Plant (Chrysler-Sunbeam)	ECS
Eau Claire Ordnance Plant	EW
Frankford Arsenal, Philadelphia	FA, F, A
Frankford Arsenal (Laboratory)	FAL
Joliet Arsenal	JA
Kelly-Springfield Tire Company (Alleghany Ordnance Plant Contractor)	KS
Lake City Ordnance Plant	LC
Lowell Ordnance Plant	LM
Milwaukee Ordnance Plant	M
Saint Louis Ordnance Plant	SL
Twin Cities Ordnance Plant	TW
Utah Ordnance Plant	U, UT

Table 3.2 U.S. Ammunition Producer Head Stamps

3-D Ammunition Inc.	IMPACT 3D
Alexander Arms	ALEX-A
Allen & Wheelock	A – W
American Ammunition, Miami, FL	A-MERC
American Ballistics, Marietta, GA	A B T
American Eagle (division of Federal Cartridge)	AM. EAGLE
Am-Tech Intl.	AMTECH
AMRON Corporation, Waukesha, WI	AMRON
Anderson Munitions, Inc., Memphis, TN	AMI
B-West	B-West[a]
Barrett Firearms	BARRETT
BBM Corp., West Springfield, MA	BBM
B&E Cartridge Co.	BE
Black Hills Ammunition	BHA
Blazer	CCI
Browning Arms Co., Morgan, UT	BROWNING
Burnside Rifle Company	BRSD
Buffalo Rock Shooting Supply	BUFFALO ROCK
Canyon Cartridge Co., Albertson, NY	CCC
Cascade Cartridge Intl.	CCI
Clinton Cartridge Co.	C.C.Co.
Colt's Patent Firearms	COLT
Connecticut Cartridge Corp.	CCC
Corbon	C-B, CORBON, GLASER
Creedmoor Cartridge Co.	C.C.C.
D.C. Sage & Company	SAGE
D&S Manufacturing, Inc.	D&S

Table 3.2 *(Continued)*   U.S. Ammunition Producer Head Stamps

Denver Bullets, Inc.	DENVER
Delta Defense Frangible Ammunition	DFA-NT[b]
DoubleTap Ammunition	DOUBLETAP
Eldorado Cartridge Corp.	E L D, STARFIRE E L D
Estate Cartridge Co.	ECC
Extreme Shock Munitions	EXT SHK
Firearms Unlimited	F-U
Federal Cartridge Corp.	F, FC, FCC, FEDERAL
Fiocchi of America	F.O.A., FIOCCHI USA
Frontier Cartridge Co.	FRONTIER
Georgia Arms (Master Cartridge)	GA ARMS, MASTER
Greenville Ammunition Supply (GAS)	Starline or Speer marked brass is used
Gromak Inc.	GROMAK
Guilford Engineering Associates Inc.	GEA
G&S Munitions	G&S
H&S Ammunition	H&S
Hall & Hubbard	H & H
Harrington Ammunition Inc.	HAI or HARRINGTON
Hornady	HORNADY, FRONTIER, HMC
High Performance Cartridge Corp.	HI-PER
Horner Munitions	H&R (not associated with gun maker)
Hydra Shok Corp.	HYDRA-SHOK
Hunting Shack	HSM
Independence (CCI)	I
Idaho National Engineering Laboratory	INEL
International Cartridge Corp.	ICC
J. B. Wise	Wise
KTW Inc.	KTW
Kilgore Ammunition Products	KFA
Longbow	L B-NTF[c]
Liberty Cartridge Co.	LIBERTY
M&D Munitions	M&D
Magtech	CBC, MRP
Master Cartridge Co.	MASTER
Maxim Munitions Corp.	MAXIM
Midway Arms, Inc.	MIDWAY
Montgomery Ward & Co.	M – W
National Brass & Copper Tube Co.	H
National Cartridge Co.	N, NATIONAL
Nevins Ammunition, Inc.	NEVINS
Ready Products Corp., Irving, CA	A-ZOOM
Remington Arms	REM, RA
Remington Arms/Peters Cartridge Co.	RP, R-P
Remington/Union Metallic Cartridge Corp.	REM-UMC, UMC, U.M.C., UMC-UEE, RyU, U
Robin Hood Ammunition Co.	R.H.
Peters Cartridge Co.	PC, PCC, PETERS
Precision Metal	PMC

Table 3.2 *(Continued)* **U.S. Ammunition Producer Head Stamps**

D.C. Sage & Co.	SAGE
Savage Arms Corp.	S.A.Co., S.A Corp., S.A. Corp, SAV, SAVAGE
Scharch Manufacturing	TOP BRASS
Sherwood International Export Corp.	SIEC IK [d]
Smith & Wesson	S&W
Smith & Wesson/Fiocchi	S&W-F [e]
Standard Cartridge Co.	SCC
Starline	*__* [f]
Speer Ammo	SPEER
Superior Ammunition, Inc.	SUPERIOR
United States Cartridge Co.	U.S., US, U.S.C.C.O.
Union Cap & Chemical Co.	U.C.C.Co
Wahib Arms	WAHIB
Western Cartridge Corp.	WCC, W C C, W.C.C.Co, SUPER SPEED, WESTERN
Winchester Repeating Arms	H, WINCHESTER, W, WIN, WR, W.R.A, W.R.A., WRA, W.R.A.Co., Super-X
Winchester/Western	W-W, W-W Super
Zero Ammunition Co.	ZERO

[a] B-West is an example of a vanity marked cartridge. B-West was not the manufacturer, but was a major importer of Eastern Bloc arms. These cartridges were produced by Russian firms and sold by B-West.

[b] NT: nontoxic.

[c] NTF: nontoxic frangible.

[d] Sherwood cartridges were another example of offshore-sourced ammunition that was marked with a vanity label. The actual manufacturer was Igman, as indicated by the IK.

[e] Not to be confused with caliber identifier, but S&W as manufacturer.

[f] In addition to Starline head-stamped brass, Starline produces vanity head stamps for other manufacturers. Some manufacturers prefer to use brass head stamped with the twin stars and dash.

- Composition of materials used to construct casing
  - Brass
  - Steel
    - Color
  - Polymer
    - Is the base or cap metallic?
  - Paper
  - Other material
- Ignition system
  - Rim fire
  - Center fire
    - Boxer primed
    - Berden primed
  - Electric
  - Pin fire
  - Other or unknown

- Cartridge casing shape
  - Straight-walled cylindrical casing
  - Tapered casing (the diameter of the casing narrows as it approaches the neck)
  - Bottleneck casing (the diameter of the casing narrows at the shoulders, creating a well-defined neck)
- Shape of the base and rim
  - Rimmed
  - Rimless
  - Rebated or reduced rim
  - Belted
- Presence of a coating on the casing
- Presence or absence of a head stamp (If there is a head stamp, identify the characters and note their relative positions to one another, using clock positions.)
  - Roman numerals
  - Hebrew characters
  - Arabic characters
  - Symbols such as stars, arrows, etc.
- Presence of a crimping mark around the circumference of the casing
- Projectile description, if present
  - Round
  - Pointed
  - Flat
  - Truncated
  - Conical shaped
- Projectile jacketing material
  - Bare lead, unjacketed
  - Copper
  - Steel, nickel, or other white metal
  - Composite jacketing material such as nylon, Teflon, or polymer
- Jacketing characteristic
  - Fully jacketed
  - Semijacketed
  - Unjacketed
  - Gilded or plated
- Projectile type
  - Ball
  - Hollow point
  - Open tip match (OTM)
  - Wad cutter
  - Semi-wad cutter
  - Sabot
  - Other (sketch appearance)
  - Shot-shell pellet
- Colored bands or colored meplat
- Presence of cannelures on projectile

Using as much information as can possibly be obtained, it may be possible to research the particular projectile back to a specific manufacturer, even if a casing is not recovered. Unless there is something rather unusual about the projectile, a generic ball projectile may be difficult to trace. On the other hand, hollow-point-type projectiles may not pose as much of an issue. All factors must be considered: the caliber class, materials used, shape, mass, even the documented performance of what occurred during the flight of the projectile until it came to rest.

Head Stamp and Manufacturing Practices for Selected Nations

The identification and marking practices of any nation is a subject all unto itself. Volumes have been written on the single topic of World War II German ordnance production, marking, and their coding system for domestic, partnered, and occupation plants. State-controlled industries tend to maintain consistent marking practices, whereas private firms tend to change more frequently. Changes in head-stamp markings may be subtle, such as a slight change in the font, the positioning of the stampings, or the inclusion of dimples or other seemingly insignificant additions that are meaningful to the factory. Since 2000, the worldwide munitions industry has seen tremendous growth, but it has been consolidated to a great degree. The consolidation within the industry has, for the most part, been virtually transparent to the consumer, since most individual labels have survived during these corporate acquisitions. It appears that most companies that have acquired others preferred to take advantage of brand-name recognition and leave well enough alone. Thus it comes as a surprise to many when they discover that their favorite cartridge company is actually a holding of a larger conglomerate. Much the same can be said for the firearms industry as well.

Albania

During the Cold War, Albania was a major repository for ammunition originating from the Soviet Union and China, in addition to domestically produced munitions. Albania has been the site of a number of accidents involving munitions stockpiles, and great investment has been made to reduce the amount of stockpiled munitions stored in the country. The state-operated arsenal, K.M. Polican, has produced 7.62×39mm cartridges, among others. K.M. Polican identifies the 7.62×39mm as the M54, as opposed to the Soviet M43 designation. The arsenal uses the number 3 to identify itself on the casing head stamp. The code 3 is marked at the 6 o'clock position, with the last two digits of the date at the 12 o'clock position. Albanian ammunition dating to at least as early as the mid-1980s used brass cartridge casings.

Argentina

Production of armaments in Argentina was under the auspices of the Directorate of Military Factories. Under the directorate, production was undertaken in various facilities located throughout the country. Military cartridges have the manufacturer head stamp positioned at the 12 o'clock position, and the stamp generally included the year by two or four digits. The caliber designation may also be present. Since 1960, Argentina has generally followed the standardized NATO color-marking system for identifying cartridge types. Concorde was a commercial brand exported from Argentina, with the head stamp OA. Charles Daley was another commercial brand that was manufactured in Argentina. Both labels are also associated with manufacture in the Philippine Islands. The following list shows the state factories and their respective head-stamp practices:

Fabrica Militar de Cartuchos San Francisco: FMMAP S.F. (1948); FMCSF (1954); F.M. S.F. (1972)

Fabrica Militar de Municiones de Armas Portatiles: F.M.M.A.P. (1939–44); renamed Fabrica Militar de Municiones de Armas Portatiles—Borghi: F.M.M.A.P. B (1944–50); again renamed Fabrica Militar San Lorenzo: FMC SL (1950–55), F.M. S.L. (1955–61); then renamed again to Fabrica Militar Luis Beltran: F.M. FLB (1961–75), FLB (post-1958)

Industria Metalurgica y Plastica Argentina, Buenos Aires: I.M.P.A.

Australia

As part of the Commonwealth, Australia was able to erect arms factories in support of the demand for ammunition during times of war. Commercially, Bertram Bullet Works manufactures brass in archaic calibers for custom loading; it can be identified by the head stamp BB. The product is exported, so it can be seen outside of Australia, especially in areas where hand loading archaic cartridges does not present legal issues. Small Arms Factory Melbourne was succeeded by Australian Defence Industries, which in turn was absorbed by Thales Australia in 2006. When the Small Arms Factory Melbourne was succeeded by Australian Defence Industries, the cartridge head stamp was switched from AF to ADI. The ADI is stamped in the 12 o'clock position, and the two-digit year is at the 6 o'clock position. Thales has continued using the ADI head stamp. Thales Australia (part of the Thales Group) manufactures 5.56×45mm and .50 BMG cartridges in Australia at its Benalla factory for consumption by Australian Defence Forces as well as for export. Both the 5.56 and the .50 live cartridges are designated as F1; the blank variants of both are designated as F3. The 5.56mm F3A1 has a chemically blackened casing for ease of identification (Thales Group n.d.). Winchester has manufactured ammunition in Australia since the late 1960s. For historical purposes, other arsenal head stamps are listed as follows:

Small Arms Ammunition Factory Melbourne: AF, AFF, MF
Small Arms Factory Number 1, Footscray: AIF, MF
Small Arms Factory Number 2, Footscray: MF2, MG
Small Arms Ammunition Factory Number 3, Hendon: MH
Small Arms Ammunition Factory Number 4, Hendon: MJ, MJB
Small Arms Ammunition Factory Number 5, Rocklea: MQ
Small Arms Ammunition Factory Number 6, Welshpool: MW
Small Arms Ammunition Factory Number 7, Salisbury: MS
Imperial Metal Works: IMI
Winchester Australia: Winchester, WCC

Austria

Austria has been a significant source of small-arms ammunition. A number of different manufacturers have appeared and disappeared during Austria's politically turbulent history. Austria started as the Austrian Empire, becoming Austria–Hungary in 1867. At the end of the First World War in 1918, Austria–Hungary collapsed and became the First Austrian Republic until it was annexed by Germany in 1938. Austria returned as a sovereign nation in the postwar period. With the exception of Hirtenberger, the other Austrian makers are archaic, most of them having faded from existence prior to World War II. Historically, these makers focused on the standardized cartridges of the state in its various political forms: the 8×56mm rimmed cartridge used by the Mannlicher M95 rifle, and the various handgun cartridges such as the 8mm Rast and Gasser, 8mm Roth-Steyr, and 9mm (9×23mm) Steyr. Postwar, Hirtenberger became one of the preeminent cartridge manufacturers in Europe and a major NATO supplier. RUAG Ammotec acquired Hirtenberger in 2003, which was known at that time as Hirtenberger AG. Ammunition of Austrian origin from before or during World War II seldom appears. A large quantity of 8×56Rmm cartridges dating from the late 1930s appeared on the U.S. commercial market in the late 1980s. These cartridges were often confused with Mauser cartridges because of the German markings on the cardboard boxes. Actually, it was German-manufactured ammunition for the Austrian M95 Steyr rifle. Relevant head stamps are as follows:

Artilleriezeugsfabrik, Vienna: AZF
Giersig & Cie, Woellersdorf: W over W, or a crowned eagle
George Roth, Vienna: GR, or G.R.
Hirtenberger Zundhutchen und Metallwarenfabrik, Hirtenberg: H, I, HP, HR, H K
   & C, W H P, X.
Keller & Company, Vienna: X K&C
Patronenfabrik, Lichtenworth Niederost: SOS
Manfred Weiss Patronenfabrik, Vienna: W, B and III

Bangladesh

The Bangladesh Ordnance Factories are operated under the auspices of the Bangladesh Army. The complex itself, comprising five factories, is located near the capital city of Dhaka in the township of Gazipur Cantonment. The organization states on its website that it obtained defense technologies from China, Germany, Italy, Australia, Belgium, and Austria (Bangladesh Ordnance Factories 2008). Head stamp BOF is observed; other elements may include the caliber or two-digit year.

Belgium

Belgium has a long history and deep tradition of arms and munitions manufacture, and is home to some of the most well-renowned names in the industry. Fabrique Nationale (FN) remains one of the oldest continuous operating armaments producers in the world. FN is part of the Herstal Group, which also includes Winchester as a holding. Commercial, government, and military consumers use FN-manufactured ammunition and firearms.

FN has used various head stamps over the years, particularly FN, FNH, FNB, and H. In the early 1990s Fabrique National entered into a relationship with Nexter/GIAT of France, presumably to supply French military small arms ammunition needs, using head stamp FNB which, according to some sources, identified Nexter as the manufacturer. Code "ch" was used during the German occupation. Other markings may appear to indicate the type of projectile that is loaded, caliber designation, and a two-year date. Stamping practices can vary by customer preference, since FN supplies ammunition to military and commercial clientele around the world.

In addition to FN, there have been other Belgian firms involved in ammunition manufacture. Anciens Establissements Pieper Hertsal (more commonly recognized by its trademark name Bayard) was founded in the 1800s and operated until the mid-twentieth century, using the head stamp AEP or A.E.P. They also manufactured firearms. There were many other Belgian firms such as Société des Cartoucheries, Belgium: S.C.B.; Francotte, May et Cie, Liege: VFM & CA V; Charles Fusnot, Brussels: FU; and Société Anonyme, Brussels: S. A. Paul Schraff Bruxelles was another remanufacturer of weapons and of cartridges, using their name as the head stamp on their shot shells. For the most part, these are more archaic and seldom, if ever, appear outside of cartridge collections. The Belgian firm Van Bruaene Rik is distinguishable because of its unique cartridge offering, the 7.92×24mm; however, other more recognized calibers are available with their variety of proprietary projectiles. The head stamp is VBR-B, affixed at the 6 o'clock position, the caliber at 12 o'clock.

Bolivia

Small-arms ammunition is solely manufactured by the state-controlled Armed Forces National Development Corporation, which markets to military and commercial consumers. The only observed head stamp is BOLIVIA.

Brazil

The firm Fabrica Nacional de Cartuchos y Municoes, Sao Paulo, was marked ESCUDO or F.N.C.M. Later this company would become Companhia Brasileira de Cartuchos (CBC), under the control of a partnership between Remington Arms and Imperial Chemical Industries. In 1979 CBC became wholly Brazilian owned. CBC uses the brand name Magtech for export cartridges; however, the CBC head stamp is still retained. Head stamp MRP for Magtech Recreational Products was also used. The V head stamp is for Velox, strictly limited to .22 rim-fire cartridges manufactured by CBC. In 2007, CBC acquired the German firm MEN, and in 2009 acquired the munitions firm Sellier & Bellot in the Czech Republic. Rio Ammunition is another brand of shot shell produced in Brazil, and has purportedly been in business since 1898. The state-controlled firm IMBEL (Industria de War Material) used the head stamp IMBEL. Fabrica Realengo, Rio de Janeiro, was founded in 1898 and operated under subtly different names until the 1980s. Its final head stamp was F R; previously it had used F.C.A.G. (1900–11) and C R F (1911–33).

Burkina Faso

Burkina Faso has also been known as Upper Volta at various times during the twentieth century. The Industrial Society Burkina Arms and Ammunition (SIBAM) is Burkina

Faso's chartered defense products provider. It is based in Ouagadougou. The marking practices are not precisely established, but markings such as SIBAM seem rational. Cartridges bearing the head stamp CV for Cartoucherie Voltaique (CARVOLT) have been observed.

Burma

Ammunition either manufactured in Burma or supplied by a third-party source to Burma has appeared bearing a triangle head stamp and a C or a cJ. Other numbers may appear that are likely a month and year of production, the year being a two-digit numeral.

Cambodia

Factory code 65 is recognized as representing Manufacture de Stung Chral. The Soviet-style head-stamp pattern was employed, with the cartridge dimension at 12 o'clock and the factory number at 6 o'clock.

Cameroon

Manufacture Camerounnaise de Munitions is the state munitions producer in Cameroon. The single observed head stamp is MANUCAM. Cartridges of 9×19mm and 7.62×51mm bearing this head stamp have been observed and include a caliber designation and perhaps a two-digit year.

Canada

Canada was a significant contributor to the overall military efforts of the Commonwealth during both world wars. After the Second World War, the Canadian ammunition industry gradually abated in presence until only SNC Technologies remained as a sole-source provider for military ammunition, and it was subsequently acquired by General Dynamics Ordnance and Tactical Systems. Commercial ammunition was produced by Canadian Industries Ltd. and sold under the name Canuck and Imperial or by Fremel Manufacturing in Ontario, using the head stamp FREMEL. Other relevant head stamps are as follows:

Canadian Industries Ltd.: IMPERIAL
Defence Industries, Montreal: DI
Defense Industries, Ltd., Quebec: DC, DI, DI Z
Defense Industries, Verdun, Quebec: V C
Dominion Arsenal, Lindsay, Ontario: D.A.L.
Dominion Cartridge of Quebec: DA, D.A., D A, D C, D, DAQ, DCCo, D.C. Co, DOMINION, MM.
Dominion Rubber and Munitions Ltd., Three Rivers, Quebec: T.R.
General Dynamics Ordnance & Tactical Systems (also SNC Technologies/Industries Valcartier): IVI
Ross Rifle Co., Quebec: R.R.Co., R.R.Co. CAN

Chile

Fabricas y Maestranzas del Ejército (FAMAE for short) is the state-controlled arms corporation and has been in operation for some 200 years. It manufactures current Western standard cartridges and shot shells for civilian markets. FAMAE has been used as a head stamp, as well as F, FME, and FM EP, and may include a two-digit date, caliber, and may also appear with the NATO cross, indicating that the cartridge has been manufactured in compliance NATO ammunition specifications.

China

China was a major supplier of ammunition into the United States until February 1994, when further importation was banned entirely by the Bureau of Alcohol, Tobacco, Firearms and Explosives (BATFE). The justification behind the action concerned the classification of Chinese 7.62×39mm ammunition as armor piercing, which led to the prohibition of future importation of that specific caliber. A comprehensive ban on further importation of all Chinese-made arms and munitions into the United States went into effect in May 1994.

Ammunition imported from China was sold under various trade names such as NORINCO (North China Industries Corporation), China Sport (NORINCO), and Jing An. Other caliber cartridges that were imported included 7.62×54mmR, 7.62×25mm, .30 Mauser (7.63×25mm), 9×19mm, 9×18mm Makarov, .223 Remington, .308 Winchester, and .45 ACP. In addition, 12-gauge shot shells originated in China, sold under the brand name Arrow, were actually manufactured by NORINCO, and these were stamped ZP over ZH. Chinese-produced casings appear in colors ranging from a reddish bronze to dark brown, brownish green, silvery bare metal, and on a very limited basis, brass. The reddish bronze casing is copper coated over base metal (also referred to as *bimetal*), giving such cartridges a distinct reddish-gold appearance.

Chinese ammunition production is through dozens, if not hundreds, of state-controlled arsenals that make cartridges for their own military purposes as well as for export. Ammunition thought to be of Chinese origin that predates the Second World War may have been foreign-sourced and imported into China from friendly nations such as the United States or Canada. Chinese ammunition dating from the late 1940s into the 1950s exhibit an irregular patterning of a mixture of Chinese characters, cartridge caliber, factory code numbers, and dates. Since China made extensive use of imported firearms, as well as domestically produced copies of foreign arms, such ammunition will be encountered in familiar calibers of the era: .30-06, 7.92×57mm Mauser, .45 ACP, 6.5mm Japanese, and the like. Chinese ammunition can been seen marked with Chinese language icons, as well as swastikas, which could be misinterpreted to mean German Nazi-era ammunition. Since ammunition production in China is a state-controlled industry, numerous factories are engaged and are identified by a factory number. Observed factory numbers are CJ, D, LY, 11, 31, 41, 61, 71, 81, 101, 111, 121, 131, 141, 215, 321, 351, 451, 661, 671, 711, 791, 21215, and a triangle. In the early to mid-1950s, it would appear that the Chinese started to standardize their head-stamp practices to match the Soviet practice, with the factory number at 12 o'clock and the date at 6 o'clock. However, certain other codes also appear in the 6 o'clock position. This would coincide with China adopting various Soviet-pattern firearms—either received from the Soviet Union or domestically produced copies of such weapons—but this may be entirely by coincidence.

Some Chinese ammunition can be seen marked with the caliber designation, particularly examples initially packaged for commercial sale. Due to the limited ability to research the practices of ordnance and munitions production within China, the firm identification of these numerical identifications to any particular factory remains elusive. For the most part, this degree of detail is probably unnecessary, as it will suffice to simply identify the ammunition as being of Chinese origin. Chinese ammunition is identified by its type, which is derived from the year it was adopted into official service. The Russian 7.62×54mmR is the Type 53; 7.62×25mm is the Type 54; 7.62×39mm is the Type 56; 9×18mm Makarov is the Type 59; and 7.62×17mm is the Type 64. The Type 64 is dedicated exclusively to the Types 64 and 67 suppressed pistol and has not been seen to any great degree in the United States. The semi-rimmed 7.62×17mm should not be confused with the Type 64; they are not interchangeable due to the variation in rim design.

Chinese-sourced ammunition imported to the United States came packaged in cardboard commercial-style boxes and in plain military-style containers, and when it was commercially sold, it was easily acquired in bulk quantity, sold in hermetically sealed galvanized steel cases. These cases open by using a can-opener-style tool and are not resealable once opened. These containers are often referred to as *spam cans* because of their similarity to the cans used to package the meat product. Cartridges contained in the spam cans come wrapped in a waxy kraft-style of paper. The 7.62×39mm cartridges in particular were loaded onto ten round stripper clips, ideal for immediate use with SKS-type rifles. The 7.62×39mm cartridges sold under the NORINCO name contained a steel core that has been reported as a "penetrator"; 7.62×39mm cartridges sold under the other names were solid lead projectiles. In the case of NORINCO cartridges containing the steel core, the outermost portion of the projectile was lead that coated this core, which was made of mild steel and was cylindrical in appearance. Undoubtedly, there is a great deal of Chinese ammunition in existence around the globe. A large cache turned up in Albania in 2008 as part of a contract bound for Afghanistan to supply pro-Western Afghan forces. The operation was disrupted by U.S. authorities when the cartridges were identified as being of Chinese origin, and were therefore embargoed. The investigation resulted in arrests and prosecutions.

Colombia

Industria Militar Colombia traces its origins back to 1908, taking the name of Industria Militar in 1954. Center-fire small-arms cartridges are head stamped INDUMIL or IM and with a caliber designation. In certain instances, the lot number and date of construction by month and year is also head stamped. Cartridge casings are marked INDUMIL and COLOMBIA. Shot shells may be identified by either marking inked on the body or head stamped; however, stars and the gauge are also used on the head stamp (Industria Militar 2011).

Cuba

Cuba's Industria Militar (Union of Military Industries) is the state-controlled enterprise concerned with the manufacture, repair, and maintenance of Cuba's various defense articles, and it claims on its website to have the capacity to manufacture ammunition (Defensa Nacional 2003). Small-arms cartridges originating from Cuba likely will be in Soviet-standard calibers and likely marked in standard Soviet-era style (a factory number and year). Pre-embargo cartridges have been seen marked "Cuba," but likely were intended to be exported to the island from U.S. sources rather than being manufactured there.

Denmark

Many cartridges from Denmark feature a crown in the 12 o'clock position of the base. In lieu of the crown, other possible information that would appear at the 12 o'clock position on the head stamp would include material supplier or when the casing material was produced. Frequently, the date is a complete four-digit year at the 9 and 3 o'clock positions, with the manufacturer abbreviation set at the 6 o'clock position. The presence of a triangle indicates a reloaded cartridge. Ammunition made for contract export has not been seen to bear the crown or other information and has been limited to the manufacturer head stamp, caliber, and production year. (Figure 3.9 depicts a 7.92×57mm cartridge made at Ammunitions Arsenalet in 1954 and sold to Ecuador.) These particular examples were loaded with steel-cased projectiles. Stamp PJJ was apparently used exclusively on 9×19mm cartridges made under German occupation. These are discernible by the wartime date that is stamped at the 6 o'clock position. Relevant head stamps are as follows:

> Ammunitions Arsenalet, Copenhagen: AA, AMA (1951–55)
> Dansk Rekylriffel Syndikat, Copenhagen: DRS (superseded by the Danish Industrial Syndicate [DISA] after World War II)
> Enger et Co.: X
> Haerens Ammunition Arsenalet, Copenhagen: HA
> Haerens Krudt Patroneret Laboratoriesvaerks: HL (1900–37)

Dominican Republic

The San Cristobal Arms Factory was started in 1948 and produced a modest number of arms and the ammunition to support them. The single observed head stamp is AC.

Egypt

Egyptian ammunition is head stamped in Arabic, having the nationality at the 12 o'clock position and a date at the 6 o'clock position. Egypt has used ordnance-grade brass in cartridge casings. British-service-caliber cartridges, as well as Soviet cartridges, are prevalent from Egypt, as both have been used in the nation over its modern history. Egyptian munitions production is controlled by the Ministry of Military Production, which operates a series of factories that are named but that generally are identified by the plant number. The two best known are Helwan and Maadi, both recognizable because of firearms known by those names: the Helwan being a handgun resembling a Beretta 92 and the Maadi being an Egyptian copy of the Russian AK rifle.

The marking practices for Egyptian ammunition have changed several times. Arabic characters are used to mark the head stamp. Until 1958, the translated term MISR appeared at the 12 o'clock position, followed by the two-digit year. In 1959 through 1971, the translated term UAR was affixed; from 1971, the term was ARE (Watson III 1984). Another head-stamp variation includes the letters A.R.E. (Arab Republic of Egypt) with a two-digit date and cartridge caliber arranged in a three-segment head stamp.

Ethiopia

Ethiopian munitions manufacturing is performed by the Homicho Ammunition Engineering Complex (HAEC), which is part of the overall Ethiopian defense industry. Like many African nations, Ethiopia received significant technical assistance in setting up this capacity, at first from Czechoslovakia and later from the Soviet Union and other Eastern Bloc nations (Griffard and Troxell 2009). There are six separate defense industrial segments, ranging from uniforms and textiles, to tanks, to aircraft. The complex has the capacity to manufacture the Russian 7.62×54mm and 7.62×39mm cartridges in both ball and blank varieties. The head stamp used by HAEC could not be definitively stated, but is believed to follow Russian/Eastern Bloc practice and would simply be HAEC accompanied by a year, caliber, or another production-related number.

Finland

Finland has a well-established reputation for building some of the finest small arms and ammunition in the world. Sako Finland (now partnered with the Beretta Holding Group) was founded in 1921 and has served military and civilian markets. The company has gone through several ownership and organizational changes through their history. They have head stamped their cartridges using S, SAKO, SAT, S A T, S.A.T., and SO. The state-owned industrial conglomerate Valmet has used VALMET, V PT, and V.P.T. Lapua is part of the Nammo Group, but it was once a state-controlled industry. They have used head stamps L and more recently LAPUA, which may be coupled with the L in a shield, as well as caliber designation of the cartridge. LAPUA- and SAKO-marked cartridges are not routinely marked with the year of production.

France

France was a major arms and armaments manufacturer until occupation during the Second World War. Numerous firms were engaged in munitions production during that period of time. After the Second World War, France occupied the portion of Germany that encompassed Mauser, and France absorbed large quantities of arms from Germany that were left over from the war, in addition to some new production. At the end of the war, Mauser was using code "svw" to identify itself, having switched from code "byf" in late 1944. Postwar, the French continued to use the svw code and svw MB to identify production from Mauser, primarily P.38 handguns and K98k rifles. Postwar production is readily identified by receiver dates; however, wartime dates are also expected. French production focused primarily on indigenous calibers such as 7.5 Lebel; however, outside calibers could be encountered post–World War II. French-sourced ammunition made during the war under occupation was marked according to German practice. As discussed under Belgium, it would appear that FN partnered with French munitions firm Nexter in 1991 and supplied small arms cartridges through this relationship, using head stamp FNB. Since 1950, the manufacturer appears at the 12 o'clock position with four-digit year at 4 and 8 o'clock.

A variation to this is to have the year at the 6 o'clock position. The presence of the NATO cross indicates a military-specification cartridge. Relevant head stamps are as follows:

Atelier de Construction de Puteaux: A.P.X., A.Px., APX
Atelier de Construction de Rennes: RS
Atelier de Construction de Tarbes: ATS, T S
Atelier de Construction de Toulouse: TE
Atelier de Construction de Valence: A.V.E., AVE., VE
Atelier de Construction de Vincennes: A.V.I.S., AVIS, VS.
Cartoucherie Française, Paris: C F, C.F., C.F, VELO-DOG PARIS, VELO-DOG
Cartoucherie Paulet, Marseille: CP, MI
École Centrale de Pyrotechnie, Bourges: ECP
Etablissements A. Pouvesle, Arcueil: APM, AP
Etablissements Rey Frères, Nimes: RY
Galand, Paris: GALAND
Gevelot S.A., Paris (Société Française des Munitions, Paris): FTCI, GG (intertwined), GEVELOT, GMx, G.S.F., G T, S F, SFM, S.F.M, SEN
Karchen & Company: KARCHEN
Manuhrin: MR
Marcel Gaupillat, Paris: G, G.P., or GAUPILLAT
Munitions Approuvées par le Comité de l'Artillerie: A R T

Germany (Prewar, West Germany, Reunified Germany)

Germany has remained a significant presence in munitions manufacture. During the rearmament period commencing in the mid-1930s through the end of the Second World War, Germany used a series of secret code systems consisting of a series of letters and numbers to indicate armaments and munitions manufacturers. There are potentially thousands of codes, as each individual manufacturer was assigned one. Codes with a P prefix are associated with munitions manufacturers, especially in the prewar rearmament period, and were gradually phased out after 1940, but they were used into the war years and are overlapped by the alphabetic code system. These codes always used lowercase letters; however, the P prefix codes used a "P." followed by a two- or three-digit number such as, but not limited to, P.25, P.28, P.120, P.131, P.334, P.405, and P.635. The identified ammunition producer codes that were used by firms within Germany, the Axis allies and puppets, and in the occupied nations include: ad, am, an, ap, asb, asr, auu, aux, auy, auz, av, avt, avu, avy, awt, axq, aym, bd, be, bf, bg, bj, bk, bne, cdp, cg, ch, dbg, dma, dnf, dnh, dom, dph, dye, dza, eba, ecc, ecd, edg, edq, eeg, eej, eem, eeo, eey, emp, eom, fa, faa, fd, fde, fee, fer, fva, ga, gtb, ha, ham, has, hgs, hla, hlb, hlc, hld, hle, htg, jtb, kam, kfg, krb, kye, kyn, kyp, lkm, mpr, mrb, ndn, nfx, oxo, oyj, pjj, suk, ta, tko, ua, uxa, va, wa, wb, wc, wd, we, wf, wg, wh, wj, wk, xa, y, ya, and zb. The identification of any individual code to the assigned manufacturer is probably unimportant outside of collector circles, but it does ascertain that the exhibit in question was produced during this period. The study of these codes as it relates to war materials, firearms, and ammunition is a topic that has justified numerous publications that delve into it specifically.

Many munitions firms disappeared at war's end. In the postwar period, several German munitions firms gradually returned and began producing for export markets as well as internal sporting, police, and military consumers. During that period of time, numerous firms had been absorbed into other firms. Regardless of the particular period, head stamps used by German firms, other than the aforementioned codes, are as follows:

Berlin-Karlsruhe Industrie Werke: BKIW (formed from DWM postwar)
Braun & Bloehm, Dusseldorf: B.B
Cartouches d'epreuve: Beschuss
Deutsche Waffen und Munitions-fabriken, Berlin, Borsigwalde: DM, DWM,* DWMB
Deutsche Waffen und Munitions-fabriken, Karlsruhe: DWMK
Deutsche Werke, Berlin: DW, DWA
Deutsche Werke Aktiengesselscahft, Berlin: D W A, DWA, D.W.A
Dornheim A.G., Suhl: G.C.D., GECADO- G.C.
Dreyse & Collenbusch, Sommerda: D&C
Dynamit Nobel, Nuremberg: DAG
Dynamit Nobel Genschow, Cologne: DNG (acquired by RUAG in 2002)
Heinrich Utendorffer Patronenfabrik, Nuremberg: HU
Hugo Schneider A.G., Leipzig: HASAG
Georg Egestorff Linden: GEL, EG.
Gustav Genschow & Cie., Durlach (also known as Durlacher Munitionsfabrik): G, D,
    GD, Ge, GE, GECO, Geco, G.G & Cie, G.G. & Co, G.G.C. (acquired by Dynamit
    Nobel in 1963)
Industrie Werke Karlsruhe: IWK
Koenigliche Munitionsfabrik, Spandau: S
Metallwerk Elisenhütte GmbH, Nassau Lahn: MEN (acquired by CBC in 2007)
Polte Armaturen und Maschinenfabrik, Magdenburg: PM
Rheinische Metallwaren und Maschinenfarbik, Dusseldorf: H
Rheinische Metallwaren und Maschinenfabrik, Sommerda: R.M, R.M.S.
Rheinische Westfalische Sprenstoff: R.W.S, RWS, R.WS, RWS/GECO (absorbed by
    Dynamit Nobel in 1931)
Sellier and Bellot, Schoenback: SB S
Steve Kornbrigel, Dornheim et Suhl: SKd, S.K.D
Utendoerfer, Nurnberg: UN
Vereinigte Zunder und Kabelwerke A.G., Meissan am Elbe: VZK

Greece

Hellenic Defense Systems S.A. (EBO-PYRKAL) was formed in 2004 with the merger of Greek Powder and Cartridge Company (PYRKAL) and Hellenic Arms Industry (Hellenic Defence Systems S.A. 2007). The company is a state-controlled enterprise. Production of small-arms ammunition in calibers 5.56×45mm and 7.62×51mm takes place at the

---

* DWM produced such a vast array of ammunition that it often placed a code denoting the caliber of the cartridge on the head stamp, instead of marking the caliber. These are often referred to as the "catalog codes."

Hymettus complex, whereas sporting cartridges are produced at the Lavrion Plant. Small-arms ammunition in calibers used by military forces are manufactured to NATO specification. Ammunition of various calibers manufactured by PRYKAL used head stamps HXP, EK, ENK, or GPC. Olympic Industries has used E.D.P. with caliber designation, NPA, and OLYMP as well as two-digit date and cartridge caliber.

Guatemala

Guatemalan munitions manufacturing is controlled by the military and is produced under the auspices of Industria Militar de Guatemala. The principle ammunition product is the 5.56×45mm, probably in the M193 (56-grain projectile) based upon observations of their product on the commercial market. The head stamp is IMG or G, accompanied by the production year and cartridge caliber.

India

During the colonial period, the British established factories in India that manufactured arms and ordnance at least as early as 1787. Presently, Indian ordnance factories are operated under the Ministry of Defense, which comprises some 40 facilities that produce defense articles (Ordnance Factory Board n.d.). Pre-independence cartridges can be identified by the British broad-arrow proof mark on the head stamp, plus the appearance of the factory head stamp and other information that may appear on British-service cartridges. Dum Dum Arsenal used DF and DI; Kirkee Arsenal used KF; and the Ordnance Factory at Varangaon used OFV. According to the Indian Ordnance Factory website, the rifle factory at Ishapore is still in operation, as is Varangaon and Dum Dum. Indian munitions manufacturing includes both NATO-standard cartridges as well as Russian-caliber cartridges.

Indonesia

PT. PINDAD is the state-owned defense industry in Indonesia. According to the company website, the company was founded in 1808 as a military equipment workshop in Surabaya under the name of Artillerie Constructie Winkel (ACW). The Dutch handed over the factory to the Indonesian Government on April 29, 1950. The factory was then officially renamed Pabrik Senjata dan Munisi (PSM), meaning "weapon and ammunition factory." PT. PINDAD was nationalized in 1983 and is managed by the Indonesian Army. The company has since changed names again several times, including PT. Pakarya Industri (Persero), PT. Bahana Pakarya Industri Strategis (Persero), and now to Kementerian BUMN (PT. PINDAD [Persero] 2011). PT. PINDAD manufactures a variety of defense articles, including ammunition in the most common calibers, including .38 Special, 9×19mm, 9×18mm Makarov, .380 ACP, 5.56×45mm, 7.62×51mm, and 12.7×99mm. Wadcutter, round nose, and blank .38 Special cartridges are produced. The 9×19mm is produced in ball and rubber projectile types, which have NSN (NATO stock numbers) assigned. The 9mm Makarov and .380 ACP cartridges are produced with ball-variety projectiles and also have an NSN assigned. PINDAD 5.56×45mm cartridges are made using blank, ball, match grade, tracer, and steel-core projectiles. The tracer and blank types have an NSN assigned. The 7.62×51mm is produced in ball, match grade, and tracer types, each having an assigned NSN. Soviet 7.62×39mm cartridges using ball projectile and blank type are also manufactured. An unusual cartridge, the 7.62×45mm, is manufactured by PINDAD. The 7.62×45mm is a unique cartridge that originated in Czechoslovakia, and used only the VZ 52 rifle, which was short-lived due to the appearance of the AK rifle. PINDAD also manufactures the 7.62×53mm, better known as the .30-06, available in ball projectile and blank types. The head stamp PINDAD is used.

Iran

The Defense Industries Group (DIG) is a subsidiary of the Ministry of Defense and is therefore a state-controlled enterprise of the Islamic Republic of Iran. The Defense Industries Group is broken down into several different industries, including the Ammunition and Metallurgy Industries (AMIG). The AMIG website claims that it is the largest industrial group within the DIG and is made up of eight different subsidiaries scattered throughout Iran (Defense Industries Group 2006). AMIG manufactures both Russian and NATO cartridges and components: 5.56×45mm M855 and M193 types, 7.62×39mm, 7.62×51mm, 7.62×54mmR, and interestingly 7.92×57mm. The DIG was previously known as the Iranian Defense Industries Organization as part of the State Arsenal of Tehran. Ammunition from the Shah era is still seen, typically stated as originating in Persia. Head stamps may appear in the form of a crown with a mix of Arabic and Farsi characters. The materials used tend to mimic Western standards. Surplus ammunition from the era includes Western calibers such as .303 British and the .30-06.

Ireland

A single head stamp has been associated with Ireland: IMI, for Irish Metal Industries. The company is still in business, but it does not manufacture ammunition.

Israel

Israel manufactures ammunition for internal use as well as for export for military and commercial customers. Early Israeli ammunition typically bears head stamps with Hebrew characters; however, by the mid-1950s the Roman alphabet and Arabic numbers began to appear. Since 1967, Israeli Military Industries has used the abbreviation IMI on commercially exported cartridges. IMI purportedly has used the head stamp TZ for military ammunition intended for use exclusively by Israeli forces. Some sources indicate the TZ head stamp started appearing as late as 1977. IMI has also head-stamped cartridges TZZ, apparently for ammunition manufactured for military customers outside of Israel. TZZ-stamped cartridges have appeared in the United States, perhaps as surplus or contract over-run ammunition sold commercially by IMI. The head stamp TA is believed to have been used to identify Tel Aviv Arsenal, appearing circa 1954. The head stamp E is unknown but has been associated with Israel.

In 2005, Israeli Military Industries changed its name to Israel Weapons Industries. Israeli ammunition has always enjoyed a reputation as being some of the finest produced anywhere in the world. Israel has been subcontracted at various times to supply NATO nations with ammunition, and NATO-specification cartridges originating from Israel will carry the NATO cross on the head stamp as evidence of such. Israeli ammunition has been sold commercially under the names IMI, Eagle, Samson, and IWI. Israel manufactures ammunition in the NATO military calibers, certain handgun calibers popular in the American market, and especially in calibers that the Israeli forces use themselves. Historically, other calibers were manufactured, including 7.92×57mm Mauser, dating back to when Israel used these surplus rifles to equip their forces before these rifles were converted to 7.62×51mm.

Italy

The nation of Italy has enjoyed a long and distinguished history of small-arms and ammunition production. Producer head stamps should not be confused with inspector stamps, using name abbreviations. Vintage Italian ammunition is archaic at this point; it is more subject to collector interest than practical use, as comparatively little survived the war and even less remained usable. Vintage Italian ammunition is primarily in the Italian firearm calibers such as 7.35×51mm Carcano, 6.5×52mm Carcano, and 9mm Glisenti; however, non-Italian calibers that were in use in the first half of the twentieth century would be expected. Modern Italian ammunition is quite common in the form of shot shells and in sporting calibers. Relevant head stamps are as follows:

Arsenal de Bologne: B P
Brombini Parodi Delfino, S.p.a., Rome: AOC, B P D, B.P.D
Giulio Fiocchi Leeco: CD., CP 99, FIOCCHI, GFL, G.F.L, G.F.L., G F L,
    (GFL is currently used in commercial Fiocchi products)
Leon Beaux Milan: BEAUX, L. BEAUX & Co.
Pirotecnia Bologna: T.M
Pyrotechnie de Bologne: A.C.B
Pyrotechnie de Capoue: AA, AA-C., C A, C.AA, C.R, D.C.E., E S, T.R, VS, ZG
Societa Italiana Munizione Leon Beaux et Cie, Milan: SIM
Societa Metalurgica Italiana, Campo Tizzoro: SMI, SYI

Japan

With the exception of naval ammunition, World War II Japanese ammunition does not appear to regularly exhibit head stamps, particularly the handgun cartridges. The military cartridges then in use—8×22mm Nambu, 8mm SR (semi-rimmed), 9mm Revolver, 6.5×50mm, and 7.7×58mm—were rendered obsolete at the end of hostilities and were subject to limited postwar manufacture. The postwar production included Norma-produced 7.7×58mm, Hornady 6.5×50mm, and limited 8mm Nambu production from Asahi Okuma circa 1960–61 as well as some postwar production in China, likely to supply various and sundry forces in Asia armed with surplus Japanese weapons. Relatively little wartime Japanese ammunition remains in existence; the majority produced was used up during the war or destroyed after Japan surrendered.

Postwar commercial production was intended to satisfy persons who used surplus Japanese rifles for sporting purposes and collector enthusiasts who wanted to shoot souvenir

Japanese pieces. Japanese ammunition was identified by its type, just like their firearms; the 7.7×58mm is the Type 99. With respect to the 7.7×58mm, caution must be exercised, as the same dimensioned cartridge was loaded for use with machine guns, the only difference being the rim design. The Type 99 rifle cartridge is rimless, whereas the 7.7×58mm Type 92 is fully rimmed like the British .303 or the Russian 7.62×54mm. The 8×22mm Nambu was officially designated the Type 14, as was the pistol chambered for it; however, the Japanese Type 94 pistol also used the same cartridge, retaining the Type 14 designation. The various Japanese submachine guns were chambered in 8×22mm Nambu as well. Due to the very limited application of Japanese cartridges outside Japan, the cartridges are typically suffixed with the term *Jap*. Colloquially, all Japanese rifles have been termed *Arisaka*. The limited number of head stamps linked to Japan are: Asahi Okuma Company, AOA; Naval Arsenal of Yokosuka (date and caliber also stamped), E; and Tokyo Seiki Co., T E and TOYO. The letter *y* has been associated with the Toyokawa Naval Arsenal.

Kenya

Kenya Ordnance Factories Corporation is Kenya's only indigenous ammunition production facility. The corporation was chartered by state order in 1997. The corporation manufactures 7.62×51mm, 5.56×45mm, and 9×19mm cartridges. Blank cartridges are produced in 7.62×51mm and 5.56×45mm but not in the 9×19mm; only live ammunition of 9×19mm caliber is manufactured. The designation SS 77/1 is used to identify ball 7.62×51mm cartridges, which likely mimics the US M80 cartridge. The company produces the SS 109 type (heavy ball) 5.56×45mm. It is claimed that all ammunition is manufactured to NATO specifications (Kenya Ordnance Factories n.d.). It is not claimed that the ammunition has been tested and accepted by NATO or that a NATO acceptance stamp should appear. A three-element head stamp has been seen, using KOF, a two-digit date, and caliber designation.

Lithuania

The Lithuanian munitions firm Giraites Ginkluotes Gamykla (GGG) manufactures NATO specification 7.62×51mm (M80) and 5.56×45mm (SS109 projectile) cartridges. The company came online in 2000 and received certification to NATO specification in May 2005. The company reports that since 2006, some 90% of their annual output is exported (AB Giraites Ginkluotes Gamykla n.d.). The head stamp is GGG, and the NATO cross is to be expected. GGG ammunition has appeared on the commercial market and appears to be the same ammunition produced to NATO specifications.

Malaysia

SMEO, formally known as Syarikat Malaysia Explosives, Ltd., is a government-controlled entity and the only munitions production firm in Malaysia. According to the company website, "SMEO was incorporated in 1969 as a joint venture Company with equity participation between the Government of Malaysia, Dynamit Nobel of Germany, Oerlikon Machine Tools of Switzerland and two local partners namely Syarikat Permodalan Kebangsaan and Syarikat Jaya Raya Sdn Bhd. However, by 1974, the Government of Malaysia acquired all the shares" (SME Ordnance Sdn Bhd n.d.). The head stamp MAL accompanied by the caliber and two-digit year has been observed. Small-arms ammunition and shot shells are manufactured. Calibers produced include .38 Special, 9×19mm, and varieties of 5.56×45mm, 7.62×51mm, and 12.7×99mm.

Mexico

The state-controlled ammunition enterprise in Mexico is the Fabrica Nacional de Municiones de Mexico, and it has marked cartridges with F de M, F.M, FM, FNC, F.N.C, and F.N.C. Remington Arms was a partner in the Mexican-based cartridge producer Cartuchos Deportavos de Mexico, which used the head stamp CDM. In 1978, the company changed names to Tecnos Industria SA. The company now markets under the brand name Aguila. The name Aguila appears as the head stamp alongside caliber designation in the traditional commercial form.

Morocco

Moroccan ammunition was manufactured by Manufacture Nationale d'Armes et de Munitions. The observed head stamp was MNAM. It is unclear if Morocco still manufactures ammunition.

Netherlands

The Dutch Government Ordnance Works, aK Artillerie Inrichtingen, was founded in 1679 and was the sole supplier of arms and ammunition within the Netherlands. The operation was later renamed Hembrug. In the early 1970s, Eurometaal N.V. was spun off from Hembrug. Assensys B.V. succeeded Eurometaal in Dutch munitions manufacturing and markets small-arms ammunition (Assensys B.V. n.d.). Head stamps used by aK Artillerie Inrichtingen were A, AI, AI C, AAI, BAI, DAI, G, and AI. Stamps 33, 37, B, D, DO, G, I, O, P, U, and X have also been observed, likely related to casing material, lot, or propellant information, and not necessarily as a definitive manufacturer head stamp. Head stamps EMZ and NWM have also been reported since the early 1980s.

New Zealand

The Colonial Ammunition Company in Auckland was the first munitions production firm located in New Zealand. Three head-stamp variations have been identified: C A C, C.A.C, and CAC. Hy-Score is a commercial brand that was manufactured in both Australia and New Zealand by the Colonial Ammunition Company. Hy-Score cartridges were in hunting- and sporting-oriented calibers such as .22 and .243 Winchester. Presently, Ordnance Developments Ltd. exclusively manufactures training and special-purpose ammunition in the form of blank cartridges as well as frangible projectile cartridges directed specifically at law enforcement, military clients, and theatrical performance. Their 9×19mm ball ammunition is the only live cartridge produced by the firm. Blanks are manufactured in 5.56×45mm, 7.62×51mm, 9×19mm, .50 BMG, and .338 Lapua. Frangible loads are manufactured in 5.56mm, 9×19mm, and .40 S&W. The company invests greatly in clear identification of its blank cartridges as a matter of safety (Ordnance Developments 2009). The company uses brass and steel in the manufacture of cartridge casings. Thales is the current supplier for New Zealand military ammunition requirements.

Nigeria

Defense Industries Corporation of Nigeria (DICON) was founded in 1963 as a chartered state enterprise. According to the DICON corporate website, the (West) German firm Fritz Werner provided technical assistance in setting up the facilities. The initial output in 1964 is reported to have been 12 million 7.62×51mm and 4 million 9×19mm cartridges annually. This output is said to have tripled during the Nigerian Civil War (1967–70). Some sources have reported that the ordnance factory is located in Lagos; however, the DICON corporate website states that the ordnance plant is located in Kaduna (Defense Industries Corporation of Nigeria 2010). A single head stamp is reported to be currently used: OFN (Ordnance Factory of Nigeria). The head stamp AFN may have been used previously.

North Korea

North Korean source ammunition is virtually nonexistent save examples in the hands of military or intelligence agency collections or a few private collections. In all likelihood, North Korea maintained a marking practice consistent with the Soviet Union and China. A head stamp bearing a solid triangle at the 6 o'clock position, coupled with either a ring or the letter A at 12 o'clock, has been observed on some examples. Variations include the use of a star, circle, or factory code 93 with a Korean character or two digits. In these variations, the arrangement of the stampings remains the same.

Norway

The Nordic Ammunition Group (Nammo AS) is a group comprised of three defense industry firms based in Finland and Norway. The group was founded in 1998 and was formed by the merger of Celsius AB, Patria Industries, and Raufoss (Nammo AS 2011). Nammo manufactures small-caliber cartridges in 5.56×45mm, 7.62×51mm, .338 Lapua Magnum, 12.7×99mm, and 9×19mm. All calibers are available in a variety of loads, including ball, tracer, subsonic, armor piercing, etc. "Nammo was the first company to introduce Plastic Training Ammunition. Since 1954, Plastic Blank Ammunition and Plastic Short Range Training Ammunition (PSRTA) have been used by military forces and law enforcement communities worldwide. Plastic Training Ammunition is available in all calibers from 4.6mm–40mm" (Nammo AS 2012). Lead-free cartridges, reduced-range training loads, and other types of specialized cartridges are available. Previous head stamps used by Raufoss were A.Y.R, AYR, R P, and R T P (may include a crown and/or date, usually segmented into four divided elements). Nammo Lapua is the commercial ammunition provider, marketing cartridges in popular calibers. Lapua uses a shield surrounding an *L* as a head stamp.

Pakistan

Pakistan Ordnance Factories is a conglomerate of 14 separate factories and three commercial subsidiaries that produce military and commercial products. The Pakistan Ordnance Factory is often simply referenced as POF. It manufactures both the Russian- and NATO-standard cartridges: 5.56×45mm, 7.62×39mm, 7.62×54mmR, 7.62×51mm, and 9×19mm (Pakistan Ordnance Factories 2010). Brass is used for the casing material on all cartridges.

The head stamp is POF, caliber designation, and last two digits of the year of manufacture. This head stamp, coupled with year of manufacture, appears to have been used since the inception of POF in 1951. During the colonial period, Britain had constructed a number of ordnance factories within Pakistan, and it would appear that the POF head stamp was at that time coupled with standard British proof marks.

Peru

FAME S.A.C., or Fabrica de Municiones del Ejército, de Lima, is a military-controlled private enterprise, having transitioned from being a state-controlled enterprise in 2009. The company was founded in 1963 by government decree with the intention of supplying the Peruvian armed forces (FAME S.A.C. 2011). Presently, the manufacturer's product mix includes 7.62×51mm, 9×19mm, .380 ACP, .38 Super, and .38 Special center-fire ammunition, in addition to 12- and 16-gauge shot shells. Their product carries the head stamp FAME and a two-digit year.

The Philippines

The Philippine Islands are home to several munitions firms that have done a lot of manufacturing of arms and ammunition on behalf of other names. Armscor is one such commercial arms and munitions manufacturer. This company should not be confused with the South African firm of the same name. The company entered the ammunition and firearms manufacturing business in 1952 as Squires Bingham Manufacturing Inc. The firm changed names to Armscor (Arms Corporation of the Philippines) in 1980. The company manufactures a variety of firearms and popular ammunition calibers (Arms Corporation of the Philippines 2005). In addition to Armscor-labeled products, the company has produced ammunition for a variety of other customers under their brand names, generally head stamping the product to suit the desires of the retailer. Commercially used names have included Concorde, Seeker, and Squires Bingham. Observed head stamps include the caliber designation, as well as AP or ACP, which could readily be mistaken as part of the caliber designation, rather than the manufacturer's abbreviation. The .22 rim-fire cartridges are stamped with a T. Floro International Corporation is another defense firm based in the Philippine Islands; however, they focus primarily on defense products, including firearms, firearm suppressors, mines, grenades, and other military accoutrements.

The Filipino Department of National Defense Government operates the Government Arsenal. Their current products include the 5.56×45mm M193 cartridge, 7.62×51mm M80 cartridge, 9×19mm, and .45 ACP, although .30 Carbine, .30-06, and .38 Special cartridges had been made in the past (Government Arsenal of the Philippines n.d.). Cartridges produced by the arsenal are head stamped RPA and the two-digit year. Compliance with NATO ammunition specifications in NATO standard calibers is claimed.

Portugal

EMPORDEF (Portuguese Company of Defense SA) operates the Portuguese defense industry on behalf of the Portuguese government and is the parent company of Portuguese munitions producer IDD (EMPORDEFF 2007). IDD appears to have superseded the former arms and munitions maker INDEP (Indústria de Defesa SA). INDEP was a recognized NATO supplier and used the head stamp FNM, the abbreviation for its prior name Fábrica Nacional de Munições de Armas Ligeiras, and two-digit year. Other historical Portuguese munitions producers were Arsenal de Exercito, which used an intertwined AE, and Fabrica de Cartuchas y Polvoras Quimicas, stamping F C.

Saudi Arabia

The most recognizable feature of the Saudi head stamp is the crossed swords and palm trees. The palm trees are stamped at the 12 o'clock position, the swords at 6 o'clock. In Arabic numbers, the caliber appears at 9 o'clock and the year of manufacture at 3 o'clock. The date calculation is based on Islamic Hirja, and they observe the lunar calendar, not the Gregorian calendar.

Scotland

The sole producer of shotgun shells in Scotland is the Caledonian Cartridge Company. The product is commercially sold throughout the world and readily recognized by inked markings on the shot shell hull.

Singapore

Singapore Technologies Kinetics is part of Singapore Technologies Engineering, Ltd. The company, in part, manufactures small arms and small-arms ammunition. A former defense products provider, Chartered Industries of Singapore, was acquired by ST Kinetics in 2000 (ST Engineering 2006). ST Kinetics currently manufactures comprehensive lines of cartridges in .50 BMG (12.7×99mm), 5.56×45mm, and 7.62×51mm. The 5.56×45mm line includes frangible low-energy ammunition (FLEA), plastic-cased blanks, and the typical ball projectiles. Chartered Ammunition Industries cartridges were head stamped HB along with the cartridge caliber designation. HG is observed as the head stamp, unusually at the 6 o'clock position, with the caliber designation at 12 o'clock.

South Africa

South African ammunition was restricted due to trade embargoes internationally because of the policy of apartheid. When this practice ended in 1994, South African ammunition began to appear on the market. Pretoria Metal Pressings, now part of Denel Group, has used the head stamps A, P, PMP, A80, D 02, DNL, and 13. PMP continues to manufacture military and commercial ammunition, including the 9×19mm, 5.56×45mm, 7.62×51mm, and the 12.7×99mm. On the commercial side, PMP manufactures a comprehensive line of handgun and rifle cartridges, including heavy rifle rounds for large game. Other numerals may appear, but these are not the head stamp, instead referencing other information. Swartklip Products became part of the Denel Group in 1992. Swartklip itself was founded originally as Ronoden, which passed to Armscor in the early 1970s. Ronoden used the head stamp RMC; Swartklip used SP. Centurion Arms produced ammunition under the brand name AmmoTech, marked AMT. A U head stamp was used to identify the Government

Factory (The Mint), Pretoria, until 1961, when replaced by SAM. A diamond after the U indicated the branch mint at Kimberly (Watson III 1984). South African Ammunition Factory used the head stamp SAAF. The firm Musgrave Manufacturers, Ltd., used head stamps M M, mus, and 0 8.

South Korea

Poongsan Corporation, located in South Korea, manufactures defense products, including small-arms ammunition for military and commercial markets. These cartridges are marketed in the United States as Precision Made Cartridges, bearing the head stamp PMC. PMC is also reported to mean Pan Metal Corporation. Poongsan ammunition was previously imported through a variety of different agents into the United States. The head stamp PS was also used, perhaps to indicate military-specification cartridges, whereas PMC denoted commercial specification. Cartridges observed with the PS head stamp also had a date stamped, appearing in two formats: one using an abbreviated month and year, such as "Nov.82", the other as a year and month, such as "80.01." Three variations in the letter M in the PMC head stamp have been observed, which may indicate the specific source that produced the cartridge. Ammunition branded as Hot Shot was another PMC brand and was commercially labeled accordingly. (Figure 3.4 reflects the PMC and PS head stamps.)

Spain

Historically, Spain has had numerous munitions manufacturers, most having disappeared by the 1970s. Santa Barbara Sistemas was a state enterprise, part of the Spanish Ministry of Defense, until acquired by General Dynamics in 2001 as part of the Combat Systems Group (General Dynamics 2001). In 2003, Santa Barbara was integrated into the General Dynamics European Land Systems (General Dynamics European Land Systems 2011). Santa Barbara continues to use the head stamp SB. The stamp "J. Costas Barcelona" appears on older, paper-hulled shot shells. Rio is another Spanish brand name that manufactures shot shells. The name Rio appears on the shot shell hulls, which are made of plastic. The historical identified head stamps and associated firms are:

Consorcio de Industries Militares: CIM
Cordo S.A., San Sebastian: CO
Empresa Nacional de Santa Barbara, Palencia: SB
Empresa Nacional de Santa Barbara, Toledo: SB-T or T
Fabrica Nacional de Palencia: FNP or P
Fabrica Nacional de Toledo: CIM FNT, A T, T A
Manufacturas Metalicas Madrilena: MMM
Pirotecnia Militaria de Seville: CIM-PS, PS, S
Secretaria de Armamento, Delegacion de Valencia: SA
Spanish Government Arsenal: S. G.
Standard Electrica, Madrid: M or T
Star Bonifacio Echeverria S.A.: STAR TRUST
Union Española de Esplosivos: U intertwined with I

Sudan

Military Industry Corporation (MIC) Sudan became operational in 1960 and is controlled by the Sudanese Ministry of Defense. The organization has five facilities and manufactures a diverse offering of military products, including small arms and ammunition, specifically 9×19mm and 7.62×54mm cartridges, although 7.62×39mm cartridge production cannot be ruled out due to the proliferation of arms chambered in that caliber, particularly the AK series of firearms (MIC Sudan 2007). One variation in the head stamp is unusual in that a four-element design is utilized: SU appears at the 12 o'clock position, a two-digit year at the 9 o'clock, and other numerical values at the 3 and 6 o'clock positions. A three-element design, using SUD instead of SU, caliber, and date may also be found.

Sweden

The Swedish state arms industry was comprised of several arsenals. Cartridges from the 1940s can appear with the three-letter code amf followed by factory numbers 24, 26, 27, 28, 29, 30, 31, or 32. Cartridges dating from the 1960s appear with codes 24, 25, 26, 026, 27, 027, 070 along with the amf marking. The presence of a 0 prefix numeric may have indicated a cartridge intended for export. Circles or stars possibly indicate reloaded casings; the appearance of roman numerals indicates production lot numbers or delivery number. Cartridges may bear other numeric codes that indicate cartridges from foreign sources such as the United States or the United Kingdom, particularly prior to World War II.

AB Norma Projektilfabrik: NORMA, norma
Arsenal de Karlsborg: EK
Dansk Ammunitionsfabrik, Otterup: 27, 027
Forenade Fabriksverken: FFV
Karlsborg Ammunisjon Faktori, Karlsborg: CG (CG now used by NAMMO)
Svenska Metallwerken Vasteras: HERTER'S, SM, 26, 026

Switzerland

In 1995, Schweizerische Munitionsfabrik (Swiss Ammunition Enterprise) was formed by the unification of the Altorf and Thun factories and was afterwards renamed RUAG Munition in 1999 (RUAG Holding 2011). Historically, head stamps used by Swiss producers are as follows:

Airmunition Industries: AM
Eidgenossische Munitions Fabriken, Altdorf: T A, D A
Eidgenossische Munitions Fabriken, Thun: B T, D T, R T, M T, MFT, T, T T, W T, (T
    W may also be associated), THUN
Fabrique de Soleure: FS
Patronenfabrik A.G. Solothurn: P
Swiss Government Marking: SW. GOV.

Taiwan

Taiwan has maintained a relatively small armaments production capacity. A historical head stamp is 60A, identifying cartridges produced by the 60th Arsenal, which was founded on September 1, 1946, on mainland China. The 60th Arsenal was reestablished in Taiwan after the machinery was moved from mainland China in 1948–49 and was subsequently renamed the 205th Arsenal (Ministry of National Defense R.O.C. 2011). The head stamp TAA is used by the 205th arsenal. The 205th Arsenal has been recognized as a NATO supplier; thus the NATO cross may be encountered on such cartridges if they are manufactured to NATO specifications. Ammunition produced by the 205th Arsenal is not directly marketed to the commercial sector, but may appear if such cartridges are sold as surplus from a customer stockpile.

Tanzania

The Tanzania People's Defense Forces operate the Mzinga Corporation as a manufacturer for small arms and ammunition. "The Mzinga Corporation in Tanzania was set up in 1971 with Chinese equipment to produce 7.62×39mm ammunition" (Anders and Weidacher 2006). The observed head stamp is MZINGA.

Thailand

Thailand has maintained a domestic defense industry that operates under the Ministry of Defense. Historically, Thailand was not a major exporter of ammunition, but the cartridges produced are NATO standard, as Thailand uses weapons originating from Western nations almost exclusively. There are several munitions operations based in Thailand. Thai Arms, based in Bangkok, used the head stamp TA. Products of Bullet Master Company, Ltd., also located in Bangkok, have been observed with the head stamp BM. Royal Ammunition is another Thai product, using the head stamp RAI coupled with a two-digit date and caliber designation. The company currently focuses on the manufacture of pistol-caliber cartridges (Royal Ammunition 2009).

Turkey

Until 1950, Turkish ammunition used a four-element head-stamp design, partitioned by lines. The crescent moon and star appear at the 12 o'clock position surrounded by the letters TC (translated to mean Turkish Republic), four-digit year at 3 o'clock, caliber at 9 o'clock, and FS appears at 6 o'clock. FS is the actual factory code, reflecting fabrication by the plant at Kirikkale/Ankara. The head stamp MKE, at the 6 o'clock position, stands for Makina ve Kimya Endustrisi Kurumu, which seems to be the more modern head stamp. The head stamp FI, coupled with TC (note previous mention) for Fabrikalar Iskenderun, may also be encountered.

Uganda

Luwero Industries Ltd. is a subsidiary of the National Enterprise Corporation in Uganda, but is owned and operated by the Ugandan Ministry of Defense. Luwero manufactures ammunition and provides repair and refurbishment services for small arms and other defense materials. The Luwero factory is located in Nakasongola and was commissioned by an act of parliament in 1989 (IHS Global 2011). The head stamp LI at the 12 o'clock position, coupled with a two-digit year at the 6 o'clock position, has been observed. Variations of this stamp may be encountered; however, the information is substantially the same. Luwero cartridges resemble those of Chinese origin, the casings having a distinctly red appearance in contrast to traditional cartridge brass or base steel.

United Arab Emirates

ADCOM Manufacturing, Ltd., based in Abu Dhabi, was acquired by the firm Tawazun Holding and renamed Caracal Light Ammunition (CLA) (Caracal Light Ammunition 2011). Burkan Munitions Systems LLC is also part of the same group. Caracal Light Ammunition manufactures 5.56×45mm, 7.62×51mm, and .50 BMG (12.7×99mm) cartridges to Western standards and using standardized designations. Ammunition variants include standard ball, armor piercing, tracer, and so forth. For ADCOM, the head stamp AD, a two-digit year, and caliber designation are used. Caracal is currently the only small-arms manufacturer in the UAE. The company also markets a pistol of the same name, and a precision rifle also appears to be in the works.

United Kingdom

The United Kingdom historically had a vast number of firms involved in producing ammunition for civilian and military consumption. The United Kingdom has long enjoyed a reputation as one of the world centers for fine sporting arms and ammunition. During the world wars, many firms were engaged in supporting the war efforts by producing munitions. After the Second World War, there was a great deal of consolidation, and many companies left the industry. In addition to private firms that were brought in, the British Ministry of Supply (later Ministry of Defense) operated a comprehensive network of Royal Ordnance Factories. In the mid-1980s, the British government privatized the remaining active ordnance factories under the name Royal Ordnance, where they passed into the hands of British Aerospace, which in turn dropped the Royal Ordnance name. The company manufactures ammunition under the name BAE Systems Land & Armaments.

Ammunition Company of Europe: ACE
Arche Wringers Ltd., Glasgow: AW
Birmingham Metal & Munitions Company, Birmingham: J
Birmingham Small Arms & Metal Company: B, B.S.A.
British Manufacturing & Research Co., Grontham: BMRC
Crompton Parkinson: C-P
Eley Brothers Ltd., London: E, EB, EC Eley, ELEY, ELEY BROs., WILKINSON
F Joyce & Cie, London: F.J.
Greenfell & Accles, Ltd., Birmingham: GA, G+A
Greenwood & Batley: G, GB
Greenwood & Batley, Farnham: GBF
Holland & Holland: HO. & HO.
Imperial Metal Industries (Kynoch Ltd.): K-33, K34, K.50, K53, K55, K56, K.57, KYNOCH
John Rigby & Sons: RIGB

Kings Norton: HG - NTN
Kynoch Factories (Imperial Chemical Industries), Birmingham: ICI, KY, WR
Ministry of Supply Factory, Hirwan: HN
Mountain & Sowden, Ltd.: M&S
Nobel Industries, Ltd. (London & Birmingham): NOBEL
P. Webley & Sons, Birmingham: WB
Royal Laboratory, Woolrich: R, RL
Royal Naval Armament Depot: RNAD
Royal Ordnance Factory Blackpool: BE
Royal Ordnance Factory Burghfield: BD
Royal Ordnance Factory, Radway Green: RG, R.G., RORG
Royal Ordnance Factory, Spennymoor: S R
Royal Ordnance Factory Thorpe Arch, Boston Spa/Yorks: TH
Royal Ordnance Laboratory, Birmingham: ROFB, R. L.
Webley & Scott, Ltd.: WEBLEY, W&S

Venezuela

Compania Anonima Venezolana de Industrias Militares, or CAVIM for short, is the state arms industry. CAVIM manufactures small-arms ammunition in the various commercially popular calibers, and their ammunition has been sold in the United States. Head stamps included CAVIM, and a caliber designation and a two-digit year may also appear. The head stamp VEN may have also been used for production under the auspices of the Ministry of Defense and not intended for commercial consumption.

Zimbabwe

Mathews Manufacturing Co., perhaps better known in numismatic circles, is said to have produced ammunition using the head stamp MMCo. Such cartridges would date to when Zimbabwe was known as Rhodesia (circa 1965–79). Currently, Zimbabwe Defense Industries (ZDI) operates as the principle supplier of small arms and ammunition within the country. Like most, if not all, African nations, Zimbabwe received foreign technical assistance in setting up ZDI. Ammunition manufactured by Zimbabwe Defense Industries was briefly available in the United States in the late 1990s, marketed under the brand name Cheetah. The observed head stamp was ZI marked in the 6 o'clock position and a two-digit year at 12 o'clock. Brass cartridge casings were used.

The Russian Federation, Soviet Union, Warsaw Pact, and Eastern Europe

In nations around the world where ammunition is produced by state-controlled arsenals, there has been an element of secrecy attached to what arsenals produce ammunition and where they are located. Ammunition produced in the former Soviet Union, China, North Korea, or former Warsaw Pact nations usually bear only a date and factory code. Most information known about factory-coded ammunition originating from the Eastern Bloc has come from declassified intelligence sources detailing where small ammunition production facilities were located. After the fall of the Berlin Wall, subsequent collapse of the Warsaw Pact, and eventual collapse of the Soviet Union itself, most nations that were once under the influence of the Soviets have capitalized on existing stockpiles of small-arms ammunition and production facilities. Much of the existing stock has been exported around the world, while the factories have either privatized or remained state-controlled enterprises and are actively selling their products worldwide. There is comparatively little ammunition predating the Cold War left over, except in the hands of collectors; thus the need to positively ascertain the identities of such specimens is limited outside the context of cartridge collectors and historians.

Ammunition originating from the Russian Federation differs little from that produced during the Soviet era. As a whole, there are no significant differences between cartridges produced by the various factories other than some variations in materials selection. With the collapse of the Soviet Union in late 1991, the former Soviet military-industrial complex faced almost certain doom. The same factories that churned out untold billions of rounds to fill state orders have turned their attention toward the worldwide market in military and civilian sectors. The cartridge casings are made of steel and appear in colors ranging from dark brown to greenish brown, olive green, and steel gray. A red or purple ring can be observed where the projectile meets the casing neck as well as around the primer opening at the base of the casing. This coloring is a waterproof sealant and is found on military-specification cartridges. Soviet-sourced cartridges not bearing this sealant have been observed; however, these cartridges apparently were not manufactured to military specification, likely intended from the outset of production bound for the commercial marketplace. These cartridges have a slightly shiny appearance, as they are treated with a thin layer of lacquer or polymer as a protective measure against the elements. Spent casings of this variety will start to rust almost immediately, even if left exposed to overnight dew, and within a matter of weeks will almost certainly degrade to the point of being nearly indiscernible and probably of little investigative value. Another common feature of Russian-sourced cartridges is a gold-colored primer. The projectiles are invariably copper jacketed.

The earliest Soviet-sourced cartridges that were imported to the United States were characterized by the most cost-effective packaging possible, generally low-grade cardboard boxes with plain white labels. Occasionally the producer's icon would appear, but most of the labels simply indicated the number and type of cartridge, along with the standard warnings about lead content and keeping out of the reach of children. The practice continued for quite some time and was overlapped by the gradual shift to more aesthetically pleasing commercial-style packaging and brand names. Despite the presence of newly minted Russian ammunition, a great deal of surplus Soviet-era cartridges remain available. Commercial ammunition in bulk appears in cardboard cartons that contain cardboard boxes of 20 rounds apiece. The cardboard cartons are generally basic and bear basic information that is stencil printed; the individual boxes are commercially marked (see Figure 3.7).

Military surplus ammunition will appear in metal tins, two tins per wooden crate. Some Russian producers will sell commercial ammunition in military-specification sealed metal tins, although this is readily recognized by the English language content stenciling, as opposed to the Cyrillic used in the surplus containers. All Russian ammunition, whether

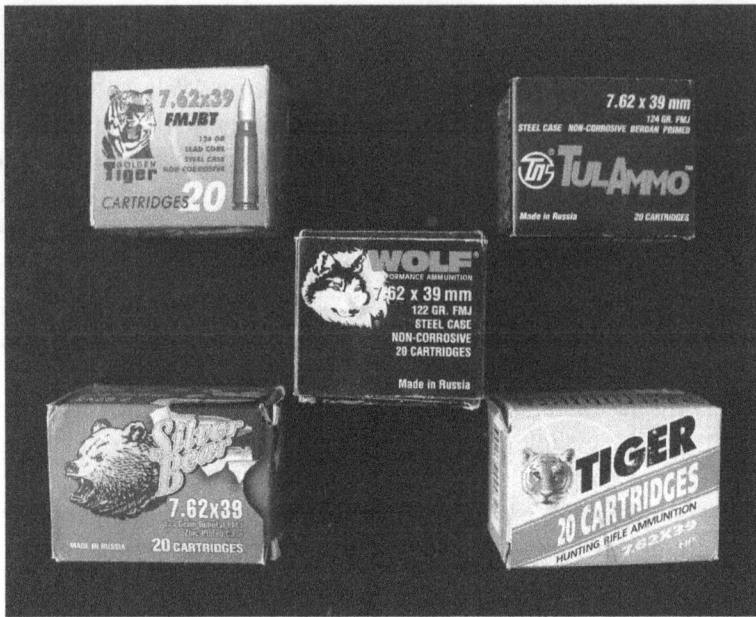

Figure 3.7 Contemporary Russian commercial ammunition boxes. Top row from left: Vympel Golden Tiger; Tulammo. Middle row: Wolf Performance Ammunition. Bottom row from left: Barnaul Silver Bear; Barnaul Tiger. (Image from author's collection.)

Cold War surplus or more contemporary production, are Berden primed, so reloading is not likely, although it is not impossible. Older surplus ammunition can be expected to use a corrosive priming compound, whereas contemporary production moved away from it and now uses a noncorrosive compound. The residuals of the corrosive compound will cause the bore to rust if not thoroughly cleaned either by hot soapy water or a dedicated gun-cleaning solvent.

Soviet Factory Codes

Soviet factory codes underwent several shifts in the twentieth century. Predating the Second World War, a mixture of Cyrillic characters, letters, and numbers were used that can create confusion between dates and manufacturers, especially when compared against wartime or Cold-War-era head stamps. Character Л at 12 o'clock represented the Lugansk plant in the 1910s to 1920s. These casings will be stamped with a year of manufacture. In roughly the same time period, the Ulyanovsk Cartridge Works used C or y at the 6 o'clock position with a date at the 12 o'clock position. The marking II was used by Podolsk. The Tula Cartridge Works used T until replaced by the number 539 during the Second World War, which was again replaced by a triangle in 1945–46. Around the time of the onset of World War II, the marking practices were somewhat standardized to a numerically based code system, with each munitions plant being assigned a particular number, although some confusion was created by the forced relocations of entire factories. The old factory numbers were retained when factory relocations took place during the German invasion, although they were established in new locations and perhaps under different names.

Numerous numeric codes have been observed and some identified: 3 (Ulyanovsk), 3B (Ulyanovsk, possibly an ancillary plant), 7 (Vympel), 10 (identity not confirmed), 17

(Podolsk initially but also used by Barnaul*), 38 (Yuryuzan), 44 (identity not confirmed), 46 (Sverdlovsk), 50 (possibly Penza), 54 (Nytva), 58 (identity not confirmed), 60 (Frunze Machine Tool Plant), 179 (Novosibirsk),[†] 184 (identity not confirmed), 187 (Tula), 188 (Novosibirsk or Klimovsk), 270 (Lugansk), 304 (Kunceskij), 528 (identity not confirmed), 529 (New Lyalya), 531 (identity not confirmed), 539 (Tula Cartridge Works), 540 (Irkutsk), 541 (Chelyabinsk), 543 (Kazan), 544 (Glazov), 545 (Chkalov), 547 (identity not confirmed), 611 (identity not confirmed), 710 (Podolsk), and 711 (Klimovsk).

From 1949 through 1957, a series of letters were used in lieu of numerical dates, stamped in the 6 o'clock position. These alphabetic values have been confused to mean a new series of manufacturer codes; however, it seems that they actually represent a year of production: А 1949, Б 1950, В 1951, Г 1952, Д 1953, Е 1954, И 1955, К 1956, and 1957 Л. A star or stars may be observed in the 9 and/or 3 o'clock positions, especially in cartridges produced between 1933 and 1940 (Watson III 1984).

The 7.62×25mm cartridges manufactured before 1942 were not head stamped because they solely originated from the Tula Cartridge Works. The 7.62×25mm cartridges that have been observed with plant number and dates were likely produced when manufacturing was expanded to other facilities in the later stages of World War II or immediately postwar. It is highly unlikely that wartime ammunition of this caliber would be encountered outside of collector circles.

The Russian Federation

Within the Russian Federation are producers that actively market cartridges around the world. These companies manufacture ammunition to military and commercial specifications. As with all military cartridges, within a specific caliber there is a variety of cartridges for particular applications. Commercial ammunition is either the ball- or hollow-point-type projectiles that are marked "hunting cartridges." The manufacturers have expanded their product lines to include many popular handgun- and long-gun-caliber cartridges, and even shot shells and ammunition-loading components, mostly primers. Center-fire and rim-fire cartridges manufactured by Russian munitions producers have not changed the materials used in the construction of the traditional military-caliber cartridges, steel casings, and copper-jacketed projectiles.

Ammunition, along with weapons, was currency that the Soviet Union used as foreign policy instruments. This fact, coupled with the popularity of Soviet-sourced weapons, the AK rifle in particular, ensures that Russian-sourced ammunition is likely to be encountered anywhere in the world where weapons that chamber their standard calibers are in use. Russian-sourced cartridges started appearing in the United States in the mid-1990s as trade restrictions were relaxed between the nations. The timing could not have been better. Imports of Chinese AK- and SKS-type rifles and their inexpensive ammunition had been cut off in the United States, and the Russian exports arrived just in time to fill the void.

---

* Additional historical information on this will appear in the material about Barnaul in the following section.
† There is controversy over whether Novosibirsk was code 179 or 188. Likely both codes were used, initially 179 but later 188, when Novosibirsk remained in service after the war and was independent of technical assistance from Klimovsk.

Presently, the Russian ammunition manufacturers hold a large market share, and there has been considerable movement within the industry in terms of collaborations and dissolutions of partnerships. The Russian-based firms presently enjoy a large market presence, and until relatively recently, it could be a challenge to discern the product of one manufacturer versus another when relying strictly upon brand names and packaging alone. There likely was a lot of collaboration between the Russian firms in the infancy of the capitalist experience in the 1990s, and this remains true in certain instances. Inasmuch as the Russian firms have entered into the commercial ammunition sector, especially in North America, they remain firmly committed to meeting the demands for the military markets as well. The proliferation of Russian-sourced weapons around the world ensures a steady market demand for the ammunition used by the arms, even in areas where localized production of ammunition is taking place, as the local product may not be to the quality or the quantity desired.

Federal State Enterprise Production's Amursk Cartridge Plant "Vympel"

The Federal State Enterprise Production's Amursk Cartridge Plant "Vympel" is located in the far east of the Russian Federation. Vympel was a latecomer to the Soviet military-industrial complex, coming online in 1982. Initially, Vympel was set up to exclusively manufacture the 5.45×39mm cartridge, but as soon as the plant realized their production goals, the Soviet Union collapsed. As with other Russian firms, Vympel entered the international ammunition market; it sells commercially in the United States under the name Golden Tiger. Vympel expanded its production line to include hunting/sporting cartridges in addition to military-specification cartridges. Production includes not only the standard Russian 7.62×54mmR, 7.62×39mm, 5.45×39mm, and 9×18mm Makarov, but also 9×19mm, 5.56×45mm, .45 ACP, .308 Winchester, .30 Carbine, and shot shells (Golden Tiger Ammunition n.d.). Soviet-era cartridges originating from Vympel were head stamped "7" but now appear with the Vympel trademark: the Cyrillic B; however, cartridges marked 7 may still be encountered, as surplus Soviet-era ammunition is still commonly found. Civilian ammunition includes both full-metal-jacketed ball and hollow-point projectiles.

Joint Stock Company's Barnaul Machine Tool Plant

According to information from the company website, Joint Stock Company's Barnaul Machine Tool Plant produced half of the ammunition used by the Russian army during the Second World War. Located in Barnaul, Russia, the ammunition manufacturing division of Barnaul is now officially called the Barnaul Cartridge Plant CJSC (Closed Joint Stock Company) and continues to manufacture cartridges for both military clients and the commercial sector (Barnaul Cartridge Plant n.d.).

Barnaul cartridges bore the Podolsk head stamp "17" from 1941 to 1942 and "17a" from 1942 to 1948. It then reverted to 17 in 1948 and remained so until 1990. Soviet military surplus 5.45×39mm cartridges bearing 1980s production dates and factory number 17 have been observed in the United States and are available by retail. During the Second World War, machinery from Podolsk, Lugansk, and Moscow was transferred to Barnaul to escape the advancing German army. Contemporary Barnaul production has been stamped with either the company logo or the Cyrillic БПЗ, somewhat resembling the English letters bp3, but more stylized. It would appear that Barnaul has transitioned (or is in the process of transitioning) to the БПЗ and phasing the company logo out, which somewhat resembles the logo used by Microsoft for Windows.

Barnaul began exporting ammunition to the United States and Europe in the mid-1990s. Military-specification ammunition sold by Barnaul uses a lacquered steel case. On commercial ammunition, Barnaul has plated-steel casings using nickel, zinc, and brass. Barnaul also uses a polymer as a weather-resistant coating. In the years immediately following the collapse of the Soviet Union, Barnaul may have collaborated with other manufacturers to enter into the commercial market. Barnaul-produced cartridges have been sold under the names Golden Bear, Silver Bear, and Brown Bear, which are differentiated by the color of the casing. Brown Bear uses a lacquer-coated steel casing; Silver Bear uses a zinc-plated casing; and Golden Bear uses a brass-encased steel casing. This marketing scheme would appear to have been inspired by three breeds of bear in Russia. Cartridges sold as Silver Bear have been observed with the Ulyanovsk head stamp; it is thus likely that there were collaborations at various times between the manufacturers to fill orders. This anomaly is probably not singular, and other exceptions could surface. Novosibirsk LVE may also have been a partner in enterprise with Barnaul. Aside from the Bear lines, Barnaul sold cartridges branded Monarch and under the name RAM (Russian Ammunition Manufacturers). Barnaul may be phasing out the Bear lines in favor of segregating their cartridges by the type of finish used on the casing material: lacquered, zinc-plated, brass-plated, or polymer-coated steel.

Barnaul manufactures the standard Russian calibers, but has expanded their line to include 5.56×45mm, .243, 7.62×51mm, .30-06, and 9×19mm as well as 12-gauge and .410 shot shells. Full and semijacketed projectiles are used in ball and hollow-point projectiles. Barnaul manufactured a semijacketed ball projectile as well. Barnaul head stamps typically reflect the trademark or Cyrillic characters and the caliber; however, a three-element design has been observed that bears a 0. The most recent addition to the lineup is the Barnaul Centaur, which uses a jacketed projectile supplied to them by the Hornady Manufacturing Company.

Joint Stock Company Tula Cartridge Works

Joint Stock Company's Tula Cartridge Works, located in the city of Tula in the Russian Federation, is one of the oldest continuously producing arsenals in the world. According to the company, Tula was founded by Russian Emperor Alexander II in 1880. Tula Cartridge Works was often abbreviated as TCW in the literature and on cartridge containers. The company also reports that up until the late 1930s, brass was the preferred cartridge casing material, but the practice was abandoned in favor of using bimetal (brass or copper over steel) casings.

During the Second World War, Tula was forced to relocate industrial machinery to escape the German invasion as it approached Moscow in late 1941. Much of the machinery was relocated to the city of Yuryuzan, some 850 miles eastward. This movement may explain the appearance of cartridges attributed to production there bearing the head stamp "38," as there likely was leftover industrial capacity that was never transferred back to the original factory site. Tula-produced cartridges have been identified by several head stamps, including T, 539, TPZ, and TCW.

In early 2010, TulAmmo came onto the commercial market as a new brand name. According to the company's website, TulAmmo USA is a partnership between Tula and the Ulyanovsk Cartridge Works. Retail packaging may be labeled with one or both names as the ammunition source. Tula distanced itself from Wolf brand ammunition, stating that "the Tula and Ulyanovsk Cartridge Works no longer support or produce any of the 'Wolf' brands of ammunition" (TulAmmoUSA 2010). This statement at least implies that Tula and Ulyanovsk may have had a business relationship as a supplier to the Wolf brand. Tula produced vanity brand cartridges in the 1990s, such as B-West, and Tula became a major exporter of Russian firearms to the United States at that time. B-West was the U.S. importing agent.

TulAmmo presently manufactures pistol calibers .380 ACP, 9×18mm Makarov, 9×19mm, .40 S&W, and .45 ACP. The rifle cartridges offered are the .223 Remington, 5.45×39mm, 7.62×39mm, .30 Carbine, .308 Winchester, and the 7.62×54mmR. Blanks are available in 7.62×39mm and 5.45×39mm and are identified by the "star" crimped casing neck. Bimetal and steel are used to produce casings, and the steel casings are coated with polymer. Large quantities of Tula-produced 5.45×39mm Soviet-era surplus ammunition have been imported into the United States in the previously described sealed tins. These cartridges are head stamped with the 539 factory code and were made in the 1970s. The projectile is the 7n6 type, having a hardened steel core.

Klimovsk Specialized Ammunition Plant

The Closed Joint Stock Company's Klimovsk Specialized Ammunition Plant, formally known as the Joint Stock Company Klimovsk Stamping Plant, produces commercial, law enforcement, and military ammunition in addition to a line of air rifles and other firearms. Originally, the Klimovsk factory code was 188, but was later changed to 711. The code sharing with the Novosibirsk plant was likely due to the fact that they were once located in the same city, Podolsk. During the Second World War, plant machinery was relocated east to Novosibirsk, where a new plant was formed (later to become the Novosibirsk Low Voltage Equipment Plant) using the head stamp 188; however, Western intelligence reports have cited Klimovsk as factory number 711 (Watson III 1984). Presently, Klimovsk head stamps their casings with their commercial logo, which resembles a teardrop shape containing a circle that appears superimposed onto a square.

Klimovsk was among the first Russian manufacturers to produce ammunition that reached the United States, circa 1995. Klimovsk produces one of the most extensive ammunition lines in the Russian Federation, which includes some of the more esoteric Russian-originated cartridges such as the 9×39mm, the 5.66mm underwater cartridge, and the 9mm Traumatic cartridge. Klimovsk 7.62×39mm cartridges have either steel- or nickel-plated casings, probably the only nickel-plated casings produced in the caliber. Klimovsk produces several brands of .22 ammunition (the Russians use the metric 5.6mm designation): Record, Temp, Standard, Standard-L, Match, Biathlon, Okhotnik, and Temp-PU (.22 short cartridge) (KSPZ 2011).

Novosibirsk Low Voltage Equipment Plant

The Novosibirsk Low Voltage Equipment Plant began cartridge production during the Second World War. Its output centered on the 7.62×54mmR as well as the 12.7×108mm cartridges, which it still produces. Since that time, it has expanded its line to include the 9×19mm, .380 ACP, 7.62×51mm, and .30-06 Springfield. The company advertises its products for commercial, military, and law enforcement customers.

Unlike other Soviet state plants, Novosibirsk used a two-element head-stamp design: the factory number 188 in the 12 o'clock position, with the year of production in the 6 o'clock position. This marking is to be expected on older surplus ammunition; however, as the company continues to mark military ammunition in this fashion, the production year must be viewed to ascertain when the cartridge was produced. Factory number 179 appears on wartime Novosibirsk cartridges. Commercial ammunition currently bears a three-element design with the head stamp LVE (Low Voltage Equipment), the year of manufacture, as well as the caliber.

Novosibirsk has marketed assorted small-arms cartridges under various names, including Kypu (observed on .380 ACP ammunition) as well as the labels Sobol, Surok, Kosach, Junior, and Rubezh. A press release on the company website (January 26, 2009) announced that these cartridge lines have been discontinued. The Sobol, Surok, and Junior lines were .22 rim-fire cartridges. The Junior brand was head stamped with V. The boxes were thin cardboard and had an animal resembling a rooster on the lid. These cartridges were steel cased and loaded with the customary bare-lead projectiles. The Junior label has been associated with Klimovsk by various sources; however, LVE claimed the line as theirs while—in the same press release—also indicating that it has been discontinued (Joint-Stock 2009).

Ulyanovsk Machinery Plant

The State Unitary Enterprise Production Association's Ulyanovsk Machinery Plant is located in the city of Ulyanovsk in the Russian Federation. The plant began operations in 1917. For head-stamp purposes, the plant was recognized as Factory 3, but more contemporary cartridges are identified by the plant trademark: two arrows pointing in opposite directions to form a circle.

In the early 1990s, Ulyanovsk marketed its own ammunition commercially in the United States under the name Sapsen, but this was apparently short-lived. This may have been a result of a partnership with Barnaul on the Bear cartridge lines, as Ulyanovsk-marked cartridges have been observed in Bear-marked packaging. In 2010, cartridges appeared on the U.S. market packaged with the Ulyanovsk trademark exclusively. In addition to producing 7.62×39mm cartridges, Ulyanovsk has also produced 9×19mm, .40 S&W, and possibly other calibers. Like ammunition from other Russian producers, Ulyanovsk ammunition was sold around the world and likely can be encountered anywhere. Tula Cartridge Works acquired Ulyanovsk some time in the mid-2000s and has capitalized on their capacity to market ammunition under the TulAmmo name.

Wolf Performance Ammunition

In the United States, Wolf is likely the most recognized Russian ammunition company. There is some dispute as to Wolf's status as a manufacturer, and various sources disagree as to whether Wolf is a brand name underneath a manufacturer or is an industrial conglomerate that brings together products from various sources and markets them under the Wolf name. In an undated memo to consumers published on its website, Wolf sought to separate itself from other Russian Federation brands and stated that cartridges head stamped by Wolf have been sold under the brand names Herter's, TulAmmo, and Ulyanovsk, but that these brands were not Wolf ammunition (Wolf Performance Ammunition 2011). A statement from TulAmmo mirrors the claims by Wolf that there is no collaboration between the two companies. Ammunition packaged as Wolf is head stamped WOLF; however, there are numerous variations in the fonts used to mark the head stamp. Wolf markets several different product lines that include using brass, bimetal, and steel casings. In the instance of steel casings, there are inconsistencies in the casing material colors, ranging in color from grey to brown to green. A hint at outsourcing by Wolf are the .22 cartridges head stamped with a crosshair that is sourced to the German firm SK Jagd und

Sportsmunitions (part of the Nammo Group), although marketed as Wolf. Wolf cartridges are available in all popular handgun and long-gun calibers, shot shells, and even primers. Wolf is sold under several lines: Wolf Military Classic, Wolf Performance Ammunition, and Wolf Gold.

As is the case with nearly all commercially available ammunition, the cartridges that have their origin in Russia are produced in other nations as well. In Germany, 5.45×39mm cartridges (sometimes referenced as .215 caliber) have been produced by Dynamit Nobel. Prior to the appearance of Eastern Bloc calibers, particularly 7.62×39mm, there was very little interest on the part of Western manufacturers to produce these calibers, principally because of a relative lack of firearms that chambered them. Winchester and Federal cartridges were probably the only companies that produced 7.62×39mm cartridges in the United States, and they were produced to Western standards using brass casings and copper-jacketed projectiles. By comparison, these cartridges were expensive, and the cost of the ammunition is certain to have impacted the market in such a way that very few manufacturers would bring such a firearm to market. The few examples of American-produced firearms chambered in 7.62×39mm were most notably the Ruger Mini-30 and the Colt AR Sporter. The Ruger Mini-30 is exactly the same as the Ruger Mini-14 model, save for the difference in caliber. The Colt AR Sporter was built off the venerable AR platform and overtly resembles any other AR weapon. However, it was the influx of relatively inexpensive Chinese imports of arms chambered for the round, as well as inexpensive ammunition, that focused American consumer interest.

The other Warsaw Pact nations produced the Soviet-standard cartridges during the Cold War, as they were obliged to "toe the line" and standardize their arms with Soviet practice. As was the case in all matters of industry under Soviet influence, whatever industrial capacity that existed was utilized; however, since these industries largely became state-controlled interests and there was little, if any, room for private property or allusions to such, facilities were given numbers to identify them.

Other Eastern European Sources

Bulgaria

Arsenal Joint Stock Company, located in Kazanlak, Bulgaria, was founded in 1878. Originally, Arsenal was located in the town of Rousse, but it relocated to Kazanlak in 1924. The company was formerly known as the Durjava Voenna Fabrika (D.V.F.) State Military Works Kazanlak and solely served the requirements of the Bulgarian Army. In the early 1950s D.V.F. was assigned factory number 10, surrounded by two segmented circles. These circles do not appear on the cartridge head stamps, but on the cartridge packaging (see Figure 3.8). The 10 almost always appeared at the 12 o'clock position, with the two-digit year in the 6 o'clock position. Between 1977 and 1989, the company was renamed as F. Engels Machine Building Plant (Arsenal JSCo n.d.). Arsenal currently markets cartridges in .22 (5.6×15mm), 5.45×39mm, 5.56×45mm, 7.62×39mm, 7.62×51mm, 9×18mm, 9×19mm, and 12-gauge shot shells. Steel, brass, and bimetal (brass plating over steel) are used to construct casings, and shot shells are typically formed of plastic. Common to Eastern Bloc practice, red or green sealant is present about the casing neck and primer opening. The rifle cartridges are offered in soft (lead) or hard (steel) cores, tracer, and blank variants.

Figure 3.8 A sealed tin containing 440 7.62×54mm cartridges produced by Arsenal Bulgaria. The twin circles surrounding the number 10 indicate the manufacturer; the second number in the sequence, left of the manufacturer identification, is the production year. The caliber is identified at the top and the number of pieces on the bottom. Soviet-era military ammunition is similarly packaged and marked. (Image from author's collection.)

Bulgarian surplus 7.62×54mmR cartridges are especially prevalent on the surplus market, often sold in 440-round brown tins (see Figure 3.8). Other Bulgarian markings include the Cyrillic Д, used in the immediate postwar years, coupled with the date and letter B. Prewar, a lion, date, and B were used. In both cases, the head stamp was divided by lines into four elements.

Czechoslovakia (Czech Republic and the Slovak Republic)

The former nation of Czechoslovakia is another nation that enjoyed a well-established armaments industry during its history, with the majority of the armaments industry located in Slovakia. In 1992, the nation dissolved itself and split into the Czech Republic and Slovakia (the Slovak Republic). Ceska zbrojovka (often simply called CZ) has been a center of armament production since its founding in 1936. When Germany annexed Czechoslovakia in 1939, they were quick to capitalize on CZ's industrial capacity and assigned ordnance code fnh to the firm. The head stamps Z or (z) have been attributed to CZ in the prewar period, and the postwar code tgf has also been attributed to them. During occupation, Argozet Brunn, also called "Waffenwerke Brunn," in Bystrica, used ordnance code dou. Munitionsfabriken Vlasim was assigned code ak. Nazi-era firearms, ammunition, and other accoutrements bearing these codes still appear. The Prague-based munitions firm Sellier & Bellot dates back to 1825. In 1945, the Czech government nationalized the company. In 1992, the company was denationalized and once again became a privately held company. Sellier & Bellot has used SB, SBP, CB&A, and S&B for its head stamp, but used bxn in the 1950s–60s. In 2009, Sellier & Bellot was acquired by the Brazilian munitions firm Companhia Brasileirade Cartuchos S.A.

Additional codes such as aym, czo, dtp, and zv have been associated with post-1945 Czechoslovakian arms activity; however, their true identities have proven elusive, leading to a great deal of speculation. There are likely other similar codes that have either been confused with wartime German codes or have been lost to history. Povazske Strojarne/

Povazska Bystrica was a state-operated company that used head stamp PS and X. In both cases, the month and year were part of the head stamp, coupled with a Star of David. This has often led to confusion in associating such cartridges with Israeli production.

East Germany

In keeping with Soviet standards and practices, East German munitions producers would head stamp cartridges in a similar fashion, although a subtle variation was identified. The 01 marked in the 12 o'clock position on East German cartridges has been identified as the ammunition lot number, not the factory number. Such cartridges were also marked 71, apparently for the year of manufacture, 1971. Observed factory numbers are 04, 05, and 22. With respect to 04-marked casings, there is a date range of 1951 through 1990. In some instances, the year was stamped inverted. Factory number 05 may also be marked simply as 5; this variation has been reported within the date range of 1962–90.

There may have been upward of 20 different munitions plants producing ammunition at one time or another in East Germany, most likely resurrected wartime factories that fell within the borders of the country after it was partitioned. Not all of these factories may have produced finished products, but may have been subcontractors producing the components such as powder, casings, and primers. East German plants produced the German 7.92×57mm and 7.92×33mm cartridges as well as the Russian 7.62×39mm and possibly 7.62×54mmR and 9×18mm Makarov cartridges. Although marked and packaged as East German, species of 7.92×33mm cartridges may bear wartime German ordnance markings, since it is likely that much of the ammunition, at least initially, was repacked from wartime leftovers, and incomplete components were manufactured into finished cartridges.

Figure 3.9 shows 7.92×57mm cartridges produced before and after the Second World War in East Germany, Yugoslavia, Turkey, Germany, Czechoslovakia, and Denmark.

Figure 3.9 (See color insert.) The 7.92×57mm cartridge remained popular even after the Second World War, and production continued because of its widespread use around the world and the large amount of surplus German arms in circulation. Top row from left: East Germany Factory 04 (1960); Prvi Partizan, Yugoslavia (1955); Kirikkale/Ankara, Turkey (1940); and prewar German production (1934)—note P126 code. Bottom row from left: Two variations of Czechoslovakian production by Povazske Strojarne/Povazska Bystrica (late 1940s); and Ammunitions Arsenalet Denmark (1954), often called the 8mm Mauser. (Image from author's collection.)

Hungary

MFS 2000 Hungarian Ammunition Manufacturing Inc. is currently the sole producer of small-arms ammunition located in Hungary. This company was originally founded in 1952 as Mátravidéki Fémmuvek. Like other nations under the influence of the Soviet Union, it produced cartridges to Soviet specifications, using factory number 21. To distinguish Hungarian factory number 21 from other uses of the numeral (as in the case of the Radom Arsenal in Poland and the CUGIR Arsenal in Romania), the date was stamped right side up, relative to the center of the casing. Unlike typical Eastern Bloc practice, the last two digits of the date were stamped in the 6 o'clock position, the factory code at 12 o'clock. In 1995 the company was privatized as part of a Canadian–Hungarian joint venture, and then called MFS, Ltd.; however, it reverted back to state control in 1998. It would later return to private ownership under the name MFS Inc., and finally it passed into the hands of RUAG Ammotec in 2008. The company currently produces popular handgun and rifle cartridges for both civilian and military customers. The head stamp MFS has been used exclusively on newly minted ammunition. Factory number 21 will be seen on military-type packaging containers of Cold-War-era surplus ammunition. Hungarian 7.62×54mmR ammunition is often encountered, manufactured in the 1970s. The casings are constructed of gray steel and have a red sealant applied around the primer and casing neck.

Femaru-Fegyver-es Gepgyar Reszvenytarsasag, often simply called "Fegyver" and based in Budapest, was the principal supplier of small arms to Hungary. The firm was founded in 1891 and remained in operation until 2004. Their products were marked FEG. Ammunition originating from FEG was head stamped F-GY or by factory number 23. With respect to date and factory number, the stamping practice coincides with those of Mátravidéki Fémmuvek. Due to the historical relationship between Austria and Hungary, historical ammunition may be associated with Austrian firms.

Poland

Poland has long maintained an indigenous arms industry, although it seems that it has served foreign powers as often as the needs of Poland. With the collapse of the Eastern Bloc, Poland began divesting itself of surplus arms and ammunition through various channels, and surplus Polish ammunition and arms are quite common in the United States. There were numerous arms factories in Poland, some predating the Second World War. MESKO was founded in 1923 as the National Ammunition Manufactory (Bumar Amunicja S.A. 2011). Under Soviet influence, MESKO products were identified by a single circle around the number 21, but more recently the stamp MESKO is used. The company is now known as MESKO S.A., part of the Bumar Group, "which consists of 20 manufacturing defense sector companies specializing in munitions, radars, rockets and armour, vehicles, 2 trade companies and 6 foreign entities" (Bumar s.p. z o.o. 2011). At present, the company offers an extensive line of small-arms ammunition produced to both NATO and Russian standards and in calibers specific to either. Ammunition offerings include the NATO 5.56×45mm, 7.62×51mm, and the 9×19mm, and the Russian 7.62×39mm, 7.62×54mmR, and the 9×18mm Makarov. As can be expected, any given cartridge is offered as soft core, hard core, tracer, pressure proof, and blank variants. Both brass and steel are used for casing construction. Reduced-ricochet variants are offered in 9mm Makarov, 9×19mm, 7.62×39mm, and 7.62×51mm. These cartridges are intended for short-range training applications and apparently have a projectile completely or partially constructed of polymer.

Another Polish arsenal is Fabryka Broni Radom. During the Cold War, Radom was identified by a single circle surrounding the number 11. Radom products have a four-element head stamp bearing the circle 11 and three other numerals; the numeral to the left of the factory number is the year of manufacture. The materials and appearance of such cartridges are atypical for Warsaw Pact practice. The Radom name has become generic for the handguns manufactured at the Radom arsenal; however, Radom also manufactured rifles. Cold-War-era firearms originating from Radom will also bear the circle 11 marking. Radom does not appear to manufacture ammunition, and remains solely in the arms business as part of the Bumar Group. Radom does market arms in the United States under the name Pioneer Arms. Their line includes handguns, shotguns, rifles, and combination guns.

Factory code 343 represents the State Arsenal at Krupski Mlyn. Factory number 54 was assigned to the Arsenal de Skarzysko-Kamienna. Factory number 234 is associated with Poland but not specifically identified to a particular firm or facility.

Romania

In 1945, Romania came under the sphere of Soviet influence and produced arms and ammunition pursuant to Soviet requirements. Around 1957, the head stamps 21, 22, and RPR (translated to mean "People's Republic of Romania") appeared. The factory code was positioned at 12 o'clock and the date at 6 o'clock. The head stamps 15, 321, 322, 323, 324, 325, and PA have been attributed to Romania but not identified. Prior to and during the Second World War, cartridges were marked CMC at the 12 o'clock position, caliber at the 6 o'clock position, and the year split across at the 9 and 3 o'clock positions. SC Uzina Mecanica CUGIR, SA, is a Romanian arms and ammunition producer and formerly used the 21 head stamp. CUGIR also represents one of the largest exporters of AK-pattern firearms to the United States using the names Romarm, Romak, WASR, and CUGIR. Commercially packaged ammunition originating from CUGIR is marketed under the name Hot Shot; however, the brand name was also used to market cartridges manufactured by Prvi Partizan in Serbia as well as from South Korea by PMC. Factory number 22 has been observed, but its identity is not known.

Ukraine

The Frunze Machine Tool Plant is now called the Sumy Frunze Machine-Building Science and Production Association, having completed privatization in 2000. According to the company website, Frunze was founded in 1896 and began producing military articles in 1914. During the Second World War, Frunze was dispersed to the cities of Tambov, Chirchik, Chelyabinsk (factory number 541), and Kemerovo. Postwar, Frunze became engaged in petroleum and chemical industries, among other endeavors (JSC Sumy Frunze NPO n.d.). Frunze likely continued ammunition production right up until the collapse of the Soviet Union. For the purposes of identification, Frunze was factory 60. Frunze ammunition in 5.45×39mm and 7.62×54mmR has been observed in the United States. The Lugansk Cartridge Plant is also located in the Ukraine. The company is currently operating under the name Joint Stock Company Lugansk Cartridge Plant. Significant quantities of Russian Cold-War-era 5.45×39mm ammunition that has come to market in the United States was produced by Lugansk in the 1970s, akin to the Tula and Frunze products. These Soviet-era specimens will bear the 270 factory number and the last two digits of the date. Observed commercial examples of Lugansk-produced cartridges marketed under the name Lugammo bear the Cyrillic head stamp Л. Calibers in production are the 7.62×39mm Soviet M43, the 5.45×39mm Soviet M74, and the 9×18mm Makarov cartridges.

Yugoslavia

Yugoslavia as a nation disintegrated amid conflict in 1990 when the newly formed Republic of Slovenia seceded from Yugoslavia, followed by the formation of three more independent states: Croatia in 1991, then Bosnia, and finally Macedonia in 1992. When the Dayton Peace Accords were signed in 1994 and peace was restored, the former Yugoslavia had broken into six independent states: Bosnia-Herzegovina, Croatia, Macedonia, Montenegro, Serbia, and Slovenia.

Munitions production in Yugoslavia resumed in 1948, having halted at the end of hostilities in 1945, and throughout the 1950s, Soviet marking practices were observed, using factory numbers 11, 12, and 14. The factory number appears at the 12 o'clock position, two-digit year at 6 o'clock, and stars stamped at 9 and 3 o'clock positions. This practice was replaced by using two-, three-, or four-digit abbreviations to designate the manufacturer. These abbreviations can appear as either Roman characters or in Cyrillic. EIGN, ATZ, and PG are known Yugoslavian manufacturers, but their identities have not been firmly established.

Founded in 1950, P.D. Igman d.d. is a munitions manufacturer located in Konjic, Bosnia-Herzegovina. The company claims to be one of the largest European producers of ammunition ranging from the 5.56×45mm to the 12.7mm and has "exported its products in more than 60 countries on all continents" (PD Igman d.d. 2005). Unlike other Eastern European nations, Igman-produced ammunition was made using brass casings, including 7.62×39mm Soviet M43 (but identified as M67), 7.92×57mm Mauser, and .50 BMG. These cartridges resemble Western-standard ammunition; however, they can be readily identified by their head stamp. Igman head stamps ИК and IK are observed and are combined with various combinations of caliber and two-digit date. Igman produced ammunition under contract for other brand names, in addition to Igman-labeled ammunition.

Prvi Partizan is a munitions manufacturer based in the town of Uzice, in southwestern Serbia. The firm was founded in 1928 as the Arms and Ammunition Factory Uzice, abbreviated as FOMU (Prvi Partizan 2009). Prvi Partizan translated from Serbian literally means "first partisan," and it currently produces one of the widest arrays of cartridges of any company in the world for military and commercial sectors. The company has invested heavily in older and even archaic European cartridges such as .303 British, 7.5×54mm French, and 7.92×33 kurz, but not to the exclusion of more contemporary cartridges with broader market appeal. The current head stamp is PPU; previously, it was the Serbian Cyrillic equivalent ППУ (see Figure 3.10) or Roman characters PP. The head stamp PP-YU is associated with Prvi Partizan and likely means "Prvi Partizan Yugoslavia," which would date specimens with this mark to having been made before 1990. Like Igman, Prvi Partizan manufactured cartridges that were sold under different brand names. Hansen Cartridge Company was a brand name in the United States but was actually manufactured by Prvi Partizan. The head stamp was HCC. During the Cold War, Prvi Partizan was called factory number 11.

Figure 3.10 (See color insert.) Ammunition under investigation in Baghdad, Iraq, in 2006. Closer inspection of the casing head stamp reveals ППУ for Prvi Partizan; the year 2002 is also marked. Other identifying characteristics to look at include the brass casing and the distinctively red primer sealant, a color unique to Prvi Partizan. (U.S. Air Force image; photograph by Technical Sgt. Adrian Cadiz.)

A smaller, less well-known name in the Yugoslavian munitions industry was Valor. The company appears to have only manufactured .22 cartridges. The letter V was used as the head stamp. In Slovenia, Arex d.o.o. manufactures plastic and metal/plastic practice ammunition in any NATO caliber up to 12.7×99mm. What is unique about this product is the use of plastic links for belt-fed weapons (Arex d.o.o. 2011). Coal Ltd., another Slovenian firm, exclusively manufactures air-gun pellets in the 4.5mm and 5.5mm class (Coal Ltd. 1996).

Firearms

4

Introduction

The variety of firearms is nearly endless, bound only by the imaginations of the designers and engineers who bring concepts and ideas to fruition. The art and science of gun making has changed dramatically over the course of the history of firearms. Firearms predating the Industrial Revolution were, for all practical purposes, handmade and hand fitted by artisan craftsmen. With the development of mass production and interchangeable parts, gun making changed dramatically, but there was still a substantial amount of handwork that went into producing the final product. Guns were made slowly and expensively of high-grade materials; parts were machined and milled from steel blanks. Wood was the choice for the *furniture*, another term used to describe the stock, grips, and hand guard. The finishing process was bluing, a form of aesthetically pleasing rust, or case hardening, which hardened a softer metal by coating it with a harder material. The fit and finish of even the average firearm exhibited attention to detail and was the mark of the artisans on the finishing line.

The outbreak of the First World War changed gun making. Such was the need for firearms that manufacturing shortcuts had to be found in the interest of economy and expediting the production process; however, the basic paradigm of firearm manufacturing did not substantially change. Firearms were still mostly made of machined parts, still exquisitely finished, and stocks and furniture were still made of wood. High-gloss blued finishes translated to duller, matte blue, but quality remained high. Great advancements had been made in industrial production, and the result was a high-quality product—a harmonious balance between mass production and quality artisanship. During the course of the Second World War, gun making would be forever changed. Once again, the wartime demand for armaments stretched even the mightiest industrial bases. Blued finishes remained the standard, but increasingly were replaced with a newer, faster, and less expensive process: Parkerizing, also called phosphate or bonderizing. Enamels were also used in lieu of other finishing practices, and in extreme instances, parts or entire firearms were left "in the white," completely unfinished. Most guns continued to be manufactured by machining processes, but increasingly parts—and then whole guns—were produced from stamped sheet metal, from investment casting, and forging. The classification of firearms, particularly less expensive handguns, often referenced as *Saturday night specials*, tended to be constructed of parts that were fabricated by investment casting (Figure 4.1). The 1960s and 1970s saw the rise of this type of firearm from foreign and domestic sources as well as hybrids that were pieced together from foreign-sourced parts and assembled in the United States.

By the 1970s, technology had evolved to the point where the use of polymers to manufacture frames or receivers became a viable option to the firearm manufacturer. Up to that time, polymers had been used in firearm construction on a limited basis, mostly for parts

Figure 4.1 (See color insert.) The Raven MP-25. Part of the identification markings are on left side of the slide (pictured); the model and caliber are on the opposite side. The serial number is stamped on the back strap of the grip. The Raven pistols are grouped as one of the Saturday night specials; untold millions were made, and they routinely turn up. (Image from author's collection.)

such as grip panels, hand guards, and other nonmechanical parts. During World War II, Germany made extensive use of the composite Bakelite as well as another composite material called Durofol. In 1959, Remington Arms introduced the Nylon 66, a .22 rifle that featured a composite stock, although the barrel and receiver were of conventional steel construction. In 1970, Heckler & Koch marketed the first polymer-framed handgun, the VP-70. Two versions were available: a standard pistol and a similar pistol capable of being made into a machine gun by attachment of a shoulder stock. The slide was finished in either a black parkerized finish or stainless steel. The VP-70 was a blowback-operated pistol, having a fixed barrel, and was double action only. Firing was accomplished by means of a striker system that was cocked and released strictly by trigger press. A push-button, cross-bar-type safety was available and placed on the trigger guard, more customary for long guns than pistols, but an effective system nonetheless. Although not commercially successful and discontinued in 1989, H&K did set the stage for a new generation of handguns.

The decision to transition to polymer frames, while controversial at the time, was a logical one given the fact that polymer was proven to be an effective alternative to metal. Polymer is lightweight, corrosion proof, and durable. Economics cannot be discounted as a factor, either. By the 1980s, many firearms that had been manufactured by traditional machine operations were being redesigned to be manufactured by CNC (computer numerical control) machinery. In general terms, it could take approximately 20 hours of traditional machine operations, typically involving skilled human machinists every step of the way, to manufacture a firearm frame. The manufacturing time was reduced to some 6 hours using CNC machinery. A polymer frame can be molded in a matter of minutes. The economic incentive is obvious when these numbers are considered. By the mid-1990s, most handgun manufacturers were producing polymer-framed firearms, and many other gun makers opened their doors as the economic viability of entering the pistol market became more realistic.

In its most basic form, the firearm can be considered a machine. A machine completes a series of operations to accomplish work; it is the work of a firearm to expel a projectile under force of an explosive. Like all machines, there are seemingly endless varieties of approaches that have been explored to build a "better," "improved," or "safer" firearm, leading to a multitude of different methods of function and operation. Firearms produced by different firms may bear similarities or may even be overt copies of another design, but even in similar designs there is likely to be some variation in parts design or purpose to avoid patent infringements. A firearm is functioned by the press of a trigger. The movement of the trigger initiates a sequence of movements of connected parts that can be called a *firing train*. Typically, the trigger acts against a part called the *sear*. Sears take on different physical appearances, and there may be other conjoined components inserted, such as a trigger bar, which may itself contain or act as the sear as a single piece, as opposed to parts that mechanically interact with one another. The sear moves to release a hammer, which then strikes a firing pin that causes the chambered ammunition cartridge to detonate. Firearms without internal or external hammers replace that part of the action with a striker that acts upon the primer of the loaded cartridge. This striker interacts with the trigger and firing train through a mechanical linkage and is typically "cocked" by spring pressure.

The action of the firearm can be prevented through a safety device, which can be effected through numerous methods and approaches. A safety need not necessarily be dependent on the operator to enable or disable it by act of manipulation. In lieu of manually operated safety features, some firearms feature an automatic safety system. Automatic safety systems have been part of firearm design for many years. A common misconception is that a firearm without a manual safety is somehow inherently unsafe, but quite the contrary is true. As long as the automatic safety has not been compromised by modifications or alterations to the firearm, it provides a positive safety feature that is not dependent upon the operator to remember whether or not it is enabled. And from a technical point of view, these "passive" safety features are at least as effective as "active" safety features when considered individually or as part of a more comprehensive set of safety features in a particular design.

Firearms have been independently designed in many nations throughout the world. The names associated with the classic designs are evidence of the diversity of thought and philosophical approaches to design that have been undertaken. The names of the designers have become synonymous with the legacy of the firearms they designed—John Browning, John Garand, Gaston Glock, Mikhail Kalashnikov, Paul Mauser, Eugene Stoner, Kijiro Nambu, and Karl Walther, to name but a few. John Browning is considered the most prolific firearm designer of all time; he developed all forms of firearms, ranging from handguns to machine guns, shotguns, and rifles of all types. Regardless of who developed and refined a particular principle of operation, these ideas are borrowed, perhaps slightly modified, and often produced—either with or without the benefit of securing the legal rights to do so. Under the surface, firearms can be so similar that differences between them are just wide enough to avoid a patent dispute. In spite of intellectual property rights, many firearms have been reverse engineered and put into production without the rights ever being secured. As a result, the same firearm could have been built in any number of places in the world, over any given period, and given any number of different names.

United States Domestic Firearm Production for 2011

According to the Bureau of Alcohol, Tobacco, Firearms and Explosives 2011 Interim Annual Firearms Manufacturing and Export Report, licensed American manufacturers produced 6,398,854 firearms during the calendar year. The type and number manufactured were reported as follows (BATFE 2011):

2,487,786 pistols
572,798 revolvers
2,293,247 rifles
862,293 shotguns
182,730 miscellaneous firearms[*]

According to this same report, 291,342 firearms were exported, comprising 116,014 pistols, 23,221 revolvers, 78,765 rifles, 54,878 shotguns, and 18,464 miscellaneous firearms. The 2011 interim numbers are well in excess of industry activity reported in 2010, when 5,459,240 firearms were manufactured. 2010 represented a very slight decline from 2009, when 5,555,818 firearms were manufactured. In the time period studied in this report, 1986 through 2010, there were only four years where annual firearm production totals exceeded 5,000,000. Those years were 2009, 2010, and 1993 and 1994, where manufacturers reported 5,055,637 and 5,173,217 firearms manufactured respectively (Bureau of Alcohol, Tobacco, Firearms, and Explosives, 2012). This spike in manufacturing activity in 1993 and 1994 is attributed by the author to market anticipation of the enacting of the "Public Safety and Recreational Firearms Use Protection Act" portion of the Violent Crime Control and Law Enforcement Act, which took effect September 13, 1994 and was allowed to sunset after 10 years. According to this same report, 2011 firearm imports into the United States totaled 3,252,404; a number not seen since 1993 when 3,043,321 firearms were imported (Bureau of Alcohol, Tobacco, Firearms, and Explosives, 2012).

Firearm Import Trends

In 2011 Brazil was the leading exporter of small arms into the United States. According to the BATFE, 846,619 Brazilian sourced firearms entered the country, consisting of just under 360,000 handguns, 381,000 rifles, and just over 105,000 shotguns.[†] Austria was the second leading point of origin, with a total of almost 523,000 firearms exported, the vast majority of which, some 515,000, were pistols. According to BATFE, in 2011 alone a total of 3,252,404 firearms were exported into the United States from twenty-eight nations (Bureau of Alcohol, Tobacco, Firearms, and Explosives, 2012). In consideration of the supplied data, when these imports are coupled with domestic production (minus exports) a total of 9,359,916 firearms entered the marketplace in the United States in 2011.

Contemporary Trends in the Federal Prosecution of Firearms Offenses

A recent study by Bowling and Frandsen (2010) focusing on federal firearm offenses found that a total of 8,595 persons were charged with federal firearms offenses in fiscal

---

[*] Miscellaneous firearms are defined as any firearm not specifically categorized in any of the firearms categories defined on the ATF Form 5300.11 (Annual Firearms Manufacturing and Exportation Report). Examples of miscellaneous firearms would include pistol grip firearms, starter guns, and firearm frames and receivers (Bureau of Alcohol, Tobacco, Firearms, and Explosives, 2012).
[†] The author has rounded the actual figures.

year 2008. The overwhelming majority of these cases were filed for violations of the provisions of the Gun Control Act (GCA), accounting for 8,320 of the total. Of those, "60% of defendants in cases filed and 61% of defendants in cases closed under the GCA were charged with a violation of subsection 922(g), which makes it unlawful for nine types of prohibited persons to ship, transport, possess, or receive a firearm." A total of 228 cases were filed under violations of the National Firearms Act (NFA). "Almost 58% of NFA cases filed in FY2008 charged a defendant with a violation of subsection 5861(d), which prohibits receipt or possession of an unregistered firearm"* (Bowling and Frandsen 2010). The remaining 47 filed cases were violations of the Arms Export and Control Act, the majority of which falling under subsection 2778(b), "which requires registration and licensing of persons who engage in manufacturing, exporting, or importing of defense articles or services" (Bowling and Frandsen 2010). These numbers represent a decrease from fiscal year 2007, when 8,935 defendants were charged, and well below fiscal year 2006, when 9,617 cases were filed. In consideration of the overall numbers, the proportions of violations remained relatively constant in the period studied.

Firearm Classification Types

There are two broad classifications of firearms: handguns and long guns. The classification is determined by the physical characteristics that embody the device. When a firearm frame or receiver is manufactured, the manufacturer affixes a serial number and determines what type of firearm the finished product is: a handgun, rifle, shotgun, etc. Once this determination is made, that receiver is defined by that determination and is subject to any potential restrictions imposed upon it in terms of its configuration. However, firearms are frequently remanufactured or reconfigured to specifications outside their original design. Such alterations may be simple, obvious moves (such as a changing a stock or rebarreling), or they could be more obscure (such as modifying the action). When examining a firearm, it is important to first ascertain what type of firearm the exhibit originally was, and then work forward to address any potential modifications that may exist, internally or externally. It is quite common for a manufacturer to produce the same basic firearm in a plurality of configurations.

Handguns

Handguns can range from very compact and concealable firearms to large-framed, full-sized models that are less readily concealable but whose design philosophy calls from something other than concealment as a primary attribute. Handguns, as such, are not subject to restrictions on barrel length, either minimum or maximum. Handguns are further grouped by their operating characteristic: revolvers, semiautomatic pistols,[†]

---

[*] For the purposes of the text, a firearm under the National Firearms Act would include machine guns, short-barreled rifles and shotguns, as well as any other weapons, destructive devices, and silencers. These types of firearms must have had the appropriate transfer tax paid and be registered into the National Firearms Registration and Transfer Record.

[†] There are pistols capable of fully automatic fire; they are discussed further under machine guns.

Figure 4.2 High Standard Sentinel R-106 revolver chambered in .22 caliber. (Image from author's collection.)

single-shot breech loaders, and derringers. In the most ordinary sense, handguns are designed to fire the smaller caliber cartridges, but more realistically, handguns are produced to fire nearly any cartridge. This simply necessitates a larger, bulkier handgun capable of handling the higher pressures generated by firing more powerful ammunition. Handguns designed for larger cartridges, including those attributed to long guns, are either pistol versions of long guns or are specialty weapons that are not common to the criminal context.

Revolvers

A revolver is "a projectile weapon, of the pistol type, having a breech-loading chambered cylinder so arranged that the cocking of the hammer or movement of the trigger rotates it and brings the next cartridge in line with the barrel for firing" (ATF 2002). The principal feature of the revolver is how it derives its name—a rotating cylinder contained within the frame of the firearm (see Figure 4.2). Within the cylinder are a series of holes called *charge holes*, where the ammunition is loaded and stored until fired. After firing, the spent casing remains encased in the cylinder until removed.

Firearms capable of multiple shots from a rotating cylinder before requiring reloading appeared as early as the 1500s; however, it was not until the mid-1800s that revolvers, in the form that would be recognizable today, started to appear. The breech-loaded, as opposed to muzzle-loaded, handgun with a rotating cylinder was patented in 1856 by Smith & Wesson, introducing the use of self-contained ammunition cartridges to the handgun arena. It is a common assumption that the capacity of all revolvers is six cartridges, but this is not true. Most revolvers have a capacity of six cartridges, but the capacity can vary from four to nine cartridges. A capacity of five can be expected in small-frame revolvers that are designed for concealed carry; conversely, large-frame revolvers such as the Smith & Wesson X-Frame have a capacity of seven. In smaller caliber revolvers, as in the case of the .22, nine-shot medium- and full-frame revolvers are common. Revolvers are manufactured in small or compact frames, medium frame, and full-sized and even oversized frame sizes. Barrel lengths vary from 2 inches upwards to 13 inches or more, with 4-, 5-, and 6-inch barrel lengths being the most prevalent. Oddball barrel lengths of five and one-half inch or eight and three-eighth inch were also produced.

While Colt and Smith & Wesson are the most recognized American nameplates attached to revolvers, such firms as Charter Arms, Clerke Technicorp, Harrington & Richardson (H&R), Hopkins & Allen, Iver Johnson, Remington, Ruger, and the US Revolver Company round off the industry. Colt preferred to name their revolver models, for example Detective Special, Official Police, Trooper, Commando, Regulation Police, New Service, Navy Model, and Python, to name a few. Smith & Wesson named their models or used a year of introduction until 1957, when they switched to using model numbers, including for automatic pistols when they were introduced. Prior to the switchover to model numbers, Smith & Wesson identified their revolvers to a year and/or a name, such as the Model of the 1905, which was also called the M&P for "Military and Police," a name resurrected recently by Smith & Wesson for a new line of products that includes revolvers, autos, and rifles. In a single instance, a Colt and a Smith & Wesson product were identified by the same model, Model 1917, which was a U.S. martial model number. Smith & Wesson maintains production and has returned to producing desirable revolver models of the past through their custom shop.

Revolvers have traditionally been associated with the United States exclusively; however, this is not the case. It can be said that the most enduring designs originated in the United States, but revolvers were developed, manufactured, and used in other parts of the world as well. Crvena Zastava, Enfield, Herman Weihrauch (sold under the name Arminus), Gamba, Llama, Manurhin, Nagant, Nambu, Miroku, Rohm, Rossi (now part of Taurus), Taurus International, Uberti, and Webley are all names associated with revolvers made in various parts of the world. Revolvers remained the mainstay of American law enforcement officers until the late 1980s, when the gradual transition to pistols began, primarily chambered in 9×19mm, the preferential automatic cartridge of the day. Revolvers continued to serve in law enforcement around the world well past that time, even in places more traditionally associated with using semiautomatic pistols.

Antique Replicas   Antique replica revolvers, generally of Colt, Remington, or Smith & Wesson pattern, are predominantly manufactured by Italian firms such as Beretta, Cimarron Firearms, Euroarms (also known as Armi San Paolo), Fillipietta, and Uberti. These revolvers can be black powder firing muzzle loaders or fire modern self-contained cartridges. In either case, the firearm will be clearly marked to indicate what type of propellant is appropriate. Colt has reintroduced the Single Action Army and New Frontier models as center-fire cartridge revolvers, not black powder guns. These modern examples very closely resemble the originals, including the use of materials such as brass, and are chambered in period-correct calibers. These examples are not ordinarily attributed to criminality, other than theft, but have been re-marked and refashioned to be sold as original pieces to unsuspecting buyers. Figure 1.3 in Chapter 1 depicts such an example.

Direction of Cylinder Rotation   The direction of cylinder rotation of a particular revolver is always of apparent investigative interest. Revolver designs have incorporated both clockwise and counterclockwise rotations, and the rotation is design specific. Without having to work the action of the revolver, there is a visual way to verify the rotation of the cylinder. By looking at either side of the cylinder, there are a series of notches. The notches have a teardrop shape to them; this indicates the direction of cylinder rotation. Teardrops located on top of the notch indicate a cylinder that rotates clockwise. Conversely, teardrops located on the bottom of the notch indicate a counterclockwise rotating cylinder (see Figure 4.3).

Figure 4.3 Taurus Model 856 in stainless steel. Note the teardrop-shaped recesses in the side of the cylinder. The position of these recesses indicates that this cylinder will have a counterclockwise rotation. (Image courtesy of Taurus International Manufacturing, Inc.)

Functionally, there does not appear to be an apparent advantage of one over the other; however, that has not deterred some from suggesting that counterclockwise rotation allows the cylinder to work itself loose and lose index with the barrel more quickly.

The cylinder "lock up" is by means of a frame-mounted latch that protrudes as the trigger is pressed, locking the cylinder into place. This piece then recesses, allowing the cylinder to rotate. The precise nomenclature for this particular part varies by manufacturer; it can also be called the *cylinder bolt* or *cylinder catch*. It is plausible that, over time by ordinary wear or as part of faulty design, positive cylinder lock up will be compromised, potentially resulting in a cylinder with excessive rotational play or slack. An easy check for this latch on an unloaded revolver is to attempt to rotate the cylinder in the appropriate direction with the hammer down, then cock the hammer and attempt to rotate it again. If cylinder rotation can be accomplished using hand pressure, a functional issue on that arm has been diagnosed.

On single-action only or double/single action revolvers, when the hammer is cocked, that is to say pulled to the rear, there will be a corresponding retraction of the trigger, thereby pulling some of the "slack" from the trigger travel. The cylinder will also rotate in its intended direction, presenting the next charge hole and the contained cartridge to the hammer for discharge. A double-action trigger press will result in the simultaneous rotation of the cylinder in concert with the travel of the hammer as the trigger travels to the rear. A very competent revolver shooter is able to press the trigger to the edge of the mechanical limits, causing cylinder rotation and the hammer to be suspended in a cocked position that is held in place by the pressure being exerted on the trigger by the shooter. If the trigger were released, the hammer would return to the down position. If the trigger were pressed again, the cylinder would rotate once more, leaving an orphan cartridge because it had not been discharged. This in and of itself may explain an accidental discharge or training accident where this technique is being demonstrated.

Mechanical Failures   If a revolver fails to fire a cartridge when the trigger is pressed, the cause may be a worn or broken firing pin; however, it is more likely to have been caused by insufficient pressure being applied by the firing pin when striking the primer or rim of loaded ammunition. The trigger spring and main spring should be inspected to verify that there have been no "modifications" made to the springs such as shortened springs, nonspecification springs, or flat main springs that have been bent or otherwise modified. These efforts are generally made to lighten the trigger press of the firearm, but manifest themselves as problems later on when the revolver fails to reliably operate.

Aside from mechanical failure, abuse, or wear, the only significant failure issue a revolver may realize is, during ejection, a casing is trapped "under the star." The star refers to the ejector star at the back of the cylinder that acts against the rim to eject by depressing the ejector rod. This type of error is operator induced and caused by the operator incompletely ejecting the loaded casings. In the event of a suspected accidental discharge, a possible scenario that involves revolvers is the possibility of a cartridge catching on the grips. In revolvers equipped with swing-out cylinders and an ejector, the cartridge in the 2 o'clock or 3 o'clock position has a tendency to catch on larger or oversized grip panels (some called magna grips), and this may go unnoticed by the operator. If the revolver were tilted downward, this could cause the cartridge to fall back into the charge hole, reloading a weapon that the operator may assume was unloaded. If a casing is caught under the star, it would not permit the cylinder to close, as the ejector would not be properly seated; however, if the cartridge were to go unnoticed and return to the charge and the ejector star returned to seat, then the gun would be once again loaded.

Accidental and Inadvertent Discharges Involving Revolvers Accidental discharges due to mechanical failure in a modern revolver are highly unlikely. In scenarios where a revolver is reported to have accidentally discharged, the circumstances by which the revolver fired should be extensively documented. Frequently, the claim of the root cause is attributed to a bump or strike of the firearm due to rough terrain, the weapon handler being pushed or shoved, or other situations where the revolver would have been subjected to an outside shock or force that would have prompted the engagement of the firing train to cause discharge. The safety mechanisms in modern revolvers make such scenarios improbable, but not impossible if the revolver has been tampered with or if there is an inherent defect in manufacture. A competent subject-matter expert should be consulted and a thorough examination of the revolver be made. The basic reconstructive study would entail loading the revolver with snap caps or other inert ammunition and attempts made to replicate the scenario as described by the handler when the accidental discharge took place. Furthermore, a trajectory reconstruction at the scene of the event should be attempted to ascertain the direction of the bore at the time of discharge. The handler should be thoroughly debriefed concerning the circumstances of handling and what was happening with the revolver at the time. The questions should include, at a minimum, the following:

Was the revolver holstered at the time of discharge?
Was this a loading or unloading operation?
What was the physical posture and position of the handler at the moment of the event (seated, standing, leaning, etc.)?
Where did the event occur (inside a vehicle, on the shooting range, kitchen, hallway, etc.)?
If the revolver was being handled at the time the discharge occurred, under what reasoning and circumstances was the revolver being handled?
If the revolver was being handled, in what manner? A demonstration may be in order with a prop firearm or another similar revolver.

Quite often, the cause of the accidental discharge of a revolver can only be attributed to handler error, as the only causation of the discharge can be attributed to trigger press. A trigger press could only be made by a finger on the trigger or a foreign object entering the trigger guard and compressing the trigger to cause the discharge to occur. It is often the case that the handler should not have been handling the arm at the time the discharge took place, or handling the arm would have been inappropriate at the time the discharge occurred. The handler may have withdrawn the weapon from the holster and then intended to reholster; however, the handler neglected to remove the finger from within the trigger guard and the finger was compressed into the trigger when the revolver entered the holster, causing the discharge. This scenario is especially made plausible if the handler unexpectedly encounters others who would view an unholstered weapon and question the circumstances, creating a sense of urgency to reholster before others take note. Scenarios as described here have occurred in vehicles, guardhouses, and even restrooms involving individuals under arms.*

Failure to Fire, Ammunition Issues   In the case of a failure to fire, ammunition cannot be eliminated from consideration, and should be explored as a possible cause. The ammunition should be inspected to verify that it appears serviceable, that is, that the ammunition has not deteriorated due to the elements, chemical exposure, and improper storage. If this does not seem to be an issue, the ammunition could be defective due to faulty primers, no powder contained, or other defect in manufacture. An often-overlooked issue that is attributed to the firearm to the exclusion of the cartridge is the hardness of the cartridge primer. If the primer is too hard, then the firing pin may not cause the primer to detonate when impacted. Such was the case during World War II with American-supplied revolvers to the British Commonwealth under the Lend-Lease program. British-specification primers were much harder than their American counterparts; thus American revolvers delivered a strike that was too light to cause detonation. To correct the problem, a slight modification was made by adjusting the mainspring to increase the force of impact on the primer.

While most revolvers are handguns, there are in fact revolver long guns as well. The only significant difference is that the long-gun revolver has an attached shoulder stock and a 16-inch barrel to comply with Bureau of Alcohol, Tobacco, Firearms and Explosives (BATFE) regulations in the United States. The Rossi Circuit Judge is a modern example, and is chambered to fire both .410 shot shells and .45 cartridges (see Figure 4.4). Certain antique revolvers are found with these features, in addition to modern production and

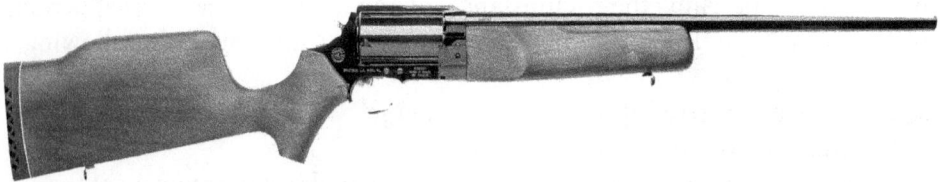

Figure 4.4 The Rossi Circuit Judge is basically a Judge revolver that has been configured into a rifle, having a shoulder stock and 16-inch barrel. (Image courtesy of Rossi Braztech International.)

---

* The term *under arms* is intended to define anyone carrying a firearm in the course of an official capacity and while directly engaged in official duties.

modern reproduction of antique-pattern arms. Revolvers are also popular as derringer-type firearms, which are subclassed as small, easily concealable firearms.

Revolvers with Swing-Out Cylinder   On modern revolvers, the swing-out cylinder is the most commonly encountered variety. The swing-out cylinder emerged quite early in modern design, having appeared in the 1880s. The cylinder is opened by way of a cylinder release latch, also called a thumb piece, which is located on the frame. Some manufacturers use a latch that is pushed to the rear; others slide forward. Cylinder release latches have customarily been located on the left side of the frame, but may be found in the area of the hammer at the top of the frame by the rear sight.

When the cylinder release latch is depressed, it permits the cylinder to be opened and the contents viewed. The cylinder is attached to the frame by the crane, which itself contains an axis pin that the cylinder spins upon. On a swing-out cylinder design, the axis pin serves the dual purpose of acting as an ejector, which allows all cartridges to be ejected by depressing the front of the pin, which is spring loaded and will return to rest when released. The ejector "star" is on the back side of the cylinder and rests within recesses there. These recesses permit the rim of the loaded casing to rest within them, and the ejector acts upon the rim. Cartridges that are rimless or that have rebated rims may load in a cylinder; however, they will not eject using the ejector, as there is no rim for the ejector to act against. As a method for permitting the function and ejection of rimless cartridges within a revolver, a moon clip or half-moon clip would be used. A revolver with a swing-out cylinder may not be equipped with a cylinder release latch. A cylinder may be released by the cylinder axis pin by depressing it. Such a design may be indicated by an axis pin that is not shrouded or covered underneath the barrel. The examiner is cautioned that this, in and of itself, is not an indication of this type of revolver exclusively. The absence of a cylinder release latch obviously eliminates the possibility of using one, and thus the cylinder would be accessible by other means.

Break-Top or Solid-Frame Revolvers   Alternatives to the swing-out cylinder are the break-top-type and solid-frame (or fixed cylinder) revolvers. The break-top revolver has a hinged frame that is opened by a release lever. Most often, the release lever is located on the left side of the frame, where the cylinder release latch would otherwise be located; however, the release mechanism can also be incorporated into the rear sight or otherwise placed at the top of the frame. In such a case, upward and/or rearward pressure would open the release. Opening the action with the lever pivots the front half of the revolver down, revealing the cylinder. Caution must be exercised when doing this because the ejector star is spring loaded and will forcibly eject the contents of the cylinder as the revolver is opened. Opening the action will also release the cylinder and it will spin freely. The British Webley and Enfield family of revolvers are commonly of this type (see Figure 4.5). Other manufacturers have been IOF (India Ordnance Factory) as well as a number of American manufacturers such as Harrington & Richardson (H&R).

Another variation of the solid-frame revolver uses either a loading gate, or the entire cylinder is removed to load and unload. In these revolvers there is no cylinder release lever; loading and unloading is accomplished by two different means, depending upon the design of the particular revolver. In one variation of the solid-frame revolver, there is a loading gate located on the right side of the frame behind the cylinder. The loading gate is a hinged door that is opened to reveal the back of the cylinder (see Figure 4.6). An ejection rod is spring loaded and encased within a shroud beneath the barrel. As the ejection

Figure 4.5 The name Enfield is not singular to the bolt-action rifle. Pictured is an Enfield Number 2, Mark 1 revolver chambered in .38/200 British Service. This is a break-top revolver; the lever to open the action is on the left side of the frame adjacent the hammer. Note the spring-loaded ejector, which opens automatically when the action is opened. (Image from author's collection.)

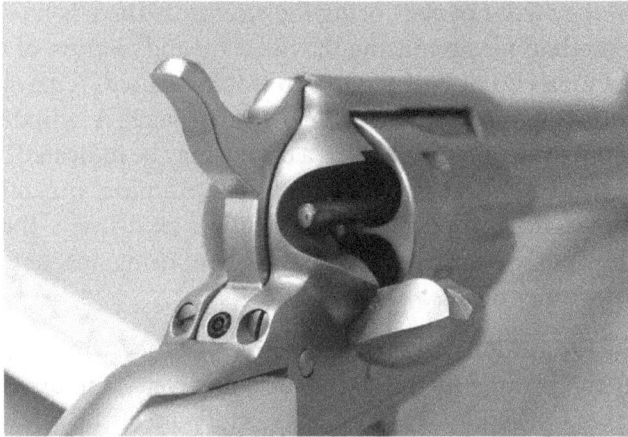

Figure 4.6 Solid frame revolver with spring-loaded ejector rod depressed. The loading gate is open. (Image from author's collection.)

rod is depressed, the contained casing is ejected. Loaded cartridges can then be loaded as empties are ejected. The visual cue is the detonated primer. With some models, it may be necessary to partially cock the hammer to release the cylinder from the trigger, allowing it to spin freely. The hammer may be cocked slightly above its resting position, or employ a one-third cock, where the hammer travel is only about one-third its full travel, or half cocked. When functioning properly, the trigger will not act to permit firing when the firearm is not fully cocked.

Another alternative to the solid-frame revolver differs in that the cylinder axis pin may also serve as an ejector rod, where the cartridges or spent casings can be pushed out of the cylinder charge holes. In this configuration, there is a small appendage to facilitate grip, and the pin is spring-loaded forward so that the pin will rebound into the shroud and not inhibit the operation of the revolver. In yet another variation of the solid-frame revolver, the cylinder axis pin is threaded into the frame and is first unscrewed and then removed from the frame or is released by depressing a spring-loaded release button located in front of the cylinder. Once removed, the entire cylinder is dropped out of the frame, and the pin can be used as an ejector rod to push casings or cartridges out of the charge holes. The cylinder is then simply returned to the frame and the pin reinserted into the frame and through the cylinder and screwed back in to secure it in place.

Revolver Safety Devices The approaches to a safety apparatus on a revolver have differed from those installed on semiautomatic handguns, that is, not having a manually engaged safety device. However, other forms of safety systems have been devised for revolvers. Frederick Felton, an employee of Colt, patented a safety device in 1895 for revolvers known as the Felton Device. The safety device prevents the revolver from discharging when the cylinder is not completely closed and locked in with the frame. The device acts as an intermediary between the trigger and the cylinder release latch, or thumb piece, denying trigger motion unless the latch is in its fully closed position (Felton 1895). As cylinder release latches are spring loaded, the operation is seamless and transparent to the user. The majority of, but not all, revolvers have incorporated the Felton Device or some similar mechanism to the design. Break-top-type revolvers, and more cheaply made examples of revolvers, should not be expected to have such a device installed. For test purposes, the Felton Device can be defeated by engaging and maintaining pressure on the cylinder release latch and then pressing the trigger. As with all mechanical safeties, their function cannot be relied upon absolutely. Barring alteration of the firearm, mechanical safety systems seldom fail. Manufacturer defect by design or material cannot be ruled out, but these are not common either.

Hammer Cock Positions Several different approaches to revolver hammer-position safety and a hammer at rest were devised. A cocked hammer presumably presents the possibility of an accidental discharge if the revolver were dropped or subjected to an outside force that may force the hammer to fall. Regardless of how positive the hammer lockup is, the perception is generally that a cocked weapon is at risk of inadvertent, accidental discharge. Other measures were devised to address the potential for the interaction of the hammer with the loaded cartridge under the hammer when the hammer was in the rest, or down, position.

Single-action revolvers, especially those loaded by a loading gate or removable cylinder, often have a hammer cock position sometimes referred to as the *safety notch*, which is designed to keep the hammer off the firing pin and the firing pin from resting on any loaded cartridges. The actual hammer position may not literally be a quarter pull relative to a fully cocked position; it may be just a few millimeters of travel to keep the hammer off the firing pin; or it may be half cocked. As with all safeties, this feature cannot be considered as a guarantee to prevent discharge due to rough handling or a trigger press. The half-cock position, also known as the *loading position*, is to be treated with equal delicacy. Some single-action revolvers go so far as to have a disclaimer inscribed on them telling the handler not to have a cartridge chambered under the hammer, essentially making a six-shooter a five-shooter.

Rebounding-Hammer Safety A rebounding-hammer safety is not a dedicated safety device per se, although it does serve two basic purposes that, for all practical purposes, can be treated as acting in the interest of safety. The first is to prevent the firing pin from resting directly against the cartridge primer, which potentially could cause accidental discharge if the firing pin were somehow to strike the primer with sufficient pressure to cause detonation. The second purpose is to prevent the firing pin from sticking in place from lack of maintenance, wear, or general deterioration of the firearm that would prevent opening the cylinder should the firing pin freeze in the striking, or extended, position. The rebounding-hammer safety is not exclusive to revolvers, it can be found on other types of firearms as well. Functionally, the rebounding hammer comes to rest in a position that prevents it from directly contacting the firing pin after the firearm has been fired.

Transfer-Bar Safety   The transfer-bar style of safety is literally a lever set within the frame between the hammer and the firing pin. The transfer bar is mechanically linked to the trigger so that as the trigger is pressed, the transfer bar moves into a blocking position between the hammer and firing pin. When the trigger is fully depressed and the trigger "breaks" (meaning the hammer disengages from the trigger and drops to fire the loaded cartridge), the transfer bar, which is fully extended upward, is impacted by the falling hammer and transfers the impact to the firing pin. Had it not been for the intervention of the transfer bar, no detonation would have occurred, since there is insufficient hammer travel for the hammer to directly impact the firing pin; the transfer bar serves as the medium between the two parts to literally transfer the force. As the movement of the transfer bar into the firing position is entirely dependent on the movement of the trigger, the safety will prevent an accidental discharge from taking place if the hammer were somehow subjected to some outside force that caused it to fall. The transfer bar will also prevent accidental discharges if the hammer were struck by force or dropped when it is "at rest" or in the uncocked position. An investigative inspection of any revolver should include checking the hammer for "push off," where considerable force is applied to the cocked hammer that attempts to force the hammer to fall when the trigger is not pressed. The potential for push off can be studied by applying force with the hands, blows with a hammer, and even dropping the revolver from height.

Hammer-Block Safety   There are several subtle variations of the hammer-block safety, but they all practically serve the same function. The hammer block is best described as providing the same level of safety as the transfer bar, yet addressing the question by taking the completely opposite approach. The hammer-block safety, like the transfer bar, is mechanically linked to the trigger. It is designed to prevent the hammer from passing through the frame unless the trigger is fully depressed. The hammer block moves down and out of the way of the hammer as the trigger is depressed, again opposite the transfer-bar system but essentially performing the same function. If pressure is relieved from the trigger, the hammer block returns to its locked (up) position, thereby denying passage of the firing pin through the frame.

The hammer block safety is mostly seen on revolvers whose firing pins are part of the hammer itself, such as is the case with Smith & Wesson revolvers, faithful copies of these revolvers, and most modern double-action Colt revolvers. Colt preferred to call the hammer-block safety the "Positive Safety Lock." This was first introduced on the Police Positive model, hence the name "Positive." The experienced revolver shooter is capable of pressing the trigger to "index" the cylinder. This involves pressing the trigger so that the cylinder rotates and the hammer cocks, but it is suspended by the threshold pressure of the shooter's finger on the trigger that is exerting sufficient force to cause this action to take place, yet insufficient pressure and trigger travel to cause discharge to take place. In such a condition, the shooter could release the trigger, which would allow the hammer to fall as the trigger is released but the hammer block would prevent the discharge from taking place.

Firing-Pin Safety   The firing-pin safety is not a singular-acting safety device per se; rather, it acts as an additional measure of protection against an accidental or unintended discharge. One approach to the firing-pin safety is the inertia firing pin. The firing pin is spring loaded so that it resists movement toward the chamber, and this resistance must be overcome by force of impact from the hammer, typically through a transfer bar. Another approach is to block the firing pin by some type of intermediary, such as the case of the transfer bar or hammer block.

Cylinder Indexing   Over time due to use and wear, and especially if more powerful cartridges are fired, revolvers can begin to lose their indexing. The index is the cylinder alignment with the barrel. If significant movement is noted in the cylinder that allows it to move freely, or if there is a report of a gun "shaving lead," the cylinder is not properly indexing. Shaving lead is caused by the misalignment between the forcing cone of the barrel and the cylinder. Slight misalignment will cause small fragments to shave from the projectile when fired; more significant loss of index can be catastrophic to the shooter and the firearm. A certain amount of slight wiggle is acceptable. The area where the barrel meets the cylinder within the frame is called the *forcing cone*, which is conically shaped to account for very slight variations in the cylinder index relative the barrel even when normal tolerances exist.

Single- and Double-Action Revolvers   Revolvers can be capable of single action only, double action only, or both single and double action. Double-action-only revolvers are not capable of firing in a single-action condition, and in fact double-action revolvers can typically be discriminated by not having a visible hammer or not having a hammer spur that would permit the hammer to be cocked. Double-action-only revolvers are generally made on smaller frames designed for concealed or discreet carry. The double action, coupled with any of the previously described revolver safety systems, is designed to enhance the safety factor of such revolvers. Revolvers that are capable of single and double action will have at least a portion of the hammer exposed that allows the shooter to manually cock it if desired. Revolvers with exposed hammers capable of single action operation will have a spur for cocking, whereas revolvers with concealed or shrouded hammers will have an exposed appendage, such as the Taurus Model 851 in Figure 4.7. The revolvers in Figures 1.3 and 1.4 have fully exposed hammers to permit single action function. Figures 1.5 and 1.6 depict a revolver in both firing conditions. In contrast, the Enfield revolver in Figure 4.5 has an exposed hammer but lacks the spur to manually cock it, indicating that this revolver is double action only.

Most contemporary revolvers are based on established designs that originated from Colt, Smith & Wesson, or Ruger. The most recent innovations in revolvers have focused on using frames constructed of polymers, compressed powdered metals, and even exotic materials such as scandium and titanium. Taurus International markets polymer-framed revolvers in its Public Defender models. Smith & Wesson's Bodyguard is constructed of a combination of a steel-reinforced polymer lower frame and an aluminum-alloy upper frame. The Ruger LCR is an interesting study in multiple materials, using aerospace-grade aluminum for the .38 Special frames, a stainless steel cylinder, and polymer firing control housing. The .357 version uses stainless steel to construct the frame (Sturm, Ruger, and Co. 2011).

Since their inception, revolvers were chambered for fully rimmed cartridges, as the casing rim supports the cartridge in the cylinder. The classic revolver cartridges include such as the .38 Special, .357 Magnum, .44 Special, .45 Long Colt, and so forth. Smith & Wesson's Governor Model revolver chambers 2½" .410 shot shells, .45 ACP, and .45

Figure 4.7 Taurus Model 851 Ultra Lite is capable of single- or double-action operation. The hammer is shrouded by the frame, but there is a small textured appendage that permits thumb cocking for single-action firing if desired. (Image courtesy of Taurus International Manufacturing, Inc.)

Colt cartridges without any modifications. Preceding the introduction of the Governor, Taurus International released a revolver called the Judge, capable of chambering both .45 class cartridges and .410 shot shells. These revolvers are something of a throwback to handguns that were designed to fire shot shells, although these were breech-loaded pistols and not revolvers. An unusual revolver produced by Smith & Wesson, the Model 547, was chambered for the 9×19mm Luger cartridge. The Model 547 was a K-frame (medium size) revolver with a six-round capacity. Unlike the N-frame Model 1917, the Model 547 did not use moon clips or other attachments to load the cartridges, but featured an intricately redesigned ejector star to accommodate the rimless 9×19mm cartridge. Revolvers are chambered up to .50″ cartridges, often called the "big bore" revolvers. The big-bore race was tempered only by legal restrictions that prohibit firearms from using cartridges greater than one-half-inch bore diameter, or .50 caliber. The .460 and .500 Smith & Wesson Magnum cartridges, .480 Ruger, and the .475 Linebaugh are likely the most well-known examples of the big-bore revolvers outside of the wildcat loads (see Chapter 2 for a discussion of wildcat loads).

Semiautomatic Pistols

The BATFE defines a semiautomatic pistol as "any repeating pistol which utilizes a portion of the energy of a firing cartridge to extract the fired cartridge case and chamber the next round, and which requires a separate pull of the trigger to fire each cartridge" (ATF 2002). Auto-loading, semiautomatic, or breech-loaded handguns are termed *pistols*. Auto-loading or semiautomatic pistols are exactly like semiautomatic long guns, firing one round with each press of the trigger until the ammunition supply is exhausted. An auto-loading pistol is supplied ammunition from a magazine that is either removable or is fixed within the frame or receiver.

German designer Hugo Borchardt devised the first pistol in 1893. Borchardt drew heavily from the work of Hiram Maxim, who had developed the automatic rifle in 1883. Maxim's design used the recoil force generated by the discharge of a cartridge to perpetuate the firing cycle of the firearm. Sir Isaac Newton's Third Law of Motion, as described in the *Mathematical Principles of Natural Philosophy*, states that for every action, there is an opposite and equal reaction. The energy released when the cartridge was detonated could be captured and used to expel the spent casing, chamber a live cartridge, and reset the firing train for a subsequent shot. This idea of recycling the energy made available through discharge paved the way to the first successful, practical machine gun, adopted by many nations under different names but universally known as "The Maxim." Capitalizing on these prior arts, Georg Luger went on to develop his own interpretation of the automatic pistol, itself a timeless design recognized not only by its classic lines, but its ominous reputation. Paul Mauser received a U.S. patent for an automatic pistol on June 15, 1897, having already received patents for the design in Germany in 1895 and other nations between 1895 and 1897. Titled "Recoil Operated Firearm," Mauser stated that:

> the main object of the invention is to provide an improved magazine repeating firearm with a moveable barrel in which recoil caused by the shot is used to unlock and open the breech to eject the empty cartridge-case and to cock the firing mechanism as well as to compress a number of springs arranged in such a manner as to effect the loading of a fresh cartridge, the relocking of the breech and the locking of the bolt, and the advancing movement of the barrel. (Mauser 1897)

John Browning entered the automatic pistol field in 1898 when he presented a design to the Belgian firm Fabrique Nationale (FN); the pistol entered production as the Model 1899. The Model 1900, which was slightly improved, entered production the following year. The FN Model 1900 was the first regular-production semiautomatic pistol to use a reciprocating slide, which is now commonplace in pistols. Browning continued to capitalize with a succession of pistols based on the 1900. Browning's approach of using a reciprocating slide was in contrast to the toggle action favored by Borchardt and Luger and the movable barrel that Mauser had pursued. The action of the toggle resembles that of the second knuckle of the finger: Pulling the knuckle withdraws the end of the finger and mimics the toggle action.

None of the German designers—Borchardt, Mauser, or Luger—had yet incorporated the concept of the reciprocating slide, although the art of pistols was obviously well underway in Germany by that time. German manufacturers largely passed by revolvers, with some notable exceptions such as the Reich's Revolvers, devised in the 1870s and still in regular service as late as World War I, with anecdotal evidence of them appearing in World War II. Mauser entered the revolver field with the 1878 pattern but did not pursue it in favor of developing pistols. Browning designs were favored on both sides of the Atlantic; all pistols manufactured by Colt were of Browning's design, and he enjoyed a long-standing relationship with Fabrique Nationale as well. In retrospect, Colt seemed to prefer Browning's blowback designs, while Fabrique Nationale preferred the breech-locked designs.

John Browning will be best remembered for designing the pistol that would become the Colt Model 1911, inspired by his Model 1900, which was chambered in .38 Rimless Smokeless. The 1911 is considered by many to be the classic pistol of the ages, and is likely the most widely copied pistol of all time. The 1911 was followed by the Browning Hi Power— also called the P35, the HP, and Model 1935—chambered in 9×19mm and first manufactured by Fabrique Nationale (see Figure 4.8). The Hi Power is one of the most popular pistol designs on the planet and, like the 1911, has been widely copied, but certainly not to the extent of the 1911. While firing the smaller 9×19mm cartridge instead of the .45 ACP,

Figure 4.8 Browning Hi Power 9×19mm pistol in military matte black enamel finish. (Image from UK Ministry of Defence; photographer Brian Douglas.)

the Hi Power featured a double-stack magazine,* providing a higher ammunition capacity than the single-stack magazine of the 1911. The basic Hi Power design has been subjected to subtle changes, and there are numerous variations. Browning died before finishing the Hi Power, and the question lingers over whether he would have pursued the locking-bar design that ultimately was used in the Hi Power or remained with the swinging-link design of the Model 1911.

As a historical footnote, the Hi Power was used by Axis and Allied forces during World War II, and was produced in occupied Belgium by FN for German consumption and in Canada by the John Inglis firm. It was also distributed to Chinese and British Commonwealth forces, especially commando units. The British formally took the Hi Power into official military service in 1954, and it has remained in use throughout the world. The Hi Power continues to be available commercially. Most modern pistol designs can attribute at least some design feature or inspiration from the ideas and innovations of John Browning.

Pistol Safety Systems   There are a number of approaches used for safety systems in pistols. The prevalent systems in use, and subtly copied, are outlined in this text, rather than outlining every approach ever used. Historically, the safety has been a manually operated function that is manipulated by a lever, switch, or button (see Figure 4.9). Use of the push-button or cross-beam safety on handguns is unusual but not unheard of. The safety button is located on the trigger guard and follows the practice where such a safety is installed on a long gun. The cross-bar safety locks the trigger out from the rest of the firing train.

Safety levers can be mounted on the frame or slide. Slide-mounted safety levers are ordinarily on the left side; however, ambidextrous levers (a lever on both sides) are available. In consideration of left-handed shooters, many contemporary designs permit the operator to relocate the safety to either side, typically without the assistance of an armorer or specialized tools. Safety levers frequently, but not always, incorporate a decocking feature. On such an equipped firearm, when the safety is depressed, the hammer safely leaves the single-action or cocked position and returns to the rest position. Obviously, if the firearm is

---

* A double-stack magazine design has cartridges stack in two staggered rows within the magazine body, whereas the single-stack design permits a single column of cartridges stacked one on top of the other.

Figure 4.9 A U.S. Military M9 (Beretta 92) pistol. The red dot at the back of the slide indicates that the safety lever is on the "off" position; the weapon is capable of firing. The old mnemonic "red you're dead" always applies. (U.S. Army image.)

capable of both single and double action, the firearm would be made ready to fire when the safety was disengaged, even though the hammer is down or in a double-action condition.

SIG Sauer installs a decocking lever almost universally on its product line. The lever is located just forward of the left grip panel, where a portion of the panel is cut out so that the lever can sit flush with the grip. The lever is spring loaded so that it can return upward once released. SIG did not generally use a slide-mounted safety catch, except in the case of its Mosquito pistol, chambered in .22. Walther's P99AS features a decocker that is flush mounted to the top of the slide and provides a tactile cocking-condition indicator.

The firing-pin-block-type safety prevents the firing pin from traveling forward without a complete trigger press. Most models with a safety lever on the slide use some form of a firing-pin-block safety. For obvious technical reasons, decocking levers are not installed on striker-fired pistols because there is no hammer to decock, as the striker is generally cocked by the action of the slide or by the press of the trigger.

Magazine Safety Another safety feature of pistols, especially prevalent in Smith & Wesson models but in use by others, is the magazine cutoff safety. The Browning Hi Power was the first pistol to incorporate the magazine cutoff safety. The magazine safety will not allow the firearm to discharge unless a magazine is inserted and seated in the magazine well (see Figure 4.10). Mechanically, this system uses a trip or a lever that moves when acted upon by the magazine body, thus permitting normal operation. With the magazine removed, the trip is relaxed and disables the firing train. An engaged magazine safety (magazine removed) disables the trigger, which can be pressed without a discharge occurring. In such a design, the trigger has no resistance at all when pressed and simply returns to its resting point when not being manipulated. Another approach causes the trigger to lock out, and the trigger is prevented from any travel when the safety system is engaged.

Firing-Pin Safety Like revolvers, firing-pin safety systems are employed in automatic pistols as an additional measure of safety that complements the other safety systems built into the design. There are several approaches to firing-pin safety. A firing pin may be physically blocked by a bar or other barrier until the trigger is pressed or the safety is relaxed by whatever means used on that design. A novel approach to firing-pin safety was to push the firing pin out of alignment with the firing-pin channel when the safety lever was engaged,

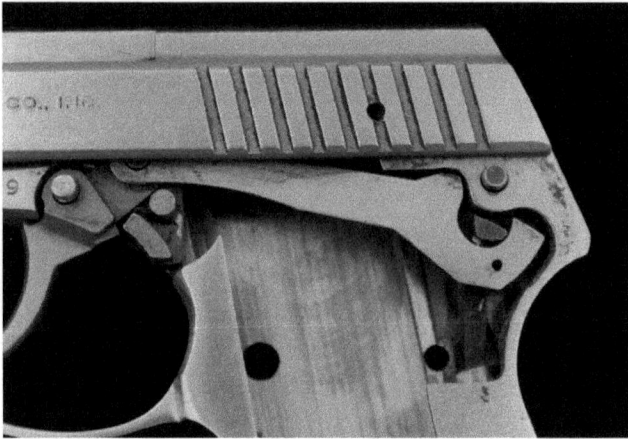

Figure 4.10 Close-up of an L.W. Seekamp pistol, showing the draw bar, which is connected to the trigger and the hammer strut. Just behind the trigger is the magazine safety lever, which disables the firing train when no magazine is inserted. (Image from author's collection.)

as in the case of the Mauser Hsc. Spring-loaded firing pins can resemble revolver-style inertia firing pins, having spring tension that resists movement until overcome by the force of hammer impact. Striker-fired double-action-only pistols generally put the firing pin under spring pressure to cock when the trigger is pressed; otherwise, the firing pin basically floats within the firing-pin channel. The Swartz safety system installed on the Colt 1911 was novel: The firing-pin safety was coupled to the grip safety panel, and had no connection whatsoever to the manual safety. The grip safety panel was connected to the firing-pin safety by means of an actuator that pushed upward when pressed. "This upward movement will slide the actuator upward which in turn will move the firing pin safety lock out of engagement with the firing pin, the pin now being free to be moved to the firing position" (Swartz 1937).

A fixed firing pin is a firing pin that is not a separate component from the bolt or action, but is permanently affixed to the breech face. A fixed firing pin can only be used in a firearm that fires from an open bolt; it is not feasible to have a fixed firing pin firing from a closed bolt. In the United States, open-bolt designs were discontinued per an ATF (Bureau of Alcohol, Tobacco, and Firearms) ruling in 1982, continuing only in machine guns, so the concept of fixed firing pins on semiautomatic weapons passed into extinction.

Loaded-Chamber Indicator   The loaded-chamber indicator is not a new idea, but it has been recently resurrected and has become a common feature, as certain venues will only permit sales of firearms that are so equipped. The indicator is not a safety device in the sense that it is a mechanical intervention to prevent the firearm from discharging. The loaded-chamber indicator is literally an appendage on the firearm that presents itself when a cartridge is loaded in the chamber or when the action is cocked. This is particularly true with striker-fired actions, where the slide has been pulled to the rear and the action is cocked, whether there is a cartridge loaded in the chamber or not.

Perhaps the earliest example of a loaded-chamber indicator was present on the Luger pistol. The Luger has a small tab on top of the slide, forward the toggle, that permitted both visual and tactile indication that the gun was loaded (see Figure 4.11). The small tab had the word *GELADEN* (German, literally meaning "loaded") stamped on the side. Recall that the Luger

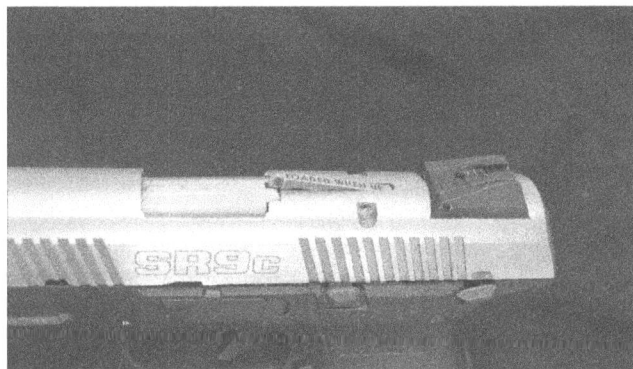

Figure 4.11 The Ruger SR9C features a very conspicuous tactile and visual loaded chamber indicator that is plainly marked. (Image from author's collection.)

was designed in the late 1800s. The Luger's replacement, the Walther-designed P.38, featured a loaded chamber indicator in the form of a small pin that protruded out the back of the slide above the hammer, giving the handler a tactile indication that the gun was loaded, a feature that Walther had already incorporated into its PP and PPK series pistols, with the exception of pistols chambered in .22, which were not so equipped. The Sauer 38h, another pre-World War II German design, featured a loaded-chamber indicator as well as a decocker, although the decocker was omitted in later war production as a matter of manufacturing expediency.

Glock redesigned the extractor on their pistols to include a small raised tab that would present itself outward when a round was chambered. The previous extractor design did not feature a raised tab; however, it did physically protrude from the slide when a cartridge was chambered, and the tab only makes it more pronounced. Taurus International has included a loaded-chamber indicator on all pistols they manufacture, mimicking Glock by incorporating it into the extractor. Ruger has incorporated a loaded-chamber indicator in their SR series handguns. It takes the form of a small tab that elevates from the slide behind the ejection port. The tab has "LOADED WHEN UP" inscribed on it, and is painted bright red on either side. Jimenez Arms, the successor of Bryco Arms, installs a loaded-chamber indicator in the form of a red plastic knob that pushes out to the rear from the slide when the firearm is loaded and the striker is cocked. Bryco Arms had incorporated this feature on its pistols before the company went into bankruptcy in 2003.

Grip Safety   A grip safety consists of a movable pressure plate that is present on the front or back strap of the pistol grip. When the hand is removed from the grip, the grip-safety system engages as the plate extends outward. When a positive grasp by the operator is made onto the grip-safety panel, the grip-safety system disables.

Grip safeties often, but not always, work in concert with other forms of mechanical safety devices. The Colt Model 1903 Vest Pocket pistol featured a grip safety on the grip's back strap and also incorporated a manually engaged safety catch. The safety of the design was further enhanced by the addition of a magazine safety. The 1903 is directly related to the Model 1908; the only difference between them is the caliber, with the 1903 chambered in .32 ACP and the 1908 chambered in .380 ACP. Although the 1903 and 1908 were said to be of hammerless design, they truly were not, as both designs had internal hammers. The Colt Model 1911 and 1911A1 featured a grip safety, also present on the back strap. The 1911 and 1911A1 also had a manual safety lever that worked independently of the grip safety.

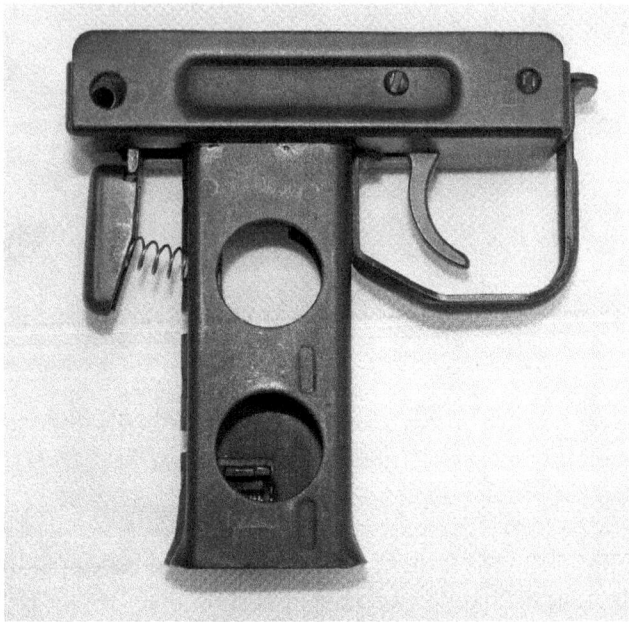

Figure 4.12 Close-up of the grip of the UZI submachine gun. The grip safety is located behind the magazine well and is spring loaded. The grip panels have been removed to show these details. (Image from author's collection.)

Perhaps the most radical application of a grip safety was the H&K P7 pistol. The P7 is a delayed blowback-type handgun that featured a prominent lever on the front strap of the grip. When the pistol was grasped, pressure was applied to the lever, which then cocked the firing pin. Of course firing could not occur until the trigger was pressed, which was quite an interesting technical approach to not use the trigger to apply the force to cock an internal firing pin. This gave the P7 a high degree of safety, and it did not feature additional safety devices, as it would have been redundant to do so. Numerous variations of the P7 were manufactured, including a dedicated training version that was specifically designed to fire plastic projectiles and a variation for special operations that featured a threaded barrel from which to attach a suppressor. It was not only the unique safety system that made the P7 an interesting, albeit complex, design. In the classic design of a blowback, the P7 featured a fixed barrel, which could only have contributed to its reputation for accuracy. The barrel had a polygonal profile, instead of traditional rifling, which was practically unheard of when the P7 came to market in 1979. In addition, the pistol had a fluted chamber, unknown for any handgun made anywhere, but a standard feature of H&K rifles and submachine guns of the day.

The Dutch Madson M/46 submachine gun incorporated a grip safety located at the backside of the magazine well. Apparently, the magazine well was intended to serve double duty as a fore grip, as the M/46 was incapable of firing unless the safety lever was depressed. The UZI, Mini UZI, and Micro UZI also featured a grip safety that was placed on the back strap of the grip, which was an addition to the manually engaged safety on the fire-selector switch (see Figure 4.12). Other firearms inspired by the UZI share the characteristic grip safety. For operators that found grip safeties to be an impediment, the safety plate was often wired or taped shut to provide constant pressure on it, thereby effectively overriding it.

Trigger Safety   A trigger safety operates in much the same way as the grip safety. A small pressure plate is installed on the trigger that prevents movement of the trigger unless the pressure plate is first depressed, which engages the trigger and permits trigger travel. The trigger safety is one of three safety features on all Glock pistols, and has since appeared on other designs as well, including the Springfield XD pistols and some Taurus pistol models. A trigger safety is readily apparent by looking at the trigger and seeing a small appendage on the leading edge of the trigger.

Safe-Action and Similar Double-Action Systems   The introduction of the Glock 17 radically changed the paradigm of firearm design in a number of ways. The Glock-patented "Safe Action" is a three-way safety system whose functions are completely passive; there are no manually operated safeties. The three safeties are sequentially disengaged as the trigger is fully pressed. The three parts of the Safe Action are the trigger safety, firing-pin safety, and drop safety. The trigger safety consists of a small tab on the face of the trigger. If this tab is not first depressed, the trigger will not operate. The firing-pin safety is a physical barrier in the form of a spring-loaded plunger that isolates the firing pin from the breech face. The drop safety is accomplished by the design of the firing-train components, preventing the trigger bar from acting with force against the firing pin in the event the firearm is dropped.

Initially, this approach was somewhat controversial among the purveyors of firearms, and the Glock was often dismissed for failing to have any form of manual safety, which often led to erroneous claims that the pistol simply had no safety device installed whatsoever. Another erroneous claim was that, due to the use of polymer, the Glock would pass undetected through metal detectors, making it easy for someone to smuggle one onboard an aircraft or another controlled area. This claim ignored the basic fact that much of the firearm, specifically the slide and barrel, was constructed of steel. Moreover, many of the internal components are constructed of metal, and there is a metal endoskeleton in the frame underneath the polymer. The controversy has abated, and the industry has come to emulate the Glock. Double-action-only, striker-fired, polymer-framed pistols are now the norm as opposed to the exception.

In 1994, Smith & Wesson responded to the appearance of the Glock with the Sigma Series, which resulted in a lawsuit brought by Glock, who claimed that Smith & Wesson infringed on intellectual property rights. Ultimately, the case was settled out of court with undisclosed terms in 1997. Taurus International Manufacturing, Springfield Armory, Smith & Wesson, Ruger, Kel-Tec, H&K, SIG Arms, Tanfoglio (sold under the name European American Armory in the United States), and a host of others now market pistols similarly configured to the Glock and are often mistaken as Glocks when viewed. The case could also be made that the impact that the Glock has had on the firearms industry has manifested itself in the form of numerous small manufacturers that have entered the marketplace. The market acceptance of polymer-framed pistols has allowed these smaller manufacturers to enter the market with polymer-framed pistols with steel slides and barrels that operate as double-action-only guns, operated either by blowback or by a Browning-type recoil action. There can be no dispute that the main reason that has stimulated this occurrence is the cost effectiveness of a polymer frame compared to one fabricated of metal. The appearance of these styles of firearms from these smaller manufacturers has prompted the larger manufacturers to respond to this segment of the market. The appearance of names such as Diamondback Firearms, Kahr Arms, Kel-Tec, and Cobra Firearms has certainly prompted companies such as Ruger, Sig Sauer, and Smith

& Wesson to respond with their own small-framed, polymer-based firearms. Kahr Arms announced in April 2011 that it had filed suit against Diamondback Firearms over "patent infringement" (Kahr Arms 2011).

Other Pistol Controls

Magazine Release   Another feature found on pistols that does not appear on revolvers is a magazine release. American tradition has called for the magazine-release button to appear on the left side of the frame close to the where the grip meets the trigger guard. Newer pistol designs may permit relocating the magazine release to either side of the frame as a concession to left-handed shooters, similar to the practice for locating the safety or safety/decocker on some models. European practice traditionally has been to place the magazine release on the heel of the grip, typically as a release latch that interacted with the magazine floor plate. When the latch was pulled to the rear, the magazine was released. Variations of these styles do exist. Beretta Model 92 handguns manufactured in the 1970s placed the magazine release button within the left grip panel, but near the heel of the grip. A more contemporary approach has been to re-form part of the trigger guard into a magazine-release paddle that is ambidextrous, eliminating the need to relocate the magazine release to suit the right- or left-handed shooter. The H&K USP series pistols—as well as the Walther PK380, PPS, PPQ, P22, and P99—all feature magazine releases built into the trigger guard. In Figure 4.8, the magazine release is a small, knurled button located behind the trigger, a customary place for this control to be located.

Slide-Lock Lever   Most, but not all, pistol designs feature a form of slide-lock lever. As the name implies, the purpose of the lever is to lock the slide to the rear so that the chamber is exposed. An exposed chamber allows visual and tactile inspection to ensure that no cartridge is loaded. Locking the slide to the rear may also be a requisite to disassembling the firearm.

During shooting operations, when the ammunition supply is exhausted, the pistol is designed to lock the slide to the rear, alerting the shooter that the gun is now empty. This is not a function of the slide lock, but rather a function of a trip or catch that is tripped by the magazine follower. Most pistol designs allow the slide to be pulled slightly rearward from this locked position to release the trip and allow the slide to travel forward and close, regardless of whether a magazine is inserted or not. Certain pistol designs, such as the Mauser HSc, do not feature a slide lock; the action is locked open entirely on the basis of the magazine follower catch, and the action cannot be closed until a magazine is inserted.

It is a common but ill-advised practice to use the slide-lock lever as the means to release the slide: When the lever is pressed, the slide slams forward. Such a practice ultimately leads to the slide-lock lever fracturing, as the small piece of metal cannot stand up against the mass of the slide wearing it down. A better practice is to simply pull the slide back and release the slide lock, then release the slide and allow it to travel forward. Inexperienced shooters will often "ride the slide," that is, they will attempt to control the forward travel of the slide by maintaining hand pressure on it, normally out of fear that they will break the gun if the slide is allowed to travel forward freely. Riding the slide is a major cause of malfunction in pistols, as the action is inhibited and may not operate properly. Using the slide-lock lever to close the action of an unknown pistol may present another potential hazard. If the pistol has been modified to have a fixed firing pin, simply releasing the slide when a loaded magazine is inserted would cause the firearm to discharge, likely continuously, until the ammunition supply was exhausted.

Single-Shot Handguns

Single-shot, breech-loaded handguns such as the Remington XP-100 and the Thompson Contender family of pistols are rarely encountered in a criminal context, unless they are recovered stolen firearms. The Remington XP-100 is a bolt-action arm, using a turndown-type bolt. The Thompson Contenders are single action, and the action is opened by manipulating the trigger guard, which doubles as the release to open the breech. Such pistols may have exposed barrels, or the barrels may be shrouded by a horizontal hand guard, somewhat resembling a small rifle. By virtue of being capable of being fired when held with one hand, and having a chamber that is integral to the bore, single-shot handguns are classified as pistols.

Derringers   The term *derringer* has generically been used to define especially small handguns and may take the form of a revolver or pistol. The earliest examples predate the use of self-contained cartridges, instead relying on percussion design. The term originated from a man named Henry Deringer, a manufacturer of firearms in the nineteenth century. (Note the difference in spelling: The modern term is spelled with *rr*, while the name is spelled with a single *r*.)

> In 1806 Henry Deringer established a firearms factory in Philadelphia, Pennsylvania and began manufacturing flintlock pistols, muskets, and somewhat later, percussion rifles for the U.S. Army. Though initially recognized as a supplier of long arms, Deringer gained renown with the production of percussion dueling pistols, which first appeared in 1825 and were primarily sought by military officers and political officials. The manufacture of a smaller version of the dueling pistol in the late 1840s and the pocket pistol in the early 1850s solidified Deringer's position as a manufacturer of quality firearms. (FBI 2001)

Pistol derringers can have single or multiple barrels. Derringers can be concealed in a wallet, mounted on a belt buckle, or carried in the pocket or a garter belt. Derringers have always proven to be the ultimate option in concealability. True to the design intent and as a matter of safety, most derringers are single-action-only firearms. Derringers can be of any caliber, from small to large, and their effectiveness is limited to very close quarters. Mechanically, nothing differentiates a derringer from a larger handgun that operates the same way, save the more diminutive dimensions. An interesting pistol that could be called a derringer was the Snake Charmer. Manufactured by Cobray, this double-barrel side-by-side pistol was capable of chambering either .410 shot shells or .45 class cartridges. The pistol was primarily designed for hikers or other outdoor enthusiasts who might find it necessary to dispatch a snake with a small, handy firearm. The barrel had a minimal amount of rifling—just enough to meet legal requirements so that the pistol could not be defined as having a smooth bore. A similar derringer is manufactured by Leinad, a single-shot .45/.410. Cobra Enterprises manufactures an over/under derringer reminiscent of the pattern manufactured by Davis Industries. Harrington & Richardson manufactures a different model called the Snake Charmer, a .410 breech-loaded shotgun.

A recent entrant into the derringer market is the DoubleTap. Designed and manufactured by Heizer Defense, it is constructed of either aluminum or titanium. The pistol has an over/under barrel configuration and will be produced in 9×19mm and .45 ACP. Publicly announced in October 2011, the DoubleTap is now in regular production (Heizer Defense 2011).

General Safety Issues

Hang Fires

A *hang fire* occurs when the trigger is pressed and discharge is intended, but the cartridge does not detonate right away. There are a few potential causes, but all are primer related. The primer could be defective or weathered, which would inhibit the primer from instantaneously detonating when subjected to force of impact. It is of critical importance to maintain muzzle discipline should a hang fire occur, as the powder charge in the casing could potentially detonate at any moment and without warning. There is no set time of delay in a hang fire other than it is generally agreed upon that a minimum of 60 seconds of muzzle discipline should be maintained before attempting to clear the action.

Squib Loads

A *squib* or *squib load* is a cartridge that was loaded with a substandard powder charge. When the primer was struck and the powder detonated, there was insufficient gas pressure created to propel the projectile down the barrel. The shooter or observer ordinarily can recognize the report of a squib load, which is a light pop when compared to the usual loud boom that the shooter or observer would be accustomed to hearing. In all likelihood, the squib projectile is lodged in the barrel and will have to be coaxed out, usually by knocking it out with a dowel rod and a mallet. Failure to clear a squib load and continuing to fire can potentially result in serious personal injury or death, and certainly the destruction of the firearm.

Slam Fire

A *slam fire* can be deliberate as part of the design of that particular firearm, or it can be the result of an accidental discharge. As described in the discussion of slide-action shotguns later in this chapter, slam fire is possible on firearms that have no trigger disconnector and permit repeating shots to be taken as the action is manually cycled and the trigger held down. In the context of an accidental discharge, particularly with closed-bolt automatic firearms, a slam fire occurs when the firing pin is pushed forward so that it presents itself as if to fire; then, as the bolt closes, the exposed firing pin comes into contact with the loaded ammunition cartridge's primer, typically with enough force to cause detonation to take place. In most circumstances, a slam fire occurs when the action goes fully into battery; however, this does not mean a slam fire could not occur as the action closes after collecting a cartridge from the magazine, which would cause discharge when the action was out of battery.

Firearms equipped with free-floating firing pins can be susceptible to slam fires, especially if there is a general lack of maintenance, which causes the firing pin to stick in the forward position. In situations where a slam fire is suspected of causing an accidental discharge, the bolt, firing pin, and the firing-pin channel should be very carefully inspected for accumulation of carbon fouling and the buildup of debris and foreign matter that could cause the firing pin to seize in place. The ammunition of preference that has been used in that firearm should also be inspected, as certain brands of ammunition tend to run dirtier than others. Another factor in sticking firing pins is the application of lubricating oil within the firing-pin channel. Typically the firing-pin channel is intended to be kept dry because the presence of oil tends to attract and trap foreign matter, which could cause the firing pin to not function properly—either not at all or to create a situation conducive to slam fire. Not all designs that originally called for a free-floating firing pin have maintained fidelity with that feature; to be sure that modifications were not made, the examiner may need to disassemble and evaluate.

Long Guns

Long guns are firearms that are designed to be fired from the shoulder. As such, long guns are customarily, but not always, furnished with a shoulder stock constructed of wood, composite materials, or metal. If a shoulder stock is not present, there is the ready capacity to attach one. There are two types of long guns: rifles and shotguns. The absence of a shoulder stock does not preclude the exhibit from being defined as a long gun, as there are numerous variations in stock design and long guns that can be equipped with a stock but are not so configured. Certain designs require the presence of a form of shoulder stock due to design characteristics. The classic AR-pattern rifle requires a stock of some form to encapsulate the recoil spring tube assembly. The Remington 1100, a semiautomatic shotgun, also has a recoil spring contained in a channel set in the stock offset at an angle from the back of the receiver. A long-gun receiver is physically larger than that of a pistol receiver, primarily because rifle cartridges are physically larger and must be able to sustain the higher pressures created by such cartridges.

Long-Gun Safety Systems

Safety features on long guns mimic those on handguns but primarily focus on mechanically disengaging or locking out the trigger from the rest of the firing train by flipping a switch, the press of a button, or the movement of a slide from the fire to the safe position. The cross bar, or push-button safety, is one of the more common long-gun safety systems. The safety button is located within close proximity to the trigger, usually on the trigger guard, and pressing the button into the safe position literally locks the trigger, preventing it from moving. Another safety feature is an interlocking mechanism that prevents the operation of the trigger unless the action is completely closed, thus preventing the firearm from potentially discharging if the action is not completely in battery, creating a hazardous condition. A trigger disconnector often is the design feature that accomplishes the lock-out feature, and also prevents the trigger from operating unless a full movement of the action has occurred. The disconnector is discussed at greater length under machine-gun conversions later in this chapter.

Rifles

A rifle is defined by 26 U.S.C. § 5845 (c) as "a firearm designed to be fired from the shoulder and designed to use the energy of an explosive in a fixed cartridge to fire only a single projectile through a rifled barrel for each single pull of the trigger." Rifles are defined by the physical characteristics of having a rifled bore and either having an attached shoulder stock or the capability of having a shoulder stock that is readily attachable, even if such stock is presently removed. In the United States, the rifle is required to have a minimum barrel length of 16 inches and have an overall length of at least 26 inches. In the instance of a rifle that is equipped with a stock that will fold away or that will collapse or telescope to a shorter length, then the overall measurement must be made with the stock fully extended or folded out. The overall measurement is a separate consideration from the barrel length, and both measurements must be taken into consideration when contemplating the legalities of the configuration of a particular rifle. When measuring the barrel length, a permanently attached muzzle fixture such as a flash hider or compensator will be included in both the barrel and overall length measurements. The BATFE defines permanent attachment as having been made by silver solder, welding, or pinning and welding into place. Use of a fixative yet easily reversible process, such as a strong adhesive, would not qualify as a permanent attachment. The firearm is a rifle based on meeting these requisite definitions and is classified as such by the original manufacturer.

Rifles may be lever action, bolt action, single shot, self-loading, or slide action, and the variety and style of rifles is nearly endless. Stocks may be fashioned of various grades of hardwood or laminates, polymer, or metal. Finishes range from deep-luster blues to chrome, stainless steel, parkerized, and enamels. Aesthetics do not concern the examiner, as they play no part in the functional characteristics of the firearm. Within the parameters of operating principles, rifles are among the most versatile; there has likely been a rifle devised for every possible method of operation. The rifle simply offers more real estate from which to develop operating systems, whether they use gas, recoil, springs, pistons, or something that combines elements of any of them.

Lever Action   The lever-action rifle is operated by the manipulation of a hinged lever that is ordinarily attached at the bottom of the receiver and behind the trigger. Customarily, the lever makes up a portion of the trigger guard. Lever-action rifles have been in continuous production since the first examples started appearing in the mid- to late 1800s. Lever-action rifles typically have a tubular magazine located underneath the barrel. Identifying information on a lever-action rifle can be found on the barrel, receiver, or on a tang that extends downward from the rear of the receiver that mates it to the stock. Lever-action rifles can be had in innumerable calibers, many of which are obsolete or archaic, given the age of a particular specimen or modern replicas chambered in an archaic caliber. The Model 1894 Winchester is considered one of the quintessential lever-action rifles.

Bolt Action   Bolt-action rifles are operated by means of manually operating a handle attached to the bolt, which cycles the action of the firearm. Several variants of the bolt-action rifle exist, including straight-pull bolts, where the bolt is literally pulled straight back as in the case of the Austrian Steyr Mannlicher M/95 rifle or the Swiss Schmidt-Rubin. A bent or straight bolt handle must be turned up and then pulled back to open the action, and these are more common than straight-pull bolts by a large margin.

Bolt-action rifles were initially developed for military applications and include the American Springfield, German Mauser, British Enfield, Japanese Arisaka, Russian Mosin-Nagant, Swiss Schmitt-Ruben, and Italian Carcano. The majority of contemporary bolt-action rifles mimic either the Enfield or the Mauser 98 design. Bolt-action rifles are ordinarily equipped with either detachable box magazines, as in the case of the British Enfield, or an internal magazine, as in the case of the Japanese Arisaka, the Mauser, or the Mosin-Nagant. Commercial bolt-action rifles may be single shot, as in the case of the Winchester Models 1900, 1902, and 1904, or they may be magazine fed, such as the Ruger M77 series, which have internal magazines. Rifles equipped with internal magazines can generally be unloaded by opening the magazine floor plate on the bottom of the stock, which will release by pressing a button or releasing a latch. Many bolt-action rifles use detachable magazines that are removed using the magazine release.

Many surplus military bolt-action rifles have been sporterized: The stocks may have been reconfigured or replaced with more ornate furniture; they may have been refinished or had their military finish polished to a higher degree of luster; and it is possible that the rifle may have been rechambered to fire a different cartridge than originally designed. Examiners are cautioned against presuming that a particular model surplus rifle is still chambered as it was originally, as many rifles have been rechambered and rebored to sporting-caliber cartridges. A seemingly unlikely user of surplus World War II German K98k bolt-action rifles postwar was the newly formed nation of Israel. These rifles were rechambered in 7.62×51mm and likely served until the 1980s in a reserve capacity. These rifles are readily identified by the large "7.62" engraving on the receiver ring and often on the stock as well.

Slide Action   Slide-action rifles operate by using a forward "pump" or slide to operate the action. Manipulation of the slide loads and unloads the firearm. Slide-action rifles are typically supplied ammunition through a tubular magazine, but box-style magazines are used as well. A few examples of slide-action rifles include the Remington Models 14, 25, 76, and 7600, and the Winchester Models 1890, 1906, and 61. The Winchester Model 1890 is often called the "gallery gun" because of its popularity in carnival and fairground shooting galleries in the bygone era. As a matter of safety in the short ranges where gallery guns were fired, the .22 Short was the preferred cartridge.

Single Shot   Single-shot rifles are breech-loaded affairs that are operated by opening the breech using some form of release lever. Such a rifle may be either striker fired or have an exposed hammer to cock in preparation to fire. Examples of breech-loaded rifles include the rolling block or the falling block, which came into existence in the mid-1800s but were superseded relatively quickly by bolt-action rifles in service with the military forces of the world. Except for hunting or recreational purposes, breech-loaded firearms were largely replaced by bolt-action rifles, principally due to improvements in ammunition technology and other inherent advantages of the bolt action.

Self-Loading   Self-loading rifles are now more universally identified as semiautomatic rifles, yet they were termed *self-loading* for a number of years. Like pistols, the semiautomatic rifle is operated through any number of systems focused around the capture and utilization of the energy created by the discharge of the weapon, either by gas operation or simple blowback.

Self-loading rifles appeared around the turn of the nineteenth to the twentieth century. Among the first to appear were from Winchester Repeating Arms, who manufactured the self-loading rifle Models 1903 (chambered in .22 Winchester Automatic Smokeless Cartridge), 1905 (chambered in .32 and .35 Winchester Self Loading), 1907 (chambered in .351 Self Loading Rifle), and the 1910 (chambered in .401 Winchester Self Loading). Remington Arms responded quickly to Winchester with the Model 8 (chambered in several Remington-developed calibers such as .25, .30, .32, and .35), which appeared around the same time as the Winchester 1907. These rifles were very advanced in appearance for the time and are still quite graceful looking, combining clean lines with classic craftsmanship. These early self-loading rifles became very popular with law enforcement users until the time of the Second World War. Postwar, semiautomatic rifles continued to escalate in popularity in the marketplace (see Figure 4.13). Modern semiautomatic rifles can take on endless forms, appearances, and configurations. Detachable magazines, internal magazines, and tubular under-barrel magazines are used to supply ammunition to semiautomatic rifles.

Figure 4.13 Benelli MR1 semiautomatic rifle chambered in 5.56×45mm. (Image courtesy of Benelli USA.)

The term *assault rifle* is a somewhat colloquial term that attempts to define a certain species of rifles. The term is applied based on various features, design characteristics, or configurations present on the rifle in question, yet the term *assault* has also been applied as *assault pistol* or *assault shotgun*. Somewhat synonymous with *assault rifle* is the term *paramilitary configuration*, apparently to describe a rifle that aesthetically may resemble a rifle that is currently or was previously in issue with a military force. The terms themselves are subjective and quite vague, apparently only used to describe contemporary firearms while typically excluding older surplus military arms of the previous generations such as bolt-action rifles, even if they were originally configured to have the same or similar features that would otherwise define them as assault rifles.

Certain models of rifles do overtly appear to resemble military arms; however, their resemblance does not necessarily reflect that the firearm in question is anything more than a rifle as the term is legally defined. Semiautomatic rifles that resemble contemporary military arms are quite common and make up a large portion of the market share of firearms that are commercially sold. Examples include AR-pattern firearms; AK-pattern firearms; the MSAR rifles, which resemble the Austrian Steyr AUG (Armee Universale Gewehr or Army Universal Rifle); copies of the Israeli Galil or Finnish Valmet; copies of the H&K G.3; copies of the FN-FAL; and so forth. The examiner/investigator is encouraged to refrain from using the term *assault rifle*, as it fails to define any rifle or its functional characteristics with any specificity and carries a certain subjective, perhaps provocative, connotation.

In the purest legal language, there is nothing that differentiates an AR-15-pattern rifle from a Remington Model 7400 rifle. Both examples are semiautomatic rifles that are fed by a detachable magazine; both chamber powerful cartridges; and both expel one round per press of the trigger. Both have shoulder stocks and otherwise meet all legal parameters required to classify a firearm as a rifle. The differences between the two are primarily aesthetic, not functional.

As in the case of an assault rifle, the terms *assault shotgun* or *assault pistol* are attempts to define a particular species of such that bear certain aesthetic or design characteristics that are thought to imply a paramilitary or nonsporting application, appearance, or configuration. Once again, the terms are somewhat elusive to define with any specificity, tending to focus on particular makes and models of shotguns and pistols based upon individual perception instead of addressing the functional capabilities or characteristics of the particular arm in question.

Characteristics that have been identified under 27 CFR pt. 178.11 to identify an assault rifle:

A folding or telescoping stock
A pistol grip that conspicuously protrudes beneath the action of the weapon
A mounting interface for a bayonet
A flash suppressor or threaded barrel
A grenade launcher

Qualities that are deemed to constitute an assault-type pistol by 27 CFR pt. 178.11:

A detachable magazine that is attached to the pistol outside of the grip
A threaded barrel
A barrel shroud that permits one-handed shooting using the nontrigger hand
A pistol that has an unloaded mass in excess of 50 ounces
A semiautomatic version of a fully automatic weapon

Qualities that are deemed to constitute an assault-type shotgun under 27 CFR pt. 178.11:

A folding or telescoping stock
A pistol grip that protrudes conspicuously beneath the action of the weapon
A fixed magazine capacity in excess of five rounds
An ability to accept a detachable magazine

Carbines and Bull Pups   A carbine is defined as a short, compact rifle. A carbine may be an original design or a physically smaller variant of a full-sized rifle. In the classical sense, the carbine was exactly the same as the full-sized rifle, except that the barrel was shortened and there may have been some other modifications to the fixtures and stock to produce a lighter and handier rifle.

Historically, the evolution of carbines goes to the days of horse-mounted armed persons, whether they were military cavalry or ranchers. A full-sized rifle was simply too unwieldy in this scenario, yet there was a need to carry something larger than a handgun. The carbine evolved and led to the gradual phasing out of the full-sized rifle in favor of these carbines for general issue in certain armed forces. The first generation of center-fire rifles using full-sized cartridges were physically large instruments having barrel lengths that could reach 30 inches and overall lengths upwards of 4 feet. These rifles had great range and accuracy, but did present the issue of being cumbersome. During World War II, the British Number 4 Enfield rifle was subjected to extensive modifications to provide British Commonwealth forces fighting in jungle environments with a lighter alternative. The result was the Number 5 Enfield rifle, henceforth known as the Jungle Carbine. Using the Number 4 rifle as the basis, the barrel length was reduced from 25 inches to slightly less than 19 inches, reducing the overall length of the rifle by approximately 5 inches. The amount of wood used in the stock was significantly reduced, and engineers went so far as to mill away as much metal from the receiver as possible. As a result, the Number 5 weighed in at some seven pounds, compared to the Number 4 at almost nine pounds. Unfortunately, the Number 5 did not perform as well as expected. It still fired the standard full-powered .303 British service cartridge. However, the recoil was heavy for such a rifle, and it was not nearly as accurate as was expected, especially given the pedigree of the Enfield action.

The Russians developed and fielded carbine versions of the standard 1891 Mosin Nagant rifle, itself a fully dimensioned rifle. Through a series of successive steps, the German Model 98 rifle was developed into the definitive carbine variant, the K98k (K for *Karbine* and k for *kurz*), roughly translated to mean Short 98 Carbine.

The United States experimented with creating carbine variants of the M1 Garand; however, they were never fully developed. The US M1 carbine, devised by Winchester Repeating Arms, was an entirely separate model from the M1 Garand and was not developed as an offshoot from the Garand. The U.S. military had not used the M16 long before Colt had developed a carbine version of it. The first carbine version, initially given the official designation of XM177, differed from the standard M16 by introducing a two-position collapsible stock and shortened barrels of various lengths. Later, it simply became known as the CAR 15. The carbine variant of the M16 has since become officially designated the M4 in U.S. military terminology. The M4 equipped with an 11½-inch barrel is further identified as the Commando; the standard M4 comes with a 14½-inch barrel, and the standard M16 comes with a 20-inch barrel.

The term *M4* was the subject of a lawsuit brought by Colt against Bushmaster Firearms. In the suit, Colt claimed trademark infringement by Bushmaster, who used the *M4* term to market their version of the rifle based on the AR-15 platform. The suit was eventually resolved when the court determined that the term *M4* was designated by the U.S. military, who applied it to a specific model of firearm, and therefore the term resides in the public domain. This decision also noted that the term *M4* had been applied by other manufacturers of similar arms and even applied to other unrelated products as a model designation (Memorandum Decision 2005).

A *bull pup* is a rifle that has very compact external dimensions. The archetypal approach used to create a bull pup is to relocate the receiver to the back of the rifle, allowing for a barrel of reasonable length to be used. The most obvious clue of a bull pup is the location of the magazine well, which is behind the grip and trigger. Bull pups may offer certain ergonomic advantages to the operator in addition to the reduced overall physical size.

Bull pups can be deceiving in their appearance, often mistaken as short-barreled rifles; however, their actual barrel length can be well within legal requirements, as the barrel is mostly contained within the stock and just a short length, or even just the muzzle, is left exposed (see Figure 4.14). Examples of bull-pup designs include the Steyr AUG, the French FAMAS, the British Enfield L85 series, and the Walther G22. Fabrique Nationale's FS2000 and the PS90 are both relatively late offerings that are bull-pup-configured rifles. Both have all-encompassing stocks that contain the receiver and barrel such that little of either is exposed. The FS2000 is chambered in 5.56×45mm, accepts the AR-15/M15–pattern magazine, and is available as a machine gun to legally authorized entities as the F2000 (see Figure 4.15). The PS-90 is the semiautomatic version of the P-90 submachine gun. The PS-90 has a rectangular shape and has the unique feature of a magazine that lies flat on the top of the receiver. A distinguishing feature is the amount of barrel that protrudes from the stock. The PS-90 standard barrel measures 16.10 inches, and the overall length is 26.23 inches. As a submachine gun, the P-90 barrel length is 10.39 inches in length (FNH USA n.d.).

O.F. Mossberg & Sons, Inc., marketed a bull-pup configuration variant of their venerable Model 500 12-gauge shotgun. The 500 bull pup was available in 12 gauge only and was equipped with either an 18½″ or 20″ barrel, with a capacity of six rounds for the 18½″ barrel and nine for the 20″ barrel. An eight-shot magazine was optional. Mossberg sold the same gun under the Maverick Arms line.

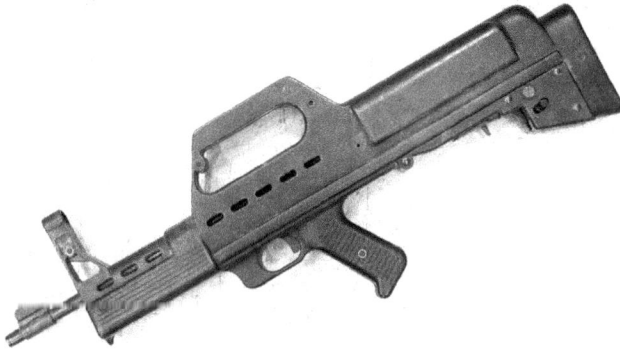

Figure 4.14 Believe it or not, this is a Ruger Mini-14 that has been fitted into an aftermarket stock, giving it the appearance of a bull-pup configuration. The barreled action is removed from the factory stock and seated into this chassis. An internal bar connects the factory trigger on the action to the user trigger. Without seeing the action, it is possible to identify the gun as a Mini-14 by looking at the front sight and the magazine well. (Image from author's collection.)

Figure 4.15 Fabrique Nationale F2000 in the hands of a Peruvian sailor. This weapon is available as a select-fire weapon or as a semiautomatic rifle (as the FS 2000) and is a classic bull pup; note the compactness of the weapon. (U.S. Marine Corps image; photographer Cpl. Brian J. Slaght.)

Contemporary Copies of Classic Military Weapons   A current trend is the reintroduction of older pattern military firearms that are semiautomatic rifles, not machine guns as the originals were. Semiautomatic rifles patterned after the Browning Automatic Rifle (BAR), Thompson submachine gun, Suomi M31 submachine gun, STEN gun, Sterling submachine gun, the UZI, the Degtyarev and the Goryunov (both are machine guns of Soviet design), the German Sturmgewehr (official designations MP-43, MP-44, and StG-44), and the German FG-42 are or have been manufactured and commercially available. Wise Lite Arms, based in Boyd, Texas, manufactures a copy of the British BREN Mark II machine gun as a semiautomatic rifle. Mere observation of a firearm of these appearances and configurations would be insufficient to deem them as contraband. Such models as described here do not include firearms that were originally manufactured as machine guns that have been remanufactured or reconfigured by the removal of parts so that the weapon would fire semiautomatically only.

These firearms are newly manufactured receivers that have been designed to function as semiautomatic weapons, although they aesthetically resemble and may even be marked to mimic period originals. One quality that often betrays these semiautomatic copies is that the originals had short barrels, but in order for the copies to be compliant as rifles, the barrel length must be at least 16 inches. The exception to this would be to designate the newly manufactured receiver as a handgun, thereby eliminating the barrel-length issue. PPS-43c, manufactured in Radom, Poland, by Pioneer Arms, is one such example, being a semiautomatic version of the venerable World War II–era Russian submachine gun, complete with an inoperable folding stock. As such, it is classified as a handgun. Another wartime-era Russian submachine gun, the PPS-41, has served as a pattern for a contemporary semiautomatic rifle, manufactured in Germany as the Model SKL-41. To date, it is unknown in the United States. As is always the case, before making snap determinations, a detailed technical inquiry is in order.

Short-Barreled Rifles

A short-barreled rifle is defined by 26 U.S.C. § 5845(a) (3) as "a rifle having a barrel or barrels of less than sixteen inches in length." A short-barreled rifle is a firearm that was originally constructed as a rifle; however, the barrel length has been modified so that its length is shorter than the requisite 16 inches. When contemplating whether the rifle has been made or remade into a short-barreled rifle, recall the basic characteristics that define a rifle: the firearm has or originally had a shoulder stock or the capability to readily attach one, has a rifled bore, and an overall length of at least 26 inches. Barrel length reduction can be accomplished by shortening the existing barrel by trimming or by replacing the barrel with one that is shorter. Short-barreled rifles can also be manufactured as such by a professional gun maker and titled accordingly, so the term should not be inferred to merely mean that it could only apply to modified firearms.

Certain firearm designs are not readily reconfigurable by replacing barrels, a question addressed by determining how the barrel is attached to the receiver. The AK-pattern firearm, for example, has a barrel that is pressed into the barrel trunnion, which is then mated to the receiver and the entire assembly riveted in place (see Figure 4.16). There simply is no practical quick replacement for such a barrel. Likewise, barrels that are threaded into the receiver are then typically pinned or otherwise locked into the receiver. It is not practical to cut the barrel down either, due to the location of the gas system relative the muzzle. On the other hand, an AR-pattern rifle can be reconfigured to a short-barreled rifle quite easily. On an AR-type

Figure 4.16 A semiautomatic copy of a Russian Krinkov, an AK variant. This exhibit is not a machine gun, it is a short-barreled rifle because of the shoulder stock and barrel that measures approximately 8.5 inches. The weapon was assembled using a parts kit mated to a U.S.-made receiver. (Image from author's collection.)

firearm, the entire upper receiver assembly can be replaced in moments with an upper receiver that has been mated with a shorter barrel (7½-, 10½-, 11-, and 14½-inch barrel lengths are common, but other lengths do exist). Replacing an AR barrel can be a time-consuming and somewhat complex undertaking for an inexperienced person; however, upper receivers with short barrels already mounted are readily available through commercial channels.

Rifles such as the FN SCAR have modular configuration capabilities that are designed for rapid reconfiguration to suit the operational needs of the user; it is simply a matter of acquiring a suitable barrel. Handguns that have had shoulder stocks affixed to them may also be classified as short-barreled rifles, as they no longer are defined as handguns when the shoulder stock is attached. When determining whether a rifle qualifies as a short-barreled rifle, the examiner must take into account whether the exhibit was originally manufactured in its present condition when examined. The method of modification should be identified and firmly established.

Certain rifles with short barrels are exempted from legal issues attached to short-barreled rifles, as they have been reclassified as curios and relics or antiques. Winchester rifles, in particular the Models 1873, 1892, 1894, and others, were frequently special ordered and delivered with barrels shorter than 16 inches before the National Firearms Act placed regulatory provisions on short-barreled rifles. The most current published ATF Curios and Relics List should be consulted if questions arise concerning the Curio and Relic status of a particular firearm.

During the late 1800s and into the first part of the twentieth century, it was not uncommon for a handgun to have the ability to attach a shoulder stock. Nearly all examples of Luger pistols, the Mauser c/96 Broomhandle, certain varieties of Model 1911, Browning Hi Power pistols, and some revolvers were manufactured with a lug to permit attachment of a shoulder stock (see Figure 4.17). In such bona fide original historical examples, no violation of law occurs insomuch as an original shoulder stock (not a contemporary copy of a vintage shoulder stock) is attached. Original shoulder stocks can generally be readily identified by an applied serial number, quite often matching the gun they are attached to as a rig, or by period inspection or property markings. Obviously, such period examples tend to exhibit patina, aging and wear associated with objects of relative age. Modern reproductions of these stocks do exist and closely resemble period pieces and may be marked in such a way to imitate originals. Certain modern handguns have aftermarket stocks produced that will attach to them, thereby creating a short-barreled rifle when they are affixed. When contemplating identification, the examiner must ensure that the exhibit is, in fact, a rifle and not a handgun.

Many firearms that are configured to appear as rifles have been reconfigured in such a way by the original manufacturer to be manufactured as legally classified handguns. Recently the Radom Arsenal in Poland (marketing under the name Pioneer Arms) began manufacturing a handgun patterned after the World War II–era Soviet PPS-43 submachine gun (also known as the Sudayev, the designer's last name); however, the new product, dubbed PPS-43C, differs from the original for two reasons. The first is that it operates as a semiautomatic, not a fully automatic firearm (machine gun). The second is that, while equipped with a shoulder stock, it is permanently mounted in a folded position so that it cannot be extended to offer the shooter a shoulder mount; hence it is defined as a handgun, not a rifle.

If there is uncertainty about the original classification of a firearm, the question may have to be answered by researching the manufacturer's A&D (Acquisitions and Dispositions) records to ascertain what type of firearm the exhibit was originally manufactured as. If the manufacturer has gone out of business, a firearms trace could be made by the BATFE to ascertain what the firearm was categorized as in the course of commerce.

Figure 4.17 During World War II, a female Canadian armaments worker tries out an Inglis-manufactured Hi Power pistol equipped with a shoulder stock. (Image from Canadian National Archives.)

In lieu of any other existing documentation, examiners must draw upon their own experience and application of the definitions of a long gun versus a handgun. Other examples of arms legally classified as "handguns" but resembling rifles include the Iver Johnson Enforcer, a copy of the U.S. government M1 carbine; sundry AK-pattern clones with inert folding stocks; and AK-pattern clones without any form of stock or attachment apparatus, such as the Romanian-produced Draco, which is classified as a pistol. In 2008, Kahr Arms started manufacturing the Model 1927A-1 Thompson pistol, patterned exactly after the Thompson Model 1927 submachine gun; however, this contemporary version is semiautomatic only and does not feature a shoulder stock or any ready means to attach one. Original Thompson submachine guns did have the capability of removing the shoulder stock, and there is a slot on the lower portion at the rear of the receiver where the stock affixes. This newly made Thompson semiautomatic weapon can be configured by the factory as a rifle, pistol, or a short-barreled rifle; it is not made available as a machine gun regardless of its configuration. Most AR-pattern handguns must be equipped with a buffer tube extending from the rear of the upper receiver; however, in such an instance, the buffer tube has been redesigned to not permit installation of either a collapsible or full rifle stock. This does not prevent the original buffer tube from being removed and a standard rifle buffer tube being installed in its place. The Czech VZ61 Scorpion pistol closely resembles the CZ M61J, a compact submachine gun with a folding wire stock. The two are marked differently on the receiver but overtly are nearly indistinguishable from one another. The VZ61 obviously lacks the stock and is not capable of full-auto fire. UZI semiautomatic carbines are easily converted into short-barreled rifles. The UZI carbine was equipped with the folding stock or with a solid wooden stock and a 16-inch barrel. The UZI barrel is removed easily

Figure 4.18 The Rossi Ranch Hand is a lever-action handgun, although it very much appears to be a cut-down rifle and could easily be mistaken for one. (Image courtesy of Braztech International, L.C.)

by unscrewing the barrel nut from the receiver and pulling the barrel out. There are plastic dummy "replica" barrels to give the look of a short-barreled weapon, but these are not legally considered barrels.

Weapon Made from a Rifle

A weapon made from a rifle is defined by 26 U.S.C. § 5845 (a) (4) as "a weapon made from a rifle if such weapon as modified has an overall length of less than 26 inches or a barrel or barrels of less than 16 inches in length." Such a weapon was originally constructed as a rifle and bore the requisite features and characteristics of a rifle before modification. Many persons are not familiar with the term, identifying any shortened rifle as a short-barreled rifle. The receiver remains intact, but substantial modifications will have been made. In nearly all cases, the overall length is reduced by cutting down the shoulder stock, perhaps leaving only the short "pistol grip" portion of the stock, and reducing the barrel length. In the case of a rifle with a functional folding or collapsible stock, the stock must be fully extended when a definitive overall length measurement is taken. As in the case for a short-barreled rifle, the examiner must conclude that the exhibit was not originally manufactured in the configuration it exists in when examined.

Many rifles can serve as the host to manufacture such a weapon, the only real limitation being set by the presence of the gas system or magazine that runs the length of the barrel, or part of the operating system that is set in the stock. The examiner should establish and document the method(s) used to effect the alterations. The Rossi Ranch Hand is one such example of a firearm that is legally classified as a handgun yet could easily be mistaken as a short-barreled rifle or a weapon made from a rifle (see Figure 4.18). Although legally classified as a handgun, the Ranch Hand is lever action, having a barrel length of 12 inches and an overall length of 24 inches. The stub stock is not intended to be used to fire the weapon from the shoulder. The Ranch Hand is available in some classic calibers: .45 Long Colt, .45 Colt, .44 Magnum, and .38/.357.

Shotguns

A shotgun is defined by 26 USC § 5845(d) as follows:

> The term "shotgun" means a weapon designed or redesigned, made or remade, and intended to be fired from the shoulder and designed or redesigned and made or remade to use the energy of the explosive in a fixed shotgun shell to fire through a smooth bore either a number of projectiles (ball shot) or a single projectile for each pull of the trigger, and shall include any such weapon which may be readily restored to fire a fixed shotgun shell.

In the United States, a shotgun must have a minimum barrel length of 18 inches and overall length of at least 26 inches. As with rifles, barrel length and overall length must be taken into consideration independently of one another when a classification is made concerning a shotgun suspected of modification or alteration that affects the requisite dimensions. Like rifles, shotguns are found in a broad array of configurations and appearances. The same basic shotgun receiver can easily be configured as a tactical-style shotgun seemingly more appropriate for military or law enforcement applications, or it can just as easily be provided a longer barrel or other aesthetic qualities that configure it for hunting, trap and skeet, or other traditionally defined sporting purposes. Shotguns can be single shot, operated by slide action, lever action, bolt action or semiautomatically.

Breech-Loaded Shotguns   The most basic shotgun is a single-barrel, breech-loaded type. The action is opened by a release mechanism, just like breech-loaded handguns and rifles. Most breech-loaded shotguns have an exposed hammer; however, internal strikers may be encountered in lieu of exposed hammers. The striker would be cocked by closing the action. Breech-loaded shotguns can have multiple barrels. A double-barrel shotgun may have barrels side by side (abbreviated S×S) or over/under (abbreviated O/U). Functionally, there is no difference between the two except in the arrangement of the barrels on the receiver and where the breech release latch is located. By virtue of design, double-barrel shotguns are breech-loaded weapons and, like single-barrel breech loaders, may feature exposed hammers that must be cocked in preparation to fire. When handling a striker-type breech-loaded shotgun, caution must be exercised when closing the action due to the uncertainty whether the strikers will discharge when cocked, even without trigger input. Double-barrel shotguns may have a single trigger or double triggers within the trigger guard, with each trigger serving to function a separate chamber, one per barrel.

Slide Action   The slide action is more often called the "pump," to the extent that if one mentions a slide action, it likely would not be recognized in conversation. The slide-action shotgun has been around since the late 1800s and shows no signs of diminishing in popularity. Slide-action shotguns have a reputation for ruggedness, reliability, and durability. Slide-action shotguns are probably the most versatile type of firearm produced—used by hunters of all types, sport shooters, and military and law enforcement operators, who can use them with either lethal or less-lethal munitions. The two most popular models of slide-action shotgun in the United States today are the Remington 870 and the Mossberg 500 series. Maverick Arms, a subsidiary of Mossberg, sells essentially the same product as Mossberg, although marked slightly differently. Winchester and Smith & Wesson marketed slide-action shotguns, but neither continues to do so. Fabrique Nationale and Benelli both market slide-action shotguns set up in various configurations for different users, and they are joined by numerous other firms such as Harrington & Richardson (including New England Firearms) and the Chinese arms maker, NORINCO.

Slide-action shotguns have a tubular magazine located underneath the barrel. With a loaded magazine, cycling the action loads the chamber. Pulling the slide to the rear ejects the spent hull and cocks the hammer. Slide-action shotguns may have internal hammers, as in the case of the Remington 870 or the Mossberg 500, or an exposed hammer that can manually be manipulated, as in the case of the Ithaca 37. In a properly functioning slide-action firearm, the slide cannot be pulled to the rear until the action is unlocked, ordinarily accomplished by pressing the trigger. In instances where discharge is not desired, an action release lever or button is furnished, typically in the area of the ejection port or trigger guard, usually located on the right side of the receiver. Depressing the action release will permit the slide to be moved to open the action. Caution must be exercised because if there are additional live cartridges available to be loaded, a live cartridge might inadvertently be chambered if the action is fully opened. A gradual pull of the slide to the rear that allows the loaded cartridge to present itself will prevent an unintentional reloading.

The simplicity and durability of slide-action shotguns has ensured their continued success and even preference over semiautomatic shotguns. The main failures of slide-action shotguns are caused primarily by operator error such as loading shells backwards, or by short stroking the action. Much like a pistol, a slide action shotgun can malfunction if the operator does not exert sufficient force or travel on the slide to completely cycle the action. Modern slide-action shotguns have separate loading and ejection ports. Ejection occurs through the port normally located on the right side of the receiver, although left-hand versions are available that eject to the left. Tactically minded operators will note that a combat or tactical load can be accomplished by introducing a live shell into the chamber through the ejection port with the shell carrier out of the way. The loading port is located on the bottom of the receiver and allows shells to be loaded into the tubular magazine. Remington's Models 10, 17, and 29 featured a combination loading/ejection port on the bottom of the receiver. The Remington SP-10 is a semiautomatic shotgun and should not be confused with the earlier Model 10, which was manufactured from 1908 to 1929.

Lever-Action Shotguns   Lever-action shotguns are rather uncommon, their commercial success likely being tempered by the appearance of slide-action and automatic shotguns, which appeared around the same time. The Chinese arms maker NORINCO manufactured the Model 87W, a contemporary copy of the Winchester Model 1887. The shotgun is 12 gauge, has a 20-inch barrel, and a capacity of five shots. Another contemporary copy originated from the Italian firm Chiappa Firearms. The Chiappa version differs in that it has 22- or 24-inch barrels and a capacity of upwards of seven shots. The original Winchester Model 1887 was capable of handling black powder loads only. A redesigned version, the Model 1901, was capable of firing shells loaded with smokeless powder and was available in 10 gauge only. The Models 1887 and 1901 were breech loaded; the shells were fed into a tubular magazine underneath the barrel when the action was opened. The Ithaca Model 66 was a single-shot lever-action shotgun that was available in 12 and 20 gauge and .410.

Bolt-Action Shotguns   Bolt-action shotguns are largely archaic; however, they are encountered from time to time. Mossberg manufactured magazine-fed bolt-action shotguns, including the Model 385 and the more contemporary Model 695. Mossberg probably sold as many Model 385s to secondary retailers branded under their name as it did Mossberg-marked products. The Stevens/Savage Models 38, 39, 58, 59, 124, and 238 were other examples of magazine-fed, bolt-action shotguns. Functionally, bolt-action shotguns operate exactly the same way as bolt-action rifles, depending on which bolt-action rifle the designer drew inspiration from.

Semiautomatic Shotguns   Semiautomatic shotguns work in the same manner as other auto-loaded firearms, by capturing the energy released by the expulsion of a projectile to cycle the action. Auto-loading shotguns started to appear around 1900 and, like so many other firearms, the concept was pioneered by John Browning. Practically every company that markets shotguns produces semiautomatic models. Remington produced a series of auto-loading shotguns, some of which were Browning design, predating the introduction of the 1100 in the early 1960s. Remington's Model 1100 and its variants is one of the longest continuously produced auto-loading shotguns. Winchester produced a series of auto loaders, also of Browning design, by obtaining the manufacturing rights; Browning designs were also manufactured by Browning Arms. For the most part, auto-loading shotguns are manipulated in exactly the same way as slide-action shotguns, save for the auto-loading operation. The Italian gun makers Benelli, Beretta, and Fabarm produced some of the finest auto-loading shotguns made, both as sporting pieces and models marketed toward tactical operators. Fabrique National markets a comprehensive line of auto-loading shotguns as well.

Shotguns have been manufactured capable of either slide- or auto-loading function. The SPAS-12 and SPAS-15, both products of the Italian firm Luigi Franchi s.p.a., were capable of acting as either a slide action or as an automatic. It is often, but erroneously, believed that the term *SPAS* was somehow derived from the "s.p.a" in the Franchi name, which is actually the business entity type. The model terminology was an acronym, SPAS (sporting-purpose automatic shotgun). Both the SPAS-12 and SPAS-15 are 12-gauge shotguns. To convert the SPAS-12 from slide to semiautomatic firing, the action was closed, a button underneath the slide was depressed, and the slide moved slightly to the rear until locked in a position marked by the letter *M* on the barrel shroud. Once locked, the shotgun behaved as an auto-loader. Another interesting feature was a magazine cutoff function that permitted the shotgun to be breech loaded by the operator. The SPAS-12 was imported into the United States for a relatively brief period of time, and was available in several configurations. An auto-loading-only version of the SPAS-12, the LAW-12, was also available, and physically looked very similar. Another variant, the SAS-12, was the slide-action-only version. The SPAS-15 bore a resemblance to the AR-10 (the AR rifle chambered in 7.62×51mm) and the USAS-12, another auto-loading shotgun of that era. The SPAS-15 was fed by a detachable magazine holding six rounds. The SPAS-15 was very briefly available for commercial sale in the United States; estimates vary from 180–300 shotguns imported before the model was banned from further importation in 1989. Even rarer than the shotguns were additional magazines; according to anecdotal information, all available spare magazines that were imported for SPAS-15 were acquired by a single arms dealer in Miami, Florida, at that time. The Benelli M3 shotgun converts from slide to auto loading with a simple quarter turn of the selector ring and is available commercially.

The Russian Saiga 12, 20, and .410 are self-loading shotguns. The term *Saiga* represents an entire product line of hunting rifles and shotguns manufactured by Izhmash, located in the Udmurt Republic of Russia. The Saiga shotgun model indicates the round it is chambered for: 12 gauge, 20 gauge, or .410. The Saiga shotgun is based on the Kalashnikov method of operation and overtly resembles the AK-pattern rifle, except for the dimensional differences. Like the AK, the Saiga shotguns are fed by detachable box magazines with capacities of 2, 5, or 8 rounds, although larger capacity magazines and drums are available. Because the operating system is based on the AK rifle, the Saiga is widely considered to be an extremely reliable auto-loading shotgun. Its gas system is tolerant of high- and low-pressure shells and continues to cycle reliably, which has been an impediment to auto-loading shotguns in general. To support different hunting, sporting, and tactical applications, the Saiga-12 uses screw-on muzzle chokes and other accessories, and the gas system is adjustable to suit the type of ammunition being used. The Saiga 12 can chamber 2¾- (70mm) or 3-inch (76mm) shells. The Saiga 12 is made in a variety of configurations, with barrels of 17 or 23 inches in length (Izhmash OJSC n.d.).

The Vepr 12 is another Russian semiautomatic shotgun. The Vepr bears a strong resemblance to the Saiga, and it is easy to mistake the two from a distance. The Vepr is manufactured by MOLOT, while the Saiga is manufactured by Izhmash. Like the Saiga, the Vepr is based on the Kalashnikov operating system and is available in several barrel lengths. Two interesting features that distinguish the Vepr 12 from the Saiga is a magazine locking mechanism that does not require that the magazine be inserted at an angle to seat in the receiver, and the bolt, which has a hold-open feature.

A very recent addition to the semiautomatic shotgun market is the Model 1216, manufactured by SRM Arms. The shotgun has a very modern, bull-pup-style appearance. The receiver is encased in a polymer shell. The magazine is rather novel. It is comprised of four tubes that rotate on an axis. The entire magazine assembly is quickly removable. The standard 1216 is 34 inches in length with an 18-inch barrel, although short-barreled versions are also available.

Three semiautomatic shotguns, the USAS-12, Striker 12, and the Streetsweeper were retroactively reclassified as *destructive devices* in 2001 by an ATF administrative ruling. ATF Ruling 2001-1 mandated that these shotguns be registered into the National Firearms Registry by May 1, 2001, and provided legal opinion that served as the basis for the reclassification. Examples of these models that were not registered by the deadline are now considered contraband arms. The Striker 12 was manufactured by Sentinel Arms and the Streetsweeper by SWD/Cobray (designed in Rhodesia and initially manufactured in South Africa). In addition to the 12-gauge Streetsweeper, SWD manufactured a .410 version. The Striker 12 and the Streetsweeper were both supplied ammunition from a drum magazine. The USAS-12 bears resemblance to the AR-pattern rifle in its general layout, outside of the dimensional differences, and is magazine fed. It was manufactured by Interord (RAMO Defense) in Tennessee. Daewoo Precision Industries of South Korea had manufactured a fully automatic version of the USAS-12; however, it is likely that there are few, if any, examples in the United States.

Drillings

A drilling firearm combines both rifle and shotgun barrels within a single frame or receiver. Drillings are unusual in the United States, but are very popular in Europe. A drilling has two or more barrels, quite frequently up to four, configured as over/under, side by side, or barrels in four positions. Drillings are breech-loaded firearms and are often quite ornate in appearance. Historically, a number of "survival guns" that are drillings were manufactured, generally firing smaller gauge shot shells and smaller caliber rifle cartridges. Such guns were included in survival kits for aircrews and in shipboard life raft kits. In the United States, a drilling is classified as a rifle since it contains a rifled barrel.

Short-Barreled Shotguns

A *short-barreled shotgun* is defined by 26 U.S.C. § 5845 (a) (1) as "a shotgun having a barrel or barrels of less than 18 inches in length." Such shotguns are often erroneously referred to as "sawed off shotguns," a technically and grammatically incorrect term. The more appropriate terms are either "sawn off shotgun" or "short-barreled shotgun." A short-barreled shotgun originally had a stock, an overall length greater than 26 inches, and barrel(s) over 18 inches in length. As with rifles, the barrel length can be changed either by replacing the barrel with one of a shorter length or by shortening the existing barrel. Many shotgun designs feature a barrel that is readily and easily removed without tools or advanced knowledge. The barrel length must be taken into account to the exclusion of the overall length of the shotgun in question. For the short-barreled shotgun definition to apply, the overall

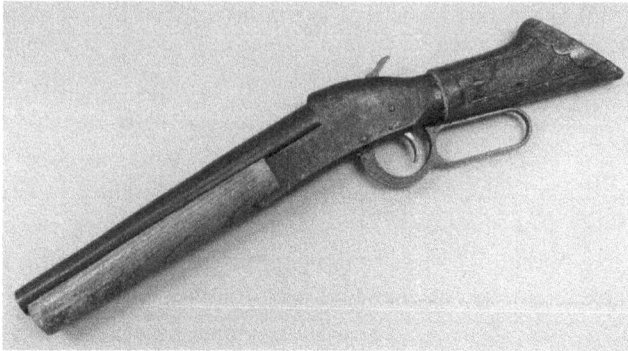

Figure 4.19 This Ithaca Model 66 12-gauge shotgun has been remade into a weapon made from a shotgun. The overall length and barrel length dimensions are used to make this determination. (Image from author's collection.)

length must remain greater than 26 inches. Practically any type of shotgun can be made into a short-barreled shotgun. Breech-loaded shotguns can be made very short, whereas the ability to reduce the length of a slide-operated or auto-loading shotgun is compromised by the presence of under-barrel tubular magazines, stocks, and other fixtures inherent to the particular design that cannot be removed. Like rifles, short-barreled shotguns can be constructed as such by a gun maker.

Weapon Made from a Shotgun

Another variation of a shotgun is the *weapon made from a shotgun* and is defined under 26 U.S.C. § 5845 (a) (2): "a weapon made from a shotgun if such weapon as modified has an overall length of less than 26 inches or a barrel or barrels of less than 18 inches in length." To be a weapon made from a shotgun, the firearm must have been originally manufactured as a shotgun and bear the requisite characteristics of a shotgun, most conspicuously the smooth bore. A weapon made from a shotgun is often mistakenly called a short-barrel or sawn-off shotgun in the vernacular, but legally, the "weapon made from a shotgun" definition is distinct. A weapon made from a shotgun will have both the stock and barrel cut down into a very compact weapon (see Figure 4.19). The stereotypical candidate of choice to manufacture a weapon made from a shotgun is a single-shot, breech-loaded shotgun. The stock is cut away to leave a minimal amount of grip, and the barrel may only just extend to the end of the stock forward of the receiver. As is often the case, when the barrel is cut down, the means of attaching the stock to the barrel is removed, so wire or tape must be used to hold the forward portion of the stock on. Auto loading shotguns and slide-action shotguns generally cannot be cut down enough to qualify as a weapon made from a shotgun due to the design particulars of these types.

Chokes   Shotgun barrels are manufactured with a level of constriction in them called *choke*. There are five basic types of choke, and they are, for all intents and purposes, the same between manufacturers. Generally, the choke is found marked on the barrel, since the choke is a function of the barrel, not the receiver. Occasionally, the term *choke* may be substituted for the term *bore*, which could create some confusion as it is not uncommon in British circles to reference the gauge as the bore. A choke may also be an external attachment installed on the muzzle of the shotgun. There are shotgun barrels that accept a choke tube at the muzzle. These choke tubes may be a type that screws on or otherwise fastens

Table 4.1 Shotgun Bore Choke Values

Choke Name	Barrel Constriction (inches)
Cylinder	None
Improved cylinder	.010″
Modified	.020″
Improved modified	.025″–.030″
Full	.035″–.040″

onto the muzzle and somewhat resembles a flash hider. Shotgun barrels can be internally threaded at the muzzle to accept a screw-in choke, which is inserted into the barrel and secured in place using a dedicated wrench. There may be subtle variations in the amount of choke prescribed by the manufacturer; these ranges are approximate. (See Table 4.1.)

The purpose of the choke is to constrict the patterning of shot as it spreads out over a given range. The tighter the bore constriction, the tighter the shot group patterning will be as the distance from the muzzle increases. Individual shotguns will pattern differently in terms of shot-pattern distribution on a target over a given distance. If shot patterning is relevant to the investigation, the examiner is encouraged to test pattern the shotgun over a series of distances using like-type ammunition to what was used in the event under investigation. Investigators seeking to make distance determinations are cautioned that individual shotguns have individual patterning characteristics. A shotgun patterning is the bias of shot-pattern dispersal, or spread, over any given distance. A shotgun may bias its patterning left or right, and up or down; this means that the majority of shot in the grouping will be focused in a center of mass that was not necessarily the point of aim that the shooter intended, assuming that the shooter was able to hit the intended point of aim.

Destructive Devices

A destructive device is defined by 26 U.S.C. § 5845(f) (2) as

> any type of weapon by whatever name known which will, or which may be readily converted to, expel a projectile by the action of an explosive or other propellant, the barrel or barrels of which have a bore of more than one-half inch in diameter, except a shotgun or shotgun shell which the Secretary finds is generally recognized as particularly suitable for sporting purposes.

Objects such as bombs, grenades, rockets having a propellant charge greater than four ounces, missiles having an explosive or incendiary charge of more than one-quarter ounce, mines, mortars, bazookas, rocket launchers, recoilless rifles, cannons, naval guns, howitzers, and other similar devices and implements are also defined as destructive devices. A subtle variation to the classification of a destructive device is the M203 grenade launcher. The M203 is a 40mm grenade launcher that is installed underneath the M16 rifle; however, any suitable weapon equipped with the MIL-1913/STANAG rail system could serve as the host weapon. The M203 can also be used as a stand-alone firearm with a specialized frame. The purchase of the receiver portion, excluding the barrel, of the M203 presently is treated as the purchase of any regular firearm, as it is not defined as a destructive device. That being said, the barrel, being 40mm in bore diameter and rifled, is a destructive device and is subject to applicable regulations therein. This rationale may leave some to question the logic, but the answer is simple: In theory, any smooth or rifled bore of any diameter could be affixed to the M203 and be usable. As long as such bore was less than one-half inch in diameter, found to be particularly suited for sporting purposes if greater than one-half inch, or fired shot shells, then the destructive-device definition would not apply.

Suppressors, Silencers, Moderators, Mufflers, and Cans

The definition of *firearm silencer* and *firearm muffler* in 18 U.S.C. 921(a) (24) is as follows:

> The terms "firearm silencer" and "firearm muffler" mean any device for silencing, muffling, or diminishing the report of a portable firearm, including any combination of parts, designed or redesigned, and intended for use in assembling or fabricating a firearm silencer or firearm muffler, and any part intended only for use in such assembly or fabrication.

Firearm suppressors are also referred to as "silencers," "moderators," "mufflers," and "cans." Regardless of what the device is called, they are all single in purpose. Suppressors have existed nearly as long as the modern firearm, for the noise generated by gunfire has always been a nuisance. Hiram P. Maxim (the son of Hiram Maxim, inventor of the Maxim machine gun) has gone into history as the father of the suppressor. However, independent research into the subject by other persons in various nations has resulted in various approaches being taken in the effort to suppress, if not eliminate, the noise generated by the report of a gunshot (see Figure 4.20). Suppressors should not be confused with other muzzle attachments such as flash hiders, muzzle brakes, compensators, chokes, and other similar items whose function on the firearm is not to reduce the noise or report of the gunshot. Suppressor design and development has been an active process around the world and is subject to ongoing improvement and refinement. There have been efforts to suppress all types of firearms that have achieved varying degrees of success. To date, even shotguns and high-powered, large-caliber firearms have been successfully suppressed.

Despite how these devices are portrayed, suppressors have a much broader scope of application. Suppressors have been stereotypically associated with professional assassins to do "hits," secret government agents, poachers, and military operators requiring stealth and surreptitious use of firearms. Anecdotal and historical evidence would appear to support

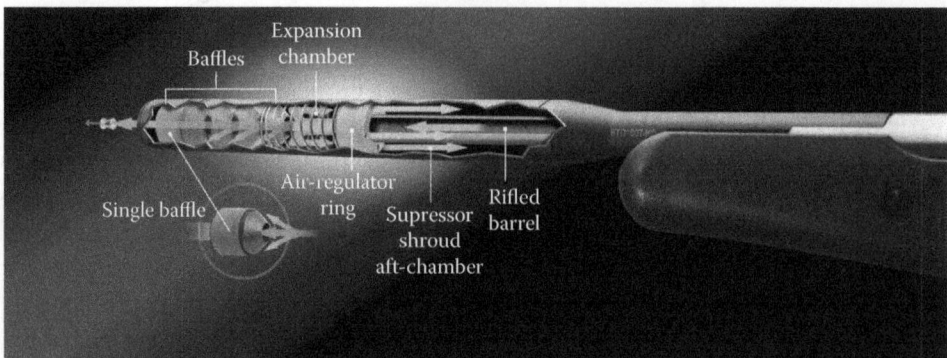

Figure 4.20 (See color insert.) Stoeger Air Gun X20 suppressor schematic showing baffles and an expansion chamber. (Image courtesy of Benelli USA.)

the use of suppressors in all the aforementioned applications; however, these are not the only uses for a suppressor. Television and movies certainly have played a role in the public perception of the suppressor, as well as how a suppressed firearm is expected to behave. Untold thousands of suppressors are legitimately possessed in the United States and used for their intended purpose—the abatement of the noise created by gunfire. Modern suppressors are often more effective at abating noise than the best ear protection, and they can be used to great benefit, such as in a training environment where the ability to hear a voice is a safety issue, in addition to prolonged and repeated noise exposure issues on the part of students and especially instructors.

A report published in August 1968 by the U.S. Army's Frankford Arsenal reported the findings of research that investigated the then state of the art of suppression of small arms. The report was coauthored by Leonard W. Skochko and Harry A. Greveris. The authors defined three sources of sound that occur during shooting, stating that, "the three noise sources are relatively independent of each other and, consequently, require different techniques for their attention" (Skochko and Greveris 1968). The three sources they identified were:

Air or propellant gas discharge preceding the projectile exit
Projectile emergence causing abrupt volumetric displacement of the air mass by the
    projectile at the weapon muzzle
Air or propellant discharge (or inflow) following projectile exit

Therefore, the object of effective suppressor design and execution is to moderate all three elements of noise creation when a firearm is discharged. What a suppressor cannot contain, however, is the mechanical noise generated by the firearm by its internal mechanisms such as the hammer, firing pin, recoil spring, and so forth that occur as a result of the workings that cause the firearm to function. To address this, firearms that are purpose built with suppression in mind can be engineered to dampen internal noise as much as possible. Another confounding variable that somewhat contradicts the stealth afforded by a suppressor is the inability to control the noise generated by ejection of cartridge casings. Such a device was patented in 2005 by Thomas Sauer under U.S. Patent 6,836,991 B1 (Sauer 2005), which specifically addressed noise, among other issues, generated by the ejection of spent cartridge casings. Another source of noise, spent casings hitting the floor after impact, have been addressed by less technical means—the attachment of a brass-catching container, such as a bag, to the ejection port.

Suppressors are mated to the host firearm by two methods: detachable suppressors and integral suppressors. A detachable suppressor is designed to be installed and removed at the whim of the operator. Attachment is typically accomplished by screwing the suppressor onto a threaded barrel, attaching it to locking lugs at the muzzle end, or joining it to an adapter attached to the barrel. Threaded attachments are finely threaded to ensure proper suppressor mounting and alignment to the bore of the host firearm. As a result, the detachable suppressor is not immediately interchangeable between different models of firearms because the threaded mounting depends entirely on what thread pattern is used to mate the suppressor to the host firearm. Many detachable suppressors have interchangeable threaded end caps that make it possible to mate the suppressor to a different host, perhaps even a host firearm that is a smaller caliber than the identified caliber the suppressor was designed to support. Alternatives to threaded barrels have included locking collars,

Figure 4.21 A High Standard H-D Military .22 pistol with integral suppressor. This is not a period piece; it was manufactured to resemble the classic H-D as it would have appeared for OSS or CIA use. (Image from author's collection.)

affixing the suppressor by the host's bayonet lug (in military rifles), and "quick detach" type setups that have become more commonplace in recent years, particularly in military applications. Quick-detach suppressors use coarser threading and generally fit over the muzzle of the weapon, if not attached to a dedicated flash hider or muzzle brake, and may include a secondary locking mechanism to secure the suppressor to the muzzle. Friction lock type linkages are not usually found on professionally manufactured suppressors, but quite often are used in homemade, clandestine suppressors as a matter of manufacturing expediency and simplicity.

An integral suppressor is one that is built into the weapon and integrated as part of the barrel. Integrally suppressed firearms have the appearance of an oversized barrel as the suppressor adds to the overall diameter of the barrel when compared to a nonsuppressed model of the same firearm. This should not be confused with firearms that have heavier match, target, or "bull" barrels that are deliberately oversized to enhance accuracy, such as the case with the Ruger 10/22 .22 rifle. This particular rifle has long been a popular platform from which to fabricate a rifle with an integrally suppressed barrel. The Heckler & Koch MP-5SD is probably the most recognizable and one of the most successful applications of the integral suppressor; however, it is not alone—integrally suppressed firearms can be found dating back to World War II. The British STEN Mark II submachine gun was modified into the STEN Mark IIS by adding a suppressor that encapsulated the barrel. Thousands of examples of the Colt Woodsmen and the High Standard H-D, both .22 pistols, were modified for clandestine and special operations by milling down the barrel, drilling holes or "ports" in the barrel, installing the suppressor body over the barrel, and securing it by way of a locking or threaded collar that had been installed to the receiver (see Figure 4.21). A Frankford Arsenal report gives an account of a silenced M1 .30 carbine (Skochko and Greveris 1968, 47). The report states that the suppressed variant was developed in Enfield, England, circa 1945, probably at the behest of the OSS (Office of Strategic Services, the forerunner of the Central Intelligence Agency).

> Since standard supersonic ammunition was used in the silenced carbine, seven holes of 0.125 inch diameter were drilled in the barrel close to the breech. This allowed the gases to be bled off through the holes, with a consequent reduction in ballistic pressure and projectile muzzle velocity. The original barrel length was also reduced to ten inches, presumably to minimize the final length of the carbine. The carbine's silencer surrounds and extends seven inches beyond the gun barrel. Inside, the silencer has a series of conical baffles, positioned throughout its whole length. The overall length and diameter of the silenced barrel are 17 and 1.4 in., respectively. (Skochko and Greveris 1968)

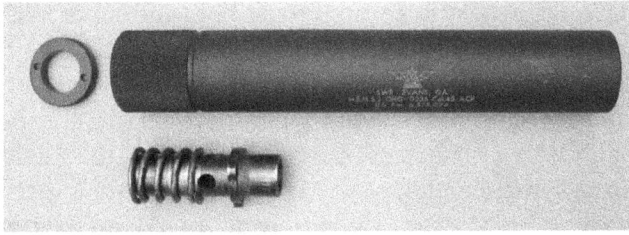

Figure 4.22 A functional suppressor manufactured by Southeast Weapons Research. It has been disassembled into its main components: the housing, recoil booster, and end cap. (Image from author's collection.)

In spite of the effort and investment made to develop the silenced M1 carbine, the report suggests that the type was not widely utilized. On the other hand, the integrally suppressed pistols (the High Standard H-D and the Colt Woodsmen) were apparently better received and quite effective.

Suppressors, regardless of the nuances of a particular design, can be broken down into three main components while certain designs call for a fourth component to be present:

1. Body, casing, or housing
2. Internal components or assembly
3. End caps, front and/or back
4. Recoil booster

The use of a linear coupling device, recoil booster, or "Nielson device" is required for certain firearm designs to permit reliable cycling of the host weapon. This is especially true on pistols with retracting barrels that move during the recoil sequence of the action, but it is not an issue on a firearm with a fixed barrel. Such a device is integral to the design and becomes relevant if the questioned device is disassembled, its presence providing technical evidence that a suspected device was at least intended by design to act as a suppressor and the designer took into account the possibility that too much gas pressure may be lost and an insufficient amount would remain to cycle the host firearm. A recoil booster typically takes on the form of a spring-loaded piston; however, variations in design could be fabricated, although functionally such a design would be equivalent (see Figure 4.22).

The internals of a suppressor vary greatly, dependent entirely upon the best thoughts and engineering of the designer. Numerous approaches to suppression have been studied and built. One approach is the use of internal "wipes," which are simply lightweight materials that allow a projectile to perforate them as it passes from the muzzle of the host, traveling through the suppressor body, and out into open air. The material used in wipes includes rubber stoppers, steel wool, acoustical foams, polyurethane bushings, and metal mesh. The drawback of such a design is that the wipes tend to wear out rather quickly and must be routinely replaced to ensure continued performance of the suppressor.

The simplest suppressor is a container that acts as a gas trap and does not use baffles, wipes, or chambering. The entire suppressor acts as the chamber. Such a suppressor may be integral or attachable to the host. The suppressor container, or body, could be filled with an aggregate filler such as metal shoelace eyelets, common fiberglass insulation, or expanding foam sealant to act as an absorbing medium. This form of suppressor may or may not contain a bore that is tapped to bleed off gas. The installation of a bore within the suppressor body certainly would be of benefit to ensure proper bore and suppressor alignment; however, it may not be absolutely required by the particular design that has been undertaken.

A third suppressor design uses internal baffles. In such a design, a plurality of baffles is employed that run the length of the suppressor. There may be the installation of some type of filler material between the baffles in the "baffle stack." In its simplest form, baffles could be metal washers stacked on top of one another within the suppressor body. Baffles of various shapes and sizes exist, again completely dependent on the designer. Baffles can be conical, shaped like the letter K (the K baffle), have a helical appearance, etc. The common characteristic of suppressor baffles is the presence of a hole that allows a projectile to pass, usually through the center. A notable exception would be a suppressor that is offset from the bore, meaning that part of the suppressor body sits beneath the barrel so as not to obstruct the sights of the host gun.

Yet another design approach is suppression by chambering, either independently or in concert with either baffles or internal fillers or wipes. Such designs rely on an intricately tuned interior that muffles the noise energy by dissipating it through a plurality of chambers. Often, acoustical chambering is used in conjunction with filler elements, either baffles or acoustically absorbent materials. In such a design, the bore is tapped to provide escape ports for the gases as they pass through the suppressor, expanding and dissipating the energy.

Suppressors can be defined as one- or two-stage designs. Two-stage suppressors have a large and small diameter sleeve or body. The large-diameter portion of the suppressor is closest the muzzle, and acts as an expansion chamber. The suppressor body then tapers down into the smaller diameter, which contains the wipes, baffles, or acoustical chambering and materials, depending on the approach of the designer. A two-stage suppressor simply confers an advantage of enhanced interior volume to permit the suppressor to work more effectively. The Bell Laboratories suppressor originally developed for the suppressed M3 Grease Gun was a two-stage design and was connected by means of a bushing. "The rear sleeve encloses a roll of wire mesh which surrounds the drilled gun barrel; the forward sleeve, which extends beyond the gun barrel muzzle, contains a stack of wire mesh discs" (Skochko and Greveris 1968). This suppressor was very large physically, extending beyond the normal barrel length of the M3, and was further enhanced by gas bleed holes being drilled in the gun barrel. A two-stage suppressor, the Sionics, was designed for Gordon Ingram's MAC-10, which not only abated the noise created during firing but also made the MAC (Military Armaments Corp.) a lot easier to shoot. A Nomex™ cover was standard equipment for the suppressor to allow the shooter to hold onto it with the nonshooting hand, serving the secondary purpose of a handle to control muzzle climb. Although built for the MAC, other hosts have been threaded to accept the Sionics suppressor.

Building a suppressor is not an exceptionally complex task, as evidenced by the fact that untold dozens of homemade examples are seized by law enforcement annually. If someone wants to see how to build a suppressor, the complete details are as close as a filed patent, found via online sources, or in the copious printed matter that discusses the construction of suppressors. Patent information is especially useful, providing the prospective builder with detailed information on suppressor design. A suppressor need not be a purpose-built device; a suppressor can be fashioned from sundry objects. Common one- and two-liter plastic beverage bottles have been used as expedient, single-use suppressors. SWD, one of the companies that manufactured MAC pistols and submachine guns, manufactured a threaded adaptor that permitted the attachment of a two-liter plastic soda bottle to a firearm. This approach may have been more novelty than functional, but the adapter itself is legally considered a suppressor. Lawn-mower and power-equipment mufflers have also been used. Quite literally, anything that is portable that is used to abate the report of gunfire becomes a suppressor by definition—even such seemingly benign objects as pillows, vegetables, or a wad of polyester batting.

Haag (1973) reported that a .22 rifle (a precise description of the host firearm was not given) had a suppressor that was attached by way of a "snug, friction fit" and contained "inner and outer tubes to be packed with glass wool." Images in the original journal article suggest that the inner tube contained a baffle stack as well. A suppressor constructed of basic hardware store materials was reported by Emanuel (1982). The host weapon was a Charter Arms AR-7. Attachment was by means of a threaded collar that capped the barrel, which had been modified by removing the front sight and the drilling of four holes. The suppressor itself was constructed of "two chrome plated sink drain pipes 9/16 inches in diameter.... The outer pipe was filled with steel wool; the inner pipe contained alternating pieces of cut plastic and a fibrous material." The AR-7 rifle, developed at Armalite by Eugene Stoner, who also developed the AR-15 (later M16), is an ideal host firearm by virtue of its design. The AR-7 was conceived as a survival gun and fits into its own butt stock when disassembled. All the parts of the rifle—the barrel, magazine, and receiver—are withdrawn from the stock, which is weatherproof, and fitted together. The barrel is mated to the receiver by a muzzle nut. The barrel is aligned by way of a small reference pin on the barrel, and the nut is screwed onto the threaded portion to secure the barrel. This subtle design cue is also found on the AR-15/M16, where a barrel-alignment guide is also found. The AR-7 is chambered in .22 Long Rifle, a relatively easy caliber to suppress, regardless of the particulars of how the suppressor will be constructed and what materials will be used. This fact, coupled with the threaded barrel attachment and a removable barrel, only enhances the potential. With the barrel removed, it is very easy to alter the barrel to conform to the needs of the builder.

Not all suppressors can be disassembled. In the case of a sealed unit, the suppressor would have to be cleaned by immersion in solvent and allowed to soak or placed into an ultrasonic cleaner. Many are designed to allow at least partial disassembly to facilitate some measure of cleaning. At least one end cap is threaded to the suppressor body and is removable, permitting internal components to be removed and cleaned. If the suppressor uses baffles, then the entire baffle stack could be removed and cleaned, as well as the interior of the suppressor body. Other suppressors may only allow for the linear coupling device to be removed, in which case the entire suppressor would have to be submerged in a suitable cleaning solvent. Commercial parts washers and ultrasonic cleaners work well for the can that cannot be taken apart. Suppressors in .22 are especially susceptible to buildup, as the priming and powder formulations for .22 cartridges tend to be very dirty.

Suppressors can be constructed of any number of metals or other materials. Aluminum has the benefit of corrosion resistance, high strength, and light weight. Weight is of special concern, as a heavy suppressor would make the weapon front heavy and affect the balance felt by the operator. Aluminum is joined by other materials such as stainless steel, ordnance-grade steel, titanium, Inconel alloys, ceramics, and even carbon fiber. Some suppressors incorporate several different materials, using aluminum for the body but other materials for the internal components. In the absence of these materials, the homemade suppressor can be constructed of PVC pipe or any tubular metal stock obtained at the local hardware store or a specialty supply shop that provides metalworking and industrial operations.

Commercial suppressors may be optimized to be fired either wet or dry, as specified by the manufacturer. Wet suppressors require the addition of some liquid to work optimally. The liquid can be water, saliva, or a commercial product designed for such application. As the weapon is fired, the liquid charge is gradually used up, such that the peak effectiveness will be reduced, requiring the suppressor to be refilled. A water charge in a wet suppressor can be expected to last several dozen rounds before needing replenishment; however, the operator will begin to notice a reduction in the effectiveness of the wet suppressor as the liquid charge is used up. A side effect of the wet suppressor is the tremendous blowback of the liquid back into the host firearm; thus it is recommended that host firearms be thoroughly dried and cleaned promptly after the shooting session. Dry suppressors do not require a liquid charge to work at peak efficiency; in fact, inserting liquid in such suppressors may be discouraged.

The effectiveness of a suppressor is measured by the sound reduction measured in decibels. The relative effectiveness or noise-abatement quality of a particular suppressor can be subject to much debate and is as much in the opinion of the observer as is in any testing that is intended to follow an objective, scientific approach. In cases where the suppressor is used in conjunction with a firearm chambered in a cartridge whose performance exceeds the speed of sound, the suppressor design may be adapted to reduce the energy enough to slow the projectile to subsonic speeds. Numerous manufacturers produce subsonic loads of cartridges that exceed the speed of sound, such as .22, 9×19mm, and 5.7×28mm. Sound travels at approximately 1,180 feet per second at 68°F (20°C) in the open air. Thus, calibers that are expected to exceed this velocity will generate a sonic boom, thereby reducing the effectiveness of the suppressor to some degree. The use of subsonic ammunition eliminates such concerns. This acoustical hurdle is not a factor for cartridges with velocities at subsonic levels.

The legality of private possession of suppressors varies greatly. In the United States, the suppressor is defined as a firearm, although the suppressor in and of itself is not capable of expelling a projectile and is not designed as such. In the United States, the manufacture, possession, and transfer of suppressors fall under the purview of the 1934 National Firearms Act. Individual suppressor components may also be defined as suppressors, even if unassembled and not accompanied by the balance of components to complete the fabrication of a suppressor. It was a common practice to sell suppressor kits in the United States until suppressor components were classified as suppressors themselves. A suppressor kit could be the housing or body, minus internal components, or it could be the components such as baffles, but absent the body.

Efforts have been made to suppress all varieties of firearms, although suppressors tend to be associated most closely with pistols. This is perhaps mainly due to the appearance and application of suppressors in popular television and movie productions. Suppression of firearms where the ammunition is contained within a sealed chamber obviously overcomes the concern of the potential for noise escaping from the chamber and thus reducing or defeating the efforts of a sound suppressor, which lends the primary application of suppressors to firearms other than revolver types. There have been attempts to suppress revolvers, but such attempts have always seemed to be dismissed as foolishness, due to the inherent characteristics of the revolver design that allows the escape of noise from the chambers in the cylinder. As mentioned previously, there is a small gap between the cylinder and the forcing cone of the barrel, permitting the escape of gases and thus noise. Bearing this in mind, certain revolver designs eliminate this small gap as much as possible and suggest that some degree of successful suppression may be possible. Such an apparatus was invented in 1973. The host weapon was a Dan Wesson Model 12 revolver with a suppressor attached. "The silencer was mounted by removing the barrel nut from the revolver and screwing the silencer in its place" (Morris 1973). In the case of a Dan Wesson revolver, it is possible to adjust the gas seal between the front face of the cylinder at the point where it meets the barrel, which would appear to be one reason for using a Dan Wesson to the exclusion of other brands; the other would be the ease of attachment to the host.

Inert, replica copies of suppressors are manufactured and overtly resemble functional suppressors. The same companies that build functional suppressors often build these inert replicas and market them as faux suppressors, fake cans, barrel safety extensions, and training devices. Regardless of the name, these items are clearly identified as inert, replica, or fake to avoid confusing them with functional suppressors. They are not properly referred to as "nonfunctional" devices because of the implication that the object in question was once functional or could be restored or made to function. Some examples of inert suppressor analogs are solid, having no bore and only the overt appearance of a suppressor and means of attachment to the host weapon, whereas other examples are "shoot through," meaning that they have a bore and can be mounted onto a host firearm that can be fired with the replica in place. Inert copies should not have the ability to be disassembled. Inspection of the end caps should reveal firm and intact welds or other permanent sealing that prevents disassembly. Furthermore, in a shoot-through type, a visual inspection by looking down the bore should reveal a smooth bore that is absent any holes or other indicia that suggest that gases resulting from a gunshot could be diverted into the body of the apparatus, as opposed to passing straight through (see Figure 4.23).

In one example the author is familiar with, a commercially available barrel extension that was factory threaded to mate with the host firearm was substantially modified to act as a suppressor. A portion of the barrel extension body was milled away, leaving a circular, winding appearance. Ports were then drilled through the bore of the barrel extension, and the device was then covered with an expandable heavy plastic material consistent in appearance with that used to form nonslip tool handles. When test fired, the apparatus held together but was, speaking generously, very marginally effective at abating the noise of a gunshot. The author classified the device as a suppressor, given the apparent attempt to fabricate a device that would abate the report of a gunshot, even if the net result in noise suppression was marginal.

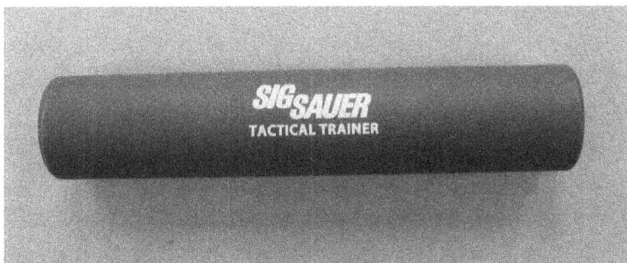

Figure 4.23 A replica suppressor manufactured by Sig Sauer. These were sold as an accessory to the Sig Mosquito .22 pistol with threaded barrel to give the suppressed "look." The suppressor replica could also be purchased separately. This replica is in no way functional; the body is solid, but it does have a bore, permitting the host gun to shoot through it. (Image from author's collection.)

Other barrel attachments may be mistaken for suppressors, including elongated flash hiders, muzzle compensators, and harmonic stabilizers. Flash hiders and muzzle compensators are readily recognized by ports or cuts that are designed to divert gases away from the muzzle instead of allowing all the muzzle gas to exit directly outward from the bore. The purpose of a harmonic stabilizer is to eliminate barrel fluctuations that result from gunfire. The harmonic stabilizer may be constructed around the barrel, giving the barrel a "bull barrel" or appearance of having an integral suppressor. Other harmonic stabilizers attach at the muzzle, which may again give the appearance of a detachable suppressor. A definitive determination may not be possible without disassembly or x-ray of the apparatus.

There are several key points to keep in mind when conducting searches or investigations where suspected suppressors or components could be encountered:

- Any object that could be used as part of a suppressor, even in absence of other parts, such as a tube, baffles, or a combination of bits and pieces that could possibly be made to fashion into a suppressor. In such a situation, the surveyor must be able to recognize objects that are particularly suited for suppressor construction.
- Apparatus, modifications, or means of attachment of a device or combination of parts to a firearm. This can be coupled with modifications to a firearm in an attempt to act as a host to an internal or attachable suppressor.
- Modifications made to inert suppressor copies or barrel extensions indicating an effort to remanufacture the product into a suppressor. Such indication may include drilling or cutting into the object or attempts to reseal it by welds or solder.
- Plans, schematics, or instruction sheets that address how to construct a suppressor. These may take the form of blueprints, mechanical drawings, or materials downloaded and printed from online sources.
- The presence of firearms with factory-threaded barrel attachments, such as flash hiders, that are or have been removed. This would include aftermarket threaded barrels, firearms with factory-threaded barrels, or barrels that have been threaded in the workshop. In such cases, this permits a ready host firearm that the constructor can use as a basis to resolve the problem of coupling a suppressor to the host.
- Leftover or discarded materials, including aborted construction projects that could have been attempts at previous suppressor manufacture.

Many currently manufactured firearms come equipped from the factory with a threaded barrel. Factory-threaded barrels were once considered taboo, but this is no longer the case. The various and sundry copies of the MAC series of firearms have always come with threaded barrels. The Intratec TEC line of handguns (TEC-9, TEC-DC9, TEC-9 mini as well as the TEC-22) were furnished with threaded barrels until the feature was removed in response to legislation targeting the pistols. Heckler & Koch produces an entire line of suppressible handguns in 9×19mm, .40 S&W, and .45 ACP, all equipped with threaded barrels. Their Mark 23 handgun chambered in .45 ACP and designed specifically for SOCOM has come with a threaded barrel since its first delivery in 1996; however, it was discontinued in late 2010. SIG-Sauer offers many handguns with threaded barrels, including the P220 Combat chambered in .45 ACP; the P226 Elite chambered in 9×19mm, .40 S&W, and .357 SIG; the P229 Elite, which is available in 9×19mm only with threaded barrel; the P239 Tactical chambered in 9×19mm; and the Mosquito, which is only offered chambered in .22. Beretta offers the Px4 Storm Special Duty chambered in .45 ACP with a threaded barrel. Remington Arms offers

Figure 4.24 Operators with the International Security Assistance Force in Afghanistan demonstrate room-clearing techniques. The operator on the left is firing an H&K UMP-45 submachine gun with an attached suppressor; the operator to the right fires an integrally suppressed H&K MP-5SD. (U.S. Air Force image; photographer Staff Sgt. Joseph Swafford.)

numerous suppressor-ready rifles based on their Model 700 bolt-action rifle, their AR-pattern semiautomatic rifle, and even the Model 597 .22 semiautomatic rifle. Smith & Wesson offers variants on the M&P pistol line with threaded barrels in every caliber.

The number and variety of firearms that come available with threaded barrels is only likely to increase. Firearms with attached flash hiders or muzzle attachments that are installed by threads make readily available hosts by removing the accessory and accessing the threads at the muzzle. In addition to factory offerings, custom barrel shops offer aftermarket, direct-fit barrels for practically all popular pistols that have barrels threaded to customer specification. Suppressor manufacturers themselves often manufacture muzzle brakes or flash hiders that are easy to install for adaptation of their product to the host firearm, especially with rifles. Suppressor-like devices have been manufactured and permanently attached to paintball guns. The BATFE addressed the existence of these devices and ruled that, as long as the device remained permanently attached to the nonregulated article (a paintball gun was mentioned specifically), there was no violation of law. When tested using a firearm, these ported devices were found to be capable of performing as a suppressor. The ruling further stated that should the device be removed, it would constitute making of a suppressor (Truscott 2005).

Figure 4.24 shows operators wielding firearms with noise-suppression devices.

Machine Guns

A *machine gun* is defined in 26 USC § 5845(b) as

> any weapon which shoots, is designed to shoot, or can be readily restored to shoot, automatically more than one shot, without manual reloading, by a single function of the trigger. The term shall also include the frame or receiver of any such weapon, any part designed and intended solely and exclusively, or combination of parts designed and intended, for use in converting a weapon into a machinegun, and any combination of parts from which a machinegun can be assembled if such parts are in the possession or under the control of a person.

Machine guns can be operated by recoil, gas action, or a combination of recoil and gas action. Certain machine guns are not capable of sustained, continuous fire per a single press of the trigger; instead, their rate of fire is moderated by a "burst" feature, typically two or three shots per press of the trigger, depending on how the fire control parts are set up. The purpose of a burst is to retard the cyclic rate of fire of the firearm, allowing the shooter to fire multiple shots while regulating the rate of fire to a manageable level in an attempt to deter excessive ammunition consumption and the inherent loss of the target due to muzzle climb that all machine guns experience during sustained fire. The argument in support of a mechanically induced modulation of fire by burst circled around the point that under sustained fire, machine guns, especially those that are handheld, typically cannot be held on target by the operator beyond two to five rounds fired. The burst option has never been universally accepted as the best approach to controlling automatic fire in lieu of a well-trained trigger finger. As a whole, machine guns fire at a cyclic rate between approximately 400 and 1,200 rounds per minute. This cyclic rate may be a theoretical value unless the firearm is fed with an inexhaustible supply of ammunition (such as with a belt-fed gun). Otherwise, time is consumed changing magazines, and thus the rate of fire per minute is greatly reduced. However, the rate of fire can be calculated on the basis of the time interval over which a given number of rounds are fired.

Machine guns can be purpose-built weapons by a manufacturer. However, many firearms can be converted into machine guns from semiautomatic firearms. The approach used varies greatly and requires some measure of understanding of the principle of operation that makes that particular firearm function. Some firearms are quite commonly converted to machine guns due to availability of receivers and parts, simplicity of design, and myriad other factors such as builder interest, availability of open-source information on the topic, and individual understanding of the operating principle behind the weapon. Machine guns can be individual weapons capable of being fired from the shoulder or in other individual shooting postures, but these can also include crew-served weapons that require several individuals to effectively maneuver the weapon. In fact, a machine gun can take the form of a rifle, shotgun, or pistol.

Submachine guns are still machine guns by definition, but they are more specifically defined as those firing cartridges *traditionally* categorized as pistol calibers, such as 9×19mm, .45 ACP, .380 ACP, 7.65 Tokarev, and even .22 caliber. The physical dimensions of the firearm are largely irrelevant, as the determination is based upon the dimension of ammunition that is used. There was nothing small or light about the Thompson Model 1927 or the simplified military version, the M1. However, they are submachine guns by definition, as they were chambered for the .45 ACP cartridge. The MAC-11, chambered in .380 ACP, offers one of the highest rates of fire of a mechanically operated machine gun, firing in the range of 1,200 rounds per minute. Other examples of submachine guns include the M3 Grease Gun, the Israeli UZI, the C.Z. Zastava M61J (often called the Scorpion), the British STEN gun, the Lancaster, the Sterling, and innumerable other models. The introduction of the intermediate* cartridge in the later stages of World War II forced a nuanced definition of the term *submachine gun*. The 5.56×45mm, although not considered a full-sized, full-powered rifle cartridge, is not defined as a pistol cartridge either; thus

---

* The intermediate cartridge is larger than the pistol calibers, but is smaller than full-size, full-powered rifle cartridges such as 7.62×51mm. The first intermediate cartridge was the 7.92×33mm *kurz* developed in World War II Germany for the MP-43, the forerunner of the modern military rifle.

there is some room for dispute over whether the term *submachine gun* would apply to an intermediate-caliber full-auto firearm.

The term *submachine gun* has been supplemented, but not replaced, by a newer term, *personal defense weapon* (PDW). Such weapons are designed to give compact fire power to users who may have a weapon requirement exceeding that provided by handguns, but cannot necessarily deploy a full-sized rifle or even a carbine. The development of PDWs was made with dignitary-protective details, aircrews, and counterterrorist operators in mind. The term *machine pistol* is often erroneously applied broadly to submachine guns. The term was probably first bestowed by German designers who developed machine guns that fired pistol-caliber cartridges as opposed to full-size rifle cartridges.

The Bergman MP-18 was one of the first practical machine pistols to be deployed. Designed by Theodor Bergman, versions were chambered in 9×19mm Luger and 7.63×25mm Mauser, both considered preeminent handgun cartridges of the day. The designation MP denoted *Machine Pistole*, a term that would be used for many years to designate machine guns firing pistol cartridges—the MP-28, MP-38, and MP-40—and even into more modern times with such firearms as H&K's MP-5 and the MP-7. The term has been applied almost exclusively to German-sourced firearms of this variety.

Figure 4.25 shows a Fabrique Nationale P-90 submachine gun chambered in the FN 5.7×28mm.

Most machine guns are configured to combine the safety switch with the fire selector switch. The M16 combines the safety switch with the selector, as do Heckler & Koch machine guns, Uzis, and others. Another approach is to separate the manual safety switch from the selector, as is the case with such firearms as the M14, M2 carbine, and Ruger AC-556. MAC machine guns have a simple two-position safety switch on the right side of the receiver adjacent the trigger. The sliding switch indicates safe or fire mode. The selector switch for semiautomatic or fully automatic fire is located on the left side of the frame and is a two-position rotating lever. The lever itself resembles the one used on the AR-pattern firearm safety/selector switch. The Beretta Model 38A submachine gun had a rather novel approach; this particular model has two triggers: one that permits single-shot fire, the other full-auto fire. This is also true for the 38A derivatives manufactured by Beretta—the Model 38/42 and the Model 5 (see Figure 4.26). Yet another option was a staged trigger, as in the case of the MG-34 and MG-42. A slight press of the trigger yielded single shots, while

Figure 4.25 The Fabrique Nationale P-90 submachine gun chambered in the FN 5.7×28mm. The P-90 is a classic bull-pup design. The PS-90 is a commercially available semiautomatic copy. The PS-90 is fitted with a 16.1-inch barrel, although conversion to the short submachine gun barrel is not difficult. (Image from author's collection.)

Figure 4.26 A Beretta Model 1938A submachine gun; observe the two triggers. (Image from author's collection.)

completely depressing the trigger would result in fully automatic firing. It is important to note that there are "dummy" or inert-replica selector switches that physically resemble an actual selector switch; factually, these are props that have no functional value whatsoever. These articles look the part, but that is it.

Machine guns can be manufactured as such in a factory environment; however, a machine gun could also be a firearm that has been remanufactured from a semiautomatic firearm. In the United States, in such an instance where a firearm has been remanufactured, the name of the firm, including city and state, that performed the work must be conspicuously marked on the receiver. Abbreviations are acceptable, as long as they are readily and easily recognized. Commercially manufactured machine guns are not generally marked "machine gun" or in any other way to convey that fact, other than recognition of the selector switch or other unique feature to a particular model that would indicate the firearm is a machine gun.

Figure 4.27 depicts an M61J submachine gun manufactured by Z.C. Zastava, Czechoslovakia. Figure 4.28 shows an UZI submachine gun. Figure 4.29 shows a disassembled Mini-UZI. Figure 4.30 shows the full-auto sear in an UZI.

Gatling guns were the earliest form of rapid-fire weapons. The Gatling gun took its name from its inventor, Dr. Richard A. Gatling. The Gatling gun is not defined as a machine gun because only one round is expelled per press of the "trigger" (or crank, as is the case of the Gatling gun). Gatling's patent, titled "Improvement in Revolving Battery-Guns" was issued November 4, 1862. The development of the Gatling gun would not have been possible without the refinement of the self-contained cartridge. The Gatling gun was not supplied ammunition by a belt or magazine like modern arms, but instead via a gravity-fed reservoir on top of the weapon. Ammunition was dropped into a hopper and channeled to the breech. The spent casings were then ejected automatically onto the ground from the barrel they were fired from. A total of six barrels were used, and the firing barrel was at the 12 o'clock position on the receiver body.

Hand-crank apparatuses have been devised to fit onto semiautomatic belt-fed weapons to simulate full-auto fire without the implications of manufacturing a machine gun. Turning the crank causes a mechanical "finger" to press the trigger. The faster the crank is turned, the faster the gun fires. Despite its antiquity, the Gatling gun concept has remained. The M61 Vulcan cannon is a six-barreled 20mm cannon that has been used as shipboard armament and has been the standard cannon used on U.S. military aircraft

Figure 4.27 The M61J is called a machine pistol, but it is still a submachine gun. It was manufactured by Z.C. Zastava in Czechoslovakia, and is chambered in 7.65mm (.32 ACP). The semiautomatic pistol version, the VZ-61, does not have the shoulder stock. Note the selector switch and the folding wire stock on the pictured example. (Image from author's collection.)

Figure 4.28 (See color insert.) The UZI submachine gun. The depicted examples were manufactured as submachine guns by Israeli Military Industries in Israel and are so marked on the left side of the receiver as UZI SMG. The English-marked selector switch is above the trigger, depicted in the lower example. Hebrew-marked selectors are also encountered. (Image from author's collection.)

since the early 1960s. The M134 Mini Gun, originally developed by General Electric and a derivative of the M61, went into service in the 1960s and remains in front-line service to date. There are two significant differences between the original Gatling gun and the modern multibarreled guns. The first is that the modern versions are driven by an electric motor, hydraulics, or pneumatic systems instead of a hand crank. The second is that the Gatling gun relied on percussion-based ignition, whereas the M61 and

Figure 4.29 A disassembled Mini-UZI. Note the compensator cuts in the barrel. The receiver is in the center of the image; this is the gun part. (Image from author's collection.)

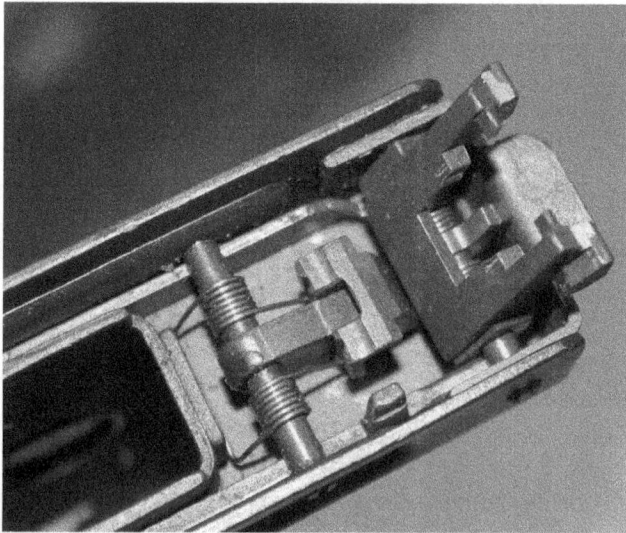

Figure 4.30 Close-up image showing the full-auto sear in an UZI; it is located at the back of the grip assembly. The sear itself is that U-shaped part that has been flipped up to expose it. (Image from author's collection.)

the M134 use electrically primed ammunition. In October 2011, Colt reintroduced the Model 1877 Gatling gun. It is a full-size, faithful reproduction of the original Model 1877, chambered in .45-70 Government and is fully functional. It is a rather elegant looking piece and includes all the requisite accessories and tackle to complete the rig. Its brass, iron, and wood construction are indicative of gun-making craftsmanship of a bygone era. It is claimed to have a rate of fire of 800 rounds per minute by hand crank (Colt's Manufacturing 2011).

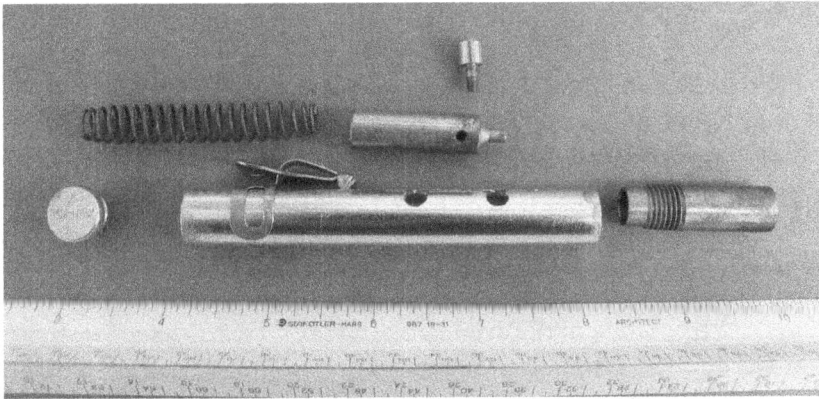

Figure 4.31 A pen gun chambered in .25 ACP. The barrel is unscrewed from the body to load and unload. A simple spring-loaded striker system is used to detonate the cartridge. The striker is pulled back and can be placed in a "cocked but safe" condition by sliding the handle into the round notches from the channel that the striker travels on. It is well made and a very interesting design, leaving one to wonder where it originated. (Image from author's collection.)

Improvised Firearms

A device need not be professionally constructed by a manufacturer of firearms to be legally classified as a firearm. The 1950s saw the rise of improvised firearms that became widely known as zip guns. Zip guns typically were fashioned from commonly available sundry materials, using simple hand tools and basic mechanical knowledge. Zip guns were often the product of high school or vocational/trade school metal shop projects or made at home in the basement or the garage. Steel pipes, steel bolts, metal-bodied flashlights, pens, tire pressure gauges, and other similar objects have all been employed in the production of zip guns (see Figure 4.31). The stereotypical urban zip gun had a barrel fabricated from metal antennae with internal components donated from mechanical clocks, cap guns, nails, and even rubber bands. Such devices usually employ a striker-type spring-loaded firing pin and have a readily accessible breech to load and unload, requiring the use of a locking mechanism that can be opened and shut, such as a threaded cap. The underlying feature that can be expected to be common of zip guns is the remanufacturing of an existing non-gun manufactured article into the firearm. This is generally preferred over fabricating each component of the firearm, which can be restricted by access to adequate machine tools, materials, and the knowledge of how to use them.

Improvised firearms of all types have been the product of cottage industries in foreign countries, particularly during times of conflict, and firearms or components of firearms were hand fabricated. Such firearms were commonly encountered in places such as Southeast Asia and, more recently, in the northern territories of Pakistan, throughout Afghanistan, the Philippine Islands, and throughout Africa. Improvised firearms may also be fashioned from existing manufactured articles, and may even include components from manufactured firearms, with the balance of components being fabricated by hand from whatever raw materials are suitable and available. Improvised firearms quite often are crude in appearance and construction, and generally are simple in the extreme in design and execution. On the other hand, such devices may be elaborately constructed and so closely resemble a factory-produced article that proper identification may be difficult.

Improvised firearms may take on the form of mundane objects that are not ordinarily identified or considered potential firearms, such as umbrellas, cameras, walking canes, heavy gloves, flashlights, cigarette lighters, cellular phones, smoking pipes, belt buckles, and briefcases. These items can be equipped to conceal and remotely fire a stored firearm. These improvised weapons may have, in fact, been professionally manufactured as clandestine firearms dedicated as part of a covert operative's kit. As such, identification of them remains somewhat elusive, except that the standard of manufacture is quite high in comparison to homemade improvised firearms. It is unusual, although not unheard of, for such improvised or clandestine weapons to appear, and it may be very difficult, if not impossible, to trace them back to the original source, as such articles are ordinarily devoid of serial numbers or other manufacturing marks that would indicate their origin. Legally speaking, improvised weapons frequently can be classified as "Any Other Weapons" due to the presented characteristics of the article.

Any Other Weapons

In the United States, a firearm may be classified in the "Any Other Weapons" category as defined by 26 USC § 5845 (e) as follows:

> any weapon or device capable of being concealed on the person from which a shot can be discharged through the energy of an explosive, a pistol or revolver having a barrel with a smooth bore designed or redesigned to fire a fixed shotgun shell, weapons with combination shotgun and rifle barrels 12 inches or more, less than 18 inches in length, from which only a single discharge can be made from either barrel without manual reloading, and shall include any such weapon which may be readily restored to fire. Such term shall not include a pistol or a revolver having a rifled bore, or rifled bores, or weapons designed, made, or intended to be fired from the shoulder and not capable of firing fixed ammunition.

Other examples of Any Other Weapons include firearms not readily recognized as firearms and generally called improvised firearms, as previously described in this text, such as flashlight guns, suitcase guns, cane guns, umbrella guns, and the like. Handguns with a vertical fore grip attached were deemed to be in the Any Other Weapons category by ATF directive (BATFE 2006a). Like machine guns, sound suppressors, destructive devices, short-barreled rifles, and shotguns, Any Other Weapons fall under the purview of the National Firearms Act of 1934. For lawful possession in the United States, the provisions of the National Firearms Act must be satisfied, with the device being lawfully registered and excise tax paid.

As is the case with most firearms, there are some designs that closely resemble an Any Other Weapon, but they do not meet the requisite characteristics. The American Derringer Corporation manufactured a firearm that was a facsimile of an Any Other Weapon, but it did not qualify as such due to its particular design characteristics. The Stinger Model 2 was chambered in .25 ACP and was constructed to resemble a pen; however, it was hinged to create a short "grip" at an angle to the bore, and it was designed not to fire until the grip was moved to be angled to the bore. In addition to the short "stock" that was created by the hinged portion of the device, the bore was rifled. With these two factors in mind, it is not an Any Other Weapon, but a pistol.

There are other examples of similar pistols manufactured by Stinger. Ironically, carbine versions of the pen pistol were marketed. Using the classification of a smooth-bore handgun as an Any Other Weapon, a shotgun-type receiver originally manufactured with a short stock or pistol grip at an angle to the bore, but not a traditional shoulder stock, is classified as an Any Other Weapon. The key elements in this particular instance rely on the facts that the original manufacturer could legally manufacture an Any Other Weapon and that the receiver was identified as such, and bore the necessary characteristics in its configuration to qualify (short grip at an angle to the bore and smooth bore).

Another similar example is a knife gun manufactured by Powell and Brown Industries, also known as Powell Knife Pistol, Inc. (PKP). The overt appearance of the firearm is that of a fixed-blade hunting-style knife or bayonet. Three appendages—the trigger, hammer, and a push bar to open the action—may go unnoticed by casual observation, but they are present and would be recognized as irregular for typical knife construction, even if the firearm portion was not readily recognized. The knife gun is a single-shot, breech-loaded affair and has been reported in calibers .22, .38, and .45. The barrel is above the blade, and the action opens on a hinge. A rudimentary notch-style rear sight is present on the hilt. The company obtained their license to manufacture in the mid-1980s; however, prototypes are certain to have existed before that. The firearm is marked on the left side with caliber, manufacturer, and serial number. The Powell Knife Gun is defined as a pistol, not an Any Other Weapon, by virtue of its short grip at an angle to the bore and the rifled bore.

In addition to such improvised firearms, starter guns and flare guns (also called Very Pistols) have been remanufactured or reconfigured into firearms. Flare guns can be used to expel fixed ammunition cartridges by using a subcaliber insert. An insert allows for a cartridge that is smaller than the bore or chamber of the host device. The insert simply fills the gap and aligns the firing pin with the rim or primer. An insert can be of any caliber, but primarily centers on handgun cartridges and shot shells. Most modern flare guns are made of plastic and will not handle the pressure created by expulsion of shot shells or even handgun-caliber cartridges from them. BATFE issued an open letter dated May 4, 2006, concerning the use of such inserts in a flare gun. The letter reported the results of experiments conducted by the BATFE Firearms Technology Branch into the use of inserts in flare guns or the firing of ammunition from a flare gun without an insert. Common modern flare guns have bores measuring 25mm, 37mm, or 12 gauge. The experiments concluded that the flare guns were destroyed in the test-fire procedure after a single shot. Furthermore, the letter reemphasized the legalities of using a flare gun as a firearm (BATFE 2006b).

Antique Firearms

An antique firearm is classified under 18 U.S.C. § 921(a) (16) as follows:

> (A) any firearm (including any firearm with a matchlock, flintlock, percussion cap, or similar type of ignition system) manufactured in or before 1898; or (B) any replica of any firearm described in subparagraph (A) if such replica—(i) is not designed or redesigned for using rim fire or conventional center fire fixed ammunition, or (ii) uses rim fire or conventional center fire fixed ammunition which is no longer manufactured in the United States and which is not readily available in the ordinary channels of commercial trade; or (C) any muzzle loading rifle, muzzle loading shotgun, or muzzle loading pistol, which is designed to use black powder, or a black powder substitute, and which cannot use fixed ammunition. For purposes of this subparagraph, the term "antique firearm" shall not include any weapon which incorporates

a firearm frame or receiver, any firearm which is converted into a muzzle loading weapon, or any muzzle loading weapon which can be readily converted to fire fixed ammunition by replacing the barrel, bolt, breechblock, or any combination thereof.

When defining a firearm as an antique, there may be instances where state law may contradict U.S. Code with respect to the particulars of what constitutes an antique firearm. The question must be examined based upon the venue where the judicial controversy is being heard. Obviously, federal laws are not applied outside the federal system, so if the controversy is confined within the jurisdiction of state-level officials, the status of an antique as defined by relevant state law would be observed. The status of a firearm as an antique presents certain legal questions that vary from venue to venue, especially if the question surrounds a person who would otherwise be prohibited from possessing a firearm but who is in possession of an antique firearm. Many venues permit an antique to be reclassified as a firearm and deprive it of antique status should the antique be used in the commission of a crime.

Air Guns

Guns that operate using a form of compressed gas have existed since at least the late 1500s. Criminal use of an air gun was central to the plot in the Sherlock Holmes mystery *The Empty House,* published in 1903 but set in the last years of the nineteenth century. The esteemed Colonel Sebastian Moran attempted to murder Mr. Holmes using an air gun with which Moran had already murdered the Honourable Ronald Adair. An aside in the story is that Holmes's companion, Dr. John Watson, erroneously identified the recovered projectile from the Adair crime scene first as a soft-nose revolver bullet and later in the story as an expanded revolver bullet. Watson is perplexed by the evidence at the murder scene: no one reported hearing any noise; Adair was alone at the time; and there was a large sum of money still in the room where the body was found. As the story draws to a conclusion, the air gun is recovered at the scene of Moran's arrest, an empty house across the street from 221B Baker Street, where Holmes resided. Moran was witnessed shooting a wax bust of Holmes strategically positioned in the window of his apartment and was hastily taken into custody by Scotland Yard detectives who were hiding in the vacant house. Holmes was quick in recognizing the weapon, saying that "I knew Von Herder, the blind German mechanic, who constructed it to the order of the late Professor Moriarty" (Doyle 1903/1992).

Air guns are popular around the world in venues where there are significant prohibitions against the possession of conventional arms, but this statement is equally true for places where firearm restrictions are more relaxed. The most modern trend in air-operated guns has been to replicate historical and contemporary military firearms, which has only contributed to broadening the interest in air guns and to advancing the state of the art. Air guns have become popular with youths in North America; however, air guns can be and are used for hunting and match competitions, including pistol and rifle matches at the Olympics.

The power of an air gun should not be underestimated; injuries can and do occur when persons are shot with even low-powered devices. Despite popular perceptions that look upon air guns as toys, many would be surprised to hear that air guns are used for hunting and have appeared in military applications. An oft-cited example of a military application of an air gun is the Austrian Girardoni, which entered service around 1780.

This air gun had a capacity of 22 rounds and an interchangeable self-contained reservoir to supply the compressed air. The reservoir stored enough air to permit several hundred shots, a considerable advantage in an era when muzzle-loaded muskets were the norm. American explorers Lewis and Clark carried a Girardoni-pattern air rifle with them on their famous expedition. That particular air rifle was apparently in caliber .462 (National Rifle Association n.d.).

The term *air gun* is a generic one, encompassing a plurality of different designs and principles of operation; they are truly a study all unto themselves outside the realm of firearms that use energetic materials to expel projectiles. Air guns may also be referred to as airsoft guns, the two terms are interchangeable. Air guns run the gambit from very inexpensive dime-store varieties of plastic construction to very precise and complex instruments fabricated of high-quality materials, made to the highest standard, and priced well in excess of comparable firearms. The brands Daisy, Crosman, and Tipman are the names most synonymous with the American air-gun industry. These names are joined by other distinguished firms such as Hammereli, Anschutz, Walther, and Benelli. Numerous firearm manufacturers license or permit air-gun makers to produce highly accurate, practically indistinguishable reproductions of their respective products.

Air guns have been broken into two broad classifications based upon the projectile they are chambered for: the BB gun and the pellet gun. Both pistol and rifle configurations are made that use one or the other. Air guns may be single-shot, chamber-loaded designs, but certain BB guns can be preloaded with a good quantity of BBs that are poured into the magazine. Certain pellet guns are fed through a rotary magazine, much like a revolver. Classically, the BB was a .177″/4.55mm diameter ball constructed of steel that was coated with copper nickel, or less commonly, bare nickel. The common .177″ pellet is a diabolo design, with a head that tapers to a waist that in turn widens to the hollowed tail or base. There is a great variety of head designs manufactured, with the most common "drugstore" variety having a flat nose. Traditionally made of lead, the material selection for pellets varies, including types made of lead in combination with polymer. There is an amazing array of pellets available for any air-gun application, and they range in caliber from the basic .177″ upwards to .50″ projectiles weighing some 200 grains, a projectile comparable to those used in self-contained cartridges!

The requisite air or gas can be introduced into the air gun by a number of methods and is design specific. Air guns can be supplied the compressed gas for function either by mechanical charging to fill an onboard reservoir or by using compressed-gas cartridges. Mechanically charged air guns use a lever, crank, pump, or other apparatus to introduce air into a cylinder; the air is then released when the trigger is pressed. This expels the air from the piston, and the pump must be worked again. Other designs employ a refillable tank that is charged by an external air compressor or pump. Especially prevalent in entry-level guns are designs that use disposable CO_2 cartridges. Disposable cartridges are loaded and then locked into place with a bracket, which also contains the nozzle that pierces the top of the cartridge and releases the gas, thereby charging the system. Such systems are not well sealed and do not hold a charge very long, and the cartridge will have to be soon replaced if not exhausted by firing the air gun. Within any given design there is a plurality of variations of how a particular air gun operates.

The paintball gun is a purpose-built device intended to expel small paint-filled spheres. Like all air guns, paint guns operate using a compressed gas. Paint-ball guns are equipped with the refillable air tank. As a matter of safety, the velocity that the paintball gun expels the paint ball should be metered before put into use. Paintballs traveling at velocities in excess of 300 feet per second are generally considered unsafe and may result in unreliability of the paint-ball gun, as the paint ball itself may shatter in the bore because of the excessive pressure. The first paint-ball guns bore little resemblance to actual firearms, but they have evolved in appearance to more closely resemble actual firearms. Paint-ball competition has become a popular sport and has expanded around the globe.

The Automatic Electric Air Gun (or AEG for short) is air-gun technology taken to a new dimension. The standard projectile for the AEG is the 6mm-diameter ball-shaped pellet fabricated of plastic. Pellets of different densities are available, allowing the AEG to have an increased range and greater accuracy. The AEG is different from other air guns in that the AEG uses a battery-powered electric motor that compresses and releases a piston to create the air charge to propel the pellet. Many designs also feature an electrically operated magazine to increase the rate of fire of the AEG. The movement in the AEG community has been directed toward mimicking all varieties and vintages of firearms, ranging from all brand names of handguns to rifles, submachine guns, and squad automatic weapons. This is a logical pursuit, given the fact that many firearms that are copied as an AEG are barred from lawful ownership. Many AEG analogs are well-built, convincing devices that use metal, plastic, and real wood to construct the parts and equal the weight of the real counterparts. In fact, many real-world accessories designed for firearms will interface with the AEG analog. It would be very easy to be deceived by the appearance of an AEG and mistake it for a real firearm were it not for the requisite blaze-orange muzzle cap to indicate that it is not a real firearm. However, this is easily removed by the end user if the apparent intent is to mask the device's identity as an AEG.

The AEG can attain an extremely high rate of fire, with some examples achieving cyclic rates in the area of 900–1,200 pellets per minute. At close range, an AEG is capable of cutting holes through drywall and shredding aluminum cans. For recreational purposes, most AEGs are used in the same manner as paintball guns: conducting simulated war games, historical reenactments, and general plinking. The AEG has also found professional uses, often being used for training purposes; these devices are comparable to using simulated munitions. The accuracy of controls, weight, and feel make them ideal for professional training environments. The added safety margin is that an AEG is not configured to accept standard ammunition, eliminating the risk of using firearms or dedicated simulation firearms and confusing training ammunition for lethal-application cartridges. Using proper safety equipment, the risk of personal injury from the pellets fired from an AEG is said to be minimal.

Air guns and the like are not legally considered firearms; they do not meet the requisite definition of a firearm, that is, being capable of expelling a projectile, or readily restored to such a state, under the force of an explosive. Air guns themselves may be subject to other regulations outside the scope of laws that address firearms, most notably concerning age restrictions for purchase.

Air guns are often represented as a firearm by persons who otherwise cannot obtain a firearm to facilitate the commission of crimes such as armed robbery or assault. There have been numerous unfortunate tragedies where a person possessing an air gun has been mistaken as being armed with an actual firearm, resulting in injuries and deaths. Quite often, the examiner is brought in to testify that the air gun used in the commission of a crime mimics that of an actual firearm, particularly if the legally mandated blaze-orange safety cap has been removed or painted over to conceal that fact that the article is a nongun. Such a case is at least as much based upon the facts obtained in the scope of the broader investigation as the opinion of the examiner.

The examiner may be called to opine as to the physical similarities between the air gun and an actual firearm, as well has how the air gun operates and what range the air gun may have if it were discharged. Another common question within the use of an airsoft gun is whether it can be classified as a "dangerous weapon" for the purposes of forming the basis to charge as a felony or misdemeanor crime. The possibility that serious, permanent injury or death can result from being struck with a BB or pellet expelled from an air gun may be very slight under the narrowest of circumstances; however, it cannot necessarily be dismissed as impossible. The Consumer Products Safety Commission issued a safety alert stating that the commission "has reports of about four deaths per year caused by BB guns and pellet rifles" (Consumer Products Safety Commission 2010), and particularly warns of BB guns that propel a BB or pellet at velocities exceeding 350 feet per second. The technical nuances of whether an airsoft gun is a dangerous weapon based upon medical datum and technical information is tempered by case law that may have set the predicate in that specific legal venue. In general, air-gun classifications should be treated as any other firearms case. It must be presumed that, unless there is testimony on record, whether the article is (or is not) deemed to be a dangerous weapon will factor into how that case will be prosecuted and what the expected outcome will be.

Ammunition-Feeding Devices

The terms *magazine* and *clip* are often said to represent the same object, and the terms are often incorrectly interchanged. Much of this confusion seems to originate from popular culture in the form of music, movies, and television, where the firearm magazine is simply referred to as the clip, lending the impression that the two terms are interchangeable, when in reality they are not. The common denominator between the two is that they provide a source of ammunition for a firearm. Magazines are inserted into firearms and supply the firearm with ammunition to the extent of its capacity. A clip may be inserted or used to load cartridges into the internal magazine of a firearm or one that is detachable. Clips generally are intended for single use and can then be discarded. Certain styles of clips are reusable and are often sought by collectors as valuable accoutrements for the firearm they pair up with.

Magazines

There are several different types of magazines. The most common is the detachable type, which may be identified as a box-type magazine because of the physical appearance of the magazine resembling a box. Not all magazines share the same basic shape for functional and aesthetic purposes. Another colloquial term to describe a magazine is *banana clip*, a reference to the curvature inherent to the design of certain magazines and its resemblance to a banana. Once again, the use of the term *clip* in this context would be a catachresis. The particular shape of an individual magazine is subject to various engineering and design issues, but also is in proportion to the total ammunition capacity. Most detachable magazines are rectangular in shape, and either angled or straight bottoms or floors can be expected.

For the most part, magazines are not interchangeable between different models and manufacturers of firearms, although there are exceptions to this. The pattern magazine used in the AR-15/M16 family of rifles is sometimes called the STANAG 4179 pattern.* Contrary to common belief, there is simply no standard NATO (North Atlantic Treaty Organization) magazine in the true sense of the word but, rather, an acceptance that a particular style of magazine manufactured to a certain specification is acceptable for use within the NATO partners. The STANAG 4179–pattern magazine is shared not only with the AR-15/M16 rifles, but is also used by the FN SCAR 16 and 16S, the British SA-80 (L85A1), the FN FNC (chambered in 5.56×45mm), FN FS2000, SIG 556, Kel Tec PLR 16 and SU-16, the Daewoo K-1 and K-2 rifles (including their civilian counterparts the DR-100 and the DR-200) as well as a plurality of other firearms originating from around the world. Within STANAG 4179–pattern magazines, there are several variants. The original AR-15/ M16 magazines were of 20-round capacity and had straight bodies with a pronounced waffle-pattern reinforcement ribbing. This version of the magazine was replaced by the more familiar 20-round magazine made of aluminum with vertical reinforcing ribs impressed into the body. The 30-round magazine has largely replaced the 20-round magazines with military, law enforcement, and civilian users.

STANAG 4179 magazines that meet U.S. military specifications are manufactured from aluminum, and the manufacturers stamp the floor plate with their information. U.S. government contractors have included Colt, NHMTG, Adventureline, Okay Industries, Center Industries, and others. Various companies may have vanity-stamped floor plates, and the floor plate may have been replaced if the magazine has been rebuilt. C-Products, a magazine supplier that specializes in this particular style magazine, supplies other companies, but they vanity stamp the floor plate with the name of the client who contracted them. Other manufacturers may be found on the magazine floor plate, but in fact these could be vanity names stamped by the manufacturer for that particular customer.

Polymer as well as stainless steel–bodied magazines have appeared from commercial sources, and these generally are found to fit and function in firearms using this pattern of magazine. The Israeli firm Orlite Engineering developed an AR-pattern magazine that used Orlite: Teflon-impregnated Zytel that covered a steel-mesh endoskeleton. STANAG 4179 magazines manufactured for the British MoD were so marked on the floor plate, constructed of steel, and had a grayish parkerized finish. Heckler & Koch developed a steel magazine dubbed the "High Reliability" magazine. H&K claims that during U.S. military testing, the magazine "improved feeding reliability of the system by 30–50% overall" (HK Tactical Defense Systems 2010). There are two variants of the magazine: anti-friction and maritime/anticorrosion. The maritime version is available in 20-round capacity. Both versions have National Stock Numbers (NSN) indicating that they are recognized within the U.S. military supply chain.

Magazines for the Ruger Mini-14 and Mini-30 rifles overtly resemble the AR-15/M16 magazine and may be finished in blue or stainless steel. These magazines are not interchangeable due to the manner that the magazine inserts and sits in the Ruger rifle, which mimics the AK rifle. Like the Ruger Mini-14, the Israeli Galil rifles superficially resemble the AR-15/M16–pattern magazine, but they are not interchangeable for the same reason.

---

* STANAG is the abbreviation for a Standardization Agreement among NATO partners. The purpose of a STANAG is primarily to establish commonality of logistics, administrative matters, terminology, and ease of interchangeabilities of materials within NATO.

Magazines for the Steyr AUG and the H&K G36 are also proprietary and do not inter-change, although a conversion package is available to convert the Steyr AUG. Galil maga-zines were manufactured either from heavy-grade steel with a heavy finish or from Orlite. Interestingly, Galil magazines had a capacity of 35 rounds but did not look much larger than other magazines.

The AK-pattern rifles chambered in 7.62×39mm are practically universally standard, regardless of their nation of origin. Certain semiautomatic versions of the AK imported into the United States may have modified magazine wells so as not to accept the standard AK-pattern magazine, but these are relatively rare. The 7.62×39mm AK pattern is seen in 10-, 20-, 30-, and 40-round capacities; larger capacity magazines can be found, but are unusual. Traditionally, they were made of fluted steel and finished in a black-enamel or parkerized finish. The top of the magazine, where it meets the receiver, is reinforced on either side and quickly bears indications of insertion and removal as the finish scrapes off. Steel AK magazines have been supplemented with polymer magazines that vary in color: black, purple, dark green, dark red, and clear. AK-pattern magazines are not inter-changeable by caliber; magazines are specific to 7.62×39mm or 5.45×39mm. East German 5.45×45mm magazines were made of bright-orange-colored Bakelite to distinguish them from 7.62×39mm magazines. These magazines had steel reinforcement filets in them. The 5.45×39mm magazines can be seen in any color.

AK-pattern magazines have been produced by practically every nation that manufac-tured the AK and can be identified by arsenal markings on the magazine body. These markings coincide with factory codes or symbols used to head stamp ammunition and mark the firearm. However, some Soviet magazines have been "sterilized"—the markings removed or never placed.

Shared Magazine Architecture

Some manufacturers share magazines across several products. Hi Point Firearms, an Ohio-based firearm manufacturer, manufactures two models of carbines, one chambered in 9×19mm and the other chambered in .40 S&W (Model 4095). The magazines are common between the handguns and carbines of the same chambering. The Beretta CX-4 Storm car-bine chambered in 9×19mm shares the magazine of the Beretta 92 pistol or the P×4 Storm pistol, depending on the model of carbine. Other firearms designers, rather than trying to design and produce their own magazine, have simply utilized another maker's maga-zine. The Kel-Tec SUB-2000 carbines are manufactured to use a specific magazine from another manufacturer—Glock, Smith & Wesson, SIG Sauer, or Beretta—and are available chambered in 9×19mm or .40 S&W. The Kel-Tec SU-16 series of rifles use the standard AR-15/M16–pattern magazine. The Kel-Tec RFB rifle uses the readily available, inexpen-sive 7.62×51mm FAL-pattern magazine. The availability of inexpensive surplus magazines built for STEN guns, M3 Grease Guns, and the Thompson submachine guns have led to their adaptation by numerous companies or to conversion kits that permit the use of these types of magazines in various firearms.

Drum Magazines

The drum is another form of the magazine. Drum magazines have a high ammunition capacity, well exceeding box-type magazines. Drums are not a new idea, dating back to the earliest days of automatic firearms. There was a drum magazine built for use with the Luger (Pistole 1908 or P/08) pistol, which collectors call the *snail drum*, and the drum especially gained notoriety with the Thompson submachine gun during the Prohibition Era. Like other magazines, the capacity of the drum can vary but typically ranges from 50 to 100 rounds. Drum magazines are currently manufactured for numerous weapons platforms such as the Ruger Mini-14 in .223 caliber, the AK and variants, firearms using the AR-15/M16-type magazine, the H&K MP-5, and others. The American AR-180 machine gun exclusively used a drum magazine; there were never other magazines developed for it. Calico Light Weapon Systems carbines have a unique helically fed, top-mounted magazine of either 50- or 100-round capacity. The magazine itself physically resembles a small cylinder.

The physical size of the magazine generally has a bearing on the total capacity, but this is not always the case. In the United States between September 14, 1994, and September 14, 2004, there was a capacity restriction placed on magazines that were sold to individuals. During this time, magazine capacity was restricted to 10 cartridges. Magazines manufactured for use by law enforcement, military, and other government entities were exempted from the restriction, and magazine manufacturers were legally obligated to mark these exempted magazines accordingly. Smith & Wesson went so far as to assign serial numbers to individual magazines, presumably for accountability. Since the physical dimensions of the magazine had to remain the same to fit the host weapon, the physical appearance of restricted-capacity magazines did not vary from full-capacity magazines. Most manufacturers simply installed plugs or internally modified the magazines to limit the cartridge capacity. Another magazine marking that has been observed is present on magazines that were manufactured outside the United States but would have been subject to restriction. Such magazines were marked to indicate they were not for importation into the United States. It is not uncommon to encounter ban-era magazines that are so marked in the civilian market today, but as the law banning civilian possession has expired, these magazines are no longer subject to any control unless possessed in a venue that prohibits it on the basis of capacity. Immediately after the ban had lapsed, stockpiles of these magazines were released for general sale and purchased not only as usable magazines, but also as novelties.

Magazine Components

The features that define a magazine are a body, spring, and a follower. Within the magazine body is a spring that compresses as a load is placed upon it, thereby providing the force required to get ammunition from the magazine into the chamber of the firearm. The spring is capped with a follower, a tab that separates the loaded ammunition from the spring and pushes the ammunition toward the chamber as the weapon is fired. Magazines are typically loaded from the top, where the feed lips are located, but some variations exist that load from the side. The feed lips are located at the top of the magazine and comprise the material on either side that narrows to ensure that the cartridge will feed smoothly from the magazine into the host firearm. Damaged or distorted feed lips are often a cause of functional failures and cannot be overlooked during a firearm exam or inspection. Thin metal that will deform, as well as plastic that will chip or crack from use, are especially susceptible to magazine-related failures such as failures to feed or double feeds. Some magazines may contain a fourth component, a reinforcing plate located at the base of the magazine between the floor plate and the spring. Drum magazines may be fed either through the top, called the *tower*, or by removing the access panel on the drum and loading into the

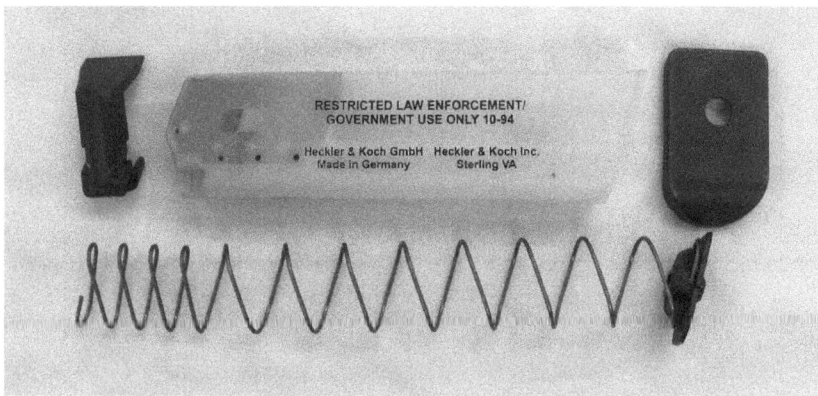

Figure 4.32 A magazine for the H&K USP 9 pistol that has been disassembled into its components: polymer magazine body, magazine follower, magazine spring and reinforcing plate, and steel floor plate. This magazine bears the restriction markings, in effect between 1994 and 2004. This construction typifies most modern magazines. (Image from author's collection.)

drum body itself. The life span of a magazine is indeterminate. For the most part, magazines are intended for reuse over a period of time, the only wear item being the spring, which is replaced from time to time. Figure 4.32 shows a magazine that has been disassembled into its component parts.

The Clip

A *clip* is a device entirely different from a magazine. A clip is simply a piece of metal or plastic that holds cartridges. The clip contains no working parts. Clips can be used to either load a firearm directly or to load a magazine. Often clips are referred to as *stripper clips*, as the cartridges are stripped out of the clip by pressing them into the appropriate magazine. In many military circles, ammunition is issued on bandoliers, cloth ammunition carriers that contain preloaded stripper clips. These clips then allow the ammunition to be quickly loaded into magazines, either directly or by use of a separate stripper-clip guide. Clips may also be used to feed cartridges directly into the firearm.

The M1 Garand rifle has an internal magazine that is loaded by way of an eight-round clip. The Garand clip was an en bloc variety where the cartridges were loaded into the clip, which was made of metal and shaped to resemble a C. When the Garand's action was open, the whole clip was inserted into the action and pressed downward until seated. The action then slammed forward, often causing the much lamented condition known as "M1 thumb" if the operator was unable to get the thumb out of the way in time. The loaded cartridges did not leave the clip but, rather, were fed out of the clip, now loaded in the Garand, until the supply of loaded ammunition was exhausted, at which time the clip was ejected from the Garand, the action would lock to the rear, and the shooter would insert a fresh clip and repeat the process. The M1 was a classic military arm, and perhaps the single greatest battle implement devised, at least according to General George S. Patton, but the en bloc clip created two significant tactical problems: The first was that single rounds could not be loaded into the gun—only a fully loaded eight-round clip. This meant that the operator would have to exhaust the loaded cartridge supply completely; there was no topping off. The second problem was that when the clip went dry, it was ejected with a loud clanging

Figure 4.33 An assortment of clips (from left): 5.56×45mm for STANAG magazine; top right: M1 Garand; bottom right: Mauser 98 pattern. (Image from author's collection.)

noise, letting everyone with an experienced ear in the area know that a Garand shooter was out of ammunition. The direct descendent of the M1 Garand, the U.S. Rifle M14, did not share this feature. The M14 was loaded by using a detachable box magazine. Figure 4.33 shows an assortment of clips.

In addition to the M1 Garand, there are other firearms that have internal magazines that are incorporated into the design and are not ordinarily removable for reloading. They are still magazines in the true sense of the definition. They contain a spring and a follower, just like detachable magazines. The internal magazine can be oriented either horizontally or vertically. The Mauser 98–pattern rifle is an example of a vertically oriented internal magazine. The magazine is mounted beneath the chamber and is loaded by either inserting individual cartridges or by pressing five rounds at a time into the magazine using a stripper clip. The SKS-type rifle,* regardless of the nation of origin, is loaded by means of a 10-round stripper clip or by loading individual cartridges by hand. The top of the receiver has a guide notch where the clip is inserted and the rounds are pressed down with thumb pressure. An unusual Chinese-made variant of the SKS was manufactured and imported into the United States. The NORINCO SKS 63 was manufactured to use detachable AK-pattern magazines instead of the internal 10-round magazine. Conversion kits to upgrade the SKS rifle from the internal magazine to a fixed larger capacity magazine are quite common. Overtly, the most telling feature that distinguishes the SKS 63 from others is the thumbhole-type stock, whereas standard SKS rifles have a more traditional rifle stock.

These two are not the only examples of firearms that can be loaded in such a manner; there are numerous rifles that have the ability to load even a detachable magazine-fed firearm through the action using stripper clips, including the Belgian FAL-pattern rifles and copies such as the Springfield SAR 48 and the sporterized SAR 4800, and the British

---

* SKS is an acronym that is translated to Semiautomatic Carbine Simonov. The rifle was developed by Sergei Simonov, a Russian firearm designer. The design was short-lived, being replaced by the AK-47. However, the basic design remained with numerous nations under various names: Chinese Type 56, East Germany Karabiner S, North Korea Type 63, and Yugoslavia M59. There are numerous subvariants that have been produced, but the basic design remains the same.

Enfield series of rifles. The presence of a stripper-clip guide on the top of the receiver indicates that the firearm can be loaded using stripper clips, even if a detachable magazine is used as the ammunition source.

Ammunition Belts

Belted ammunition was once limited to use in machine guns, but that is no longer the case. Ammunition belts have been made of cloth or metal. There are two versions of metallic ammunition belts: continuous belts and those with disintegrating links. Continuous belts remain wholly intact as they pass through the receiver of the firearm. Linked belts break apart into individual links or into short sections when fired through the firearm, which depends entirely on how the belt was designed. There are a number of semiautomatic firearms that have been manufactured that are fed by an ammunition belt. These firearms are built using newly fabricated receivers capable of semiautomatic fire only, and then married to parts kits from donor firearms or from newly made components. Semiautomatic versions of the Browning M1919, Browning M2, as well as the German MG-34 and MG-42 have been manufactured. Ohio Ordnance is a manufacturer that has come to specialize in such firearms, manufacturing a semiautomatic version of the current U.S. military–issue M240, a belt-fed 7.62×51mm machine gun. The SHRIKE 5.56, developed by Ares Defense Systems, is a system that bolts onto the lower receiver of any AR-pattern firearm, converting it from a magazine-fed weapon into one fed by a belt. The current-issue U.S. military M249 SAW or Squad Automatic Weapon can be fed either by belt or by inserting a detachable magazine. This was also the case with the H&K 21 general-purpose machine gun.

Internal Magazines

Rifles, shotguns, and even handguns can be fed through a tubular magazine. Tubular magazines are mounted underneath the barrel of the firearm and are loaded either by a loading gate underneath the tube or through the loading or ejection port of the firearm. Such firearms can be operated by slide action, by pump, or as semiautomatics. In the case of a rifle, the insert in the magazine is removable to facilitate loading and unloading. At the muzzle end of the magazine is a knurled knob. When twisted, a log tube containing the magazine spring and follower is withdrawn from the magazine. Cartridges contained in the magazine can be emptied by turning the firearm down. In the case of a shotgun, the tubular magazine cannot be unloaded in the same manner. The action may have to be cycled or the tubular magazine can be emptied by partially opening the action to engage the shell latches in the magazine tube. Rifles fed by tubular magazines typically only feed cartridges with round-nose projectiles; it has been a long-standing fear that dropping spitzer or pointed-nose projectiles onto one another in a tubular magazine where the cartridges would stack and allow primer-to-projectile tip contact could result in a discharge if the primer were impacted strongly enough.

Firearm Alterations and Modifications

All firearms can be subjected to tinkering and modifications. Modifications and alterations to firearms can take the form of simple replacements of stocks or internal components to those thought to improve some deficiency in the original firearm design, or to modify the firearm to behave in a manner inconsistent with the original specification as manufactured. Alterations and modifications may also be for purely aesthetic purposes.

Finish, Refinishing, Colors

Cosmetic changes to firearms are quite common. Guns are refinished to suit an owner's preference or to restore or rehabilitate a worn firearm. Refinishing was quite common on war trophies, where owners were tempted to enhance the plain finish of their souvenir gun from a basic military finish to bright chrome. Modern firearms made of aluminum can be anodized any color, and polymer firearm parts can be molded in any color desired. This carries an inherent risk of a misidentification where an assumption could made that a blue gun could be mistaken as a training firearm, when in fact it was simply finished in blue. Likewise, colors such as orange or red could be interpreted to mean that a firearm was intended for less-lethal application or other purposes. However, it cannot be assumed that changes to finish or to parts were made by the preference of the possessor.

Other popular refinishes include contemporary camouflage patterns or any other colors that may mislead someone from initially believing that the object is a firearm and not a toy or a model. Orange tips are legally required on firearm analogs such as toy guns, airsoft guns, and models; however, such tips can be removed or painted over. A refinished firearm may resist identification due to the markings being smoothed if the firearm is overpolished in the refinishing process. The proliferation of polymer has only added to the rainbow, since polymer can be made any color. See Figures 4.34 and 4.35 for examples of colored handguns.

Figure 4.34 (See color insert.) The Taurus PT738.380 ACP pistol. This example has a frame molded of pink polymer with a blackened stainless steel slide. The colored appearance of an apparent firearm cannot be relied upon to distinguish that the object is not a real firearm. (Image courtesy of Taurus International Manufacturing, Inc.)

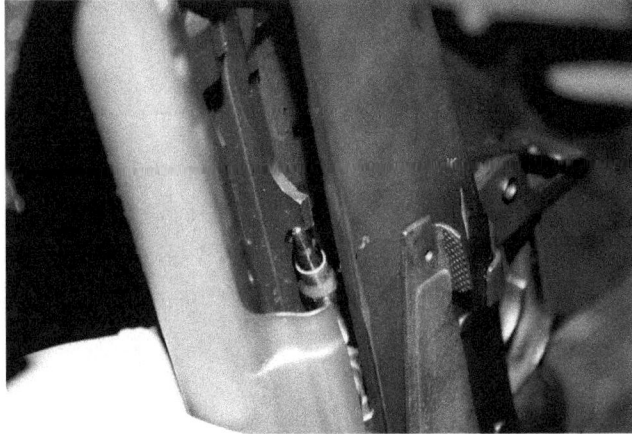

Figure 1.1 A semiautomatic rifle action is opened, revealing the presence of a live cartridge in the chamber. Such a find can be of great investigative interest. (Image from author's collection.)

Figure 2.3 A cross-section of various size shot shells, from left to right: 12-gauge 3½″ Super Magnum, 12-gauge 3″ Magnum, 12-gauge 2¾″, 16-gauge 2¾″, 20-gauge 2¾″, Aguila 12-gauge 1½″, and .410 3″ shell. (Images from author's collection.)

Figure 2.9 Caseless-telescoped cartridges. (Image courtesy of Paul Shipley, AAI Corporation.)

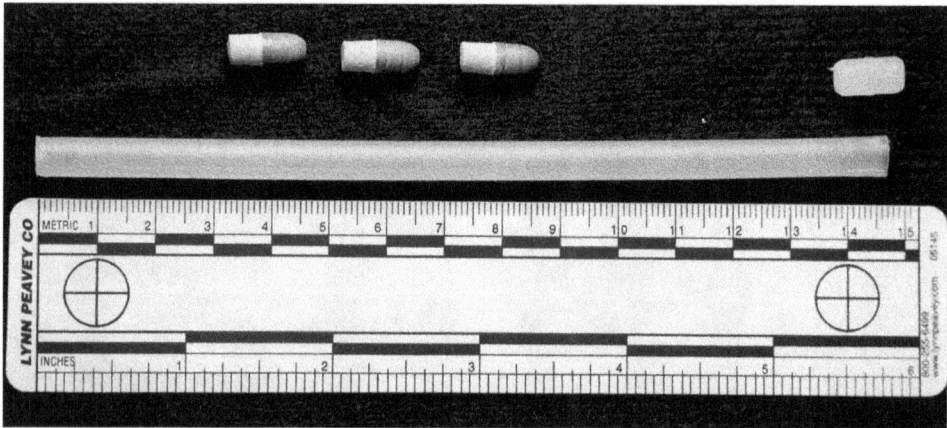

Figure 2.10 Daisy V/L caseless cartridges. The bare lead projectile is seated against the yellowish, granular propellant charge. The plastic tube is the container the cartridges were shipped in. (Image from author's collection.)

Figure 2.12 Example of 5.56×45mm M855 "Green Tip" cartridges on stripper clips. Note the annealed casing neck. (U.S. Marine Corps image; photographer Cpl. Lydia M. Davey, USMC.)

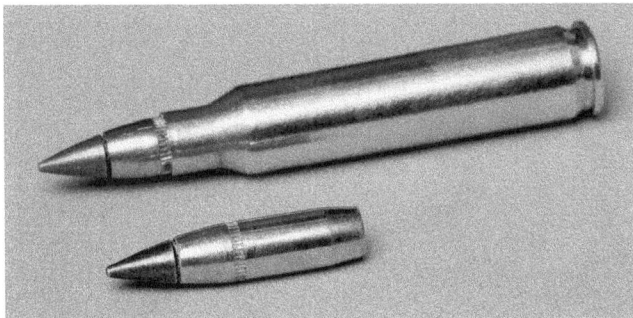

Figure 2.14 The 5.56×45mm M855A1 EPR. Note the gap between the tip and projectile body and the cannelure on the projectile itself. (U.S. Army image; photographer Todd Mozes.)

Figure 2.18 A cutaway image of the Hornady Critical Defense cartridge. (Image courtesy of Hornady Manufacturing.)

Figure 2.20 The pop culture phenomenon of a world taken over by zombies prompted Hornady to release a line of "zombie killing" ammunition, a rather jocular bit of marketing. (Image courtesy of Hornady Manufacturing.)

Figure 2.28 Examples of various dummy, drilled, and inert cartridges (from left to right): safety orange marked 9×19mm, drilled casing 9×19mm, solid black plastic 5.56×45mm, corrugated casing 5.56×45mm, and a 12-gauge shell marked DUMMY. (Image from author's collection.)

Figure 2.34 The Glock 17T is dedicated exclusively for training. The frame is blue polymer to clearly identify it as a training aid. Also depicted is the magazine with blue floor plate for identification. Note the Simunition cartridge loaded in the magazine. (Image from author's collection.)

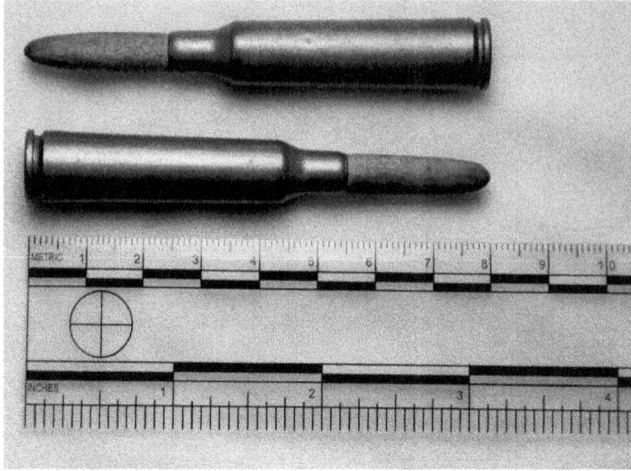

Figure 2.37 Swedish 6.5mm cartridges loaded with wooden projectiles. (Image from author's collection.)

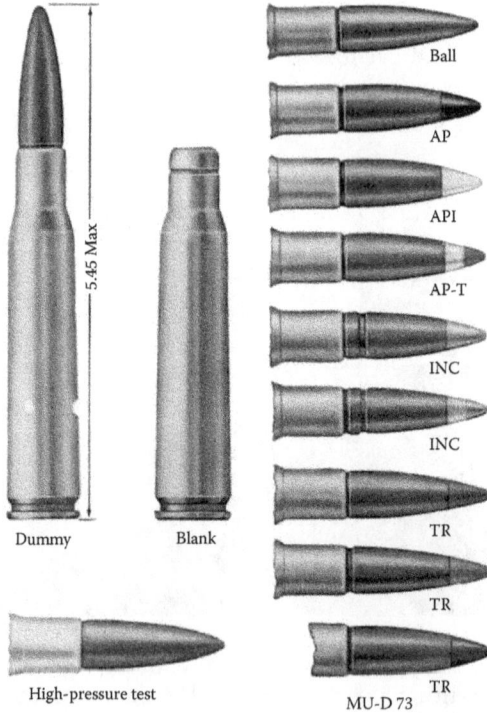

Figure 2.38 U.S. Military .50 BMG (12.7×99mm) color-coding scheme. (From U.S. Army Technical Manual TM 9-1300-200, 1993.)

Figure 2.39 U.S. Marine Lance Corporal Richard Mueller mans a turret-mounted M2.50 machine gun. The ammunition load consists of armor-piercing incendiary (gray bands) and armor-piercing tracer (red over gray bands) cartridges. (Image from U.S. Marine Corps; photographer Gunnery Sgt. Scott Dunn, USMC.)

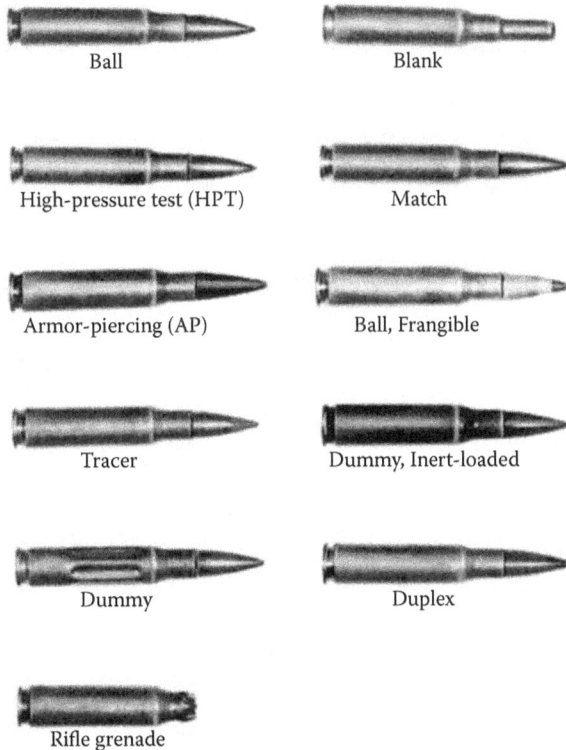

Ball

Blank

High-pressure test (HPT)

Match

Armor-piercing (AP)

Ball, Frangible

Tracer

Dummy, Inert-loaded

Dummy

Duplex

Rifle grenade

Figure 2.40 U.S. military 7.62×51mm color-coding scheme. (From U.S. Army Technical Manual TM 9-1300-200, 1993.)

Figure 3.2 A study of two Chinese-sourced cartridge head stamps. On the left is a typical example of 7.62×51mm ammunition of Chinese origin using block characters. The 61 indicates the factory number; 92 indicates that this cartridge was manufactured in 1992. The example on the right, also 7.62×51mm, is a forgery. The head stamp indicates production by Royal Ordnance Radway Green in 1960, complete with NATO standard cross. L2A2 was the British military designation for the 7.62×51mm NATO ball cartridge. Both examples are of similar construction; the casing is copper-plated steel. (Image from author's collection.)

Figure 3.4 An assortment of 5.56×45mm/.223 cartridge head stamps. Top row from left: Precision Metal Corp.; Poongsan Corp. Korea (October 1981); Federal Cartridge (2005); Hornady Manufacturing; International Cartridge Corp.; Hornady Manufacturing. Bottom row from left: Lake City Ordnance Plant (2002); Lake City Ordnance Plant (nickel plated) (2005); Lake City Ordnance Plant (1975); Winchester Cartridge Corp. (2001); Barnaul Cartridge Plant Russia; Wolf Performance Ammunition. (Image from author's collection.)

Figure 3.5 An assortment of commonly encountered 7.62×39mm cartridge head stamps. Top row from left: Klimovsk Specialized Ammunition Plant Russia (2000); Barnaul Cartridge Plant Russia (newer logo); Barnaul Cartridge Plant Russia (earlier logo); Barnaul Cartridge Plant Russia (1995); Tulammo Russia; Federal State Enterprise Production Amursk Cartridge Plant Vympel. Middle row from left: Four varieties of Wolf Performance Ammunition head stamps; Igman, Bosnia-Herzegovina (1981). Bottom row from left: Klimovsk Specialized Ammunition Plant (1993); Pretoria Metal Pressing (1988); China State Factory 71 (1991); China State Factory 31 (1971); Winchester Cartridge. (Image from author's collection.)

Figure 3.6 An assortment of 7.6×51mm/.308 cartridge head stamps. Top row from left: Wolf Performance Ammunition; China State Factory 61; Giraites Ginkluotes Gamykla (2004); Federal Cartridge. Bottom row from left: Gevelot S.A.; Lake City Ordnance Plant; Remington Peters. (Image from author's collection.)

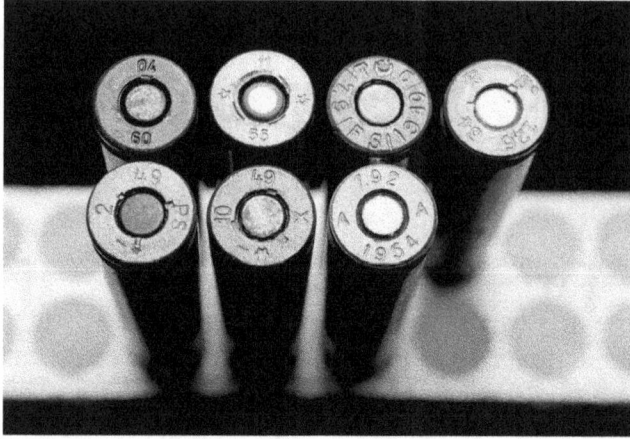

Figure 3.9 The 7.92×57mm cartridge remained popular even after the Second World War, and production continued because of its widespread use around the world and the large amount of surplus German arms in circulation. Top row from left: East Germany Factory 04 (1960); Prvi Partizan, Yugoslavia (1955); Kirikkale/Ankara, Turkey (1940); and prewar German production (1934)—note P126 code. Bottom row from left: Two variations of Czechoslovakian production by Povazske Strojarne/Povazska Bystrica (late 1940s); and Ammunitions Arsenalet Denmark (1954), often called the 8mm Mauser. (Image from author's collection.)

Figure 3.10 Ammunition under investigation in Baghdad, Iraq, in 2006. Closer inspection of the casing head stamp reveals ППУ for Prvi Partizan; the year 2002 is also marked. Other identifying characteristics to look at include the brass casing and the distinctively red primer sealant, a color unique to Prvi Partizan. (U.S. Air Force image; photograph by Technical Sgt. Adrian Cadiz.)

Figure 4.1 The Raven MP-25. Part of the identification markings are on left side of the slide (pictured); the model and caliber are on the opposite side. The serial number is stamped on the back strap of the grip. The Raven pistols are grouped as one of the Saturday night specials; untold millions were made, and they routinely turn up. (Image from author's collection.)

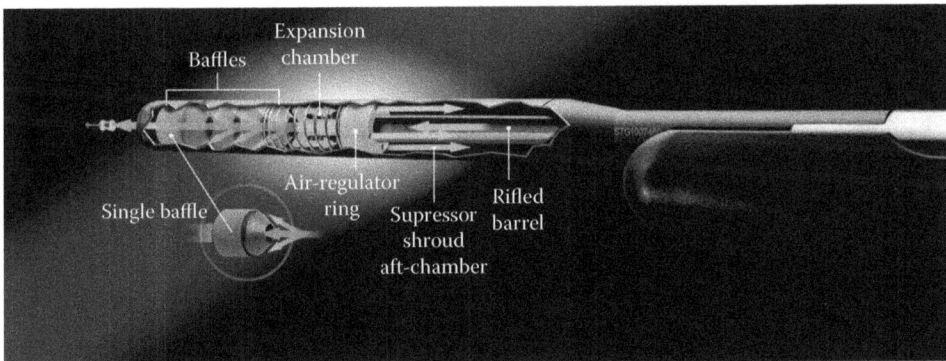

Figure 4.20 Stoeger Air Gun X20 suppressor schematic showing baffles and an expansion chamber. (Image courtesy of Benelli USA.)

Figure 4.28 The Uzi submachine gun. The depicted examples were manufactured as submachine guns by Israeli Military Industries in Israel and are so marked on the left side of the receiver as Uzi SMG. The English-marked selector switch is above the trigger, depicted In the lower example. Hebrew-marked selectors are also encountered. (Image from author's collection.)

Figure 4.34 The Taurus PT738.380 ACP pistol. This example has a frame molded of pink polymer with a blackened stainless steel slide. The colored appearance of an apparent firearm cannot be relied upon to distinguish that the object is not a real firearm. (Image courtesy of Taurus International Manufacturing, Inc.)

Figure 4.35 The Glock 22P. The 22P is an inert training aid used to familiarize the user with the pistol without the risk of handling a firearm that could become loaded. The pistol accepts a magazine, and the trigger is fully functional, but the barrel is solid. The red frame is meant to alert the user that this example is inert. (Image from author's collection.)

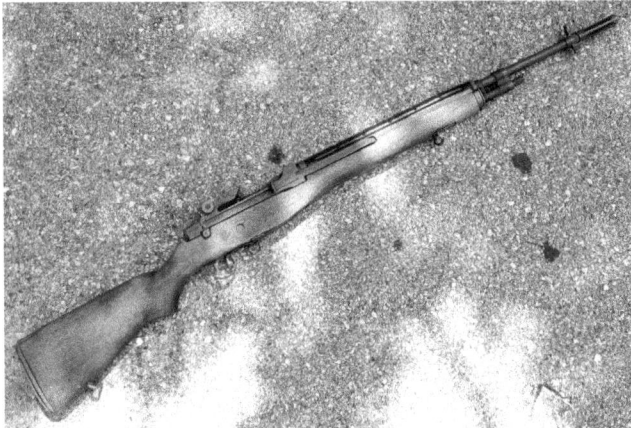

Figure 4.55 A U.S. Rifle M14 manufactured by Harrington & Richardson is a machine gun. This example bears the circle *P* proof mark and eagle cartouche on the stock. (Image from author's collection.)

Figure 4.56 Close-up of the receiver markings of the H&R M14. Note the selector switch knob on the right side of the receiver. Semiautomatic copies of the M14 generally do not have the notch cut in the stock for the selector, although surplus stocks in circulation may or may not be cut for the selector. (Image from author's collection.)

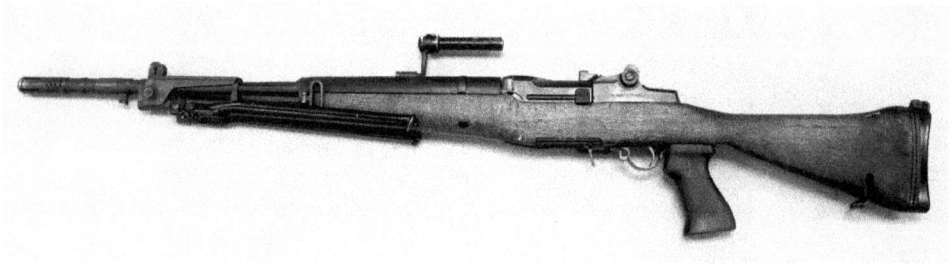

Figure 4.57 The M14-pattern rifle manufactured by Beretta as the BM59. There are definite similarities, but note the distinct differences between the M14 and the BM59. This example has a bipod, a grenade-launcher spigot on the muzzle, a grenade-launching ladder-type sight, and a pistol grip instead of a standard rifle stock. (Image from author's collection.)

Figure 4.58 Close-up of the BM59 receiver. Note the selector switch on the left side and the "P.B. BM59" marking on the action-lock button. The marking practices used by Beretta on the BM59 mimicked other manufacturers of the M14. Barely visible on the receiver below the sight-adjustment knob is the importer's mark "Springfield Armory," not to be mistaken for the government arsenal. (Image from author's collection.)

Figure 4.65 The FN SCAR 16S, chambered in 5.56×45mm. Also pictured is the FN two-tone STANAG magazine. (Image from author's collection.)

Figure 4.66 A contrast between the upper receivers of the FN SCAR 16S (top) and an AR-15 (bottom). The bolt carrier groups have been removed to contrast the differences between the two. The SCAR is operated by gas piston; the AR uses the traditional gas impingement. Installed on the AR upper receiver is a Knight's Armament RAS free-floating rail system, an aftermarket flash hider, a MaTech rear sight, a Knight's Armament vertical fore grip, and an Aimpoint Comp M4 using a Knight's Armament mount. (Image from author's collection.)

Figure 5.3 This 1913 vintage Colt 1911 has had its original serial number ground from the frame, and it has been renumbered. It is apparent that this was historical and likely done at the factory or at a government arsenal. The United States Property marking has also been obliterated from the frame by grinding and then stamping a series of X's across that part of the frame. Students of the 1911 will note the crude stamping as well as the die being inconsistent with that used by Colt. (Image from author's collection.)

Figure 4.35 (See color insert.) The Glock 22P. The 22P is an inert training aid used to familiarize the user with the pistol without the risk of handling a firearm that could become loaded. The pistol accepts a magazine, and the trigger is fully functional, but the barrel is solid. The red frame is meant to alert the user that this example is inert. (Image from author's collection.)

Firearm Chassis

A *firearm chassis* is a variation on the basic stock or furniture setup of a firearm, especially rifles and shotguns. Various chassis have been developed that allow a user to remove the supplied stock and reset the barreled receiver within the new chassis. Such an alteration may give the firearm the appearance of a completely different weapon and may lead to an erroneous identification. Chassis systems may also mask the actual make, model, and serial number of the firearm, and it may require disassembly to ascertain the true firearm data.

An example of this is the Commando Mark I, manufactured by Volunteer Enterprises. It was made of a composite material with a pistolized grip and vertical fore grip. The butt stock was made of wood and resembled that used on the Thompson submachine gun. The firearm portion of the Commando Mark I was a U.S. military M1 .30 carbine barreled action that was set into the chassis. A metal upper hand guard was placed over the barrel. Original carbines used a wooden hand guard; the metal examples appeared postwar and were used on commercial copies of the carbine. The Commando Mark I was marked on the side as such, causing many people to mistake this marking as indicating the manufacturer. As the chassis was not a firearm, no serial number was affixed.

In addition to the Mark I, other variants were produced, some using M1 carbine actions, other relying on semiautomatic copies of the Thompson submachine gun. Contemporary semiautomatic copies of the U.S. military M14 rifle can be found with a variety of stock configurations, such as the M1A manufactured by Springfield Armory, among other models, including a copy manufactured by the Chinese NORINCO company. The SOCOM 16, SOCOM II, and other versions feature the basic M1A design but with a slightly shorter barrel (16″) in a composite stock, in contrast to the traditional wooden stock commonly associated with the military M14 and the civilian M1A. The U.S. military M14 has received a new lease on life, designated the Mark 14 Mod 0 Enhanced Battle Rifle. A major portion of the enhancement is the stock, now featuring a pistol grip for improved ergonomics, a telescoping butt stock, and the addition of the military standard MIL-STD-1913 picatinny rails to attach accessories. The barrel length is reduced from 22″ to 18″. The chassis system was developed by Sage International. The Mark 14 Mod 0 is simply a modernized M14: The original receiver is still used, although aesthetically they do not have much of a resemblance (Armstrong 2007).

The Ruger Mini-14 is available from Ruger in a chassis configuration that gives it a distinctively different appearance than other Mini-14 rifles. Aftermarket chassis are also available that radically alter the appearance of the Mini 14. Another Ruger product, the 10/22, a .22 semiautomatic rifle, has a multitude of chassis conversion options. The Ruger 10/22 can mimic the appearance of a Thompson submachine gun, a Krinkov, and even an MG-42. Owners choosing to reconfigure the appearance of an H&K USC or SL-8 carbine can find the necessary parts to give the USC the appearance of the UMP submachine gun. The H&K SL-8 rifle can be modified to give the appearance of the G-36 automatic rifle. H&K modifications may be accomplished using factory H&K components or aftermarket parts patterned after factory UMP and G-36 components.

Chassis conversions are not limited to long guns; handgun chassis conversions are also available. HERA-ARMS, GmbH (High-grade European Research for Small Arms), manufactures chassis systems for various model Glock handguns as well as Model 1911–pattern firearms and the SIG 2022 that convert these handguns into carbine-style firearms, complete with shoulder stocks and accessory rail systems. Such a conversion of a handgun may present legal issues in certain venues. The investigator is reminded that these chassis conversions are strictly aesthetic and do not change the operational nature of the firearm, other than questions related to the configuration or reconfiguration of the firearm. Figure 4.14 demonstrates another chassis system that can substantially alter the appearance of the firearm.

Mechanical and Functional Modifications

The alteration, addition, or replacement of factory parts is generally for two purposes. The first is for purely aesthetic or ergonomically beneficial purposes and in no way affects the actual function of the firearm. A nonfunctional modification is the addition of bridge mounts or other means for attaching accessories. All popular makes and models of firearms have a wide array of aftermarket parts available, including lasers, flashlights, optical scopes, bipods, grips, and other similar objects.

The second purpose is to cause the firearm to behave in a manner inconsistent with the specifications as originally manufactured. These modifications work under the supposition that the firearm can be improved upon by the modification of a factory part, or by replacement of a factory part with an aftermarket one. Modifications of this nature include changing components in the firing train such as firing pins, springs, triggers, and other "tuning" kits. Such alterations or modifications may be directed at the trigger to adjust the amount of resistance felt in the trigger press, the trigger travel, or trigger reset. Modern firearms are not designed in the same manner that they were even 40 years ago, very recent in the context of firearm history. Previously, when fine trigger adjustments were desired, especially (but not always) for competition guns, a competent gunsmith would make fine alterations to the relevant interior components, such as polishing feed ramps in an effort to enhance the reliable feeding of ammunition from the magazine. Sears, trigger bars, and other components in the firing train were polished, trimmed, or otherwise had the "marriage"* of trigger parts reduced to minimal tolerances. Such actions did have the desired effect; however, the modifications came at a cost, quite often compromising the inherent safety features built into the firearm per the specification as determined by the designer and manufacturer.

In previous generations of firearms, modifications were typically undertaken by a gunsmith. Modern firearms, by design, generally do not require the services of a gunsmith to achieve performance modifications such as trigger adjustment. The Heckler & Koch USP pistols are available with upwards of 10 different factory trigger configurations, ranging from a traditional double/single-action trigger to a double-action-only trigger, with or without a manual safety/decocking lever. With such versatility, there is no need for "gunsmithing" the action of this firearm; the desired configuration is simply obtained from the factory or the trigger system is swapped out by an armorer. Fine trigger stop (trigger travel) adjustments can be made by using the provided tools to suit the user.

In the case of Glock handguns, the trigger travel is set by the design of the system; however, the trigger press resistance is adjustable by the simple replacement of the trigger spring and connector to one of five different (approximate) weights offered by Glock: 3.5, 4.5, 5, 8, and 12 lbs, the latter being identified as the "New York" trigger. A trigger modification on a Glock is an exceptionally easy plug-and-play approach to reconfiguring the trigger resistance of the pistol; nonetheless, nonstandard efforts to adjust the trigger press are frequently seen, most typically taking the form of the cruciform portion of the trigger being cut short to reduce the trigger "break" (the position of the trigger in its travel where the pistol will fire) and to reduce the felt trigger resistance. Another modification in conjunction may to be to cut the trigger spring or alter the connector to provide additional trigger alteration. A Glock will typically continue to work when such modifications are made; however, the operation will become erratic.

Another example of trigger alteration is the "release trigger." Shotguns used for trap and skeet shooting can have their triggers modified to work in reverse, that is that pressing the trigger does not cause the discharge; rather, releasing a pressed trigger causes the firearm to discharge. The rationale behind this modification was to allow the shooter to press the trigger and let it go; theoretically, this would have two effects: It could increase the speed that the shooter gets the shot off, and it would reduce or perhaps eliminate trigger control errors causing errant shots. Such a modification obviously creates a safety concern for uninformed handlers.

Machine Gun Conversions

Aside from "performance enhancements" that are made to firearms, the other purpose for internal modifications is the conversion of semiautomatic weapons into one capable of fully automatic fire (see Figure 4.36). As described earlier in this chapter, firearms are generally operated by a trigger that acts upon a sear, which causes release of the hammer, causing the firing pin to impact the primer of a cartridge, resulting in detonation. Some designs omit hammers or sears in lieu of other components that functionally perform the same work but

---

* The marriage is defined as the physical interaction of components, especially in the fire train of a firearm. Shaving and polishing the contact surfaces has the effect of reducing the marriage of the components, or how they interact with one another, to cause certain effects such as lightened trigger press, full-automatic fire, or reduced trigger travel and trigger reset.

Figure 4.36 The H&K 93 rifle, chambered in 5.56×45mm. The trigger group is modified to permit operation as a machine gun. The conversion required no external modifications, so it is not obvious. (Image courtesy of Jared Ford.)

may reduce the number of parts or better suit the philosophy of the designer. In cases of semiautomatic firearms, part of the internal makeup of components may include a disconnector. A disconnector prevents the trigger from acting on the sear, literally disconnecting one from the other between shots, thus allowing for single shots only instead of continuous firing while the trigger is held down. In so-configured firearms, the disconnector is often a target for modification to override its intended purpose, to stop the trigger from reengaging the sear after the action cycles post discharge. The sear, if so equipped, could also be a target for modification if alteration of the disconnector alone would not suffice to cause unrestricted firing.

Firearms that operate from an open bolt, such as early-production MACs, TECs (KG-9), Uzis, and others, are relatively easy to convert. The nature of an open-bolt weapon is that the bolt is typically held open by a trigger bar or sear and is captured upon recoil by a form of trip, disconnector, or an interrupter. By parts modification or replacement, an open-bolt firearm quickly goes from a semiautomatic to a full auto. As long as there is no ability to interrupt the bolt's forward travel after recoil, in theory the gun will continue to cycle as long as the trigger is held down. Firearms that operated from an open bolt typically had firing pins fixed to the bolt face, which only simplified the relative ease of conversion. Since open-bolt guns tend to be operated by straight blowback, with the sheer mass of the bolt providing sufficient impact force by way of the firing pin to detonate the chambered cartridge and repeat the process, a machine gun was easily made. In 1982, ATF requested that manufacturers of open-bolt firearms reconfigure their designs to closed-bolt operation in an effort to eliminate such easy conversions. Open-bolt designs did not disappear from the landscape, however. Machine guns designed for sustained fire, or squad support roles in a military context, have been configured to fire from an open bolt to avoid the possibility of a "cook off," which is when a cartridge detonates not by primer detonation, but by the chamber heat being so high that it causes the powder within the cartridge to combust.

Closed-bolt weapons may require a slightly more detailed approach, but nonetheless, successful conversions can be accomplished. Like their open-bolt siblings, closed-bolt firearms can be converted by replacement of parts that restrict operation to semiautomatic function with full-auto-capable parts, or are otherwise modified or reengineered to enable full-auto operation. The possibilities of conversion are nearly endless. One technical hurdle that must be overcome is that most semiautomatic firearms use a spring-loaded firing pin contained within the bolt, or in the case of a handgun, the slide. Fixed firing pins have largely disappeared from the landscape, since they were only used in open-bolt guns. As a general statement of safety, spring-loaded firing pins contained within a bolt that are not regularly cleaned can result in a slam-fire condition that is prompted by the accumulation of debris buildup around the firing pin, which may inhibit the fluid spring action and cause the firing pin to fix at the breech face. Of course it is always possible to override the firing pin by modification, such as removing the spring and blocking the firing pin by other means to create a fixed firing pin bolt. In another possible scenario, the use of a spring of insufficient strength could have the effect of erratic function, or too strong a spring may also fix the firing pin in position. These unintended consequences are likely to be arrived at by persons trying to reengineer a firearm to somehow enhance the performance of same, but without regard for the true implications of their handiwork.

Conversions are often attempted, and may be successful, by using mundane items such as wire ties, shoe strings, fishing line, paper clips, hair pins, and other sundries as a means to alter the behavior of the firearm's action. Conversion can be effected by using fishing line, shoe strings, or twine by looping both ends of the line and attaching one end to the charging handle and the other end to the trigger. This method would only be effective on arms whose charging handle reciprocates back and forth with the action. The theory is that the trigger is depressed by line tension created as the operating handle travels forward. Such a conversion would not likely leave any usable physical evidence behind when the line was unwrapped from the firearm. However, the presence of the line, assuming a firearm were seized with the line intact, gives clear indication of the intent of the modification. In such a scenario, the examiner would have to identify the line as the "sear" because that is the mechanical function that the line is assuming.

As is often the case, semiautomatic firearms, particularly rifles whose design is shared with machine guns, differ little mechanically from their full-auto counterparts. Conversions of all varieties can be accomplished on nearly any semiautomatic firearm. If weapons are seized and there is suspicion that alterations may have been made to convert firearms into machine guns, it is strongly suggested that knowledgeable personnel be brought in as soon as practical. Items of particular interest in the scope of a search concerning weapons offenses should include the following:

Literature: books, computer printouts, or electronic media that outline conversion techniques, especially the modification or construction of parts

Additional firearms components and parts, especially those showing evidence of milling, grinding, polishing, and cutting from their original design

Recognized firearms components that may be considered contraband: full-auto sears, machine gun receivers, and other components and parts that could be used to construct a machine gun or convert an existing firearm to a machine gun

Discarded materials or components from aborted prior attempts, experiments, and tinkering

Abnormally high wear on a firearm, based upon its relative age, indicating a higher than expected round count

Field-Testing Procedure for Automatic Weapons

The generally accepted procedure for field testing a weapon suspected of being a machine gun is relatively simple. It must be stressed that this procedure is not an absolutely conclusive evaluation that can be performed in lieu of a more detailed analysis by a competent examiner. However, the field test can be treated as an indicator that a firearm may be a machine gun by definition. If the firearm in question was manufactured as a machine gun originally, but has been reconfigured mechanically as a semiautomatic firearm, this procedure will not work. The examiner must be able to identify the exhibit as a machine gun by sight or design characteristic. As with all firearm handling, the first consideration is safety: The handler must be certain that the exhibit is unloaded. As previously discussed, the ammunition source must be identified, isolated, and removed. The chamber is then cleared; do not attempt to catch any ammunition or casing that is ejected from the chamber. Repeat the clearing procedure several times to ensure that the firearm is unloaded.

Closed Bolt-Action Field-Test Procedure

Ensure that the firearm is unloaded, i.e., the ammunition source has been removed and the chamber is clear. Clear the action several times and inspect visually and tactilely.

Ensure that the firearm safety is in the off position. If the firearm has multiple fire positions, each position should be tested independently.

Allow the bolt or slide to return to battery (forward). Do not "ride the slide" by keeping hold of the bolt or slide, allow it to freely move under the action of the firearm mechanism.

With slide forward or bolt closed, press the trigger and maintain pressure, do not relax the force being exerted on the trigger.

While fully pressing the trigger, fully pull the slide or bolt back with the nonweapon hand and release it. The bolt or slide should return to a closed or battery position.

Release the trigger, and then pull the trigger again.

If the firing mechanism trips, often with an audible report such as a click, when the trigger is pressed again, the firearm is testing as a semiautomatic.

If the firing mechanism does not trip, the firearm is testing as a machine gun.

Open-Bolt-Action Procedure

Ensure that the firearm is unloaded, i.e., the ammunition source has been removed and the chamber is clear. Cycle the action several times and inspect visually and tactilely.

Ensure that that the firearm safety is in the off position. If the firearm has multiple fire positions, each position should be tested independently.

Open the bolt fully to the rear; it should lock in place.

Press the trigger and maintain pressure on it; do not release the trigger.

While still holding the trigger down, pull the bolt to the rear to cock it again.

If the bolt locks again in place, the firearm is semiautomatic.

If the bolt freely reciprocates with the trigger pressed, it is a machine gun.

Case Studies of Commonly Converted Firearms

MAC-Pattern Firearms

The MAC (Military Armaments Corp.)-pattern firearm has been a target for conversions since the firearms were released into the commercial market, despite the fact that they were legally available as machine guns until 1986. The MAC was originally conceived by Gordon Ingram, and this family of firearms is often colloquially referred to as "Ingrams," although they were officially titled as the MAC. There were two models of the MAC: the MAC 10 chambered in .45 ACP and 9×19mm, and the MAC 11, which is chambered in .380 ACP, as well as a 9×19mm version, the MAC 11/9. Mechanically, the MAC 11 is exactly the same as the MAC 10, although its overall dimensions are diminutive when they are compared. The MAC design has changed hands numerous times over the years, the guns being manufactured by such firms as SWD, RPB, Jersey Arms, and most recently MPA (Masterpiece Arms). The MAC has been available as either a machine gun or as a semiautomatic pistol. The only significant design change to the MAC was the switch from open bolt to a closed-bolt operation in 1982, per ATF requirement. Nonetheless, all MAC firearms are popular for conversion to machine guns, primarily due to their relative availability, simplicity, and numerous established methods for conversion.

The most obvious MAC conversion can be accomplished by copying the factory work. A hole is drilled in the receiver in the correct spot, and a full-auto sear, sear spring, trip, disconnector, and selector (sear pin) are installed. This is remanufacturing the firearm by replacing and adding the parts to complete the conversion.

A simple and popular method of converting a MAC entails filing down the bolt trip so that the bolt would not catch it after cycling when the trigger was held down, allowing the bolt's forward travel to perpetuate sustained fire. The standard trip was not constructed of hardened steel; only the version intended for installation in the machine gun was. The alteration permits the firearm to fire full auto only, since there is nothing for the hammer to catch to stop the cycle of action from occurring. Alternatively, the bolt can be modified so that the marriage between it and the bolt trip is altered, the same effect as described by the prior alteration. However, this approach differs in that the modification is made to the bolt rather than the trip. On the bottom of the bolt on the right side is a groove that marries with the bolt trip, and removal of material in that space completes the conversion. This may be done in concert with alteration to the bolt trip itself if the person undertaking the conversion is unsure that it will be successful or wants to be more thorough. Modification to the bolt as described may be desirable since, in absence of modifications to other components, a replacement bolt may be installed and the modified bolt hidden away, thereby erasing evidence of the conversion should the firearm be subject to inspection. Conversely, when full operation is desired, the firearm is quickly disassembled and the modified bolt is installed.

An extremely simple conversion that can be attempted in lieu of these mechanical modifications is to install a wire in the receiver by wrapping it around the disconnector to retard its positioning, hence rendering it essentially inoperable and accomplishing the same goal as a mechanical reconfiguration of the disconnector. This alteration would cause the firearm to fire in full auto only. Because the MAC was purpose built as a machine gun, there is little risk that the conversion will cause a catastrophic failure to the firearm. An even easier and simpler, yet fully reversible method to converting an open-bolt MAC pistol into a machine gun is to take a pencil eraser, small coins, or similar articles and fashion them into a buffer of sorts. This buffering is then stacked behind the trigger to stop the trigger in the sweet spot, which is a cusp where the trigger cannot be tripped by the disconnector, creating a very reliable machine gun. This functionally has the same effect as retarding the disconnector, but instead affecting the trigger/disconnector relationship from the trigger side.

Heckler & Koch

Heckler & Koch firearms were, at one time, a very popular platform for those seeking to convert a semiautomatic weapon into a reliable, functioning machine gun. The renowned quality and durability of H&K firearms ensure that converted models are capable of sustained use as machine guns. The popular targets for conversion were the H&K Models 41 (7.62×51mm), 91 (7.62×51mm), 911 (7.62×51mm), SR9 (7.62×51mm), 93 (5.56×45mm), 94 (9×19mm), and the SP-89 (9×19mm). H&K briefly imported a semiautomatic version of its G.3 machine gun that was marked G.3. This nomenclature was later changed to HK 41, mechanically differentiated only by the omission of the components to permit fully automatic operation. All of the aforementioned models are essentially the same gun—all operated by the delayed roller lock bolt system—the only significant difference being the firearms' physical dimensions and the caliber. Hence certain components are not universally interchangeable between the models. It has been over 20 years since H&K was able to import the semiautomatic versions of its machine guns; however, several companies manufacture H&K clones, such as Cohaire Arms in Mesa, Arizona, and PTR 91 Inc. (formerly JLD Enterprises). Other companies such as Hesse manufactured receivers in the United States and built new firearms from parts kits that originated from various areas of the world. In addition, H&K copies were imported into the United States from places such as Greece (the SAR 3 and SAR 8 through Springfield Armory), Spain (CETME), and Portugal (INDEP).

There are numerous approaches to converting these H&K models into machine guns, some requiring modifications to the firearm receiver. Other methods employ a combination of receiver modifications coupled with converting the trigger pack within the grip assembly. Yet others are a hybrid of receiver alterations and replacement or alteration of semiautomatic parts with machine gun parts. Due to numerous design changes engineered into the firearms over the time they were sold commercially, the result is a dizzying array of possibilities. When faced with a potentially modified H&K-type firearm, the examiner must first understand the workings of the firearm and identify the parts and configuration combination to ascertain what has been altered, if anything, to convert the firearm into a machine gun. As a general rule, the trigger group is not a "gun part" until mated to a machine gun modified receiver; it may be possible to encounter machine gun marked trigger groups installed on semiautomatic H&K firearms or H&K clones.

The earliest imported H&K G.3 (or HK 41)-type rifles are identified by their characteristic markings: "Golden State Arms Co., Santa Fe Division." These rifles started being imported into the United States in the early 1960s. The model and other particulars are stamped on the left side of the receiver above the magazine well. These early imports potentially could be made to fire in full auto by counterclockwise rotation of the fire selector lever to the blank spot where the full-auto position would normally be, roughly at the 5 o'clock position. By design, the fire selector lever would normally be rotated clockwise* to set the fire condition of the rifle. This conversion is accomplished by removing the stock and the grip assembly to gain access to the trigger housing that contains the safety/selector switch. The switch is then rotated counterclockwise and the parts are reassembled. If such

---

* The H&K pistol grip assembly can be marked in either letters, numbers, or icons to indicate the position of the safety/selector switch. White *S* indicates safe; red *E* indicates semiauto fire; red *F* would indicate full auto. The numbers reflect white "0" safe, red "1" single shot, red "20" or "25" for full auto. The H&K icon-marked pistol grips show bullets in a barrel to indicate safe, single, full, or burst-fire switch position, with white indicating safe and red indicating fire positions.

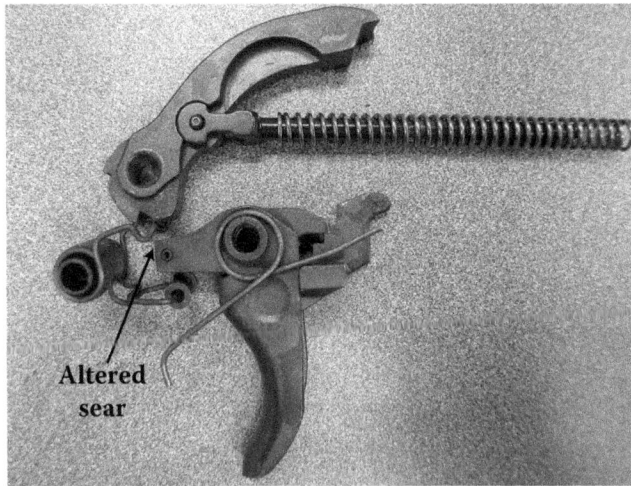

Figure 4.37 Close-up image of the modified trigger group from the H&K 93. The sear has been modified by beveling the edges, and an H&K factory full-auto hammer has been installed, replacing a semiautomatic-style hammer. Note the "XX" marking, indicating that the hammer is machine gun specification. (Image courtesy of Jared Ford.)

an example is observed, it may be possible to determine if the firearm has been subjected to this alteration by observing the characteristic circular wear mark on the side of the grip assembly. Under normal, routine use, the selector lever would not be rotated past the safe position and certainly not in a counter-clockwise motion. Otherwise, an internal inspection will reveal the positioning of the selector switch.

A method of conversion for H&K firearms that has come to light has been reported by Firearms Examiner Jared Ford, Oregon State Police Forensic Services. The firearm in question was an H&K 93 rifle, chambered in 5.56×45mm. The conversion process entailed switching the hammer from the purpose semiautomatic hammer to the full-auto specification hammer, and then polishing the sear where it engages the hammer to reduce the marriage between the two parts (see Figures 4.37 and 4.38). The adjustment of the marriage between the sear and the hammer was probably sufficient to accomplish conversion because when a semiautomatic hammer was installed in place of the full-auto hammer, the firearm functioned as a machine gun. Full-auto fire was accomplished when the selector switch/safety was in the "1" position, the semiautomatic position; thus the firearm was incapable of semiautomatic fire as modified. The advantage of this approach is that it likely will leave little, if any, residual evidence behind of the alteration when the parts are removed and replaced with the standard semiautomatic parts, thus completing the deception (Ford 2010).

The Heckler & Koch USC (Universal Self-Loading Carbine) and SL-8 carbines operate using different systems and are not interchangeable with the H&K models that operate using the delayed roller lock bolt system. The USC carbine is chambered in .45 ACP and operates on the blowback principle; the SL-8 functions on the now-familiar short-stroke gas piston. Conversions effected on these firearms may be made by modification to the receivers and internal components. There are some similarities between the USC and the UMP-45, which is a submachine gun, as well as the SL-8 and the G.36. The G.36 is a machine gun. Aesthetic conversions are performed to make the USC resemble the UMP, likewise with the SL-8 to resemble a G.36 and even accept G.36-pattern magazines. In

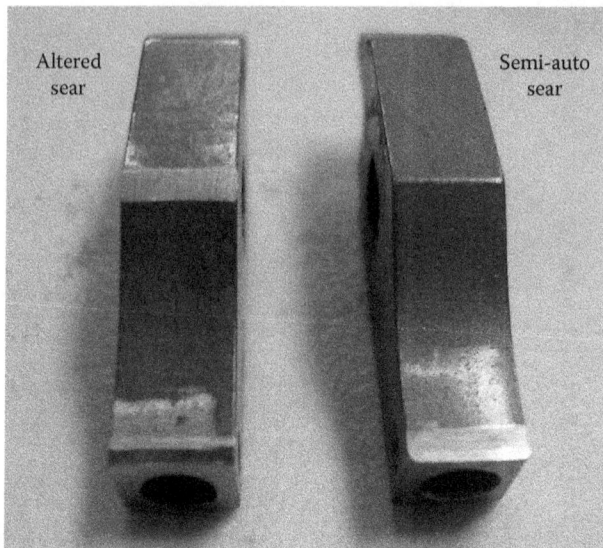

Figure 4.38 The modified sear from the H&K 93. Both pictured sears are for semiautomatic guns. The sear on the left has been modified by making additional cuts to alter the marriage of the sear in relation to other trigger parts. The altered sear was installed in the trigger group. (Image courtesy of Jared Ford.)

theory, conversions of these firearms into machine guns can be accomplished by modifications to the receiver and internal parts.

AR-Pattern Firearms

The AR-pattern firearm is a very popular platform for full-auto conversions. Converted AR firearms may take the form of pistols, short-barreled rifles, carbines, or full-sized rifles. There are numerous methods of effecting a reliable conversion of an AR from a semiautomatic weapon into a machine gun. The attractiveness of the AR platform as a vehicle for conversion is multifold. First, an AR lower receiver is relatively inexpensive and readily available, especially when compared to other firearms that could be considered as candidates for conversion. Second, the availability and low cost of replacement parts allows for a good deal of experimentation and reworking of components. Third, the modularity of the design to fit multiple configurations and calibers appeals to a wide audience.

In an effort to stem postfactory alterations, Colt originally undertook a number of different measures that have not been adapted across the plethora of manufacturers who now produce an AR firearm, and there are many. One measure was a redesign of the bolt carrier group. A full-auto bolt carrier is a completely round rear section with a channel on the bottom that interacts with the hammer, allowing it to reset and act against the full-auto sear. The revised semiautomatic bolt carrier did away with this, instead having an open horseshoe-shaped rear portion. The missing part of the horseshoe is where the hammer travel occurs; thus there is no interaction between the bolt carrier and the hammer. A full-auto-type (M16) bolt carrier may not be required if a bolt carrier converter is used. The converter was simply a piece of steel in the shape of the bolt carrier group that filled in the missing material in the M16-style bolt carrier. The converter is installed onto the bolt carrier and affixed into place by screws, allowing a semiautomatic bolt carrier to act as the full-auto variant. These converters are seldom, if ever, encountered anymore because of the widespread availability of M16-style bolt carriers.

A second modification was to not mill out the portion of the lower receiver where the auto sear would sit, often called the *sear block*. Like the bolt carrier, this practice is not observed by all manufacturers. A third alteration, while not affecting the fire-control parts, involves an oversized (.315") pivot pin hole. The pivot-pin hole is the hole at the front of the lower receiver, just underneath the barrel. This nonstandard-diameter pivot pin was designed to deter users from replacing the Colt commercial specification upper receiver with a military one. Certain manufacturers may occasionally offer upper receivers with the oversized pivot-pin hole; however, the Colt nonstandard pivot-pin diameter was overcome by use of an offset pin that would permit a mil-spec (.250") upper to mate with a Colt-manufactured lower receiver with the oversized hole.

Replacing Semiautomatic Parts with Machine Gun Parts

The most obvious way to convert a semiautomatic AR to a machine gun is simply by replacing the semiautomatic fire-control parts with machine gun control parts. This involves changing the bolt carrier (to possibly include swapping the firing pin to a large flange type, but this is not really necessary to complete the conversion), hammer (there is an extra spur on a full-auto-style hammer),* trigger, selector/safety switch, full-auto sear, and disconnector.

The author is aware of instances where AR-15 rifles have had five of the machine gun parts installed, minus the full-auto sear, and they were capable of full-auto function, albeit erratically. In theory, disabling or removing the disconnector from a modified receiver, especially without an auto sear in place, will cause the firearm to function as a machine gun, even if it discharges in unpredictable bursts before the action trips and ceases fire. On a gun set up for burst fire, as opposed to full auto, the hammer will be a different design from a semiautomatic-style or full-automatic-style hammer. A burst-style hammer is ratcheted; the burst-type hammer spring will also be different, as the hammer-spring legs are offset. The standard M16 burst-fire control parts are set up to fire in three-round increments per press of the trigger, and then the sear resets and the next press of the trigger will fire three additional rounds and so forth until the ammunition supply is exhausted, rounding off to 10 three-round bursts per 30-round magazine.

Burst parts and full-auto parts are not interchangeable per se to complete a conversion; however, the presence of these parts within the firearm may constitute a machine gun, or at least intent to manufacture one, and likely could be the result of the unfamiliarity of components to the builder. This relative ignorance may result from someone who obtains parts that were mislabeled, or the builder may have been misled by following instructions obtained from sources who themselves know no better. The physical installation of a full-auto sear will require the lower receiver to have a hole drilled in it to accommodate the sear pin that holds the sear in place. This hole is located above the SEMI-marked fire position above the safety selector switch. The mere presence of this hole—regardless of the presence of other fire-control parts, physical modifications, or other means of apparent alteration—qualifies the receiver as a machine gun under U.S. federal law (see Figure 4.39). Templates

---

* The hammer style can be instrumental to the success or failure of the conversion, as the notch on the hammer is designed to prevent full-auto fire should the sear or disconnector fail to work in semiautomatic operation.

Figure 4.39 A close-up of the right side of an AR-15 lower receiver. Note the three positions of the selector switch: SAFE, SEMI, AUTO. The hole above the SEMI position is for the auto sear pin. Although no sear is installed in this gun and the pin hole is empty, the mere presence of this hole would define this receiver as a machine gun. (Image from author's collection.)

Figure 4.40 The drop-in auto sear. The DIAS is easy to manufacture and conceal. Literally dropping into the lower receiver, the DIAS converts the host AR weapon into a very reliable machine gun. (Image from author's collection.)

and jigs that provide the precise location to drill the hole and the hole diameter are readily available from various sources.

Drop-In Auto Sear

In lieu of swapping semiautomatic parts for full-auto parts and the requisite modification to the lower receiver, the introduction of two alternative-style sears has long been known as a means of conversion, and both are quite common. The first type of sear is the DIAS (drop-in auto sear) (see Figure 4.40). As the name implies, the drop-in auto sear literally drops into the AR lower receiver and acts as the traditional full-auto sear. The DIAS inserts into the rear of the AR lower behind the hammer and in front of the buffer tube. An M16 (full auto)-style bolt carrier is used in conjunction with the DIAS. It is important to note that properly registered DIASs can be legally possessed in the United States. That having been said, there is an untold number of contraband DIASs in existence, and they routinely turn up and likely are clandestinely manufactured.

The DIAS is a simple and reversible conversion. All that is required is to properly fit it to the lower receiver and install the unit. Some AR lowers do not readily accept the DIAS. The portion of the lower receiver where the DIAS would be installed has a "sear block." The material is not milled away, and so the lower must be modified to accept the DIAS or a standard full-auto sear. The DIAS has to be trimmed to fit the appropriate width of the receiver well, given subtle tolerances either way on the part of the receiver and the DIAS. In the United States, the possession of an unregistered DIAS could conceivably be construed as possession of a machine gun. The DIAS was not considered a machine gun until November 1, 1981; thus, if the example in question was manufactured prior to that date, there is no prohibition against having one unless it is possessed in conjunction with an AR-pattern firearm. Proving that the DIAS is a grandfathered example may be very difficult; however, proving that it is not may be equally difficult.

Lightning Link

Another form of conversion is the installation of a *Lightning Link* (see Figure 4.41). Like the drop-in auto sear, the Lightning Link is rapidly installed and removed. Like the DIAS, the Lightning Link is installed at the rear of the AR lower receiver and acts as the sear, allowing for full-auto operation. The Lightning Link works by retarding the travel of the disconnector, moving it out of reach from the hammer, thereby preventing the hammer from engaging with the disconnector after the trigger is pressed. Unlike some other approaches to conversion, the Lightning Link allows controlled fire: When the trigger is released, the hammer will be stopped in the cocked position by the Lightning Link. Like the DIAS, there are lawfully possessed Lightning Links that were manufactured prior to the 1986 prohibition of further manufacture of machine guns, and like the DIAS, possession of an unregistered Lightning Link could constitute possession of a machine gun.

The Lightning Link was manufactured by SWD in Atlanta, Georgia, as the M15AC. This was the same SWD that manufactured MAC-pattern firearms, including machine guns. As the Lightning Link itself is considered a machine gun, even if uninstalled, there

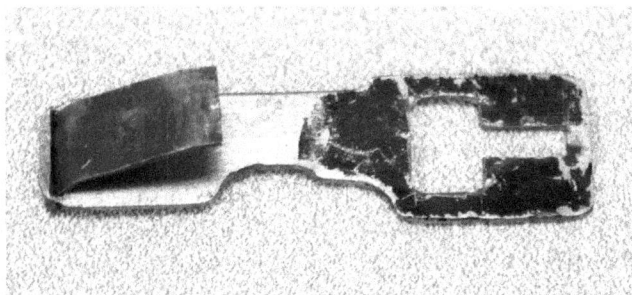

Figure 4.41 The lightning link can be manufactured using hand tools and sheet metal. The lightning link is easily concealed and requires no modifications to the host weapon to install. It would likely be overlooked by an untrained eye during a search. (Image from author's collection.)

would be a serial number and other pertinent information affixed. The Lightning Link can be a well-manufactured item of choice materials, but it can just as easily be crudely sawn from sheet metal and roughly fashioned, so clandestine manufacturing can be easily accomplished. Like all auto sears, instructions and blueprints are easily obtained for guidance in manufacturing a Lightning Link. There is no 1981 exemption to the Lightning Link, as in the case of the DIAS.

Open-Bolt Modification

Another possible conversion, albeit more complicated than those discussed previously, involves converting the AR firearm to operate from an open bolt by substantial modifications to the fire-control parts and the addition of a striker to the bolt-carrier group that acts upon the firing pin, with the modified hammer acting on the striker, as opposed to the firing pin itself. As would be expected in any open-bolt striker system, the hammer takes on the role of a sear and simply holds the action open until the trigger is pressed, releasing the bolt and causing discharge and subsequent cyclic fire to occur. Such a conversion may entail the removal of the bolt catch assembly from the lower, as using the bolt catch to release the bolt carrier into battery would cause the firearm to discharge. The bolt catch itself is traditionally a problematic piece in the AR, being susceptible to releasing the bolt if the weapon is jarred hard enough or if the firearm is banged on a hard surface. Open-bolt M16s are not unheard of; the Colt Light Machine Gun fires from an open bolt.

Indicia of a Converted AR Firearm

Indications that an AR has been converted, or that there has been an attempt to convert, should be noted, because this may be prima facie evidence that the firearm in question is a machine gun:

> Presence of a sear hole in the lower receiver located above the SEMI fire position on the safety selector, regardless of whether or not there is a third position marked AUTO or BURST
>
> Evidence of grinding, cutting, or milling at the rear of the lower receiver behind or around the area of the hammer
>
> Installation of any combination of identified machine gun–specific fire-control parts, even in the absence of a sear (remember that a sear can take several forms)

Receiver Markings

The most obvious AR-pattern machine gun is one that is factory built or a firearm reconstructed to imitate a factory-built machine gun. Contrary to popular belief, not every machine gun manufactured by Colt is marked M16; there are examples marked AR-15 (see Figure 4.42). Nor is it true that only M16-marked rifles were manufactured for the U.S. military; AR-15 rifles can be marked PROPERTY OF U.S. GOVT. Early examples of Colt-manufactured AR15 (in this instance, the "-" is omitted from the receiver marking) machine guns may be marked PATENTS PENDING, dating the firearm to when Colt had just secured the rights to manufacture from Armalite. Such examples are marked as Colt and Armalite, as the Model 01 in .223 caliber. Factory-manufactured machine guns will be readily apparent, regardless of the markings on the receiver, by the three-position safety/selector switch that is marked SAFE, SINGLE, AUTO (or BURST in lieu of AUTO). Some variations, such as the H&K 416, use icons to denote the same information.

Figure 4.42 For anyone who has ever doubted that Colt ever manufactured an AR-15 for the U.S. military, here is the proof. This is an early Colt-built gun, bearing the dual AR-15 and M16 markings. (Image from author's collection.)

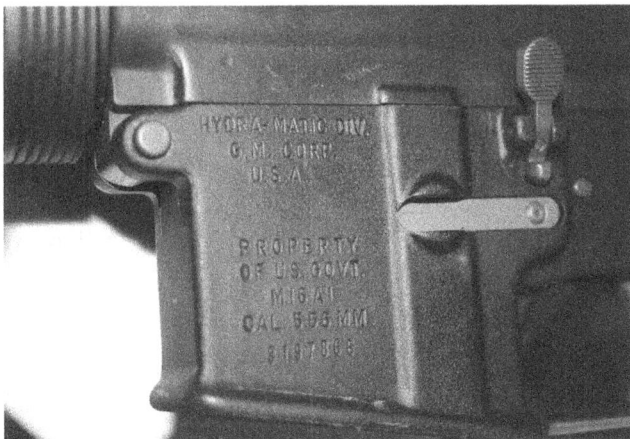

Figure 4.43 An M16 produced under contract by the Hydra-Matic Division of General Motors. Hydra-Matic–marked examples are seldom seen. (Image from author's collection.)

The AR-15/M16 is not solely the realm of Colt; many other companies have manufactured these machine guns, and so any number of manufacturer names can be expected. In addition to Colt, M16s have been purchased by the U.S. government from FN Manufacturing in South Carolina as well as earlier production (late 1960s to early 1970s) by the Hydra-Matic Division of General Motors (see Figure 4.43), and Harrington & Richardson in about the same time period as Hydra-Matic.

The examiner cannot rely on the general features of the exhibit under study because the design is so modular that anything on the weapon could have been upgraded or changed, so the receiver markings are what are materially relevant on a factory-manufactured machine gun or a firearm remanufactured into a machine gun by a third party. As with semiautomatic examples, AR-15/M16–pattern rifles are manufactured as machine guns by other companies, so variations in manufacturer markings and model numbers are expected if such an example is encountered.

Intratec Pistols

The KG-9 pistol originated from the Swedish firm Interdynamic. The design was brought into the United States and manufactured by the firm Intratec in Miami, Florida, until the company ceased operation in 2001. Classified as a handgun, this pattern was produced in calibers .22 Long Rifle and 9×19mm Luger. Both calibers were marketed under a variety of different model names, the .22 as the Scorpion and the Sport; the 9mm version as the TEC-9, TEC-DC9, and KG-99. The term *TEC-9* has come to generically identify all the models, even the .22 versions, regardless of the proper model designation that appears on the firearm. A more compact version of the pistol chambered in 9mm was sold as the Model AB-10 (where the term AB is an acronym for "after ban"). Another model, the MP-9, was touted as a commercially available machine gun, but apparently the model was plagued by development issues and there are likely very few, if any, in existence.

Although legally defined as a handgun, the Intratec pattern exhibits characteristics more customarily associated with a short rifle or carbine, short of having a shoulder stock. It is physically large in contrast to other handguns, and its overall appearance is not consistent with that traditionally associated with pistols, including a magazine well that is forward of the pistol grip and a cocking handle instead of reciprocating slide. All models were fed by detachable magazines, which were available in various ammunition capacities. The frame is constructed of polymer, with the upper portion constructed of steel and having a matte parkerized or stainless steel finish. A threaded barrel was initially available but was later removed when the firearm became subject to various legislative initiatives that classified it as an assault weapon. To complement the threaded barrel, barrel extensions were sold that resembled the TEC barrel, having a metal shroud around the barrel itself. These extensions may be mistaken for a suppressor, although they were not originally manufactured as such.

Since a threaded barrel was listed as a feature on a potential assault weapon, postban models featured a barrel without the threading on the muzzle. The model designations also changed, as the firearm was banned by identifying it specifically by model. A different model designation effectively circumvented the affecting legislation. On the 9mm models, the safety is integral to the operating or "charging" handle; the .22 models had a safety sliding switch on the left side of the receiver. Another difference between the two was the operating handle on .22 guns consisting of two tabs, one on either side of the receiver.

The firearm operates by blowback, and early models fired from an open bolt and had a firing pin that was fixed to the bolt face. This feature was later changed so that firing was by closed bolt with a floating firing pin. The simplicity of the design made the KG-9 very susceptible to conversion from a semiautomatic firearm to one capable of fully automatic operation. In the case of the open-bolt version, the bolt was held open by the sear. Polishing of the sear could directly affect the marriage between the sear and the bolt, denying the sear the ability to capture and retain the bolt if the trigger were held down, allowing continuous fire. It is possible that ordinary use could eventually cause the sear to wear down and inadvertently cause full-auto operation.

Modification to the sear alone may be sufficient to achieve full-auto fire, but sear modification may be coupled with filing down the disconnector as an additional measure to ensure success. A thorough conversion could be made to an Intratec by much the same method as a MAC—by altering the bolt body in concert with (a) retarding or defeating the disconnector and (b) sear modification. The bolt body would have material removed from it in the area where it interacts with the disconnector, once again affecting the marriage between the two parts.

Intratec manufactured other models beyond the TEC family. Handguns marketed as the CAT and Pro"tec"tor series had a more traditional pistol appearance. Intratec also manufactured a derringer, the TEC-38, chambered in .38 Special. The TEC-38 had two barrels in an over/under configuration.

Glock Pistols

The simplicity and durability of the Glock design has made it a handgun that has successfully been converted to a machine gun using several approaches. Because all Glock models are built exactly the same, any single method of conversion will work on any model, regardless of caliber. Glock does, in fact, manufacture a full-auto version of the Model 17 9×19mm, the Glock 18. There is also the compensated model, the Glock 18c. The internal components of the Glock 18 were deliberately redesigned to thwart attempts at using diverted components to convert other pistols into automatic weapons.

The Glock 18 has an external selector switch located on the left side of the slide in rear-slide serrations. This selector switches between single shots and full auto. Glock 18s are marked and serial numbered as any other Glock product, and are overtly the same as the other standard-frame models. Modifications performed with the intent of converting a Glock into a machine gun typically focus on changing the geometry and/or the marriage between the firing pin and the cruciform portion of the trigger bar. U.S. Patent 5,705,763 illustrates a device that is installed in the place of the slide cover plate. The device alters the marriage between the trigger bar and firing pin to permit full-auto fire (Leon 1998). Various other examples of similar devices have been fabricated and distributed, typically as a direct replacement for the factory slide cover plate. Some versions of these conversion plates (which are definable as a sear by virtue of function, as a standard cover plate serves no mechanical function) have an external switch that acts as a selector switch. Other examples may not be equipped with a switch, instead making the modified Glock capable of full-auto fire only. These conversion plates should not be confused with the orange armorers cover plate that is designed for inspection of the marriage between the trigger bar and firing pin to determine if replacement is necessary. Some conversions may require the removal of some of the polymer material from the frame around the trigger bar.

The modified Glock 17 shown in Figures 4.44 and 4.45 demonstrates another approach. The trigger mechanism housing has been modified by cutting away some of the polymer on the top and left side. The left side of the cruciform portion of the trigger bar has been bent upward, allowed by the relief cut made in the trigger mechanism housing, and beveled on the edge. The slide had a hole drilled adjacent to the firing pin, and a metal pin was seated that would correspond to the modified part of the trigger bar. The net effect would be that this pin would retard the marriage between the cruciform portion of the trigger bar and the firing pin, ensuring reliable full-auto function. The trigger bar, as modified with the upward facing tab, has the functional role of an auto sear. As the trigger is pressed, the movement would cause the upward bent portion of the trigger bar to meet with the pin in the slide, which would push the trigger bar down under trigger pressure.

A similar conversion has been observed, albeit more complicated than the one described. This other conversion adds a selector switch to the left side of the slide. The

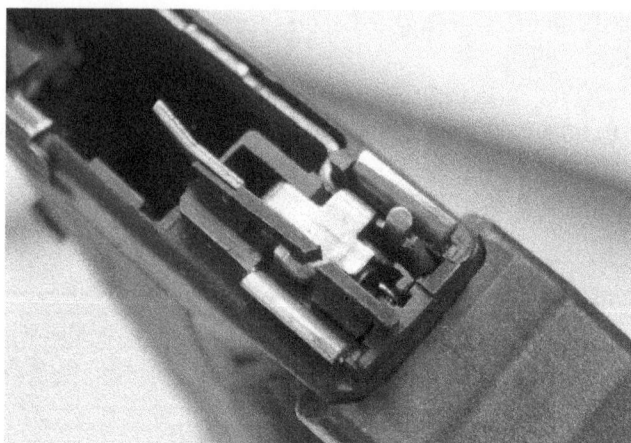

Figure 4.44 Details of the trigger mechanism housing of a Glock 17. The modifications made to convert this pistol into a machine gun are evident. A shoulder stock has been attached to the pistol, and is partially visible on the right side of the image. (Image from author's collection.)

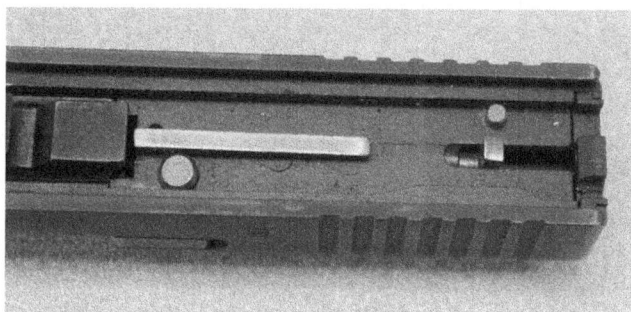

Figure 4.45 Details of the slide from the same Glock 17 shown in Figure 4.44. Observe the added round metal piece to the right of the firing pin, which would correspond to the trigger-bar modification in the frame. As an additional measure, the elevated rail in the slide has been polished down, probably to ensure that it would not push the trigger bar down into the disconnector. This was a labor-intensive conversion. (Image from author's collection.)

switch controls a small trapdoor that extends or retracts, thereby affecting the trigger bar. Extensive machining is involved in this conversion, requiring more skill and tools than the first conversion, with practically the same result.

Advisory bulletins alerting law enforcement to the existence of converted Glock handguns have been published since at least as early as January 2004, although it is likely that conversions have been undertaken long before any examples surfaced. With the proliferation of handguns of similar design such as the Smith & Wesson M&P pistol line and the Springfield Armory XD pistol series, it is likely that conversions will be attempted on these as well using similar methods.

SKS (Simonov Carbine Self-Loading)

The SKS is often mistaken as an AK-pattern firearm, and there are some outward similarities between the two. However, they are distinctively different. The SKS predates the

AK by a few years, and they are both of Russian origin. The SKS was widely copied by Soviet satellite states and was known by numerous designations. The SKS can be found in a plurality of configurations, but the basic firearm is the same. The SKS is, mechanically speaking, a relatively simple firearm that is built with fairly loose internal tolerances. Conversion of an SKS into a reliable machine gun can be accomplished by several different methods.

One method involves the construction of a drop in auto sear with supportive parts modifications made internally. One sear design is constructed of flat metal and has the appearance of a two-pronged fork or a U that is slightly hooked on the forked end. The sear is installed in front of the trigger with the forked end adjacent to the trigger and underneath the factory sear. Some disassembly of the rifle is required to accomplish the conversion, but when the drop-in auto sear is removed, it will leave no evidence behind. The simpler method entails blocking the sear disconnector, which permits the trigger to control the sear and defeats the disconnector, a classic approach to converting a semiautomatic firearm with an internal hammer that relies on a disconnector to catch and retain the hammer after each shot is fired. Shims can be fabricated or foreign objects installed to create this effect.

An untold number of SKS rifles manufactured as machine guns were imported from China in the late 1980s and entered the United States. The error was not detected until some time after the rifles had been distributed through commercial channels. These examples will have a selector switch located in front of the trigger guard. This should not be confused with the safety switch, which is beside the trigger guard. Despite efforts to confiscate these firearms, there are likely many still in existence. SKS rifles were also regularly brought into the United States by American servicemen returning from the Vietnam War as souvenirs, and some of these may have been configured as machine guns. Under ideal conditions, such a weapon would have a provenance attached to it, such as declaration paperwork. SKS-pattern rifles have been commercially imported into the United States from Albania, Russia, and the independent states within the former state of Yugoslavia. Production of this rifle was undertaken by other nations as well.

AK-Pattern Firearms

AK-pattern firearms, like AR-platform firearms, are popular for machine gun conversions. Like the AR, the AK is derived from a machine gun, which has been reconfigured into a semiautomatic rifle by modifying the fire-control parts and changing the receiver design to deter simple alteration into a machine gun. Like other semiautomatics, this means that the receiver itself could not have been initially manufactured as a machine gun and simply reverted to a semiautomatic. The receiver itself must not have been deemed to have been manufactured as a machine gun in the first place. The receiver may have been constructed to the standard of a machine gun; however, there are design changes incorporated that reflect that the receiver in question is not that of a machine gun.

Regardless of where the AK was manufactured, the basic receiver remains the same, and the same conversion could be effected on practically any standard receiver, regardless of origin. The conversion can be made using a full-auto parts kit that includes the bolt carrier, sear spring, sear, sear pin, full-auto disconnector, and full-auto hammer. Semiautomatic AK-pattern receivers lack the hole where the sear pin would be installed that sets the sear in place in the receiver. This hole is located at the front edge of the magazine release. With the safety/selector switch down, the presence of this hole is obscured. As with the AR-15, an AK receiver that is drilled for the sear is considered a machine gun, regardless of whether there is an auto sear installed or not. On an AK, the combination safety/ selector switch is located on the right side of the receiver. The safe position is the uppermost position, regardless of how the receiver is marked. With the safety engaged, the bolt carrier is physically blocked from traveling to the rear, so the firearm cannot be unloaded, charged, or otherwise manipulated. An AK capable of fully automatic fire ordinarily will have three positions on the safety selector. Semiautomatic operation is the third, or lowest, selected position. The switch center position is the full-auto position. If full-auto parts were to be substituted for semiautomatic parts, substantial internal modifications would have to be made to a semiautomatic AK receiver to accomplish a reliable conversion to allow the receiver to accept the full-auto parts.

Other conversions reconfigure parts to interfere with the marriage between the hammer and the disconnector as the primary focus. An effective and reliable conversion can be as simple as taking a wire tie and stuffing it into the crevice in front of the trigger and twisting it around the sear, thus permitting full-auto operation. Like the AR-15, a drop-in auto sear (DIAS) exists for an AK-pattern firearm. Like all other conversions, there are numerous approaches that are only limited to the ingenuity of the person undertaking the task; thus the particular route that is taken means less than the end result.

There are lawfully possessed AK machine guns in the United States. These examples take the form of those constructed by licensed manufacturers or dealers. The AKs may have been constructed domestically, or they could have originated from a foreign source.

M1 Carbine

The World War II–era U.S. M1 carbine (not to be confused with the M1 Garand) was initially produced as a semiautomatic rifle. During the war, the full-auto M2 carbine was introduced as a way to upgrade the firepower available to the carbine. M2 carbines were manufactured as such, but there was also conversion of existing M1 carbines both during and after the war. U.S. military kits T17 and T18 were issued to convert an M1 into an M2 in the field (see Figures 4.46 and 4.47). Visually, the differences between the M1 and M2 are almost indiscernible, save for the presence of the selector switch located on the left side of the receiver ahead of the ejection port and behind the hand guard, opposite the operating handle. The M1A1 was differentiated by having a folding metal stock for paratrooper use. Postwar, the designation M3 was also used, the only difference being that the rear sight was removed and a sight mount installed for use with optics.

These firearms may or may not be marked as M2 or M3 when encountered, and such markings may be overstrikes or restrikes of the original M1 receiver legend to indicate modification. M2-style stocks were slightly larger than the original M1 stock to help compensate for full-auto fire and are notched for the selector switch. An M1 stock would have to be modified to accommodate the selector switch. To reconfigure an M1, using original-specification components, into an M2 requires the following parts—selector, selector spring, selector lever assembly, hammer, and disconnector (including spring and plunger). Note that there is variation between an M1 and M2 hammer, sear, operating slide, and the trigger housing (War Department 1947).

A successful conversion of an M1 into a machine gun need not require M2 compo nents as described; one need only understand how the M1 operates and how to go about

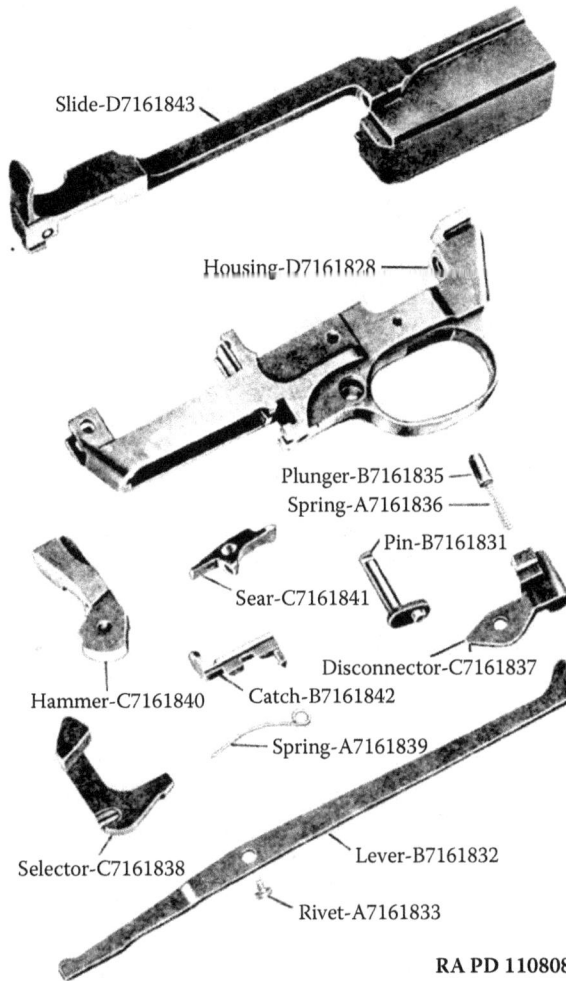

Slide-D7161843

Housing-D7161828

Plunger-B7161835
Spring-A7161836

Pin-B7161831

Sear-C7161841

Disconnector-C7161837

Hammer-C7161840   Catch-B7161842

Spring-A7161839

Selector-C7161838   Lever-B7161832

Rivet-A7161833

RA PD 110808

Figure 4.46 M2 Full Auto Parts diagram. (Source: U.S. Army Technical Manual TM-9-1276.)

modifying the action. Another approach to the conversion process could be successfully made using some M2 parts, with improvised components making up the balance. A conversion using this approach was undertaken where the trigger housing, operating slide, and the disconnector assembly were of M2 specification. The only part lacking was the selector switch. A homemade selector switch was fabricated using hardware-store-type materials. The selector switch was affixed to the right side of the receiver low on the stock, in contrast to the standard M2 selector, which is positioned on the left side of the receiver adjacent to the ejection port. In this instance, the homemade selector communicated with the disconnector lever through a hole drilled in the side of the trigger housing. The end of the selector that interacted with the disconnector was fashioned into a cam of sorts, which had the effect of lifting the disconnector lever, hence causing it to override the disconnector when engaged. This conversion was identified on a postwar commercial M1 copy, the Enforcer, manufactured by Iver Johnson (Carr 1992). Aside from the internal modifications, the presence of the pronounced switch and stock modification to accommodate it should give immediate rise to suspicion.

Figure 4.47 Close-up of the receiver of an M2 carbine manufactured by Inland Manufacturing. This example was manufactured as an M1, then later converted to an M2; the receiver markings are obscured by the rear sight. Note the selector switch on the left side of the receiver near the breech. (Image from author's collection.)

Assorted Other Conversions

The author is familiar with a conversion method for a SIG Sauer pistol. Unlike the Glock full-auto conversion, which fires from a closed bolt when converted as previously described, the converted SIG operates as an open-bolt machine gun. The conversion process involves affixing a small-diameter ball bearing to the breech face over the firing-pin hole opening. The alteration renders the pistol inoperable as a semiautomatic handgun, and the trigger becomes of no functional value, the pistol is capable of full-auto fire only. To operate, the slide is locked to the rear, a loaded magazine is inserted, and the slide release is depressed. When the slide moves forward into battery, a cartridge is fed into the chamber from the magazine, and when seated in the chamber the affixed ball bearing, now acting as a fixed firing pin, impacts the primer under the force of the closing slide, causing detonation. The pistol then cycles, ejecting the spent casing, and then the slide comes forward into battery, again loading a live cartridge, which again automatically detonates. In this particular conversion, there is no way to stop the pistol from firing until the ammunition supply is exhausted.

A similar conversion was reported for a Ruger Mark I .22 pistol. An additional striker was spot welded or soldered to the breech face. When the slide release was depressed, the pistol would function as an open-bolt machine gun. This basic conversion method could conceivably be made on other similar firearms that could be made to function from an open bolt, although the term *open bolt* is used somewhat loosely, as functionally there is no difference. Another similar method of conversion of a pistol to an open-bolt machine gun is the Walther P.38 and its post-war copies. The P.38 would appear to be an unlikely candidate for conversion from a semiautomatic pistol into one capable of functioning as an automatic weapon. Although the P.38 is a full-size-frame pistol constructed of machine steel, its magazine capacity is a mere eight rounds of 9×19mm ammunition, and there was no apparent ability to attach any form of stock or support other than the short pistol grip. A conversion kit, reported to originate from (West) Germany, was available through mail order during the 1970s and perhaps into the 1980s. The "kit" consisted of a replacement firing pin. The instructions called for the disassembly of the P.38 slide by removing the stock firing pin and installing the replacement. The conversion is completed by inverting the slide stop lever spring located underneath the left grip panel. In theory, the safety/decocker could act as a selector switch, or the firing pin could be fixed and, once again, you have an open-bolt machine gun.

Colt 1911 and 1911A1

Colt experimented with machine gun variants of the 1911/1911A1 pistol, likely in the 1920s and 1930s. Full-auto handguns were quite popular in that era, and it is perfectly logical for Colt to have followed up with their own design for one of the most popular handguns then in existence to counter competitors' machine pistols, especially from Germany. John Dillinger is reported to have used or at least possessed a Colt 1911 machine pistol, purportedly chambered in .38 Super. The pistol had a vertical fore grip resembling that from a Thompson submachine gun and an extended magazine. It is unclear whether the firearm had been converted or was originally manufactured as a machine gun by Colt. The modification would be quite simple, and many 1911 frames have inadvertently been made into machine guns while performing a trigger job. As with all conversions, there are several schools of thought, including the construction of an auxiliary trigger that is part of a vertical fore grip, as well as typical modifications to the trigger and sear. Fixing the firing pin and using the slide release as the trigger has also been contemplated, as in the case of the previously described SIG pistol. Examples of full-auto Colt 1911 machine pistols purporting to be factory made as such are said to exist. Machine gun variants of the basic 1911 likely existed in foreign versions of the handgun as well, either as factory work or by conversion after manufacture.

Browning Hi Power

In 1978, a method of converting the Browning Hi Power was brought to light. The conversion was accomplished by replacing the stock sear with a longer replacement (the original is 1 5/8″ replaced by one that is 1¾″), which gives the replacement sear sufficient reach to remain in constant contact with the trigger lever. As long as the sear remains in contact with the trigger, and the trigger is depressed, then fire will be sustained until the ammunition supply is exhausted. Such a converted pistol suffers the same malady as other full-auto pistols: a very high rate of fire with extreme muzzle climb and controllability issues. The other functional issue is that the magazine safety is defeated by the alteration and is rendered inoperable (Zahn 1978).

This report was followed up with another in April 1981, where another method of converting the Hi Power had been unearthed. In this conversion, the forward travel of the trigger lever is inhibited by the insertion of an L-shaped piece of steel forward of the trigger lever. The additional part seats into the pistol without need for modification to the firearm whatsoever and no tools are required to install or remove. The exemplar firearm had been modified by way of increasing the size of the ejection port by milling an angled portion of the slide away (Zahn 1981). As with the previous conversion, the magazine safety feature is inoperable when this modification is affected. This should come as no surprise, given that the safety works on a plunger that is attached to the trigger and is pressed against an inserted magazine, or relieved when no magazine is inserted. It is interesting to contemplate if these conversions were inspired by someone trying to engineer a trigger enhancement on the Hi Power by altering the magazine safety, which affects the trigger press. It seems unlikely that a select-fire version of the Hi Power would never have been manufactured, since full-auto handguns were very much in style in the 1930s when the Hi Power came to market. It also seems unlikely that select-fire Hi Powers were not manufactured, even if by a third party, and that such examples were registered before the 1986 machine gun ban took effect in the United States.

Early examples of the Hi Power had a slot in the grip back strap to attach a shoulder stock, turning the Hi Power into a carbine, which would have enhanced the controllability of the pistol as a machine gun to a certain extent. An interesting historical footnote of the Hi Power was that it was used concurrently by both the Allied and Axis forces during the Second World War. During German occupation of Belgium, the Hi Power continued to be manufactured by Fabrique Nationale, with German military forces absorbing all output. On the Allied side, the Hi Power was manufactured by Inglis in Canada and was issued to the Commonwealth Forces. Nazi Hi Powers carry the Nazi proof mark, showing an eagle over the legend "WaA140" as well as firing proof showing the eagle and swastika; these markings are clustered together on the left side of the receiver and frame, but variations to this practice can be found. Otherwise, the Nazi Hi Power carries the standard Browning slide legend. Hi Powers manufactured by John Inglis in Canada are generally marked "Inglis" "Canada," and "Browning." Another marking that may be observed is the British model designation, "MK.I."

Marlin/Glenfield Model 60

A seemingly innocuous firearm, so far as full-auto conversions are concerned, is the Marlin/ Glenfield Model 60, a semiautomatic .22 rifle fed from a tubular magazine. Introduction of a foreign object into the bolt such as a toothpick or a similarly sized object has the effect of fixing the firing pin in place. When the bolt hold-open latch is released, the force of impact of the bolt (with fixed firing pin) will detonate the chambered cartridge and continue cycling the action until the ammunition supply has been exhausted. In this type of conversion, there is no way to stop the gun from firing, and the factory trigger plays no role. Such a condition is called a *runaway gun*. Once again, the remanufacture of a firearm may cause components to have to be technically redefined. The trigger may no longer act as such, the function of the trigger being assigned the bolt or slide release, which could be defined as the trigger because, from a technical standpoint, that component is now fulfilling that role on the firearm. Other firearms that are commonly encountered as conversions include the Ruger Mini-14 and 10/22 as well as the UZI. In theory, any semiautomatic firearm is a candidate for modification or tinkering to effect the successful and reliable conversion, so methods described in converting other firearms serve as a general guideline to potential avenues taken in converting other semiautomatics.

STEN Gun

The STEN gun was developed in England during World War II as an expedient mass-production submachine gun. The STEN was designed from the outset to be rapidly manufactured using minimal facilities, basic materials, and by nearly anyone following basic instructions. The STEN was chambered in 9×19mm and was fed ammunition from a detachable box magazine. The magazine well was attached to the receiver and fed rather unusually to the side of the gun. The STEN operated by blowback, firing from an open bolt and having a fixed firing pin on the face of the breech block. Aside from the breech block assembly, the only other part in the tubular receiver was the recoil spring. The cocking handle was attached to the breech block and traveled along a channel cut in the receiver; a safety notch was provided. The barrel screwed into the barrel nut at the front of the receiver, and is easily removed and replaced. The trigger group was welded to the bottom of the receiver and one of several crude butt stocks attached.

Some 4 million STEN guns were produced during the war in England and Canada, and they were widely distributed even well into the postwar era. Many nations even produced their own versions postwar. There are lawfully possessed STEN guns in the United States, but there are likely many contraband wartime examples lying around as well. Untold numbers of demilled STEN gun parts kits were imported into the United States until the practice was discontinued. Per ATF regulation, the receivers were cut up and rendered useless, but the remainder of the assemblies—the stock, barrel, barrel nut, trigger-housing and firing-train components, magazine housing and its associated parts, and the breech block and its associated parts—were left intact. Many STEN gun parts are newly manufactured, so having a vintage parts kit is really not necessary to complete the repair, restoration, or the build. Surplus wartime magazines—often unissued and still in period wrapper—remain readily available.

The STEN gun remains a popular machine gun to fabricate because of its simplicity. Using a parts kit, a new receiver can be fashioned from seamless metal tubing and the rest of the parts assembled onto this new receiver. STEN gun receiver templates are readily available for purchase. These templates consist of a seamless metal tube that is covered by a wrapping onto which the template is printed. The template provides the builder with instructions on where to cut material away, drill holes, and perform other manufacturing operations to complete a receiver that is ready for assembly. The final product is at least as good as wartime production.

If not for the actual construction of a STEN gun, the study of constructing such a device, coupled with an understanding of the principle and method of operation—being a straight blowback, open bolt, firing pin fixed to the breech automatic weapon—makes for an interesting case study. Certainly, the features of the STEN gun have served as an inspiration, or perhaps indirectly, even a teaching guide.

Determining if a STEN gun is a vintage example or newly manufactured requires careful evaluation. It was the wartime marking practice to place on the magazine housing the manufacturer information and serial number. Another serial number could appear on the barrel nut; however, these vintage parts could be donors from a parts kit. The best chance lies in examining the receiver to see if the material is marked by a manufacturer or the material type is identified. If the receiver is finished, it may be necessary to remove the finishing material to ascertain if modern material markings are present. Another possibility is to ascertain if parts on the exhibit are newly minted or vintage, but once again, the question focuses squarely on the receiver and determining whether it is contemporary or vintage. STEN machine guns could be original wartime pieces, in which case the original markings would apply. Other possibilities include a "tube gun," where the receiver would be marked by the manufacturer with their information and a serial number, or a reactivated machine gun, which again would require manufacturer information to be conspicuously marked.

Devices Designed to Simulate Fully Automatic Firing

In an attempt to increase the rate of fire of a semiautomatic firearm without manufacturing a machine gun, one approach that has been taken involves external devices that are affixed to the firearm but are not mechanical modifications, additions, or contrivances to the action itself. In general, these devices are designed to take advantage of an action that has come to be broadly known as *bump fire*. Bump fire is taking advantage of the reciprocation (forward motion) of the firearm action after recoil has taken place to assist the shooter in manipulating the trigger in a manner that is much faster than one could manually manipulate a trigger in the absence of such a device. The increased rate of trigger press translates to an increased rate of fire, although the firearm continues to discharge a single shot per press of the trigger. In effect, the trigger finger remains stationary in space while the firearm moves forward, reinitiating the firing train as soon as the action is physically capable of doing so. The bump-fire concept need not necessarily require an attachment; a shooter can attempt bump fire by simply stacking fingers inside the trigger guard of a firearm and pushing the firearm into the "finger stack" with the nonshooting hand. Theoretically, as fast as the shooter can push the firearm into the awaiting fingers, discharge will take place. From a practical standpoint, this is not always possible due to the ergonomics of certain firearms and the relative safety factor of controlling aim and fire while wildly pushing forward the firearm as fast as possible. These add-on devices, as described, are merely ways of accomplishing the same effort while permitting a more conventional shooting style to be observed. Audibly, such an equipped firearm does sound as if a fully automatic weapon is being discharged, and visually it would appear the same, although from a purely technical point of view, the trigger is still being manipulated in a semiautomatic mode.

Methods to take advantage of the bump-fire principle have emerged in the form of stocks and attachments to firearms whereby the operator merely has to hold onto the firearm, press the trigger, and the attachment does the rest. Devices such as the Hell Fire, Hell Storm, Trigger Activator, and the Akins Accelerator, all functionally synonymous, are designed to accelerate the manual trigger manipulation to mimic fully automatic fire by employing the bump-fire technique. However, because they are mechanical devices, they were deemed to be machine gun components in 2006 by the BATFE (Sullivan 2006). Furthermore, homemade analogs of such devices constructed of basic materials such as wooden dowel rods, plywood, rubber bands, and the like, are fashioned into bump-fire devices by those attempting to fabricate their own. These contrivances are also declared machine gun components inasmuch as their affixing to a firearm is an attempt to modulate the rate of fire. More recent attempts at capitalizing on bump fire focus on stocks and other non-mechanical devices, which, for the time being, have not been declared machine guns or machine gun components. The bump-fire principle is an example of the regulatory cat-and-mouse game that is played between the authorities who regulate firearms and the tinkerers and inventors looking to find ways around and through those same regulations. In this case, the intent is to create a firearm that mimics the behavior of a machine gun without actually manufacturing one.

Firearm Receiver

In the United States, the part of a firearm that is legally controlled is the frame or receiver. The frame or receiver is the base to which all other parts are assembled when a firearm is constructed. A frame/receiver is required to bring together the collection of parts to form a functional firearm; without the frame/receiver, there is no firearm. What is considered the actual frame or receiver varies from firearm to firearm, but generally it is the part that encompasses where the ammunition is fed as well as where the fire-control parts are, but there are always exceptions.

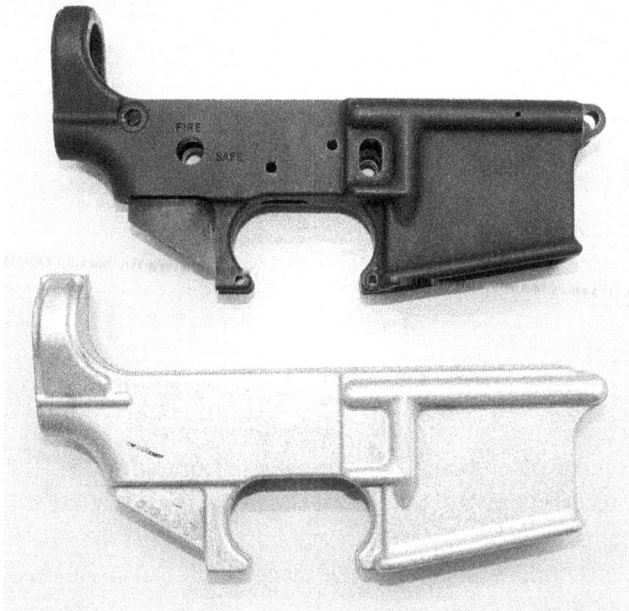

Figure 4.48 Regardless of whose name ends up stamped on the receiver as the manufacturer of record, most AR-15/M16 lower receivers start off as an aluminum forging (the same is true for the upper receivers as well). The example on top is ready to make operational and only needs the internal parts to operate; the forging shown on the bottom is the legal equivalent of a paper weight and is not considered a receiver at this stage of manufacture. AR receivers can be made of carbon fiber or other metals, but aluminum remains prevalent. (Image from author's collection.)

A revolver frame is the part that all the other parts are attached to. The barrel is seated into the frame; the cylinder is attached to the frame by the yoke. Pistol receivers are the part that encompasses the grip and what the slide attaches to.

Shotguns and rifles with single-piece receivers are the part that all the other parts attach to. Firearms with two-piece receivers, an upper and lower, will be model specific. In the case of AR-pattern firearms and their variants, the lower receiver, which encompasses the grip, magazine well, trigger, and the fire control parts, is the firearm. Conversely, in the case of firearms such as the FN-CAL, FN-FAL, L1A1 (British variant of the FN-FAL), H&K G.3 and MP-5 and variants, and the FN SCAR 16 and 17, the upper portion is considered the receiver, which does not include the trigger pack.

In the case of belt-fed firearms, whether they are machine guns or semiautomatic copies, the side plate is considered the receiver. The Browning M2, M1917, M1918, M1919 (all variants), and Maxim/Vickers-pattern firearms all represent examples where the right side plate represents the actual receiver. The slide plate is literally affixed to the receiver box. An incomplete build, that is, one not having the side plate attached, could be mistaken to be the actual firearm, since it physically appears to be complete. Since the side plate is the receiver, this piece is where all markings are mandated to be affixed, but that would not exclude the potential for identifying markings on other parts if the article in question is a vintage machine gun that has "come out of the woodwork" or a newly minted receiver that has been assembled into a lawful firearm using surplus parts to finish assembling the firearm.

ATF Ruling 97-2 defined the 1919A-4 semiautomatic firearm and 1919A-6 semiautomatic rifle to be classified as "portable firearms," despite the fact that each weapon weighs approximately 30 pounds and is approximately 54 inches in length. The significance of the ruling was regulatory in nature and had nothing to do with the function of either described firearm (BATFE 1997).

There are venues outside the United States that control any component used in the construction of a firearm and consider them the same as a receiver, including barrels, slides, magazines, and trigger assemblies. The SIG Sauer P250 is a unique pistol that permits the user to create two firearms from a single frame. The definition of frame/receiver in this instance is a little flexed, as the actual receiver is the mechanical assembly that sits within the full- or compact-sized grip; the user converts the assembly at will. In the case of the SIG P250, the mechanical assembly insert is serial numbered, but displayed through a windowed portion of either frame.

Incomplete receivers may be encountered. Figure 4.48 depicts a incomplete receiver and a finished receiver. Typically, these incomplete receivers are manufactured to 80%, requiring an additional 20% of work to become a functional firearm. The 80% receiver kits are not firearms or receivers per se, at least in the sense that they are not "readily made to function," inasmuch as some work processes remain to finish the project. These receivers may take the form of template covered work pieces such as tubular receivers, side plates, forgings and castings, and so forth.

DEWATS, Unserviceable Firearms, and Inert Firearms

The term *DEWAT* stands for DEactivated WAr Trophy. The definition of a DEWAT has changed significantly in the last 60 years. Traditionally, a DEWAT firearm was one that was rendered inert by some modification such as removing the bolt, grinding off the firing pin, and plugging the barrel with lead. As such modifications are reversible, they are no longer considered to be acceptable to legally define a firearm as deactivated, since firearms so modified could be readily restored to function. A firearm that is a DEWAT using reversible methods could still be construed as a firearm and must be permanently rendered inoperable for it to be definable as a nongun.

An unserviceable firearm is one that is incapable of operating as a firearm; it cannot discharge a projectile nor is it readily restored to a condition that would permit it to operate as a firearm. An unserviceable firearm has been substantially modified to render it useless and unserviceable, such as welding the receiver shut. Such measures may be taken to preserve a firearm that is of some interest, historical or otherwise, but due to legalities, possession of the firearm in question is prohibited.

In contrast, an inert firearm or nongun is one that has had the receiver removed and replaced with a dummy receiver (see Figure 4.49). Actual firearms parts such as the stock, barrel, and such can then be installed onto the dummy receiver. A typical dummy receiver cosmetically resembles a functional receiver, perhaps even to include markings and a few operable features such as cocking handle, trigger, and safety switches, but there are no internal parts and hence the article is not considered a firearm.

Manufacturers

Throughout the history of firearms, there have been many manufacturers ranging from individual gun makers in small shops to industrial conglomerates. Some of the oldest

Figure 4.49 This piece would appear to be a World War II–era German MP.40 9×19mm submachine gun; however, it is, in fact, a nongun dummy. It is very convincingly manufactured of metal and plastic components; the magazine is removable; and the charging handle and trigger are functional. It is even marked in the wartime manner with code *ayf* (ERMA = Erfurter Maschinenfabrik B. Geipel), dated 1940, and bears a serial number. Such analogs can easily be mistaken for a real firearm. (Image from author's collection.)

industrial corporations in existence are gun makers. Most industrial gun makers got their start as sole proprietorships, partnerships, family businesses, or unified trade guilds as evidenced by their names—Colt, Mauser, Beretta, Walther, Smith & Wesson, FN Herstal, and the like.

It is impractical to try to publish a record of firearm manufacturers, as companies come and go all the time, and such a list would rapidly become obsolete. Alongside the long-recognized names of gun makers, there have been many companies that entered the arms business. Some entered under lucrative government contracts in times of conflict. Others were collectivized and forced into the industry under the duress of the political system under which they existed or came to exist. Many nations operate or did operate state-controlled arsenals that solely filled the military obligations of that nation and its partners, but often sold weapons commercially. Many nations never developed an indigenous armaments manufacturing base, instead choosing to purchase their weapons from the establishments of Europe and the United States. As with all industries, not every company will survive, and many nameplates have fallen into bankruptcy and disappeared from the market. Firearms, however, have no real life expectancy, and thus even the oldest firearms from a long-forgotten nameplate can reappear when discovered in barns, chests, closets, drawers, and other places where they were laid and then forgotten. Many of these names are not readily recognized as firearms manufacturers. In addition, there are historical nameplates that may create some confusion on the part of the person coming into contact with a firearm.

The United States operated several arsenals that manufactured martial firearms, particularly rifles designed by government employees or designs purchased by the government. The last, Springfield Arsenal, closed in 1968. Despite the output from several U.S. government arsenals, the demands for firearms and the availability of the private industrial base led the federal government to subcontract manufacturing to any firm willing to bid on the work. As a result, U.S. military firearms can be found having been manufactured either by government arsenals or by a diversity of private firms. For example, Browning-pattern machine guns were manufactured by various divisions of General Motors, including A/C Spark Plug in Flint, Michigan, the Brown-Lipe-Chapin Plant of the Guide Lamp Division in Syracuse, New York, and the Frigidaire Division in Dayton, Ohio. The M3 submachine gun, nicknamed the "Grease Gun," was manufactured by the Guide Lamp Division. M1903 rifles were manufactured by the government arsenals at Rock Island and Springfield, and were later supplemented by production from Remington Arms and L.C. Smith Corona. M1 Garand rifles were manufactured by the U.S. government arsenal at Springfield, but were joined by International Harvester, Harrington & Richardson, and Winchester Repeating Arms. New England Westinghouse and Remington Arms were contracted to manufacture Russian Mosin Nagant 1891 rifles for the Russian Tsarist government during World War I, although many were apparently never delivered and ended up back in the hands of the U.S. government.

Brand names are often the result of acronyms and abbreviation formed from the physical location of the manufacturer, cofounders names, or other pieces of information, especially true with Italian and Spanish firms.

Historically, Spanish gun makers have been concentrated around the city of Eibar. Since many Spanish-sourced firearms came into the United States before comprehensive regulations that standardized marking practices, Spanish nameplates can be very confusing, which is compounded by the large number of firms that operated, especially during the 1920s and 1930s. The Spanish firm Astra was formerly known as Unceta y Cia and includes references to Eibar, Campo Giro, and Esperanza y Unceta, names used during different eras. Other Spanish manufacturers include Garate, Anitua, y Cia (GAC); Azanza y Arrizabalaga (who primarily copied Browning-pattern pistols); shotgun maker Arrieta S.I.; pistol maker Star Bonifacio Echeverria (often just known as "Star"); Tomas de Urizar y Cia (probably best known for the Velo-Dog brand revolvers) used a plurality of different names, retaining references to Eibar or Barcelona in some instances; Armas Garbi; Aguirre y Aranzabal (AYA); Alkartasuna Fabrica De Armas Guernica; Arana y Cia; Arizaga Eusebio; Hijos de Victor Aramberri; Francisco Arizmendi (trademark was a crest with FA over a five-point star); and Apaolozo Hermanos, located in Zumarraga, near Eibar. The Spanish gun maker Gabilondo y Cia, in Elgoibar, Spain, markets guns under the name Llama. The name Francisco Ascaso was associated with a Spanish revolutionary personality, and an unofficial handgun was made using his name as the model on Astra machinery. The majority of Spanish pistols are copies of the Colt 1911 or other similar small-frame Browning-pattern automatics, although Spanish firms did develop their own designs.

Brescia, Italy, is the home of numerous Italian gun makers, particularly those who specialize in cartridge and black powder antique replicas of American revolvers. Firms based in Brescia include Rigarmi di Rino Galesi (RAG); Fabricca Armi Pietta, whose focus is black powder reproductions of classic firearms; Armi San Marco; Armi San Paolo and Vincenzo Bernardelli, manufacturer of shotguns, rifles, and the Model 68 pistol. Gardone Valtrompia is another hub of Italian gun making. Gardone is home to the world's oldest gun maker Pietro Beretta, or more officially Fabbrica D'Armi Pietro Beretta S.p.A. F.lli.; Poli Armi (Fratelli Poli), manufacturer of fine sporting shotguns; FAMARS (Fabrica Armi Mario Abiaticco e Remo Salvinelli); and Angelo Zoli, another sporting arms manufacturer in Gardone. Gardone-based gun maker Angelo Zoli went out of business in the late 1980s. Other firms are scattered about Italy, such as Armi Tecniche de Emilio Rizzini, which is located in Marcheno, and Chiappa Azzano (branded as Armi Sport, Kimar) in Kimar.

The Russian arsenal MOLOT is an acronym for Vjayskiye Poljany Arsenal. The Russian arsenal Izhevsky is often confused with Izhmash, the two words being somewhat similar. Izhmash is a separate entity that continues to manufacture weapons for military and commercial markets. Izhevsky, or Izhevsky Mekhanichesky Zavod, or now just Izhmech, was privatized in 1994, using the name Baikal to market products commercially. The Baikal pistol was one of the first Russian-sourced firearms to enter the United States through commercial channels, and was simply called the "Baikal" and less often referred to as the Model IJ-70, although it was actually just the standard Russian Makarov pistol. Makarov-pattern pistols have been imported from East Germany, China, Poland, and other Eastern Bloc nations. Russian arms can generally be found with the arsenal marking on the receiver. Often the date will be stamped on the receiver as well; the character r precedes the year and is apparently a Russian abbreviation for *year*.

Manufacture by Multiple Firms

Firearms designed by one firm may have been manufactured by another, as in the case of the Browning BDA (Browning Double Action). The BDA slide legend can be quite confusing. There are several variations of markings, some showing Fabrique Nationale Herstal but "Made In Italy"; others bear the Beretta trademark on the right side of the slide and the FN reference is not present. This anomaly is due to the pistol being manufactured by Beretta under contract from FN, who designed the handgun but apparently never actually manufactured it. Some variations have the Browning *B* medallion on the original factory grips, yet others bear the Browning buck-head logo. Aftermarket grips are frequently seen on this pistol. On the left side of the slide is the importer information for examples imported through routine channels into the United States, e.g., "Browning Arms Company Morgan, Utah & Montreal P.O." The Browning BDA was chambered in .380 ACP. Another pistol, also called the Browning BDA, available in .45 ACP or 9×19mm, is essentially the SIG P220. This variation is distinguished by the markings on the front edge of the right side of the slide: "SIG-Sauer System Made In West Germany." The left side of the slide, also toward the muzzle, is marked: "Browning Arms Co. Morgan, Utah & Montreal P.O." The factory-installed grips are marked BROWNING.

There have been many firms that have engaged in commercial firearms manufacture through the modern era, and comparatively few survive to the present day. Occasionally an old nameplate will be resurrected. Each name has its own history, which is often quite extensive and largely outside the purview of the firearm examiner. These companies often changed names to reflect new ownership, consolidation with another firm, or a restructuring. This information can be significant if the time period of when a specific firearm was manufactured is important, but for the most part it is largely irrelevant to the examiner except for historical collector appraisal and conversational purposes.

Historical Footnotes on Older, Common Nameplates

Iver Johnson

Iver Johnson was founded in 1871 as Johnson, Bye, and Company. In 1883, it was renamed Iver Johnson and Company. In 1891, the company relocated to Fitchburg, Massachusetts, from Worcester, Massachusetts, and was renamed again to the Iver Johnson Gun & Cycle Works, where it remained for the better part of a century. In 1973, Iver Johnson, now known as Iver Johnson's Arms Inc., acquired the Plainfield Machine Company of New Jersey and moved to Middlesex, New Jersey. In 1983, the company relocated once again to Jacksonville, Arkansas, and purchased Universal Firearms, which was Iver Johnson's biggest competitor at the time, producing commercial versions of the U.S. military M1 carbine. By 1986, the company was in bankruptcy, and it was acquired and reorganized a couple of times through 1993, when it again shut down (Royal Canadian Mounted Police 2011). In 2004, the name was resurrected as a new Florida-based corporation called Iver Johnson Arms, Inc., but with no ties to the former organization. The new company manufactures 1911-frame handguns and a slide-action shotgun. The new company website indicates that it is entering into production a multibarreled derringer-type pistol and, once again, an M1 carbine. The company also reports it has produced new receivers to fit onto surplus FN Browning pistol parts, chambered in .25 ACP (Iver Johnson Arms n.d.).

Marlin Firearms

Marlin Firearms was founded in 1879 by John M. Marlin, who was employed by Colt at one time. John Marlin was an innovator and developed his own line of firearms. In 1915, the company was acquired from the Marlin family and renamed Marlin Rockwell Corp., which was broken into two business units: a commercial and sporting division (Marlin Firearms Corp. n.d.) and a military products division. The sporting company folded in 1923, was acquired at auction, and renamed the Marlin Firearms Co. Marlin was a large supplier of small arms and small arms components during both world wars. In 2000, Marlin acquired the H&R 1871 name, which included Harrington & Richardson, New England Firearms, and Wesson & Harrington. In 2007, Marlin and its holdings were acquired by Remington Arms of Ilion, New York. Marlin Firearms is currently a holding of the Freedom Group (Marlin Firearms n.d.).

Harrington & Richardson

This company was founded in 1871 by the formation of a partnership between Gilbert Harrington and Frank Wesson as Wesson & Harrington. The name was changed in 1875 to Harrington & Richardson. In 1888, the name was again changed to Harrington & Richardson Arms Company, Inc. Harrington & Richardson expanded from their plant in Worcester, Massachusetts, to facilities in Gardner and Rockdale, both in Massachusetts, and even briefly expanded into Canada. The company was awarded numerous contracts during both world wars to produce firearms, and was awarded contracts postwar to manufacture the M14 and the M16 on behalf of the United States. This success did not remain, and the company folded in 1986. In 1987, New England Firearms came into being and operated from the Gardner facility. In 1991, H&R 1871 was formed, but this has no ties to the original company (Royal Canadian Mounted Police 2011). In 2000, the company was acquired by Marlin and still produces firearms as part of the Freedom Group under the names New England Firearms, H&R 1871, and Harrington & Richardson (H&R 1871).

High Standard

The High Standard Manufacturing Co. was started in 1926 in New Haven, Connecticut, but did not start producing firearms until 1932. Around 1940, the name was again changed to

High Standard Manufacturing Corp., Inc. The company moved several times throughout Connecticut until it closed its doors in 1984. In 1993, the company relocated to Houston, Texas, and has since acquired AMT-AutoMag* and has "affiliated itself with Interarms and US Cartridge" (High Standard Manufacturing Company 2010). At present, the company produces a number of pistols of AMT and prior High Standard models, 1911 model pistols, as well as AR- and AK-pattern rifles. The AK rifles are branded as Interarms.

Savage Arms

Savage Arms shotguns and rifles were sold under the private brand names Aldens (Chieftan); Belknap; Canadian Industries Ltd. (C.I.L.); Coast to Coast; Cotter & Co.; Gamble Skogmo, Inc. (Hiawatha); Simmons (Quail Fargo); and Talo (Golden West). The J. Stevens Firearms Co. became a subsidiary of Savage Arms in 1920; the two names are essentially interchangeable, as are their products after this date. The name Springfield is associated with the Savage/Stevens Co. Prior to 1968, J. Stevens manufactured arms under the name Acme Arms Co. Crescent Firearms appeared as the American Gun Co. and Armory Gun Co. Smith & Wesson marketed a slide-action shotgun under the name Eastfield.

Firearms Imports & Exports

Firearms Imports & Exports, simply known as F.I.E., imported vast numbers of inexpensive revolvers, pistols, and shotguns into the United States until the company went into bankruptcy in 1990. The handguns in particular are frequently encountered by law enforcement. F.I.E. imported either complete firearms or various components used to construct firearms on frames and receivers produced in the United States by F.I.E., and quite often the formal manufacturer or source nation is not readily identified. As a general rule, F.I.E. suffices as identification of the manufacturer, unless it is clearly discerned that the company acted only as the importing agent and there is some evidence contained on the firearm that conclusively indicates otherwise.

Titan, an F.I.E. subsidiary, manufactured receivers that were mated to Italian-sourced slides, barrels, and other parts to construct the Titan series of pistols, including the Titan, Titan II, and Titan 25. The Titan slide marking, indicating "Made In Italy," can be easily misunderstood to believe that the firearm itself is completely of Italian origin, and the marking was simply a matter of having enough real estate to affix the stamping there, but this is not the case. F.I.E. also participated in the manufacture of the Titan Tiger revolver.

Certain revolvers marked F.I.E. may be mistaken for Rohm- or RG-pattern guns, as they are overtly very similar in appearance. Some Hermann Weihrauch (Germany)-sourced revolvers were imported by F.I.E., as well as black powder single-action revolvers from Tangfolio, Riva Esterina, and Luciano Giacosa of Italy, and shotguns by Maroccini. Other shotguns have been sourced from Brazilian makers Companhia Brasileira Cartuchos (CBC) and ER Amantino & Cia. Spanish sources included shotguns from Aguirre y

* AMT was formerly called Arcadia Machine and Tool and was located in Irwindale, California, prior to its acquisition.

Aranzabal (AYA) and Unceta y Cia, which was formerly known as Esperanza y Unceta but is perhaps better known as Astra, located in Eibar, Spain.

RG Industries

RG Industries has produced another series of firearms that are often difficult to identify. The name *Rohm* is also often used to identify this brand. RG is an abbreviation for Rohm Gesellschaft, the German firm that manufactured these small, inexpensive firearms and exported them into the United States. It seems probable that RG Industries was established to continue the production of Rohm-pattern firearms in response to the 1968 Gun Control Act, which would have prohibited further importation of the RG brand.

RG Industries in Miami produced finished firearms using components supplied to them by the German factory. Receivers marked "Rohm," "Germany," and bearing German proof marks were imported into the United States before 1968. U.S.-manufactured receivers will be conspicuously marked "RG Industries" per Federal requirements. RG Industries in Miami apparently commenced manufacturing activities in the early 1970s and remained in business until the mid-1980s, when it was litigated out of business.

In most instances, RG and Rohm both appear on the firearm, the RG logo appearing on the grip panels and Rohm generally marked on the receiver or barrel. All Rohm- or RG-named firearms used a model number with an RG prefix followed by a two- or three-digit letter/number combination. Revolvers included the .22 revolvers RG14, RG14S, and the RG23. The RG30 revolver was available as a .22 LR or Magnum, or in .32 S&W. The RG31 revolver was chambered in either .32 S&W or .38 Special. The RG38, RG39, and RG40 were only available in .38 Special. Two derringers were manufactured: the .22 Magnum RG16 and the .38 Special RG17. Automatic pistols were the RG25 (likely named because it was chambered in .25 ACP) and the RG26 (also in .25 ACP). RG copied a prior pistol design from Erma Werke, Model RG42, also chambered in .25 ACP. For all practical purposes, identification of any of the products as an RG or a "Rohm pattern" would likely suffice.

The Rohm brand became somewhat synonymous in defining the term *Saturday night special*, which was meant to mean any inexpensive firearm, but the term has evolved to take the meaning of any firearm that has a criminal following. The RG14 holds a place in history as one of the most notorious guns of all time: John Hinckley used the model in his attempt to assassinate the late former President Ronald Reagan.

Hermann Weihrauch

Hermann Weihrauch was a German arms maker. Although Weihrauch was the actual manufacturer, the company branded their products using several different names, including Arminius, Burgo, and Fabico. The Arminius brand is accompanied by a warrior wearing a winged helmet stamped on the frame. Most Herman Weihrauch firearms models start with an HW prefix, but this would exclude arms dating back to the prewar and Second World War era. As discussed in the section on Firearms Imports & Exports (F.I.E.), the two companies collaborated as manufacturers and later exporters to the United States. German-produced arms can be expected to have German proof marks. The majority of postwar Weihrauch firearms are small- to mid-caliber revolvers. The company has produced vanity firearms on behalf of European American Armory (EAA) and Herter's. Aside from traditional powder firearms, Weihrauch is also a well-known manufacturer of precision air guns.

Department Store Firearms

Commonly encountered shotguns and rifles are those once sold through department stores, mail-order catalogs, and general stores. In the United States prior to 1968, it was possible to mail order firearms from a retailer and have it shipped directly to a residence by U.S. mail, and many catalog sales retailers marketed guns in this way. When encountered, such shotguns and rifles frequently only bear the name of the retailer that sold the arm, not the company that actually manufactured the article. Many gun makers manufactured department store guns under contract for the retailers, and these firearms were not unique or specific patterns for the retailer. Instead, they were often basic versions of firearms that these manufacturers were already producing. These firearms were well-built, quality weapons, but they would have lacked excessively fancy finishes or high grades of wood. Instead, they were focused on providing a basic, utilitarian tool at the best possible price.

Any student of twentieth-century shotguns and small-caliber rifles who can readily identify cornerstone models by such firms as Mossberg, Marlin, Stevens/Savage, and Winchester would have no problem identifying the department store equivalents, as they are exactly the same gun. The majority of department-store-branded firearms lack serial numbers, and it is for this reason that they are virtually untraceable. Occasionally an example will appear that has a name, address, or some other form of personal information etched onto it, even if the information is archaic. It is a common mistake to identify the part or inventory number as a serial number on these firearms. Some examples carry a stock number, which is typically located on the barrel and ordinarily takes the form of a series of letters or numbers that would identify the manufacturer to the retailer, but this information would not be relatable to the casual observer; it would simply appear as a model. In certain instances, model numbers may coincide between the retailer and the actual manufacturer.

Sears & Roebuck

Sears used the private brand names Eastern Arms, Norwich Arms, J.C. Higgins, Ranger (see Figure 4.50), and Ted Williams when they sold firearms, and the firearms were so marked by the particular brand name. Sears-branded firearms can be traced to the original manufacturer using the stock number. The pattern used by Sears was a two- or three-digit

Figure 4.50 A Sears brand Ranger shotgun. Above the trigger is the stock number, in this case 102.25, which would indicate the actual manufacturer as J. Stevens/Savage Arms. (Image from author's collection.)

Table 4.2   Sears Brand Firearms Cross-Reference Table

Stock Number	Manufacturer
10	Marlin Firearms Co.
11	J. Stevens Firearms Co. (operating as a division of Savage Arms)
18	Savage Arms (may be marked as Springfield)
20, 21	High Standard Manufacturing Co.
30	J. Stevens Firearms Co.
31	Savage Arms
42	Marlin Firearms Co.
49	J. Stevens Firearms Co.
54	Browning Arms
66	J.C. Higgins or High Standard
73	Savage Arms
80	High Standard Manufacturing Co.
88	High Standard (J.C. Higgins) .22 revolver
97, 98, 101	Savage Arms
102	J. Stevens Firearms Co./Savage Arms
103	Marlin Firearms Co.
104, 105	Harrington & Richardson
121	Universal Arms
153	Laurona Arms (Spain)
200	Winchester Repeating Arms
201, 202, 203, 204, 205, 206, 207, 209, 210, 212, 213	O.F. Mossberg & Sons, Inc.
234	Savage Arms
273	Olin Corp. (Winchester Western Division)
281	Antonio Zoli Arms Co. (Italy)
282	Companhia Brasileira de Cartouches (CBC)
340	Ithaca Gun Co.
390, 400, 401, 402, 404, 405, 414, 420, 446, 447, 448, 449, 455, 458, 465, 467, 468, 472, 474, 483, 484, 486, 487, 488, 489, 491	O.F. Mossberg & Sons, Inc.
583	High Standard Manufacturing Co.
667	J. Stevens Firearms Co.
684	Winchester Repeating Arms
870	Voere, Austria

Sources: (Thompson 1978; LaVoy 1979; Royal Canadian Mounted Police 2011; Rosenberg 1972).

number followed by a period, then a sequence of numbers that indicated the stock number, e.g., 340.123456 would indicate a firearm made by the Ithaca Gun Company. There are potentially hundreds of different sequence numbers associated with Sears-branded firearms. The purpose of the information in Table 4.2 is not to provide a comprehensive list of all sequence numbers that could exist but, rather, to focus on providing the consumer of the information with the prefix letters and numbers needed to identify the manufacturer. The majority of firearms—comprised primarily of shotguns and some single-shot .22 rifles—sold by Sears under any given brand will not have a factory serial number affixed, as they were exempted until 1968. The most notable exception is the Model 88 revolver produced by High Standard, which would have a serial number per law.

Montgomery Ward

Montgomery Ward sold firearms under the brand names Lakeside and Western Field. As was the case in the other department stores, these firearms were manufactured by other companies and were sold primarily by catalog order, but purchases could be made over the counter. The code or number that would reveal the identity of the actual manufacturer would appear in the model number as recorded on the receiver or barrel, such as EMN-171, denoting a product by Marlin (see Table 4.3). Montgomery Ward sold two handguns, an Iver Johnson .22 revolver, the Model 75, and a French-made .22 pistol, the Model 5. The name Montgomery Ward, Ward, Wards, the initials M.W., or some variation thereof is typically marked on the firearm and can be used to ascertain that the firearm was sold through Montgomery Ward. Some Ward-branded firearms were serial numbered, even before 1968.

Western Auto Supply Company

The Western Auto Supply Company was another department store firearm retailer. Western Auto used the brand name Revelation on their rifles and shotguns. Like the other companies, codes comprising letters, numbers, or letter and number combinations can be used to cross-reference the actual manufacturer (see Table 4.4). Western Auto–marked firearms were manufactured primarily by Savage Arms, Marlin Firearms, and Mossberg. Like Montgomery Ward, Western Auto marketed .22 revolvers, the Model 76 and the Model 99, both manufactured by High Standard. It is important to note that although the revolvers may have appeared before 1968, they were required to have a serial number affixed to the frame. Western Auto–branded ammunition can be found from time to time, having more of a collector value than a shooting value.

Kmart

Kmart was another department store retailer of arms and ammunition. Kmart did not offer the plethora of models of other similar chains. Kmart is known to have sold two models of shotgun, the 151 and 251, both manufactured by the Brazilian firm Companhia Brasileira de Cartouches. The 251 was also labeled the "Junior." Both were single-shot, break-open actions and were not serial numbered. Kmart branded ammunition is also found from time to time, and has more collectible value than that of useful ammunition.

J.C. Penney

J.C. Penney retailed their brand of firearms under the name Foremost. As in the other department stores, the Foremost line was made up of utilitarian rifles and shotguns that came from contracted gun makers. It should come as no surprise that the same companies that furnished the arms to Sears, Montgomery Ward, Western Auto, and Kmart also supplied J.C. Penney. It would appear that J.C. Penney did not have the more complex model-numbering system of other retailers and assigned basic four-digit model numbers, as seen in Table 4.5.

Table 4.3   Montgomery Ward Manufacturer Reference Table

Stock Number	Manufacturer
ECH	Colt's Patent Firearms, Inc.
EFW	Firearms International, Washington, DC
EGP	Golden State Arms, Pasadena, CA
EHM	Fr. Heym Waffenfabrik, Munnerstadt, West Germany
EJN	Jefferson Corporation, North Haven, CT
EKN	Kessler Arms Company, Silver Creek, NY
EMJ	Miroku Company, Kochi City, Japan
EMN	Marlin Firearms Co.
ENH	Noble Manufacturing, Haydenville, MA
ERI	Remington Arms Co., Bridgeport, CT
EY	Iver Johnson, Fitchburg, MA
FR	Kessler Arms
M	O.F. Mossberg & Sons, Inc.
SB	Savage Arms
SD	J. Stevens Firearms Co.
XNH	Noble Manufacturing
14	Savage Arms
15, 16	O.F. Mossberg & Sons, Inc.
19	unknown manufacturer
30, 31	Savage Arms
33	Marlin Firearms, Inc.
33A (.22 slide-action rifle)	Noble Manufacturing
35 (12-, 16-, or 20-gauge slide-action shotgun)	J. Stevens Firearms Co.
35A (rifle)	O.F. Mossberg & Sons, Inc.
36 (.22 single-shot rifle)	O.F. Mossberg & Sons, Inc.
40 (.22 single-shot rifle)	Marlin Firearms, Inc.
40N (12-gauge slide-action shotgun)	Noble Manufacturing
43, 45, 46, 47, 48	O.F. Mossberg & Sons, Inc.
50	Marlin Firearms, Inc.
58 (.22 rifle)	unknown manufacturer
59, 60	Savage Arms
61	unknown manufacturer
72, 72C, 79	O.F. Mossberg & Sons, Inc.
80A, 81	Savage Arms
500, 730, 732, 734	O.F. Mossberg & Sons, Inc.
750	Fabrique Nationale, Herstal, Belgium
765, 766, 767, 768, 771, 772, 775, 776, 777, 778, 782, 792	O.F. Mossberg & Sons, Inc.
808	Savage Arms
822, 830, 832	O.F. Mossberg & Sons, Inc.
836	Savage Arms
840, 842, 846, 850, 865	O.F. Mossberg & Sons, Inc.
880	Colt's Patent Firearms, Inc.
890, 891	Marlin Firearms, Inc.
894, 895, 895A	O.F. Mossberg & Sons, Inc.
5000, Hercules	J. Stevens Firearms Co.
Premier	O.F. Mossberg & Sons, Inc.
Side by Side	Crescent Firearms Co.
Sporter	Savage Arms

Sources: (Westenberger 1972; Royal Canadian Mounted Police 2011).

Table 4.4   Western Auto Manufacturer Reference Table

Stock Number	Manufacturer
BD, R	O.F. Mossberg & Sons, Inc.
39	Marlin Firearms, Inc.
76, 99 (.22 revolvers)	High Standard Manufacturing Co.
100	O.F. Mossberg & Sons, Inc.
101	Savage Arms
105	Marlin Firearms, Inc.
107	O.F. Mossberg & Sons, Inc.
110, 115, 116	Marlin Firearms, Inc.
117	O.F. Mossberg & Sons, Inc.
120	Marlin Firearms, Inc.
125	O.F. Mossberg & Sons, Inc.
135	Savage Arms
150	Marlin Firearms, Inc.
160	Savage Arms
200	Marlin Firearms, Inc.
205, 207, 210, 220	O.F. Mossberg & Sons, Inc.
225, 230, 250, 260, 300	Savage Arms
310, 312, 325, 330	O.F. Mossberg & Sons, Inc.
335	Marlin Firearms, Inc.
336, 350, 355	Savage Arms
356	J. Stevens Firearms Co.
394, 400	Savage Arms
425	High Standard Manufacturing Co.

Contributing source: (Royal Canadian Mounted Police 2011).

Table 4.5   J.C. Penney Manufacturer Reference Table

Stock Number	Firearm and Manufacturer
2035	.22-caliber slide-action rifle made by Marlin Firearms, Inc.
2066	.22-caliber semiautomatic rifle made by Marlin Firearms, Inc.
3040	.32 Winchester Special or .30-30 Winchester lever-action rifle made by Marlin Firearms, Inc.
4011	12-gauge slide-action shotgun by High Standard Manufacturing Co.
6400	.22 Hornet, .222, or .30-30 Winchester bolt-action rifle made by Savage Arms
6500	Bolt-action .30-06 rifle made by Firearms Company, Ltd. (United Kingdom)
6610	.22-caliber single-shot rifle made by J. Stevens Firearms Co.
6630	12-gauge bolt-action shotgun made by Marlin Firearms, Inc.
6647	.410 single-shot shotgun made by Savage Arms
6660	.22-caliber semiautomatic rifle made by Marlin Firearms, Inc.
6670	.410 slide-action shotgun made by J. Stevens Firearms Co.
6870	12- or 20-gauge, or .410 slide-action shotgun made by Savage Arms

Firearm Designs Manufactured by Multiple Manufacturers

Certain firearms have proven so popular that their production has passed from manufacturer to manufacturer, the designs often outliving the manufacturers. The MAC-pattern firearms, both as a handgun and a machine gun, were originally manufactured by the Military Armaments Corp. in Georgia. The design has passed from company to company in the wake of MAC going bankrupt. Nameplates that have manufactured the MAC are RPB, MAC (located in Texas, not Georgia), SWD, FMJ/LEINAD,* Jersey Arms Works, and Masterpiece Arms. The name "Cobray" is often associated with the MAC firearm, and the company logo, a stylized cobra snake, often appears stamped on the firearm. The MAC has been manufactured in .45 ACP, 9×19mm, and .380 ACP; however, other calibers could be expected to appear, and conversion kits exist to allow other calibers to be used by the basic MAC receiver. Numerous variants of the basic models exist, including several carbine versions. Aftermarket modifications by individual owners are too vast to cite; any number of accessories or attachments are available.

The U.S. Model M1 carbine is another firearm that has been produced commercially postwar by Plainfield, Iver Johnson, Universal Firearms, National Ordnance, and Auto Ordnance.† These copies may appear in various configurations, from the original military wooden stocks, paratrooper stocks, or more modern composite furniture. Iver Johnson and Universal manufactured a handgun variant as well, identified by its pistolized stock and shorter barrel. The model, called the Enforcer, is frequently mistaken for a short-barreled rifle, but is actually classified as a handgun and meets all requisite requirements as such. For the most part, the .30 carbine M1 remains the preferred caliber, but there have been experiments to rechamber M1 carbines to fire a different cartridge. U.S. military contract M1 carbines were manufactured by Commercial Controls Corporation (Rochester Defense Corp.), Irwin Peterson, International Business Machines Corp. (IBM), Inland Manufacturing Division (General Motors), National Postal Meter Co., Quality Hardware & Machine Co., Saginaw Steering Gear (General Motors), Rock-ola, Standard Products Co., and the Underwood-Elliot-Fisher Company. Innumerable subcontractors were also involved, and military-surplus small parts may bear a plurality of markings to indicate these firms. The manufacturer's name and the serial number will appear on the heel of the receiver behind the bolt and is often obscured by an adjustable rear sight. The size of the sight may make it impossible to read the information without removing it. Ordinarily, the rear sight was staked into the dovetail that the sight was placed in and can be punched out with minimal effort.

Walther PP and PPK

One of the most enduring firearm designs is the Walther PP and PPK. The PPK is simply a scaled-down version of the Walther PP, which came onto the market in 1929. The PPK followed in 1931 (see Figure 4.51). The PP was designed by Walther as a police pistol for carry by uniformed officers. The PPK was intended to be carried by the plainclothes officer. Both models are nearly identical except for the PPK's smaller dimensions.

---

* FMJ (Full Metal Jacket), located in Ducktown, Tennessee. LEINAD-manufactured MACs also bear the same information on the receiver.
† The name Auto Ordnance is currently in use by Kahr Arms. The name is used to market copies of the original Thompson submachine gun, as well as other historical firearms.

Figure 4.51 A Walther PPK manufactured during World War II. Both the slide and frame are serial numbered and match one another. The finish imperfections and presence of milling marks clearly identify this example as a wartime piece and are common features of wartime German firearms, especially during later stages of the conflict. (Image from author's collection.)

Figure 4.52 A copy of the PPK's sibling, the PP. This example was manufactured by the Hungarian arms maker Fegyver as the model R61. The slide legend, S.A.P.S., is a property mark, "South African Police Service." (Image from author's collection.)

The PP and the PPK were manufactured until 1945, when production was interrupted. Postwar, the French company Manurhin manufactured the PP series under license from the newly formed Carl Walther. Manurhin even manufactured a copy of the Walther P.38. Copies of the PP were also manufactured by the Hungarian arms maker Fegyver. Examples of the Hungarian PP, designated the R61, were used by the South African Police Service (see Figure 4.52). These can be readily identified by the S.A.P.S. engraved on the slide, along with the police crest. Fegyver used several different model numbers, including Walam and AP66, to identify the pistol series, and they may also be marked as manufactured by Femaru. The Italian firm Galesi produced PP copies. The American arms importer, Interarms, of Alexandria, Virginia, concurrently imported

Figure 4.53 A contemporary PPK/S manufactured in the United States by Smith & Wesson under license from Walther. Observe the differences between this pistol and the one depicted in Figure 4.51, although both are PPK pistols. (Image from author's collection.)

PP-type pistols while at the same time manufacturing their own version domestically. This can lead to some confusion when attempting to determine whether Interarms was the manufacturer or the importer. The identifications markings must be clearly read and understood. Iver Johnson manufactured two models outwardly similar to the PP: the TP-22 and the TP-25. The Turkish firm MKEK manufactures a PP copy called the Kirikkale. In 2002, Smith & Wesson entered into an agreement with Walther to manufacture the PPK in the United States as the PPK/S (see Figure 4.53). This was part of an apparently broader business relationship between the two companies, as Smith & Wesson now imports Walther's products exclusively.

Like all prolific models, the PP series pistols have innumerable variations, and they can be found chambered in most of the popular small-pistol calibers of the twentieth century, from .22 to 9×18mm Makarov. There can be no doubt that the popularity of the PPK can be attributed to the fictional British secret agent James Bond, who took issue of the gun from the British government in exchange for his Beretta pistol. Ironically, he did so grudgingly and only under direct order. There is no dispute that the PP and the PPK were both very advanced and influential designs, but they were not singular. The Mauser HSc and the Sauer 38h were also quite advanced; however, neither was ever able to gain the following of the Walther duo. Mauser briefly reintroduced the HSc to market in the 1970s, but it did not fare well in the commercial sector and was discontinued.

Model 1911 and 1911A1

Although designed by John Browning through a series of successive pistols, the 1911- and 1911A1-pattern pistols were originally manufactured by Colt (see Figure 4.54). Colt has continually produced the pistols, but they have been joined by an ever-increasing number of firms that manufacture them as well. Colt was joined during both world wars by various contractors in order to fill wartime demands. Postwar surplus 1911 and 1911A1 pistols were reworked and reengineered by all varieties of gunsmiths and custom gun builders in their interpretation of improvements to the basic design. U.S. military 1911

Figure 4.54 The Colt Model 1911 Series 80 chambered in .45 ACP. The Series 80 was manufactured from 1983 to 1988. Overtly, there is nothing different about the Series 80 from other 1911 pistols save for the introduction of an internal firing pin safety. Regardless of the manufacturer and exact date of manufacture, the 1911 design is classic and immediately recognized. (Image from author's collection.)

and 1911A1 frames were not marked by manufacturer; instead, the manufacturers placed their name on the slide. Since it is not uncommon to find mismatched slide and frame combinations, the manufacturer may have to be identified by the inspector stamps and serial number. World War I companies involved in military production included Colt, the U.S. military Springfield Arsenal, North American Arms, and Savage Arms. World War II subcontractors included Remington Rand, Ithaca Gun Co., Union Switch & Signal, and Singer Manufacturing. Licensed copies from Colt were manufactured in Argentina and were marked "Ejército Argentina," or "Marina Argentina" if for naval use.

During the first half of the twentieth century, Colt filled overseas orders for Canada, England, Russia, Mexico, and Norway. Colt has made changes to the design over the years and even manufactured wartime replicas, complete with packaging, instruction manual, and markings that were authentic to the era. Colt has commercially produced many models based on the 1911. The proliferation of companies now manufacturing a 1911-pattern handgun is enormous, affirming the continued popularity of the design. At present, there are likely no fewer than a dozen companies that manufacture a 1911 handgun, including major manufacturers such as Colt, Smith & Wesson, and SIG Arms, and others such as Armscor of the Philippines (under the names Charles Daly and Rock Island Armory), Auto Ordnance, Browning Arms, Essex, High Standard, Kimber, Magnum Research, Metro Arms, Nighthawk Custom, Para Ordnance, Remington Arms (not to be confused with Remington Rand), Shooter Arms Manufacturing (marketed in the United States by ATI, or American Tactical Imports), Springfield Armory, Taurus, Wilson Combat, and others. The Spanish gun makers Astra and Star extensively manufactured the design, and their products are so marked.

Variations of the basic 1911 frame include cosmetic additions, widened frames to accommodate double-stack magazines as opposed to the original single-stack magazines, and a wide variety of calibers. The popularity of the design does not appear to be waning, even in light of the subsequent developments in handguns.

Beretta 92

Versions of the Beretta 92 have been manufactured in Brazil, Egypt, South Africa, and the United States. Beretta manufactures the 92 not only in Italy, but also in the United States through its Beretta USA subsidiary. Close attention must be paid as to the true origin of a particular Beretta-manufactured 92 if it is of interest to determine whether the pistol was of Italian or U.S. origin. Taurus International produces its own version of the 92, the PT-92, and numerous variations of it using a factory that Beretta established in Brazil. The Egyptian copy of the 92 is called the Helwan, which has been imported into the United States. The South African version, the Z88, was manufactured by two firms: Vektor and Lyttleton Engineering. American Tactical markets a version, the AT92C, which is manufactured in Turkey by MKEK. The Beretta 21, unique for its tip-up barrel, is made not only by Beretta, but also by Taurus as well. Overtly, they are indistinguishable. The continuity of design cues by Beretta allows for even vintage Berettas to stand out. The 92 design is dimensionally larger, but carries all the classic Beretta aesthetics going back to the Model 1915. The top portion of the barrel is exposed between the breech and the front sight, with the corresponding portion of the slide left open. The Model 951R and the Model 93R are both machine pistols, capable of functioning as machine guns. The 93R has a selector switch where the safety/decocker is located on the 92 series. A folding forward grip is also affixed, and a shoulder stock completes the rig.

M14, M1A, and Variants

The M14 was conceived as a replacement for the M1 Garand as the primary infantry battle rifle for the U.S. military. For all practical purposes, the M14 was simply a reiteration of the basic M1 action that was rechambered in .308, which itself was a shortened .30-06 cartridge. The M14 was slightly improved: It was loaded from a detachable 20-round magazine instead of relying on an 8-round en bloc clip. In theory, the M14 could have filled several roles, replacing the Garand as the main infantry battle rifle and the Browning Automatic Rifle as the squad automatic.

The M14 was manufactured under contract by Harrington & Richardson, Thompson-Ramo-Wooldridge (TRW), and Winchester Repeating Arms. The M14 was the last rifle to be manufactured by the U.S. arsenal in Springfield, Massachusetts. M14s manufactured for the U.S. government are machine guns (see Figures 4.55 and 4.56); however, they were generally configured to fire semiautomatically only. Surplus M14 rifles were available for civilian purchase for a brief period of time in the mid-1960s, but further sales were stopped by 1968. The interest in the M14 has prompted production of numerous commercial copies. The Chinese gun maker NORINCO exported a version, and a domestic version was produced by Federal Ordnance and Smith Enterprises, and it is still made by Springfield Armory in Illinois.

Semiautomatic versions of the M14, generically coined the M1A, overtly resemble the M14, and early commercial examples made extensive use of surplus military parts to assemble them, including stocks. The receiver markings, found on the top of the back end of the receiver, include the model manufacturer, caliber, and serial number. On a commercial gun, these resemble the military markings very closely and may cause an erroneous identification of an M1A or M1A1 as an M14.

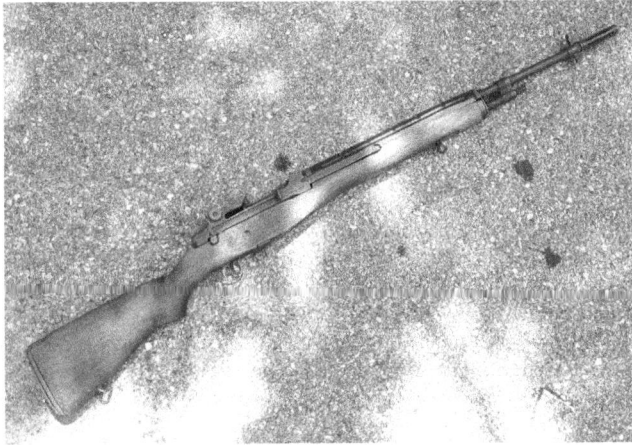

Figure 4.55 (See color insert.) A U.S. Rifle M14 manufactured by Harrington & Richardson is a machine gun. This example bears the circle *P* proof mark and eagle cartouche on the stock. (Image from author's collection.)

Figure 4.56 (See color insert.) Close-up of the receiver markings of the H&R M14. Note the selector switch knob on the right side of the receiver. Semiautomatic copies of the M14 generally do not have the notch cut in the stock for the selector, although surplus stocks in circulation may or may not be cut for the selector. (Image from author's collection.)

Current production by Springfield Armory is in a plurality of configurations. The U.S. military has retained use of the M14, although current iterations of it hardly resemble the original M14. The rifle has been subject to regular enhancements and was frequently revisited because nothing else quite met the need. The M14, in U.S. military terminology, has recently gone by several names, and several more modifications include the M21 and M25 Sniper variant, Mark 14 MOD 0 EBR (Enhanced Battle Rifle), and the M39 EMR (Enhanced Marksman Rifle). Although officially slated for replacement once again, it is likely that the M14 will leave the service in name only.

Perhaps some of the most interesting variants of this series of rifle are the Italian contributions. Few are aware that Garand production was initiated in earnest in Italy by Beretta and Breda in the immediate postwar era, using surplus machinery and technical assistance from the United States. The Breda-manufactured examples carry a BRM (Breda

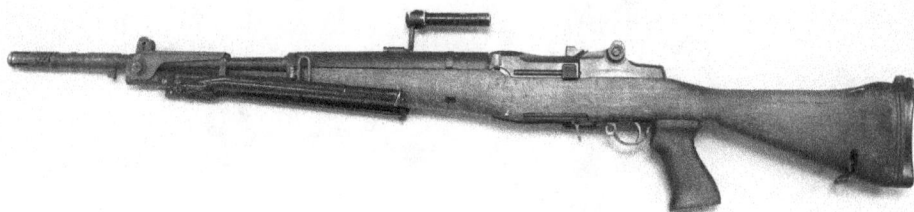

Figure 4.57 (See color insert.) The M14-pattern rifle manufactured by Beretta as the BM59. There are definite similarities, but note the distinct differences between the M14 and the BM59. This example has a bipod, a grenade-launcher spigot on the muzzle, a grenade-launching ladder-type sight, and a pistol grip instead of a standard rifle stock. (Image from author's collection.)

Figure 4.58 (See color insert.) Close-up of the BM59 receiver. Note the selector switch on the left side and the "P.B. BM59" marking on the action-lock button. The marking practices used by Beretta on the BM59 mimicked other manufacturers of the M14. Barely visible on the receiver below the sight-adjustment knob is the importer's mark "Springfield Armory," not to be mistaken for the government arsenal. (Image from author's collection.)

Meccanica Romana) marking on the heel and on the left side of the receiver. Beretta's products carry the atypical PB marking. Beretta modified the Garand into a select-fire rifle, the BM59 (see Figure 4.57). Several variants were manufactured, usually identified by the type of stock, either a standard wooden full stock or a folding style. Beretta exported the BM59 and a semiautomatic variant, the BM62, to the United States during the 1960s and through the 1970s. Several variants of the BM59 produced in the United States also exist, as either select-fire or semiautomatic rifles (see Figure 4.58). The Beretta- and Breda-produced examples apparently were widely sold around the world and could turn up nearly anywhere. Other markings may also be observed, such as the crown over "FKF," indicating Danish use. The BM59 was also manufactured in Indonesia. Overtly resembling an M14, only examination of the receiver markings will clearly define the BM59.

Armalite Rifle

The AR pattern has become one of the dominant firearms in the U.S. commercial marketplace. Contrary to popular belief, the AR designation was not an abbreviation for "assault rifle," but "Armalite Rifle." The popularity of the model has prompted nearly every major gun maker in the United States (or with a U.S. market presence) to produce their own version, and innumerable smaller manufacturers have joined in to capture a piece of the market.

The AR-15 is traditionally associated with Colt; however, names such as Charles Daly, Ruger, SIG Sauer, and Smith & Wesson have added the model to their product line. Certain firms specialize in the AR, such as Bushmaster, DPMS (sold under the name Panther Arms), Rock River Arms, Knight's Armament, Lewis Machine and Tool, and Sabre Defense have made it their cornerstone product. Currently, there are some 400 manufacturers of record putting their name on AR-pattern firearms, a broad spectrum ranging from major manufacturers (as previously described) to sole proprietorships and every size in between. Companies outside the United States have also joined the market, and foreign names can be added to the list of domestic manufacturers. Astra Arms, formerly located in Spain but since relocated to Switzerland, manufactures an AR, designated the StG4. Schmeisser of Germany simply calls it the AR; Oberland Arms of Germany markets it as the OA-15; and the Italian firm ADC sells it as the Bodyguard. The Military Ordnance Corp. in Sudan manufactures a rifle that bears overt similarities to the M16 chambered in 7.62×51mm called the Terab. There is likely no other firearm that has the aftermarket following of the AR; nearly any type of functional or aesthetic modification can be expected.

AR stocks can be fixed or collapsible. The standard fixed stock is constructed of polymer and resembles a typical rifle stock. Earlier style full stocks have a distinctly shiny plastic appearance. Later examples, especially prevalent when the M16A2 was developed, featured a duller matte finished stock of better quality material. Collapsible stocks require the use of a different recoil buffer tube that is notched to accommodate the different lengths of extension. The original Colt collapsible stock that was initially tested on the XM177 (CAR-15, or Commando) during the Vietnam War featured a two-position buffer tube mounting a "fiberlite" stock, which is rather plain in appearance in comparison to the wide assortment of stock styles now available. Currently there are innumerable styles of collapsible stock, ranging from very basic patterns to those with storage compartments, interchangeable cheek pieces, carry-sling mounts, and other customizable options. A faux collapsible stock is available, giving the appearance of the real thing, but it is fixed in position and nonfunctional. Until recently, a buffer tube was a fixture on an AR and was part of the operating system, since the recoil buffer and spring were contained within the tube that extended from the rear of the lower receiver. Pistol versions of the AR featured the buffer tube, typically covered with a rubber or foam coating, but denying the immediate ability to attach a stock. Recent proprietary designs from Rock River Arms and Sig Arms have eliminated the need for the buffer tube, as the recoil system has been redesigned to fit inside the upper receiver.

Classically, the standard barrel length for an AR-15 rifle was 20 inches. The original barrel was a thinned, tapered style. The M16A2, often called the H-BAR, featured a heavier barrel profile. Other common barrel lengths include 10½, 11½, 14½, and 16 inch. Barrels longer than 20 inches are available and typically are used on ARs intended for precision shooting and hunting. Barrels shorter than 11½ inches are made and typically reserved for handgun applications, but can be mounted to any standard upper receiver. One of the latest iterations of the M16 platform in military circles is the Mark 18 Mod 0, which is equipped with a 10.8-inch barrel. A 14½-inch barrel is standard on U.S. military M4 carbines and carbine AR-15 rifles; a 16-inch barrel is the minimum allowed by U.S.

law without registration as a short-barreled rifle. In addition to the standard barrel and the H-BAR, other barrel profiles are manufactured, including a stepped barrel that is notched near the muzzle, as well as other profiles of various diameters. The traditional triangle-style front sight post may be omitted in favor of a gas-block-style mount that permits the installation of alternative styles of front sight in lieu of the traditional triangular front sight post.

Destructive testing of M16 barrels was undertaken by the U.S. Army Armament Research, Development, and Engineering Center. Comparison testing was conducted using the M16A2 equipped with a 20-inch heavy barrel and the M4A1 equipped with 14½-inch stepped barrel. The final report was released September 1996 and revealed that the M4A1 barrel failed after firing 596 continuous rounds in 30-round increments. The M16A2 barrel failed after firing 491 continuous rounds, also in 30-round increments. The test subject M16A2, while originally set up with a 3-round burst configuration, had full-auto-fire control parts installed for testing purposes. Ten seconds between bursts to change magazines was allotted. The temperature at the time of catastrophic barrel failure was also measured as part of the evaluation. The temperature of the M16A2 barrel at the moment of failure was 1599° Fahrenheit; the M4A1 barrel was 1639° Fahrenheit (Windham 1996).

Barrels manufactured for installation on new receivers destined for the commercial market between September 1994 and September 2004 were not threaded on the end to accept any form of flash hider or compensator. In certain instances, a manufacturer may have opted to install a faux flash hider, which in every respect resembled the A1- or A2-style flash hider but did not feature the cutouts to make it functional; it was there for aesthetic purposes only. Otherwise, barrels are threaded on the muzzle, and an endless variety of flash hiders or compensators are available. The attachment is installed using one of two types of washer, a crush type or a peel type. The peel type gets its name from its resemblance to a fruit skin that has been peeled. The crush washer is circular and is literally "crushed" into place as the attachment is threaded on.

As ammunition specifications changed in U.S. military use, the rifling twist rate* was changed. The original rifling twist rate was 1:12, then changed to 1:9, and is currently set at 1:7 for 5.56×45mm chambered firearms to optimize firing heavier projectiles (62 grains and above). The rate of twist for other calibers will be different. Current U.S. military barrels are set up in 1:7 twist, but 1:9 twist rates are standard for commercial-specification barrels. Barrels may be lined with either chrome moly or chrome. Ordinarily, the barrel information will be inscribed on top of the barrel between the muzzle and the front sight.

The barrel is shrouded by a hand guard that covers from the receiver to the front sight. The original A1-style hand guard was triangular in shape. The A2-style hand guard was round and featured heat shielding. Full-length, midlength, and carbine-length hand guards are available, depending on the setup of the barrel. Contemporary hand guards are more likely to feature rail attachment systems for accessories. These rails, the M1913, are often referred to as the "picatinny rail," a term derived from the Picatinny Arsenal in New Jersey.

The AR-pattern firearm is made up of two receivers, an upper receiver and a lower receiver. The two are joined by a pivot pin and a takedown pin. Several styles of upper receiver exist, including the standard, original style featuring an integrated carry handle.

---

* The given ratios of barrel rifling are the number of rotations the projectile makes over a given distance of barrel. A 1:16 indicates that a projectile will rotate once for every 16 inches of barrel travel.

Very early examples and some new-production retro-style upper receivers feature the charging handle within the carry handle. More contemporary upper receivers are called *flat tops* and have the top of the upper receiver incorporating a rail system for the attachment of various styles of sights and accessories. Regardless of the appearance of the firearm, the focus remains on proper identification of the receiver. The upper receiver is not considered the firearm receiver in the United States; the lower receiver is the actual firearm receiver, thus the lower receiver markings are the most relevant, including serial number, manufacturer, model, caliber, etc. Traditionally, the AR upper and lower receivers started their lives as aluminum forgings. The capacity to produce such forgings falls upon a few firms within the United States, which then sell them to various manufacturers that machine the forging into the firearm receiver. Variations of the basic forged aluminum receiver exist, including variants made of carbon fiber, cast-metal types, and those machined from billets of other metals. The beauty of the AR lies in its design approach. The basic lower receiver can practically be rebuilt perpetually as long it is not physically damaged. Pistol versions are quite common, featuring very short barrels and no shoulder stock, per U.S. regulations.

Colt has apparently responded to increased competition from the new generation of modular weapon systems as well as other developments in the AR-15/M16 platform. For military and law enforcement clientele, Colt has started marketing new models: the Advanced Colt Carbine Monolithic (ACC-M), Colt Infantry Assault Rifle (IAR), Colt Sub Compact Weapon (SCW), and a gas-piston-operated variant called the Colt Advanced Piston Carbine (APC). The Advanced Colt Carbine Monolithic is similar to other single-piece upper receiver assemblies offered by other manufacturers, featuring a free-floating barrel. This carbine introduces ambidextrous fire control, magazine release, and fire selector switch. The Colt Infantry Assault Rifle incorporates the Colt Monolithic upper receiver, ambidextrous safety switch, and a unique heat sink to increase barrel life under prolonged fire. The Colt Sub Compact Weapon is a departure from the standard M16 architecture, featuring a unique bolt carrier and buffer design that reduces the overall length. The Sub Compact Weapon also incorporates the Monolithic upper receiver. The Colt Advanced Piston Carbine combines the gas-piston system with the single-piece upper receiver, almost a standard across the spectrum of the AR-15/M16 architecture (Colt Defense 2011).

The AR is sold in countless variations and in a broad spectrum of calibers (see Figure 4.59). There is likely no other firearm that has ever been manufactured offered in the array of calibers as that of the AR-15. Colt has made the rifle in .223, 9×19mm, and 7.62×39mm, but there are now literally dozens of available options for caliber on the market, from the most popular handgun and rifle calibers to those that are rather obscure. The most basic AR caliber conversion takes the form of replacing the upper receiver with one with an appropriate barrel affixed, changing the bolt carrier group, and perhaps replacing the magazine. The only obstacle presented is when the cartridge dimensions cannot be accommodated into the AR-15 magazine well, such as in the case of attempting to use .308 in a 5.56/.223 lower receiver. The magazine well is simply too short to accommodate the magazine dimensions required for the larger rounds. To get around this technical issue, some conversion kits convert the AR into a single-shot rifle or use a magazine or ammunition-feeding interface (such as a belt) that feeds into the upper receiver through an alternative route than the magazine well in the lower receiver. In the case of longer cartridge lengths, the AR lower receiver is slightly elongated to provide the accommodation.

Figure 4.59 Two examples of the endless AR configurations that can be encountered. Both examples are semiautomatic rifles. (Image from author's collection.)

Until recently, the buffer system was integral to the function of the AR-pattern firearm, regardless of caliber or whether it was configured as a rifle or pistol. Contained within the buffer tube assembly are a spring and buffer. Several versions of the buffer exist, including a standard buffer and a carbine buffer, often called the *H* or tungsten buffer for "heavy." The carbine buffer is designed to offset the shorter carbine barrel and gas pressure, thereby helping to manage the rate of fire in a machine gun. The 9×19mm and .22 AR firearms use a mechanical buffer instead of the traditional buffer and spring, because neither cartridge will develop sufficient gas pressure to work using gas impingement. The 9×19mm version is essentially a blowback-operated firearm. Recent developments by Rock River Arms and SIG have resulted in an AR-influenced design that does away with the traditional buffer tube assembly. As previously mentioned, Colt's Sub Compact Weapon has a dimensionally reduced recoil buffer and redesigned bolt carrier group. This allows pistolized or shortened carbine versions to be very compact and provides for an even greater flexibility in stock designs for rifles. Rock River describes their design as featuring "a purpose-designed bolt carrier, adjustable gas piston, and over-the-barrel spring and guide rod placement" (Rock River Arms 2011). In company literature Sig Sauer does not specify what design changes were made to develop the Model 556 pistol, but it is likely that they are similar to the Rock River.

In recent years, the major movement in the AR is the push toward using a gas-piston operation over the original gas impingement. It is ironic that Eugene Stoner envisioned a gas-piston-operated AR variant as a cheaper alternative; however, the gas piston was discarded in favor of the gas impingement. The principle disadvantages of the gas piston include a weight penalty to account for the additional apparatus, as well as additional parts. A standard aluminum AR-15 with 16-inch barrel and collapsible stock weighs approximately 6½ lbs. empty. Gas-piston AR rifles can weigh upwards of 7½ lbs. As a machine gun, gas-piston operation generally equates to a slightly lower rate of fire, which itself is not necessarily a handicap. Gas impingement is lighter because the gas tube does the work of the mechanical linkages of a gas piston; however, certain powder formulations and less-than-routine maintenance can clog the action with fouling carbon and debris buildup. The AR firearm must be routinely cleaned and lubricated to maintain proper function.

The AR was meant to be built to a high standard, using very advanced materials, and made with very close internal tolerances in mind, all of which have resulted in a durable product. Most weapons-related failures of the design relate directly to preventive mainte-nance issues, but they can also be attributed to substandard parts. Unknowledgeable build-ers can also be found at fault for "unreliable" or "under reliable" firearms. Gas ports that are out of specification create an overgased or undergased firearm that results in inherent cycling issues, as will a nonspecification or damaged gas tube. An AR that has been assem-bled using substandard or nonspecification parts will directly affect the reliability and func-tion of the component in question. Another common cause of malfunction is the magazine, in particular the magazine spring. The spring will cause insufficient pressure to be exerted when worn, causing misfeed or failure to feed. Broken extractors are a common issue on ARs that have fired quantities of steel-cased ammunition, which wears the part prematurely.

The basic AR-15/M16 platform has served as a basis for an entire series of firearms. The AR receiver has been configured into all forms of firearm, from handguns to carbines to stan-dard rifles and even precision rifles. Regardless of the individual characteristics of any partic-ular firearm, the basic AR platform remains the same. Countless aftermarket parts suppliers and the various and sundry manufacturers tout some facet of the design as being superior. However, the parts supply chain and manufacturing is, generally speaking, so homogenized that these professed advantages of one over another are not very clear or well articulated.

Kalashnikov-Pattern Firearm

Without question, the most ubiquitous firearm on the planet today is the AK pattern (see Figure 4.60). It is generally estimated that in excess of 35 million AK-pattern firearms have been manufactured across the planet since the AK-47 first entered service with the Soviet Union in 1949. The term *AK* is an abbreviation for Automat Kalashnikov, in tribute to its designer, General Mikhail Kalashnikov. The term itself has become somewhat generic and has come to mean any firearm bearing the general appearance of an AK-47 as a machine gun. This genre of firearm is often erroneously referred to as the AK-47; in fact, most of these

Figure 4.60 Ten AK-pattern rifles recovered from a clandestine weapons cache in Baghdad, Iraq, September 2008. The diversity of stocks, grips, and attachments in this small sample clearly shows that such a rifle can be found in nearly endless varieties. (U.S. Army image; pho-tographer Staff Sgt. Brian D. Lehnhard.)

firearms in existence are likely a copy of the AKM. The AK-47 had a receiver milled from solid steel and was only produced by Soviet state arsenals until 1959, when it was superseded by the AKM and production of the AK-47 ceased. The *M* of AKM indicates "modernized": The receiver was manufactured of sheet metal that is stamped into form. The plurality of models goes on from there, as the Kalashnikov system of operation served as a basis for other firearms from squad automatic weapons* to compact personal defense weapons. In addition to the Soviet-produced examples, many nations manufactured their own copy of the AK. The AK has inspired other designs, including the Swedish Valmet and the Israeli Galil.

The proliferation of the AK has led to it having something of a cult status. The silhouette of an AK is featured prominently on the national flag of Mozambique as well as on the emblems of other nations and movements; it is mentioned by name in modern music. The AK is so recognized from television and movies that anyone, regardless of interest or knowledge of firearms, seems to be able to recognize it. An AK firearm, ammunition, magazines, and the limited accoutrements available for it are considered hard currency in many parts of the world. What is ironic is that the AK never came to personify the political system of communism to the same degree that it became an icon for innumerable revolutions.

The features of the AK betray its age. The construction techniques were typical for firearms of the 1940s, using steel and wood. The bore was chrome lined, a typically Soviet practice for automatic weapons. The rear sight is the obsolescent tangent sight system, more common during the Second World War on rifles, but still effective and simple. Outwardly, the AK resembles the German-developed Sturmgewehr rifle, having a layout that became common in postwar military rifles. Although the AK bears physical similarities to the Sturmgewehr, the method of operation between the two is different.

Like other paramilitary style firearms in the United States, four basic categories of AK exist today:

Factory-manufactured machine guns originating from any nation engaged in production of AK-pattern firearms

Factory-manufactured semiautomatic copies of the AK sold commercially in the United States that are imported from foreign nations

Factory-manufactured semiautomatic copies of the AK sold commercially in the United States that are imported from foreign nations in a "sporting" configuration, which are then reconfigured to a classic military appearance by conversion

U.S.-manufactured receivers mated to a mixture of foreign- or domestic-sourced parts to complete the firearm build

AK-pattern firearms imported into the United States are semiautomatic weapons. Semiautomatic AKs, both pistols and rifles, have been imported from China, Romania, Hungary, Bulgaria, the Russian Federation, and Egypt. In addition to complete firearms, AK parts kits have appeared from the Czech Republic, East Germany, Hungary, the nations

---

* A squad automatic weapon is a machine gun that is capable of being deployed by an individual and is designed to fulfill a sustained-fire role. Often abbreviated as SAW, the squad automatic weapon is larger than the standard infantry rifle but smaller than other machine guns. The term *general-purpose machine gun* predated the use of the term SAW. Examples of a squad automatic weapon include the U.S. M60, M240, and the M249. The World War II German MG-34 and MG-42 and the later variations of them can be defined as SAWs, although that is not how they were defined at the time, when the term *general-purpose machine gun* applied.

Figure 4.61 This image is a close-up showing the markings on the left side of an AK barrel trunnion. In this case, the manufacturer is Zastava, located in Kragujevac, Yugoslavia (now Serbia). This marking dates the manufacture of this AK to prior to the breakup of Yugoslavia. Many Eastern Bloc AKs were marked on the barrel trunnion, and not necessarily on the receiver itself due to space considerations. Many thousands of Zastava AKs were demilled and sold as "parts kits" to the U.S. market, allowing builders to install these on a new receiver. The receiver must be researched for markings, as these other markings would be erroneous for identification. (Image from author's collection.)

within the former nation of Yugoslavia, and Poland (see Figure 4.61). A parts kit is defined as all of the components to build a complete firearm, minus receiver. When parts kits were imported into the United States, the receiver was completely destroyed by cutting it into pieces by a BATFE-prescribed and -approved method. Using these parts kits, companies and individuals have assembled AK-type firearms on newly manufactured receivers made in the United States. Such hybrid arms can create a certain amount of confusion because of the contradiction in markings, serial numbers, and other features found on the firearm. As is the case with all firearms, the receiver markings are the relevant ones to be used for identification. It may be of trivial interest to recognize where the other components may have originated. It was common practice for the original manufacturer to mark every part with a serial number, which matched the original receiver. However, since the original receiver was destroyed and the parts retained, the original serial number cannot be expected to match a newly made receiver. Most U.S.-made receivers bear their legal markings on the bottom of the receiver in front of the magazine well; others are marked on either the right or left side of the receiver, usually above the trigger.

Regardless of when and where the AK was manufactured, the basic principle of operation remains the same. The AK operates on a long-stroke gas piston, and the gas system is very simple. Gas pressure bleeds from the barrel through a gas block above the barrel and returns by way of a gas tube, where it acts against the gas piston. The size of the gas port is quite generous (over ½ inch in diameter) and is therefore not susceptible to clogging. The piston is threaded into the bolt carrier as a single piece, allowing it to also act as an operating rod. This bolt carrier assembly contains the bolt, which rotates within a channel milled into the assembly. The bolt carrier assembly reciprocates along rails at the top of either side of the receiver and is returned to battery by way of a recoil spring. The recoil spring locks to the receiver by two ears that lock it to a corresponding lug at the back of the receiver, and sits directly in a recess at the back of the bolt carrier assembly. To access the interior of the firearm, the sheet-metal dust cover is removed from the top of the receiver by depressing a button of rectangular appearance at the rear of the receiver, allowing the dust cover to be lifted off. To replace, the dust cover locks into place above the chamber and is secured by sliding it back over the release button.

The basic AK can be found in a vast array of configurations, driven in part by market forces but primarily by various legislative efforts that affected the firearm. Stocks may be wood, polymer, or metal. Wooden stocks are found in several lengths. Later Chinese exported AKs, designated MAK-90 (MAK for Modified AK), were sold with a thumb-hole-style stock in an attempt to "sporterize" the basic rifle to comply with legislation. Earlier Chinese exports had either a standard wooden stock or a folding-style metal stock. Polymer stocks can be either fixed or folded to the side. Metal stocks can fold underneath or to the side. Nonfunctional folding stocks are also found and are designed to be compliant with venues that prohibit that particular feature. The folding stocks are rendered nonfunctional by permanently pinning or welding them into place. It is theoretically possible to reengineer such a configuration; however, from a practical standpoint, it seems unlikely.

More recently, M16-style stocks have appeared that can be mated to the AK. Barrels may be found with various styles of functional or nonfunctional compensators or "flash hiders" attached to the muzzle, and the bayonet lug may either be present or absent. Some models are completely devoid of any muzzle device, whereas others feature a cylindrical design with horizontal openings on both sides and several off-center holes at the top, a very efficient design. Another design is slanted with a bias to the upper right, which is the direction that an AK will climb under sustained full-auto fire. Some muzzles are threaded, allowing for any of the plurality of muzzle attachments to fit. To release, there is a locking detent pin that is depressed, allowing the muzzle attachment to be removed.

The combination safety and selector switch are on the right side of the receiver. Table 4.6 shows AK selector switch markings by nation of origin. A semiautomatic AK has two positions. The safe position is the topmost position and prevents the action from opening. The lowest position is the semiautomatic fire position. The presence of a middle position, whether marked or not, should be a point of suspicion that the firearm may be altered to fire in full auto. The Israeli Galil and the U.S.-made commercial copy, the Golani Sporter, have a safety switch on the left side of the receiver above the trigger, which works in concert with the safety lever on the right side of the receiver. The two switches are mechanically linked so that manipulation of one affects the other. The Israelis were impressed enough with the AK pattern that the Galil is patterned after it, as are the ARM and the MAR (Micro Assault Rifle), which first appeared in 1995 and are copied from the AKS-74U. The South Africans were sufficiently impressed with the Galil that they manufactured their own copies of it, the R4 and the R6, which appeared in 1975.

The AK was not originally designed with the feature to hold the bolt open when the last round was fired from the magazine, nor was there a catch installed that would allow the operator to lock the action to the rear for the purposes of safety or inspection. Izhmash AK-pattern rifles commercially imported into the United States feature a small bolt catch on the right side of the receiver adjacent the trigger guard. Certain U.S.-made receiver AK firearms feature a bolt hold open that differs in that the feature is automatic, but the bolt is easily closed with even modest pressure. Some professional builders and home tinkerers have modified the safety switch by cutting a small notch to act as a bolt hold open.

Table 4.6   AK Selector Switch Markings by Nation of Origin

Country	Selector Characters
Bulgaria	AB
China	ЕД
	连
	单
	L
	D
Czechoslovakia	30
	1
East Germany	D
	E
Finland	...
	.
Hungary	∞
	1
Korea	련
	대
Poland	C
	P
Romania	A
	FA
	FF
	R
	S
Russia	AB
	0Д
Yugoslavia	R
	J
	1

Source: (BATFE 2006c).

An unusual commercial variation of the AK is the Romanian CUGIR Model PAR-1, which is slide action, not semiautomatic. The action had been modified to turn the fore grip into a functional slide. Otherwise the rifle was the same, still accepting detachable magazines and having the same AK-style fixtures.

AK firearms classified as handguns do not have a shoulder stock and should not be confused with the shortened version of the AK generically called the Krinkov. The Krinkov has a metal frame folding stock and very short barrel, measuring approximately 8½ inches in length. As with all variations of the AK, this short version was manufactured by many different nations under various model numbers; it was formerly known in Russia as the AKS-47 and later the AKS-74U, which is chambered in 5.45×39mm. Copies intended to mimic the design often feature an elongated tube, in essence a faux suppressor, to get the barrel length up to the minimum for a rifle to avoid the implication of making or having a short-barreled rifle. Such an attachment must be permanently affixed by prescribed methods. Lawfully possessed and registered short-barreled rifles in semiautomatic, mimicking the Krinkov, have been manufactured. As a historical footnote, Osama bin

Laden apparently favored the weapon; some of the most famous images of him reflect one, likely an AKS-74U, close at hand. Figure 4.16 shows the Krinkov configuration.

The Chinese call their AK the Type 56, as does North Korea. The Czechs labeled theirs as the M70. The Albanian AK series of weapons use the model prefix ASH, followed by a number designation to indicate the specific configuration. The Bulgarian model designation is similar, using the prefixes AKK, AK, and AKS. East Germany used the prefix MPi followed by numbers to describe specific models within the family of firearms. The Sudanese variant is the MAZ. In Poland the design is made as the AKMS, manufactured at Lucznik. In addition, NATO-chambered 5.56×45mm variants of the AK, dubbed the Beryl and Mini-Beryl are produced as well. During the Cold War, the AK-pattern weapons were the PMK series, including some sporting variations. With the appearance of the 5.45×39mm cartridge, the Poles produced the Kbk.wz.88, the Kbk.wz.89, and the Kbk. wz.90 (two versions, the Tantel and the Onyx). The Romanian arsenal at Cugir started producing AK-pattern weapons around 1960, calling them the AIM. Cugir still produces the AK pattern in the 5.45, 5.56, and the 7.62×39mm. The 7.62 versions are the Models 63, 65, and 90. The 5.45 is the Model 86. The Egyptian-made AK is called the Maadi, MISR, or ARM, a direct AKM duplication. The Hungarian versions are the AKM-63 and the AMD-65, both copies of the AK-47 (see Figure 4.62). The NGM is the Hungarian copy of the AK-47 chambered in the NATO 5.56×45mm. The Finnish manufactured the m/60, m/62, m/76, m/78, and the m/90; all are called the Valmet. The m/60 and m/62 are copies of the AK-47; the m/76 can be found chambered in either 7.62×39mm or in 5.56×45mm. The m/76 was really just an improved m/62—using stamped and formed metal and adding tritium sights for shooting in diminished light. The m/76 was exported to other nations, including Qatar and Indonesia. The m/78 is the squad automatic rifle variant resembling the Russian RPK. Iraq manufactured its own AKM in 7.62×39mm, called the Tabuk. Regardless of the specific model name or number applied to the weapon by the producer or by the nation or persons carrying it, the term Kalashnikov is recognized in practically any language.

Figure 4.62 A Hungarian AMD-65 in the hands of a member of the Afghan National Police. Although it is an AK-pattern rifle, the AMD-65 is distinguishable by its forward vertical fore grip that is an exact copy of the pistol grip and by the selector-switch markings: The center position is the symbol for infinity, and the lowest position is the numeral 1. A semiautomatic version of the AMD 65 is manufactured as the SA2000M or SA65M, produced as such by Fegyver or using a U.S.-made receiver. (U.S. Air Force image; photographer Staff Sgt. Joseph Swafford.)

Commercially there is a huge aftermarket following to accessorize the AK-type firearm. Most of these accessories are aesthetic in nature and include all styles of grips, rail systems, and butt stocks.

The Russian Dragunov sniper rifle, as well as the various copies of it such as the Yugoslavian M79, is a direct copy of the AK but with its dimensions enlarged to accommodate the 7.62×54mmR full-sized cartridge. The Czechoslovakian VZ-58 overtly resembles an AK; however, they are unrelated weapons. The VZ-58 was developed in Czechoslovakia independent of Russian influence. However, the proliferation of the AK prevailed, and the VZ-58 became a footnote in history. The VZ-58 is available as a semiautomatic rifle and is imported from Czechoslovakia.

Sporting rifles using the Kalashnikov operating system have emerged onto the market. The VEPR is manufactured by the Russian firm MOLOT, a joint stock company formed from the Soviet-era Vjayskiye Polijany Arsenal. These sporterized versions bear a passing resemblance to the military-style AK, the principal differences being the style of stock that is used and other aesthetic qualities that give it a unique appearance despite its common heritage to the AK under the skin. The VEPR is chambered in the popular U.S. and Russian calibers: .223, .308, 7.62×39mm, and 5.45×39mm.

The Izhmash Saiga sporting rifles have been used as a base to reconfigure the firearm into one resembling an AK. Such operations are undertaken by professional custom builders, gunsmiths, and even home builders. The reconfiguration process involves removing the sport stock and other aesthetic modifications and may involve some remanufacturing to modify the receiver and barrel to accommodate the design differences between Saiga-specific components and those of the AK. The quality of reconfigured Saiga weapons or firearms built on U.S.-made receivers using parts kits varies greatly.

Heckler & Koch G.3

The G.3 is a full-sized rifle chambered in the powerful 7.62×51mm cartridge (see Figure 4.63). Aside from the Russian AK family of weapons, the G.3 would likely rate as one of the most prolific firearms in the world. It entered service with the German

Figure 4.63 Sergeant Edward W. Deptola, U.S. Marine, fires a G.3-pattern rifle on the range at Naval Station Manda Bay in Kenya. (U.S. Marine Corps image; photographer Lance Cpl. P. M. Johnson-Campbell.)

Bundeswehr in 1959 and has been adopted by more than 50 nations during its long career. The G.3 was one of the preeminent rifles in the Western world, well known for its accuracy and reliability. The G.3 rifle has always been associated with Heckler & Koch. The design originated in Germany during the war years, but it was fully developed at the Spanish firm CETME. Later the design reverted back to Germany, where H&K put it into production. The G.3 was sold commercially in the United States initially as the G.3, later the Model 41, and still later the Model 91. The 91 was banned from further importation into the United States in 1989.

Attempts to sporterize the design were made by installing a thumbhole-style stock and removing the flash hider, then designating the firearm as the SR9. Another rare variant of the G.3/HK91 is the SR9(T). The ultraprecision PSG-1 and MSG90 rifles are both directly descended from the G.3, albeit with significant ergonomic modifications to suit the needs of a precision-rifle operator. Since that time, various U.S.-based firms have gone about manufacturing clone receivers and using surplus parts kits supplied from around the world to build up a rifle. H&K clones made in America by PTR Industries (formerly JLD Enterprises) and Cohaire Arms can be almost indistinguishable from a distance, and close-up inspection is needed to confirm. It has been reported that JLD Enterprises obtained their G.3 tooling from INDEP, Portugal, who manufactured the G.3 under license from H&K.

Several styles of stock and fore grip appear. The stock can be of a collapsible style or a full-size fixed type. The stock and hand guard can be of wood or green or black polymer. The G.3 is one of the most widely copied designs in the world, and many examples have been made outside of Germany by firms other than H&K. Within Germany, G.3 production was shared by the Rheinmetall; their production is recognized by a pentagon on the left side of the receiver on the magazine well. Most G.3-pattern rifles will be marked in this location, although some may be marked on the right side of the receiver on the magazine well, or on the receiver itself in the area of the selector switch. In addition to manufacturer information, in some instances other markings may be noted such as markings indicating use by a particular nation or military force. It was customary for H&K to indicate the month and year of manufacture in a four-digit format, such as 11/63 for example.

The G.3 was not only manufactured in Germany, and not only by German firms. Across the world, the G.3 is manufactured in Greece by Hellenic Defense Systems (EBO), in the Sudan by the Military Defense Corp. as the Dinar, in Pakistan by the Pakistan Ordnance Factories as the G3P4, in Turkey by MKEK as the G.3, and in Iran by the Iranian Defense Industries Organization.

Previously, production took place in Portugal by Fabrica de Braco Prata (FMBP) and by INDEP, short for Industrias Nacionais de Defesa; in Sweden by Forenade Fabriksverken, or FFV; in France by Manufacture Nationale d'Armes de St. Etienne (MAS); in the United Kingdom by Royal Ordnance and H&K Ltd.; in Nigeria by Defense Industries Corp. (DICON); in Mexico by state-controlled arsenals; in Saudi Arabia; in Sweden by Carl Gustav; in Norway by Kongsberg Vapenfabrik; in Bangladesh by Bangladesh Ordnance Factory; in Thailand; and in Burma by state arsenal. Semiautomatic versions of the G.3 have been made in the United States by various manufacturers, some manufacturing complete firearms, others manufacturing the receiver, which is mated to parts kits that could have originated from almost anywhere. Until such weapons were barred from further importation into the United States, semiautomatic G.3 clones were sold in the United States, primarily from Portugal and Greece, in addition to the H&K Models 41 and 91.

Figure 4.64 Cameroon sailors undergo familiarization training with the FN FNC rifle circa March 2010. (U.S. Navy image; photographer Mass Communications Specialist 1st Class Gary Keen.)

H&K's standard-setting MP-5 submachine gun is also produced in Sudan as the Tihraga, in Turkey in several variants as the MP-5, in Greece, and in Iran as the MPT9K.

FN FAL and the FN FNC

The FAL-pattern rifle, meaning Fusil Automatique Leger, was designed by the Belgian firm Fabrique Nationale (FN). Another version, the LAR (Light Automatic Rifle), was also manufactured. A scaled-down version, the FNC (Fusil Nouveau Carbine), was developed to chamber the 5.56×45mm cartridge (see Figure 4.64), while the FAL used the larger 7.62×51mm cartridge. Like the G.3 and the AK, the FAL has seen massive distribution around the world, and FN was joined in manufacture of the FAL pattern by manufacturers in many different countries. Two patterns of the FAL exist: one using English dimensions, the other using metric dimensions—the principle difference being that there are two patterns of magazines that do not interchange. American manufacturers produced quantities of semiautomatic FAL clones, primarily using foreign-sourced parts kits coupled with receivers made either within the United States or imported, especially from IMBEL in Brazil. Some semiautomatic FAL clones were manufactured in Brazil and imported into the United States as complete firearms, marketed through Springfield Armory, and sold under the model name SAR-48. The FN-FNC rifles in the United States originated from Belgium until further importation was banned. Many FNC rifles in the United States were lawfully converted to machine guns using legally registered sears.

The FAL and FNC, like the AR pattern, comprise two receivers (an upper and a lower) that are joined together. Until 1981, the ATF considered the lower receiver of an FNC to be the firearm; however this opinion was reversed in 2008 by ATF Ruling 2008-1.

The FNC rifle consists of two major assemblies, the upper assembly and the lower assembly. The lower assembly houses the trigger, hammer, disconnector, safety/selector, and an automatic trip lever in the automatic version. It also incorporates a pistol grip and a magazine release. The upper assembly houses a barrel that is attached to the upper assembly by means of a barrel extension. It also houses the bolt carrier with gas piston affixed, gas tube and hand guard, bolt, operating rod, and spring. The two assemblies are mounted together with a front

Cartridges and Firearm Identification

and rear takedown pin. Since 1981, ATF has classified the lower assembly as the receiver for purposes of the GCA (Gun Control Act) and NFA (National Firearms Act).

ATF has reconsidered its classification of the lower assembly of the FNC rifle as the receiver. The upper assembly of the FNC rifle is more properly classified as the receiver. The upper assembly of the FNC rifle houses the bolt and provides a connection point for the barrel. Moreover, the upper assembly is classified as the receiver on similar types of firearms, to include other FN rifles, such as the FN FAL and FN SCAR. Reclassification of the upper assembly as the receiver will also allow the continued installation of a lawfully registered sear into an FNC rifle because no modification to the receiver, which is the upper assembly, is required to properly install the sear. (Sullivan 2008)

Modular Weapon Systems Concept

The concept of the modular weapon system was first touted by the inventor of the AR-15 rifle, Eugene Stoner. Stoner put forth the idea that a firearm could be conceptualized and engineered in such a way that it could be reconfigured to suit a wide variety of applications. In theory at least, the concept makes perfect sense; it allows a force to utilize a single pattern of firearm, thus reducing logistical issues, and permits a simplified training regimen on the single weapon; hence every operator could conceivably operate the weapon. At the heart of the modular weapon system is the basic receiver. To facilitate user or mission requirements, the basic receiver could be outfitted with various barrels, stocks, and other accessories to produce any possible configuration, ranging from a basic infantry rifle to a carbine, precision rifle, or a general-purpose machine gun. Stoner's first effort was the Model 63, which saw limited service with U.S. Navy SEAL teams during the Vietnam conflict; however, the concept did not advance further. The concept, however, is not dead. Arguably, the M16 has evolved into a modular weapon system, as the basic receiver design and layout has been used to produce not only a basic infantry rifle, but also innumerable precision rifles such as the Knights Armament SR-25, the carbine variant M4, and even a squad automatic in the form of the Colt Automatic Rifle, which has replaced the Colt Light Machine Gun, although the change appears to be in name only, as both appear to be functionally identical. An interesting variation from the basic M16 design, the Colt Automatic Rifle fires from an open bolt, the quintessential feature of firearm designed for a sustained-fire role (Colt Defense 2003).

In 2005, Lewis Machine and Tool, a manufacturer of AR-pattern rifles, announced the release of its Monolithic Rail Platform (MRP). Designed for the AR-15/M16 weapon platform, the MRP is a single-piece aluminum upper receiver that mates to any AR lower receiver and offers the capability to quickly change barrels. Barrel lengths range from 10.5 inches to 18 inches. The upper receiver itself is available in two different sizes, a standard length and a CQB (Close Quarter Battle) length. In addition to reconfiguring the barrel to suit different needs, the receiver architecture supports multiple calibers and can be changed by installing the appropriate barrel and replacing the magazine and bolt carrier group (Lewis Machine & Tool 2011). The MRP is available in the traditional gas-impingement operating system or a gas-piston driven system. The entire length of the upper receiver is made up of the U.S. military standard M1913 "picatinny rail," allowing for any accessory with the rail interface to be mounted and used. The MRP concept was a very radical concept and one that has had much influence on the progression of the AR-15/M16 platform.

Lewis Machine & Tool (LMT) further enhanced their AR-15/M16 platform when they announced in January 2009 that they were offering a "dual operating performance system." According to a press release by the company, this is the first such system that permits "operators to quickly change a direct gas impingement to a piston system operation to fit any scenario" (Lewis Machine & Tool 2009). This dual operating system stems from the Monolithic rail platform concept, yet maintains commonality with other AR-15/M16 platform firearms.

Lewis Machine & Tool achieved a major milestone with their Monolithic rail platform concept when it was awarded a contract to supply the British Ministry of Defense with a .308 (7.62×51mm) precision rifle, designated by the British as the L129A1. The commercial Model LM308MWS (.308 chambered Monolithic rail platform) served as the basis for the L129A1, likely with some minor additions or alterations made to suit the preferences of the customer. There can be no doubt that the LMT design was in direct competition with other manufacturers marketing their versions of modular weapon systems.

Fabrique Nationale (FN) has capitalized on the modular concept with the advent of a completely new rifle. In February 2006, the prototype for a new rifle system—called SCAR (Special Operations Forces Combat Assault Rifle)—was introduced. FN developed the SCAR specifically to compete for a U.S. Special Operations Command contract for a new rifle. Testing and selection of candidate rifles to meet a specification called for by the U.S. Special Operation Command had begun several years prior, with the FN SCAR being selected in 2004. In May 2007, the SCAR had moved forward into limited production and further testing, which was expected to be followed by actual deployment of the weapon system at the end of 2007. In October 2007, FN announced that the SCAR would be available to law enforcement customers in 2008. At the end of 2008, the SCAR-16S was shipped to FN dealers in the United States. It was not until July 2010 that the SCAR-17S entered the U.S. civilian market. The delay is undoubtedly attributed to FN meeting contractual demands from the military.

The SCAR appears to be influenced heavily by the AR-15/M16. The SCAR has two receivers: The lower is constructed of polymer, and the upper receiver is one piece and constructed of aluminum. It is apparent that FN chose to use the M16 as an inspiration in developing the lower receiver unit because the layout of the controls is the same; even the same style grip is used. However, the grip is fastened using a hexagonal head screw as opposed to the mil-spec flathead screw. It is presumed that the military-version SCAR utilizes a flathead screw, like mil-spec M16 grips. Unlike the M16, the SCAR features a magazine release and safety/selector switch that are ambidextrous. The standard M16-pattern magazine is utilized, and any such magazine or drum will function with the SCAR, although FN offers a magazine. The FN magazine is subtly different, using the now widely accepted no-tilt follower, but the body is constructed of a heavier grade steel and comes in several colors. Only the exposed portion of the magazine is colored; the portion that sits within the magazine well is not.

Internally, the differences are obvious; the SCAR is highly simplified in contrast to the AR-15/M16. The civilian SCAR features a flash hider of ornate design that was developed specifically by FN. The military version flash hider is threaded to accept a quick-detach sound suppressor. The barrel is chrome lined and is free floating.* The SCAR is gas-piston

* A free-floating barrel is one that is only attached at the receiver; it does not contact any other part of the firearm such as the stock.

operated and has a user-adjustable gas-pressure valve. The entire length of the upper receiver comprises the M1913 rail system for the attachment of accessories. Additional rails are located on the left, right, and lower part of the receiver. It is apparent that the SCAR could readily support rapid caliber interchangeability with the replacement of the barrel, bolt carrier, and magazine to support the dimensions of the different cartridge. Presently, the SCAR 16, which is chambered in 5.56×45mm, and the dimensionally adjusted SCAR-17, chambered in 7.62×51mm, are available. The S suffix on the line indicates a semiautomatic carbine, but the SCAR is available as a select-fire weapon to authorized entities. The latest addition to the SCAR line, the SCAR-H PR, was announced in an October 12, 2011, press release by Fabrique Nationale that stated:

> Derived from the innovative FN SCAR weapon system, the new SCAR-H PR precision rifle is a tailored design for long-range precision fire applications while also providing capability to fight close in. The SCAR-H PR features a 20″ heavy barrel and a two-stage trigger module (Match type) allowing high accuracy. The folding butt stock and the cheek rest can be adjusted, respectively in length and in height, without tools. The operator can therefore optimize the rifle to his requirements (such as body size and body armor). (Fabrique Nationale Herstal 2011)

The SCAR features almost universal adjustments to suit the ergonomics of nearly any operator. The stock will collapse to six different positions and also folds to the side. The stock features an integrated cheek piece that is height adjustable. The charging handle can be installed on either the left or right side of the receiver. The charging handle itself is the punch to disassemble the bolt.

FNH USA was gracious enough to provide the author with a SCAR 16S for testing and evaluation. The test subject was finished in the popular flat dark-earth color, a shade of tan. The supplied magazine was equipped with a no-tilt follower and was finished in tan with only a small portion of the top of the magazine revealing the black undercoating (see Figure 4.65). Test firings consumed approximately 1,000 rounds of ammunition of mixed manufacturer .223 and 5.56×45mm cartridges. No malfunctions occurred during the test cycle. The test firearm was not cleaned, and lubrication was not added during the test cycles. Addition of debris in the form of loose sand into the firearm did not impede performance either. When the test cycle was concluded, the amount of carbon accumulation

Figure 4.65 (See color insert.) The FN SCAR 16S, chambered in 5.56×45mm. Also pictured is the FN two-tone STANAG magazine. (Image from author's collection.)

within the firearm could be best described as "negligible." The fully adjustable ergonomics to suit the operator was greatly appreciated by different shooters.

The Bushmaster ACR (Adaptive Combat Rifle) represents another approach to the modular weapon system concept. Bushmaster Firearms International, a longtime producer of AR-pattern firearms, was acquired by Cerberus Capital Management in 2006 and joined Remington Arms, DPMS, Advanced Armament Corp., Marlin, H&R, Dakota Arms, and Parker Gun Works under the corporate umbrella of the Freedom Group, headquartered in Madison, North Carolina. The ACR was a "collaborative effort between Bushmaster, Magpul, and Remington" (Bushmaster Firearms International n.d.). The ACR concept originated with Magpul Industries, who called it the Masada. Magpul is well known within the firearms industry for creating some very innovative and effective products such as the no-tilt follower for the AR-15/M16-pattern magazine as well as the polymer P-MAG, an alternative to the standard aluminum or steel AR-15/M16 magazine.

The ACR lower receiver is constructed of polymer and closely resembles the general layout of the M16, albeit with a more streamlined and modern appearance. The ACR can be fitted with two different stocks, a fixed or folding type, both with an adjustable cheek piece. The folding stock is also capable of telescoping within six positions. The ACR has a multicaliber bolt carrier group that supports 5.56×45mm and the 6.8 SPC cartridges, and it is not unreasonable to presume that other calibers of similar dimensions could be accommodated with minimal modification to the basic design. The ACR operates using the short-stroke gas piston, and the piston is adjustable. The upper receiver is a single-piece extruded aluminum unit. The barrel can be quickly changed, and three different barrel lengths are offered: 10½, 14½, and 18 inches. The ACR represents some of the most contemporary thought processes in the next generation of rifle. In many ways, the ACR and the FN SCAR bear strong similarities in the approach and philosophy that underscore the overall design and features (see Figure 4.66). The ACR is available commercially as a semi-automatic rifle or as a select-fire weapon to authorized entities.

Figure 4.66 (See color insert.) A contrast between the upper receivers of the FN SCAR 16S (top) and an AR-15 (bottom). The bolt carrier groups have been removed to contrast the differences between the two. The SCAR is operated by gas piston; the AR uses the traditional gas impingement. Installed on the AR upper receiver is a Knight's Armament RAS free-floating rail system, an aftermarket flash hider, a MaTech rear sight, a Knight's Armament vertical fore grip, and an Aimpoint Comp M4 using a Knight's Armament mount. (Image from author's collection.)

Another entrant into the arena of the modular weapon system was Heckler & Koch. The Models 416 and 417 are essentially the H&K versions of the M16, chambered in 5.56×45mm and 7.62×51mm, respectively. In consideration of other modular weapon contenders, the 416 and 417 appear somewhat conservative. H&K was among the first manufacturers to begin utilizing a gas-piston operating system instead of gas impingement.

The HK-proprietary gas system uses a piston driving an operating rod to control the function of the bolt, preventing propellant gases and the associated carbon fouling from entering the weapon's interior. This increases the reliability of the weapon and extends the interval between stoppages. It also reduces operator cleaning time, heat transfer to the bolt and bolt carrier, and wear and tear on critical components. (Heckler & Koch USA n.d.a)

Further elaboration on the system is made in information published by the company about the HK417.

The HK417 uses the unique operating rod gas system pioneered by HK in the HK416 and G36 weapons systems. This system uses a solid "pusher rod" operating rod in place of the more common hollow gas tube normally employed in AR15-style rifles. (Heckler & Koch USA n.d.b)

True to the form of H&K products, these rifles are produced to the highest quality standards. Initially 416 and 417 were available only as machine guns, and thus sales were restricted to military, law enforcement, and certain other authorized entities, but this caveat only affected the receiver portion of the firearm, which is the lower receiver. As early as 2007, very limited quantities of 416 upper receivers made their way onto the commercial market in advance of general commercial sales, much to the consternation of H&K. In keeping with the general movement of the modular weapon system concept and the direction that AR-15/M16 platform weapons have gone, the 416 and 417 are adaptable to a wide variety of applications, from carbine to a precision rifle. A departure from contemporary thinking is the decision on the part of H&K not to chrome line the bore.

In a 2011 press release, HK-USA announced the availability of the civilian model of the 416, the MR556A1, available as a semiautomatic rifle but otherwise indistinguishable from the 416 (Heckler & Koch 2011a). The upper receiver can now be purchased by itself and will interface with any standard AR-15/M16 platform weapon as a retrofit. The 7.62×51mm version, called the MR762A1 was officially introduced by H&K to the market in January 2012. Both rifles are manufactured in the United States by H&K in Newington, New Hampshire. The 416 rifle has been taken into service by the armed forces of a number of nations under client-specific model designations, including the U.S. Special Operations Command, which apparently took the rifle into service in 2004 but may have withdrawn the weapon from service in 2007.

The U.S. Marine Corps has taken an H&K IAR (Infantry Automatic Weapon) variant of the 416 into service. Designated M27, it was approved for fielding in the summer of 2011. "The M27 IAR replaces the heavier M249 SAW (Squad Automatic Weapon), which has been used by the Marines in Infantry Squads since the mid-1980s in the automatic rifle role. Both weapons fire the 5.56 mm NATO cartridge" (Heckler & Koch USA 2011b). The development of the IAR variant of the M16 by Colt Defense LLC was likely to compete for this contract.

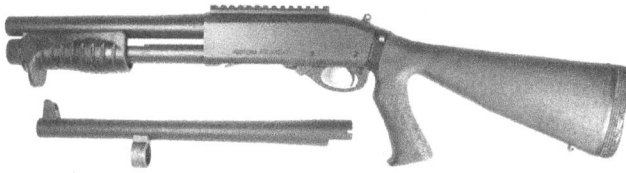

Figure 4.67 The Remington Modular Combat Shotgun. The MCS is based on Remington's 870 shotguns, with some upgrades to suit its purpose as a combat shotgun. (Image from author's collection.)

The Remington Modular Combat Shotgun (MCS) is a submodel of the Remington Model 870 shotgun (see Figure 4.67). Remington has long marketed a plurality of sub-models of the 870 shotgun for law enforcement, government, and military users; however, the MCS concept took this idea a step further. The MCS was developed with military and law enforcement operations in mind to provide a shotgun-based system that afforded the user maximum flexibility. The MCS is supplied as a kit that furnishes the operator with the standard 870 receiver and three barrels: 10, 14, and 18 inches in length. The system is furnished with two stocks: a traditional full-length stock and a short pistol grip. Both are quickly attached using a proprietary interface system unique to the MCS. Other accessories included in the kit include screw-in chokes, a cleaning kit, and capacity-expanding magazine extension tubes with the requisite magazine springs for use with these extension tubes. The entire kit fits into a single case for ease of portability.

The philosophy behind the MCS was to provide maximum versatility to the user to meet a variety of operational needs: a standard-duty shotgun with an 18-inch barrel, a close-quarter battle shotgun with a shortened barrel, a breeching tool, or an accessory weapon. One significant design difference between a standard 870 and the MCS is the addition of the M1913 accessory rail atop the receiver. This rail system allows the user to attach the normal accessories such as sights, flashlight, and so forth, but was installed more specifically as an interface between the MCS and a host M4 rifle or other model weapon of that pattern. A special attachment is used that connects to the MCS rail and the lower rail on the front hand guard of the host firearm, the result resembling an M4 with an M203 grenade launcher attached. The MCS is yet another iteration of the current trend of modularizing weapons platforms; however, it was not the first of the concept. Knights Armament, a manufacturer located in Titusville, Florida, manufactures a similar system using the Remington 870 mounted to an M4, calling it the Master Key.

Firearms Identification Methodology

When the examiner is uncertain about the particulars of a firearm, every piece of obtainable information from the weapon itself can be utilized to create at least a list of candidate firearms to provide a positive identification.

1. Is the firearm a classic design? Does the firearm mimic a pattern/design of a known firearm such as the 1911, AR-15, etc.?
2. What materials and methods are used to construct the firearm?
   Machined steel receiver and components

Stamped sheet metal using rivets or welding to assemble

Polymers used in construction

Metals other than steel such as aluminum

3. What, if any, markings exist on the firearm? All markings, however slight, should be documented as to what they are and where on the firearm they are placed. To facilitate this, the firearm may need to be partially disassembled, the grips removed, etc. Since many markings cannot be replicated on a computer, such markings may need to be hand drawn.

Manufacturer name, logo, address, and other geographical information

Presence of logos, icons, or trademarks on the grips

Presence of a ZIP code (the ZIP code was not implemented until 1963)

Make and model name or number

Identification of caliber, if possible

Indication that the firearm is or is not suitable for use with smokeless powder

Importer stamp(s)

Proof marks

Markings that indicate a secondary manufacturer or remanufacturer

Identification of a Firearm from Photographic or Video Images

The firearm examiner is often called upon to identify the particulars of a firearm using images from video or still cameras. Cameras have become so prolific in modern life that the average person is likely to be under some form of surveillance more often than anyone would like to believe. Cameras installed in public spaces and especially on private property (such as retail outlets) as well as cameras on cellular phones, computers, and other electronic gadgets have provided valuable evidence to police investigating all types of crimes.

The appearance of firearms in video or still images is a common theme. A surveillance video may reveal the brandishing of the firearm during the commission of the crime, or persons may photograph or video themselves or others brandishing firearms as a matter of posterity, especially in a criminal context. Identifying an individual with a particular firearm may prove to be valuable criminal intelligence, especially if it can be placed in connection with crimes that have occurred. Another aspect is identifying a prohibited person in possession of a firearm, which in and of itself is a crime.

Yet another facet is identifying the particulars of a firearm caught in an image. There is no doubt that the experienced examiner can identify certain firearms by virtue of design particulars that are specific to that certain model. The central problem to identifications made from imagery is being able to say, with any degree of certainty, that the depicted article is a firearm. With the availability of inert replicas, airsoft guns, and other props that are modeled with extremely fine detail and accuracy on an original firearm, this may prove too difficult to pass legal muster. That being said, if the exhibit under investigation is so unique that it has no analogues, the case could be made that some detail is sufficiently compelling that the examiner is confident in making an identification.

For such fine detail to be revealed, a great deal of photographic enhancement may be needed. This is not unrealistic, given the resolution of contemporary digital imaging devices of even mediocre quality. The scale of the object can be just as important as its shape and distinguishing features. Scale is best ascertained using a photogrammetric

Figure 4.68 An assortment of arms found in a clandestine cache in Iraq circa November 2003. Note the different styles of AK rifles depicted. The bolt-action rifle at top center is a British SMLE (Short Magazine Lee Enfield), likely dating back to World War I; the lower center is a Mauser 98, likely 1920s or 1930s vintage. (U.S. Air Force image; photograph by Suzanne M. Jenkins, USAF.)

approach, which uses the size of an easily quantified object within the same image as a comparison against the article of interest. Easily quantified objects include coins, paper currency, and other standardized, industrially produced articles. When comparing a suspect exhibit in an image to a seized exhibit, characteristics between the two would have to be compared to make any reasonable effort to attach the two as being one and the same. Particulars that the examiner can cue in on are wear patterns toward the muzzle from holster wear; unusual, nonstandard, or irregular grip styles or a feature of the grips such as position of the grip screws; cracks, fractures, or other damage to the grip panels; or other nonstandard, aftermarket, or unusual accessories to modify the firearm.

When witnesses and victims are interviewed during the course of an investigation where a firearm is involved, it may be difficult for the average person to identify even general details about the firearm. There are any number of reasons why a person's attention may not have been focused on the firearm or why that person's focus may have squarely been on the firearm. Persons who have had a gun pointed at them tend to focus exclusively on the firearm, even to the exclusion of other obvious details. One practice that investigators can use is to have a series of flash cards in their kit that depict a typical example of each type of firearm, such as revolvers, automatic pistols, various styles of shotguns and rifles, etc. These examples can be artistically rendered silhouettes or photographs of general examples of firearms taken on a neutral background. Such photographs should provide a broad perspective of the general appearance of the weapon and exclude such finer details as make and model. The depiction should allow any person, regardless of language or familiarity with arms, to identify the type of weapon simply by its overall design characteristics.

In the absence of using visual aids—and even if the witness identifies the firearm as a particular type—it may be advisable to verify the terminology that the witness is using. The terminology will be correct in the witness's mind, but that does not make it an authentically correct term. One common example is a person who identifies the perpetrator as wielding a machine gun. From the perspective of the witness, the perceived firearm was a machine gun. However, is the witness's definition of the term based upon perception drawn from television, movies, video games, or news media? It is common for a layperson to identify a weapon as a machine gun by virtue of appearance, having seen a similar-looking weapon in news reports or in popular entertainment media.

Firearm Markings

5

Serial Numbers

When the Gun Control Act was enacted in 1968, one of its provisions was the requirement that all firearms be affixed with a unique serial number. Prior to the act being signed into law, shotguns and .22 rifles were exempted from this requirement, but this did not preclude a manufacturer from affixing a serial number on any firearm as a matter of normal business practice. Despite the lack of legal requirement, many manufacturers voluntarily affixed serial numbers to arms, the major exception being shotguns and rifles branded by department stores. Since 1968, all articles that are legally articulated as a firearm that is manufactured or imported into the United States are legally required to have a serial number affixed to the frame or receiver. As part of the serial number requirement, the act further specified that serial numbers be unique; manufacturers could no longer replicate serial numbers. Furthermore, importers of firearms could not replicate a serial number on any other firearm they had previously imported, and in the case of a foreign-manufactured firearm where foreign characters are used as part of the serial number structure that was originally affixed by the manufacturer, the firearm would be assigned a new serial number made up of Arabic numerals.

In locations where firearms are manufactured, relevant local laws may classify a firearm as any component that is used to assemble a firearm such as a slide, barrel, revolver cylinder, significant smaller parts or subassemblies, and even grip panels, in addition to the frame or receiver. Such firearms can be expected to have a serial number, or partial serial number, marked on these components, as they are legally deemed controlled parts. It has been a long-standing practice by many manufacturers to affix a serial number or a partial serial number to firearm components other than the frame or receiver as a matter of routine, whether there was a legal mandate to do so or not. There were several rationales given outside of legal requirements to serialize even small parts.

One rationale was that serial numbering the small parts was part of a manufacturer's internal quality control measures. When a firearm was assembled, it was then tested and inspected for proper function, typically by firing several "proof" loaded cartridges from the firearm. If the firearm passed this firing test and the degree of accuracy, function, and safety was guaranteed, the firearm was then moved to finishing. In this instance, most manufacturers chose not to finish the firearm, instead preferring to leave the firearm "in the white."* This would allow for reworking of parts found out of tolerance or specification that could be salvaged, or simply replacing the part altogether with a new part. Only when the complete firearm passed inspection would it be moved forward for final finishing. Since all parts intended to be finished cannot be accessed when the firearm is assembled,

---

* The term *in the white* means that no finishing processes or operations such as applying coatings or polishing have been performed. The firearm is constructed of parts that are in the natural material color.

it would have to be disassembled once again to access the individual components. Despite the interchangeability of parts, in order to ensure that the same gun that was tested was in fact the same composition of parts completed as a firearm that left the factory, certain parts would be serialized so the same firearm using the exact same components was reassembled after the finishing operations were complete. It was not uncommon for the finished firearm to receive an additional step of quality control after finishing, ensuring that the finished product met the requisite standard. The practice of serializing even small parts was once prevalent across the industry; however, by the 1970s the practice had abated considerably, especially in the United States.

European gun makers in particular were fond of serial numbering magazines, although the practice is observed in arms from other parts of the world as well. Unlike American practice, where magazines are considered consumable products and not necessarily intended for reuse (in a military setting), other entities consider the magazine accountable equipment and part of the firearm, and simply disposing of the magazine once it is empty is unacceptable. Serial numbering magazines also provided a measure of equipment accountability and inventory tracking.

Serial Number Structures and Practices

Prior to the enactment of the Gun Control Act, most manufacturers chose to use numerically sequential serial numbers. When a new model firearm was rolled out, they would start with serial number one and continued to issue numbers until the production run was finished. Since replication of numbers was henceforth prohibited, even different model firearms made by the same manufacturer would have a separate serial number series used. Modern firearm serial numbers are more akin to vehicle identification numbers, most likely containing a great deal of information beyond the mere sequence number. Certainly a sequential number is in the serial number; however, manufacturers could include the numerical model number and other attributes that would be revealed only if the number was deciphered. Serial number practices are left to the manufacturer as long as they are compliant with applicable law concerning repetition and placement on the frame or receiver (see Figure 5.1). Manufacturers may opt for separate serial number ranges and sequences, using prefixes and suffixes, which are specific to a certain model, including special serial number ranges for situations such as limited production runs and commemorative and presentation arms.

An often-encountered error on Smith & Wesson revolvers is the erroneous identification of an affixed numerical sequence on the cylinder crane as the serial number. It is true that Smith & Wesson revolvers can have serial numbers affixed to the frame where it meets the crane, which would be the same number that is affixed to the bottom of the grip. The number on the crane is likely to be a production or product number, which is generally four or five characters in length. A person mistaking it for a serial number may try to obliterate it. Practically all Smith & Wesson revolvers had the serial number affixed to the bottom of the butt of the grip. If oversized grips were installed that obscured the bottom portion, then the grips would have to be removed to verify the serial number. More vintage Smith & Wesson revolvers could conceivably bear a serial number on up to five parts of the gun: the cylinder under the ejector star (see Figure 5.2), the bottom portion of the barrel underneath the ejector rod, the frame near the cylinder crane, the butt, and possibly inscribed by pencil on the inside of either grip panel. An alternative to the bottom of the butt would be the front strap of the grip underneath the trigger guard. Due to the

Figure 5.1 Taurus PT-145 PRO. Note that the serial number is marked in three places: the slide, the barrel, and at the rear of the receiver above the grip. Only the frame number can be used to identify the firearm. (Image courtesy of Taurus International Manufacturing, Inc.)

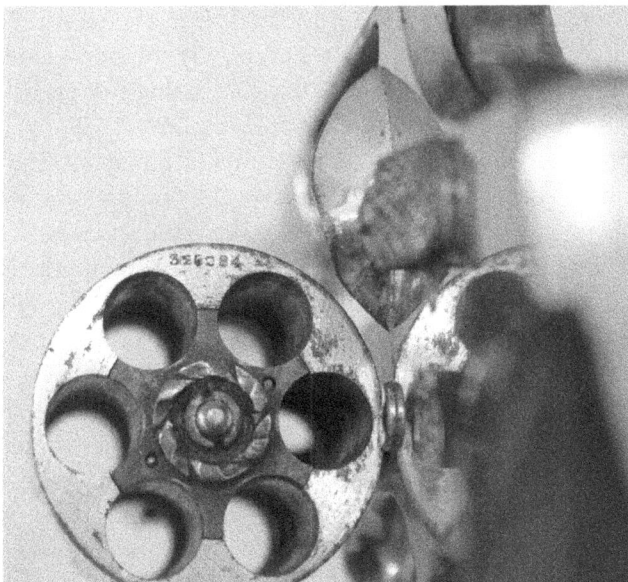

Figure 5.2 Pictured is the cylinder of a nickel Smith & Wesson Model 1905 Military and Police model. The serial number is stamped on the cylinder, as well as on other parts. This is an all-matching gun; no parts have been replaced. (Image from author's collection.)

prevalence of switched grips and the switching of parts, the frame-affixed number, the one at the bottom of the butt, must be relied upon as the actual serial number; however, if there is sufficient confidence that the other parts are original, then these parts could be referenced to assist in reconstructing an obliterated, defaced, or altered frame number. Smith & Wesson auto pistols most generally have the serial number affixed to the frame only, either on the side or the bottom of the frame. In the case of second- and third-generation auto pistols, a serial number may be found stamped in the frame underneath the left grip panel.

Vintage Colt revolvers were profusely marked with serial numbers, with the frame number being affixed to the bottom of the frame in front of the trigger guard. Adjacent to this number could be a serial number affixed to the trigger guard, and another on the barrel. On vintage Colt derringers, as well as certain other revolver models, the serial number is affixed to the bottom of the butt and on the grip portion of the frame underneath the grip panel. Modern-era Colt revolvers generally could be found to have serial numbers affixed in three places: the cylinder crane, the frame behind the cylinder crane, and inside the left side plate, requiring its removal to be seen. The actual serial number is that which is affixed to the frame; however, this does not prevent the other numbers from being used as a reference if the examiner is comfortable enough that the gun is completely original and there is confidence that a partially restored serial number will mesh with these other numbers. Colt reproductions of vintage revolvers mimic period revolvers, except that the serial numbers are affixed to the bottom of the frame only, in front of the trigger guard. Investigators are cautioned that markings that appear on the bottom of the butt or on the back strap, including police, military, security, or national property markings and asset numbers, could be mistaken for a serial number. Vintage Colt revolvers can have the serial number affixed to the frame on the grip, visible only when the grips are removed. If the investigator is uncertain about the vintage of the specimen under inquiry, all areas should be examined for the potential presence of a serial number. Colt-manufactured 1911 and 1911A1 pistols typically had a serial number stamped on the back of the slide, visible only when the hammer was cocked, in addition to the frame affixed number. The 1911 and 1911A1 pistols made by government contractors typically did not include a serial number anywhere other than on the frame. Other Colt pistol models often have serial numbers affixed inside the slide in addition to the frame-affixed number.

For the most part, affixing serial numbers is now a matter of finding sufficient real estate on the frame or receiver, particularly with respect to handguns and more so on small-framed handguns. Long guns do not generally face the same space issue. Regardless of the manufacturer, most serial numbers on revolvers follow the general practice of being affixed on the bottom of the butt, on the interior of the frame adjacent to the cylinder crane, or on either side of the frame, but not including a removable side plate. AK-pattern firearms generally have the true serial number stamped on either side of the receiver, although military AK rifles tended to be marked on the left side of the barrel trunnion. AK receivers manufactured in the United States tend to have their identification information, including serial number, stamped on the bottom of the receiver in front of the magazine well. The majority of AR-pattern firearm builders prefer to stamp the serial number and other identification information on the left side of the lower receiver on the magazine well. Since the upper receiver is not considered the firearm, no identification markings should be expected to be found there, other than that of a manufacturer who opts to place their name or trademark for marketing or product identification purposes. Polymer-frame firearms do not have the serial number stamped directly into the polymer. Instead, the number appears on a metal plate that is molded into the polymer or is a portion of metal that is "windowed" to provide a spot for the serial number. These plates can appear on the bottom or on either side of the receiver.

A firearm that has another firearm within or attached to it will bear separate serial numbers. For example, a suppressed firearm will have a serial number affixed to the host weapon, but the suppressor, itself a firearm, will have a unique serial number attached. There may be instances where the suppressor may be serial numbered to match the host gun, but this is purely dependent on the manufacturer to make such an accommodation. A host firearm that has a registered full-auto sear installed will also have two serial numbers, the sear itself will have to be accessed to verify its own serial number, although a registered receiver with a full-auto sear will have only the receiver serial number, since the auto sear is a part of that receiver and not separately installed.

Firearms Absent Serial Numbers

Historically, most firearms were affixed with a serial number by the manufacturer prior to the Gun Control Act requiring the practice. In the case of firearms without serial numbers, omitting the serial number was likely done as a matter of cost savings by eliminating that part of the operation and the bookkeeping associated with it. It is the practice of many manufacturers to postpone serializing the firearm until it has satisfied all in-house quality control measures and is ready to package. Again, this is logical as a matter of economy and paperwork reduction. Once a firearm is issued a serial number, it becomes accountable. Should there be complications from production, those firearms would have to be written off and documented as such, thereby creating additional bookkeeping burdens.

Untold thousands of weapons were retained as souvenirs and carried or mailed home by soldiers, sailors, marines, and airmen at the end of the First World War, the Second World War, and even into more modern times from the wars in Vietnam, Korea, Iraq, and Afghanistan. With the occupation of Germany and Japan and the liberation of the other nations that had been previously occupied by the Axis powers, armaments factories were captured, and many GIs' souvenir of choice was firearms. A great many firearms were captured on the battlefield and retained, but many were also taken from captured factories. Firearms captured at the factory may have been assembled but not reached the point where a serial number would have been affixed when production was interrupted. Bona fide firearms of the era without serial numbers can be treated as exempt from the serial number requirement, as they are of historical value. The same arguments cannot necessarily be made for war-trophy arms brought home from other conflicts post-1968, except perhaps in the case of antique arms, those made before 1899. Numerous examples of unmarked Model 1911 and 1911A1 pistols bearing no serial numbers, and in some cases no markings whatsoever, presumably manufactured during the World Wars under military contracts, reside in the hands of collectors. In collector circles, such arms are called *lunch-box* guns, as they are thought to have been smuggled out of the factory in a worker's lunch box. As with captured foreign arms brought home as souvenirs, these firearms are generally exempted from serial number requirements due to their age and collector status.

Firearms bearing no serial numbers, which have been established to have been manufactured after 1968 but that bear no indications of attempts to remove, obliterate, or alter the factory serial number should be treated with a degree of suspicion. One plausible answer to such a firearm suggests that the firearm may have been stolen from the factory or that the frame, receiver, and possibly other components were stolen or obtained and used to construct a firearm from a receiver or frame that was never issued a serial number. Another plausible explanation is that the firearm is a homemade article and a serial number was not affixed. A homemade firearm should generally be instantly recognizable due to the relatively crude construction compared with professionally manufactured firearms.

Firearms Bearing Multiple, Different Serial Numbers

Firearms are encountered that bear multiple serial numbers. In such cases, the serial number of record is the number that is affixed to the frame or receiver, to the exclusion of numbers that appear on the barrel, slide, stock, or other non-gun parts. Such a firearm may have been restored or rebuilt using parts from other donor firearms installed onto another frame or receiver to make a complete firearm. Such examples may be of investigative interest if there is suspicion that parts were replaced in an attempt to thwart an investigation by swapping a barrel or other replacement to remove the potential for making ballistic comparisons against other evidence in the case. In keeping with the serial number provisions of the Gun Control Act, firearm importers have often affixed new serial numbers to imported arms, and the original factory serial number is not used for recording purposes. When examples of this practice are observed, the differences are readily apparent by way of the location on the receiver where the new number was affixed, as well as the manner in which the affixing took place, such as laser engraving or electroscribe.

Renumbered Firearms

Surplus vintage military firearms have often been transferred around the world or were retained by a nation when captured from the opposition in the course of conflict. Such firearms were often "re-arsenaled" to remanufacture usable firearms using donor firearms that would be scrapped. In some instances, the original serial number would be defaced and a new number applied. Surplus World War II–era German K98k rifles that have emerged from Russian stockpiles are routinely seen with defaced serial numbers and new numbers applied by electroengraving. This is even true with firearms of Russian origin that underwent refurbishment postwar. Firearms were also refurbished in arsenal postconflict by the nation when the firearms were returned to inventory at the armory, which can also result in a mismatch of serial numbers on different parts, if numbered parts other than the receiver were used. A nation or military force that came into possession of foreign arms may have decided to attach their own serial number or other inventory control numbers to the firearm. As is the standard rule, the serial number affixed to the frame or receiver is the proper identified serial number, regardless of the attachment of other numbers to other parts. If a firearm was issued a new serial number as part of the process of refurbishment, typically the original serial number is obliterated or defaced by peening or overstamping (see Figure 5.3).

   Firearms with obliterated or illegible serial numbers are eligible to have a new serial number assigned by the Bureau of Alcohol, Tobacco, Firearms and Explosives (BATFE). This service is available to owners seeking to reclaim their firearms recovered after theft or loss and where the serial number was compromised. Obviously, the serial number would have been restored and identified to the legitimate owner. There is a process involved that requires the coordination of the law enforcement agency in possession of the firearm and

Figure 5.3 **(See color insert.)** This 1913 vintage Colt 1911 has had its original serial number ground from the frame, and it has been renumbered. It is apparent that this was historical and likely done at the factory or at a government arsenal. The United States Property marking has also been obliterated from the frame by grinding and then stamping a series of X's across that part of the frame. Students of the 1911 will note the crude stamping as well as the die being inconsistent with that used by Colt. (Image from author's collection.)

the local BATFE field office. Older firearms with government-assigned serial numbers carried a serial number prefix of IRS, for Internal Revenue Service; the prefix was later changed to ATF, for Alcohol, Tobacco, and Firearms. Older firearms that did not have factory-affixed numbers and arms where the serial number has become illegible are also eligible for this service.

Examiners should familiarize themselves with serial number practices and structure by manufacturer. In many cases, the manufacturer can be contacted directly, and inquiries can be made to them about serialization. This may assist in finding a serial number's location and/or what format the serial number will take. Occasionally, a firearm may have a production number affixed that has no immediate attachment to the serial number, but it may be possible to use that production number as a serial number derivative to determine, with the assistance of the manufacturer, what the serial number is.

BATFE Serial Number Traces and E-Trace

The Bureau of Alcohol, Tobacco, Firearms and Explosives (BATFE) is the central repository of firearm transaction records in the United States and its territories. When a firearm is sold through commercial channels by licensed entities, the transaction is documented through an Acquisitions and Disposal log, typically referenced as the "A&D log." The retail purchaser of a firearm through a licensed dealer is legally required to complete a standardized form, Form 4473, also called the "yellow form," which identifies the purchaser, the firearm, and the license holder who is disposing of (selling) the firearm. Until relatively recently, Form 4473 was completed on paper; however, an electronic version of the form is now available and is used in lieu of the paper document. Firearms dealers are required to permanently maintain these records until they leave business, at which time the records are to be forwarded to the BATFE. Certain states also require additional forms to be completed in addition to the 4473. In such instances, the state or locale would maintain these firearms transactions records. Law enforcement agencies can—either directly or through their local BATFE Field Office—request a firearms trace through the BATFE National Tracing Center

(NTC). The trace request form can be completed and transmitted by fax, mail, or electronically through the e-Trace program. Only law enforcement agencies may apply and be accepted to become e-Trace users. A trace query is made against a firearm by serial number only, not by an individual's name or other parameters. There are tremendous investigative benefits to tracing firearms seized pursuant to criminal investigations, including:

Confirming ownership of a found firearm where an individual is attempting to claim the weapon but cannot provide substantial documentation to prove ownership
The identity of a firearms purchaser from a licensed dealer
The dealer from which the firearm was obtained
When the firearm entered the chain of distribution and sale, which could implicate or eliminate a suspect firearm from consideration in a crime
Trends and patterns of individuals purchasing firearms and transferring them to prohibited individuals (so-called *straw purchasers*) and firearms traffickers
The potential to provide other investigative leads in firearms-related crime
The firearm-type classification as originally manufactured: rifle, handgun, etc.

Methods Used to Apply Serial Numbers

Serial numbers can be affixed to the firearm using any number of different methods that are at the discretion of the manufacturer. The method used by a particular manufacturer is generally considered to be the method that is most cost effective for that particular firm and the manner in which the affixed numbers can change. The federal requirements for placing a serial number were changed effective January 30, 2002, and articulated in 27 CFR 178.92(a)(1), which required that serial numbers be stamped, engraved, or cast to a minimum depth of 0.003 inch and in print no smaller than 1/16 inch. This standard also included other required information such as the manufacturer name, city and state, importer, caliber, model, remanufacturer, and so forth. The most common modern methods for affixing a serial number are:

Laser engraving
Stamping
Pin stamping or dot peening (also called dot-matrix engraving)
Roller marking
Electroscribe

In order to mark the serial number, mechanical force is applied to stampings and roller marking as the means to affix the serial number. This compresses the material, resulting in a deformation of the crystalline structure of the material well below the surface where the force was applied. As the material is compacted by the force applied to affix the serial number, it does so in such a way that it imparts the imprint of the force, resulting in a character that potentially is restorable even if the surface is significantly damaged. Laser etching or engraving does not cause this phenomenon to occur, yet laser-engraved serial numbers are restorable by the same methods. *Laser engraving* is a generic term used to refer to any number of similar methods that use a laser beam to mark an object. Laser-marking devices have distinct advantages over mechanical marking processes because there are no mechanical parts to wear through repetitive use. Lasers also can work in very small areas and leave very clean markings behind.

Each of the methods described has distinctive characteristics that show the examiner how the serial number and other information is affixed. Pin stamping, dot-matrix stamping, and dot peening are essentially the same type of operation. This style of marking consists of a series of dots formed in the shape of the represented letter or number. Roller markings are among the oldest methods of affixing data to the firearm. A die is fabricated containing the necessary information and it is machine-rolled across the surface of the firearm, leaving behind the impression. Stamping can take the form of die stamping by machine or by hand stamping using a set of dies and striking the die onto the surface.

Manufacturers may use more than one single method to apply identification information such as manufacturer, model, and caliber. This is particularly the case in two scenarios: firearms that are imported into the United States and firearms that are built using a parts kit that is installed on a different receiver or frame. In the case of an imported firearm, the original markings would be affixed by the manufacturer using any of the described methods. However, when the firearm is imported into the United States, the importer must also affix its specific information somewhere on the frame or receiver. In the case of firearms that are manufactured offshore and then imported into the United States by a subsidiary of the same company, such as Sig Sauer in Germany and Sig Arms in the United States, the method of marking will be to a high standard and will not be different in order to maintain the aesthetic quality of the firearm. Firearms imported by third-party organizations generally do not receive the same quality of craftsmanship and attention to detail. Quite frequently, imported arms are pin stamped or electroscribed just enough to satisfy legal requirement. The electroscribed markings tend to be of the poorest quality and are often difficult to read under the best of conditions. Importers are required to stamp their corporate name and the city and state where they are located. The firearm caliber and nation of origin also appear. Commonly recognized abbreviations are acceptable when affixing this information.

In addition to the person-readable information that is impressed into the firearm, some manufacturers have adopted the QR (Quick Response) code that will also appear alongside the human-legible information. The QR code can be read by a plurality of readers, including camera-equipped smart phones, tablet devices, and other devices that have the requisite application installed on them. The QR code is capable of holding a tremendous amount of information and could include not only the serial number, make, model, and caliber, but also could include other production information as well (see Figure 5.4). Smith & Wesson has bar-coded serial number information on the frames of their Sigma series pistols since releasing the line to market. The bar code is located adjacent to the person-legible serial number on the bottom of the receiver on a metallic plate embedded in the polymer. In a case where a Sigma series pistol has an obliterated plain-character serial number but the bar code remains reasonably intact, it may be possible to have the serial number identified with the assistance of the manufacturer.

Obliteration, Alteration, or Defacing of Serial Numbers

The Merriam-Webster dictionary defines the term *obliterate* as, "to utterly remove from recognition or memory; to remove from existence, destroy utterly all trace, indication, or significance of; to cause to disappear, to make undecipherable or imperceptible by

Figure 5.4 A Remington Model 770 rifle; note the dot-matrix pattern QR code. (Image from the author's collection.)

obscuring or wearing away." The term *alter* is defined as, "to make different without changing into something else; to become different," and the term *deface* is defined as, "to mar the appearance of; to injure by effacing significant details." Generally, the three terms may appear to be interchangeable; however, this should not be considered to be the case. Despite the common definitions given, there is context and shade that should be read into each of them. *Defacing*, for example, may suggest an incomplete removal, while *obliteration* would suggest that the serial number has been completely removed and no trace has been visibly left behind. Contrast this with the term *alteration*, where a serial number is somehow changed in substance while remaining intact. Examiners are encouraged to use their words carefully in written reports as well as in verbal testimony and to maintain a continuity of verbiage and not interchange the use of words throughout the process.

Methods Observed Used to Alter, Obliterate, or Deface Serial Numbers

The destruction of a serial number on a firearm seems illogical at best. In every venue, the mere possession of a firearm without a serial number is a crime, and the evidence of attack is generally quite apparent, even by casual observation. It stands to reason that possessing such a firearm would only call more attention to the situation than if the firearm had been just left alone, notwithstanding the circumstances under which the firearm comes under investigation. Despite the irrational nature of the urge to obliterate, deface, or alter the serial number, this practice has persisted. There are numerous ways to go about removing the serial number, and the majority involve some form of mechanical force, although this should not be construed to mean that this is the only possible means.

Serial numbers may be attacked by several methods—primarily by the use of tools and instruments such as grinders, punches, drills, pointed or edged objects, or perhaps even chemical abrasives in an attempt to remove the serial number from the frame or receiver. The presence of grinding marks, concave surfaces, scratches, peening marks, or other surface irregularities should draw the immediate attention of the person seizing the firearm, even if the attempt is superficial and does not appear to fully obscure the serial number. Because the weapon's finish has been attacked, the area generally becomes susceptible to rust, which will only further conceal the information.

Figure 5.5 A scratching type attack on a pistol. As can be seen, the obliteration was only partial, and the number was restored chemically with relative ease. (Image from author's collection.)

The tool used to mechanically attack a serial number has great bearing on the realized success of the attack. The tool of choice is frequently not ideal for the task, and is often softer than the material that it is being forced against, resulting in degradation of the tool perhaps more so than the surface under attack. This is especially the case where hardened finishes or especially harder materials were used to construct the firearm. The deposition of the tool material onto the attacked surface may leave the obliterator with the supposition that the attack was more successful than it actually was. In such a scenario, using a soft brush and oil will generally clean the attacked area, revealing the serial number underneath the deposited material.

Serial number attacks are grouped into several broad categories, based upon the type of attack conducted.

Scratching

The surface is attacked by a narrow-bladed, wide-bladed, or pointed object. Ordinarily the scratching is done along the horizontal axis of the firearm by virtue of ease of manipulation and tool-to-surface interaction (see Figure 5.5). Common tools used in such an attack include flathead screwdrivers, awls, punches or ice picks, electric scribes, chisels, scrapers, engravers, and even nails. The general characteristics present in such an attack are an irregular, nonparallel series of lines that tend to wander erratically across the surface. The appearance of multiple tool marks should be assessed, as quite often the tool used to attack the firearm is made of a softer material, thereby wearing the attacking tool down.

Gouging

A gouging type of attack attempts to physically remove the material that the serial number is stamped upon, whereas a scratching attack is intent on damaging the material. Gouges are typically done using a chisel or similar tool, often in concert with a hammer. Gouging may result in an uneven rectangular-shaped void left where the serial-number-bearing surface was removed by first damaging the surface around it and then forcing it out, or by simply forcing the tool into the material to a given depth and attempting to pry or roll off the number-bearing surface.

Grinding, Sanding, or Other Abrasives

In such an attack, a concave impression is left behind, providing clear evidence of a grinding wheel being used. In most instances, the grinding was made on the same plane as the grinder's rotation, leaving long, parallel lines. Grinding where the attack occurs perpendicular to the rotation of the working tool edge is not often seen. Such attacks can be called angular or cross-grain attacks, depending on how the tool was applied to attack the surface. Handheld rotary tools, bench grinders, and electric drills with grinding or polishing bits or wheels are all candidate tools for this type of attack. The style of bit, wheel, or tool edge will have great bearing on the final appearance, ranging from smooth to an extremely coarse depression resulting from the attack. In the case of hand files being used as the instrument, the *file cut* is the measure by which the coarseness of the file's teeth is identified. The range of cut from smoothest to coarsest is very (or dead) smooth, smooth, second cut, bastard, and rough. Correspondingly coarser teeth have the net effect of shaving away more material with fewer strokes. If an electric tool is used, it is likely that only individual characters will be attacked, leaving the spaces between the characters alone. Since the tool, such as an electric scribe, must be pressed into the surface, the amount of pressure and the strength of the working bit will result in impressions of irregular depth and tend to result in a stippled appearance of small impressions of irregular, often overlapping spacing.

Drilling

A power drill using any sized bit can be used to remove material. The result of such an attack is a series of irregularly spaced circular, sometimes overlapping, holes of various depths. A hand drill or drill press could be used in such an attack.

Punch

A punch may be used to peen out a serial number by repetitious blows to the affected area. The compression of the punch into the material causes its displacement and disfigures any markings, thereby potentially rendering them illegible.

Peening

A surface that has been peened has been mechanically attacked using blows from a hammer. A ball-peen hammer is specifically designed to bend and shape metal; however, this does not preclude other types of hammers from being used. Peening differs from punching, as the term is used in the context of a blunt striking object being employed as opposed to a shaped instrument that serves as an intermediary to deliver the force to the surface under attack.

Filling

A marking that has been filled has had the area covered with another material. Cold filler may be used, and this can include using a finish material similar to the factory-applied finish. Fillings may also be accomplished by soldering, welding, or other similar techniques where such materials can be used to plug the depressed surfaces bearing the markings. As an extra measure, this area can then be filed or sanded to smooth over the attack in an attempt to cosmetically conceal the remnants of any markings. Filling can be done on its own or in conjunction with another destructive method, with the filling being used to further reduce the chances of a successful restoration. A variation of filling is to use intense heat generated by plumber or cutting torches to deform the surface to the point of rendering it illegible. However, thermal attacks are not widely used due to the inherent risk of destroying the firearm by deformation when an excessive amount of heat is applied for prolonged periods.

Chemical

Caustic chemicals can be used to eat or etch away at the surface bearing the serial number. Chemically destructive methods are exceptionally uncommon due to the relative lack of availability of the caustic agents required and the knowledge of chemistry needed to know what agent should be used.

Mechanical Serial Number Restoration Techniques

The object of serial number restoration is to get below the damaged areas to reveal the compression markings deep in the metal. When viewing the damage at the surface, examiners should determine what material was used to construct the firearm and then ascertain what method was used to attack the serial number. The examiners should inspect all surfaces of the firearm, using their accumulated knowledge to determine if the serial number or a partial serial can be found on another part of the firearm. Although this number cannot be guaranteed as being the true serial number affixed to the frame or receiver, it may give the examiner a basis from which to work. In the case of a partially restored number, the presence of other numbers may fill the gaps or voids where restoration was not successful. This statement is based upon the presumption that there is concurrence between the damaged serial number and other numbers located elsewhere on the firearm.

Mechanical restoration techniques are often overlooked by many examiners who prefer to go straight into the chemical restorative techniques, but the potential value of mechanical restoration cannot be overlooked. Although mechanical restorations may not suffice to totally restore a serial number, there is no guarantee that chemical methods will succeed either. Mechanical restoration lies in the ability of the examiner to understand how to use basic hand tools to repair the damage done to the surface bearing the serial number. Depending upon the type of attack, as defined previously, the surface is left in an irregular condition at a microscopic level. If debris and oxidation are present, these areas should be cleaned thoroughly with a good-quality gun cleaner and a nylon-bristle brush. Dedicated cleaners marketed expressly for firearms need not be the only option available; other commercial cleaners suitable for use on metals are acceptable. In fact, it is advisable to use a commercial cleaning agent that does not leave a residue behind. If the product does leave a residue, the area should be thoroughly wiped out.

The examiner should first attempt to level out the material in the attacked area. It may be that by simply refolding or flattening the attacked material that obliterated characters may start to become more legible. Very fine files, i.e., the dead smooth and the smooth, but not likely as harsh as a second-cut file, are used to flatten material that has been made jagged and irregular. This process in and of itself may start to reveal numbers. Some examiners prefer the use of polishing compounds such as jewelers rouge used with a low-velocity buffer instead of, or in concert with, the file. Rotary-tool buffing heads are ideal in such situations, inasmuch as the buffer is only mildly abrasive. As with all finishing procedures, finer tools are used in succession, starting with the coarsest first. Emory boards and waterproof, fine-grit sandpaper can also be useful. A delicate hand is required when wielding tools against the attacked surface. Remember that the amount of material that must be removed or adjusted is in the thousandths of an inch or fractions of a millimeter. Heavy-handed application of chemical agents or tools can simply remove any remaining vestiges of the serial number altogether. Remember that attacks may be to various depths into the material; it is not uncommon to see a deep penetration on one side the attack, and see the depth lessen across the attack, where the pressure was reduced or the working head of the instrument became less effective or had less contact over the course of the attack.

The examiner may find it helpful to create a bit of contrast in the characters that are raised during the restorative effort. Commercially available document-correction fluid can be applied with a fine brush to bring out hard-to-read characters. Be careful to avoid over-applying the fluid, or it will get into all the crevices. If a mistake is made, it is easily wiped out with a towel or a nylon brush. Instead of correction fluid, various colors of chalk can be used.

Traditionally, most firearms have been constructed of various grades and types of metal, most commonly machine-grade steel, aluminum, or stainless steel, but more recently, cast alloys and even compressed powdered metals have appeared. The proliferation of the polymer-framed gun cannot be ignored either. Various restorative techniques for use on metal have long been known, and restorative techniques for plastics have not been unheard of, even before the appearance of the polymer-framed firearm. Surplus military arms often had a serial number stamped into the wooden butt stock. As discussed in Chapter 4, there may be instances of conflicting information. In the case of a "force match," a prior serial number would have been crossed out or otherwise defaced and a new number applied. This number cannot be accepted as the serial number because it is not the frame-affixed number; however, it can serve as a basis to assist the examiner in recovering the frame or receiver number. Age, refinishing, sanding, and routine wear and tear on the wood may have rendered some of these markings somewhat illegible. An old gunsmith trick to resurrect old and faded stock impressions or cartouches is to use steam applied directly to the area where the markings are observed. This method is well established as a good way to restore the collector value of the stock by resurrecting desirable marks.

Chemical Serial Number Restoration Techniques

Chemical restoration techniques have long been the staple of examiners trying to successfully restore serial numbers. Restorative agents can be mixed on an as-needed basis by anyone with a sufficient knowledge of chemistry, or they can be purchased as a complete kit from various forensic-supply houses. It is recommended that—before applying any restorative reagents—the users read and become familiar with the Material Safety Data Sheet for the substance they will be using. There are variations in the materials used to manufacture firearms, so different restorative agents have been developed to suit these particular materials, including various grades of steel, aluminum, and even polymer. These reagents are well known and long established within the field for those familiar with restoration of serial numbers. Common restoration reagents are used to attempt to restore an obliterated serial number:

Acidic ferric chloride: ferric chloride, hydrochloric acid, water
Fry's reagent: cupric chloride, hydrochloric acid, water
Hydrochloric acid: hydrogen chloride, ethanol, and water

Iron (III) chloride: ferric chloride and water
Nitric acid: nitric acid and water
Turner's reagent: cupric chloride, hydrochloric acid, ethyl alcohol, water
Sodium hydroxide: sodium hydroxide and water

Recommended reagents for a particular material:

Brass: sodium sulfate
Cast iron: ammonium persulfate
Copper: ferric chloride, nitric acid, hydrochloric acid
Steel: hydrochloric acid or Fry's reagent
Aluminum: sodium hydroxide, hydrofluoric acid
Nickel: ferric acid or hydrochloric acid
Polymers: ethyl ethanoate (also called ethyl acetate)

Before applying the restorative reagent, it is suggested a dam consisting of hobby-type putty or plumbers putty be constructed around the area where the reagent will be applied. This dam will serve to keep the reagents from leaking out from the application site. The reagent can be applied using a dropper or by swabs. The working time for the reagents ranges from instantaneous to upwards of half an hour and perhaps longer in more difficult cases. The examiner is encouraged to err on the side of caution to prevent overuse of the reagent, which will defeat the restorative effort. A chemical neutralizer should be included in any restorative kit and should be at hand. The corrosive effects of the chemicals on the treated surface cannot be ignored, and the restored information must be documented in short order to prevent it being obscured and lost once again. It may take several applications of the reagent to effect a successful restoration. Further application of reagent should be by swab, allowing examiners to rub the reagent into the area they wish to restore; application of pressure may achieve better results.

Nondestructive Serial Number Restoration Techniques

Digital Imagery

Before application of restorative chemicals, it may be of benefit to attempt to recover at least partial information from an attacked serial number or other information using a nondestructive technique. The author has had success in using the digital camera and photo-enhancement software. Good-quality digital photographs taken of the attacked surface can be digitally enhanced to potentially discern letters and numbers through the marred surface. Shifting the contrast, color, or reverse color (negative color) can be quite productive in reading numbers that are illegible under normal, white light. A succession of photographs can and should be taken over the course of the restorative effort, and they can always be referred back to at any step of the process to see if the image can somehow be enhanced to provide additional clues or information. Using photo-enhancement software is not generally well accepted in forensic circles, due to the implications that suggest that photos may be altered; however, if the adjustments are documented and can be replicated, this should satisfy legal questions that may arise. Outside of using software, digital cameras, even inexpensive models, often have the capability of capturing images using creative-effects exposures such as black and white, sepia tone, reverse color, and negative. More sophisticated cameras and experienced users can adjust the color balance, contrast, and brightness onboard the camera to create the same effect.

X-Ray

X-ray may prove to be another viable nondestructive technique to read through the damaged surface. X-ray equipment is a standard fixture at hospitals, some clinics, and morgues. Another possible source for X-ray equipment would be an explosive ordnance disposal (EOD) unit. Frequently EOD units carry portable X-ray machines that are specifically designed to handle smaller objects. The advantage of such a setup is immediate accessibility by law enforcement, and the equipment can be brought, deployed, and used almost anywhere. Newer X-ray machines are digital, allowing for immediate capture of the image and ease of portability and transfer of the captured digital images. The use of X-ray for serial number restoration has long been contemplated and, like most techniques, probably will not be ideal in all situations. Perhaps the best application for X-ray would be for a serial number that has been filled in. X-ray would make the filling material transparent and could reveal the surface underneath.

Ultrasonic Cavitation

A spin-off technology from NASA called "cavitation" was first reported in 1979. This method was called the "NASA-Chicago State" process, and the report stated that the "process is advantageous because it can be applied without variation to any kind of metal, it needs no preparatory work and number recovery can be accomplished without corrosive chemicals; the liquid used is water" (Lewis Research Center 1979). The study utilized cavitation, the rapid formation of bubbles created by vibrations in the liquid medium. Cavitation can be created in an ultrasonic cleaner such as those used to clean jewelry, firearms, and other items that are small or have a lot of hard to reach places. As noted in the report, "Because the vapor bubbles contain high energy, they etch, or pit, the surface of the metal they strike" (Lewis Research Center 1979). These bubbles are not created by heat, but by energy. It was reported that the cavitation-induced pitting of the surface around the area where the serial number was applied resulted in a gradual restoration of the serial number. The test subject depicted in the report showed very favorable results. The most likely single greatest hurdle to using this technique is acquiring an ultrasonic tank large enough to accommodate long guns. If this technique were to be used, then all other evidentiary exams—DNA, fingerprints, and so forth—must be completed before immersing the article in the tank for restoration of the serial number.

Magnetic Particle Inspection Process or Magnetic Flux

The Magnetic Particle Inspection (MPI) process (also *magnetic flux*, or *magnaflux*) is another possibility for a nondestructive attempt at restoration of serial numbers. MPI was originally intended as a method to study and detect defects or flaws in a ferromagnetic object in applications such as welding, firearms manufacture, and shipbuilding. For obvious reasons, this method is not effective on nonferrous materials. This method calls for the creation of a magnetic field on the item to be studied, such as the firearm frame. This field, the magnetic flux, flows across the article. The flow then "leaks" from the article, particularly where there is disturbance on the article, such as an attacked surface that once bore a serial number or other information. The operation can be performed in a "wet" environment by immersing the study piece in water. In the water is a suspension of ferrous particulate that is attracted to the area of attack. The procedure can also be performed in a dry setting, using aerosol magnetic particulate. In an ideal scenario, this operation would rebuild a serial number that had been attacked. According to Utrata and Johnson (2003), using fluorescent magnetic particles resulted in better restoration, coupled with using ultraviolet light sources to enhance contrast: "Better serial number recovery seems clearly possible using DC instead of conventional AC electric power" coupled with good magnetic connections to the study piece. The surface should be prepared using fine grit sandpaper or mild polish, and the magnetic particle should be attempted first, before turning to chemical reagents.

Variations of Temperature as a Restorative Technique

Exposure to cold is a nondestructive restorative approach as reported by Cook and Rhoden (1978): "After the area is polished (mirror finish) it is wiped over with dry ice, causing a frosting of the metal. Again, the number may become visible on the frosted surface." Exposure to high levels of heat was also explored by Cook and Rhoden, who wrote that "by heating the metal to a cherry red, the number may be in the heated area." However if the area is overheated, then recovery will not be possible.

Best Practices for Serial Number Restoration

1. Determine where a serial number or other defaced information would have been affixed on the questioned article, usually readily obvious by inspection.
2. Have a scratch pad and writing instrument at hand. Attempt to first fill in information that is legible before any methods are employed. Leave ample space between the characters on the writing surface so that as characters are developed, they may be filled in. If a character is changed, simply write the amended character underneath the prior entry.
3. Prior to any restorative efforts being made, photograph the exhibit thoroughly, with close-up emphasis on the attacked area.
4. Visually examine the area using good, direct lighting. Discern any partial or full digits that may be legible.
5. Clean the attacked site of debris or oxidation by using a nylon brush or other nonabrasive item such as shop towels and any common gun oil or solvent.
6. After the initial exam with the unaided eye, it may also prove useful to use a loupe or magnifying glass to assist in identifying the characters before performing additional work at restoration. Excessive magnification may prove counterproductive as it can become more difficult to discern the numbers.
7. Ascertain what material comprises the attacked surface to assist in decision making about what chemical reagent to use, should that route be chosen.
8. Apply desired restorative technique.
   a. Nondestructive methods
   b. Gentle sanding or filing
   c. Chemical reagents
9. Bring in another set of eyes. Without telling the other person what your results were, ask them to try to discern the information and compare what they see with your results.

10. Continuously document results step by step by photograph or detailed note taking. Be familiar with the terms *obliteration*, *alteration*, *defacing*, and *modification*. Use the term most appropriate to the situation that you are facing.

Photographic Considerations

Firearms bearing obliterated, altered, or defaced serial numbers should be thoroughly documented photographically. A photographic copy stand with a nonreflective background is in order. The author has had great success using black, white, and blue backgrounds, the best results usually being achieved using black for the purposes of contrast. Photographs should include both sides of the firearm to identify the specific design characteristics. Photographs should be taken of identification markings that were not subject to attempts to obliterate, alter, or deface. Finally, photograph the area of attack in specific detail. Remember that these photographs should be taken prior to any restorative work being undertaken, including cleaning the site. Once the site has been cleaned, and before restorative work begins, the area should be photographed again. At various intervals during the restorative work, photographs can be taken. These photographs are intended to assist the examiner, in addition to providing a form of documentation. At some point, the photographs may be useful to view to assist in reconstructing the serial number.

Frozen, Fixed, and Environmentally Exposed Firearms

Firearms are often discarded into bodies of water as a convenient means of disposing of the evidence. Occasionally, firearms are dropped and lost in areas that later flood or may be hidden and left exposed to the elements. Eventually, many such firearms are found when the creek bed dries or someone is mowing and strikes the firearm in the thick brush. Inevitably, these firearms are rusted shut. The first and most immediate concern is rendering the firearm safe. If the firearm is so corroded that the action cannot be opened, it is highly improbable that the ammunition contained inside will have fared any better. Be that as it may, evidence personnel and agency rules and regulations generally have something to say about firearms that are loaded, regardless of their apparent condition. The examiner recognizes the type of firearm and where the ammunition source is contained. As during a routine seizure, the object is to identify and isolate the ammunition source. In the case of pistols, if the magazine cannot be removed because it is frozen into the firearm, removing the magazine floor plate, even destructively, is the easiest way to access the ammunition. Once the floor plate is removed, simply remove the reinforcing plate (if there is one present) and the magazine spring, and then allow the contained cartridges to come out. This will not clear the chamber, but it will isolate these cartridges. Tubular magazines are easily cleared by removing the magazine follower and spilling any contained cartridges out by using gravity.

A firearm recovered in water and which has evidentiary value should be collected in a container that can be filled with the water that the firearm came out of. Most handguns will fit into a general-purpose plastic bucket with a fitted lid; 5-gallon buckets are ideal in these scenarios. Long guns can generally be stored in 4-inch PVC pipe that is cut to length and capped off on either side when the firearm is inserted and the pipe filled with source water. Minimal, if any, exposure to open air should be permitted. Under ideal conditions, the container should be filled with water and the firearm deposited into it right away. The container can then be sealed and removed from the recovery site. Waterborne firearm recovery is no different than firearm recovery under other circumstances. The possibility of evidence remaining, however slight, should be treated as a real possibility. It is entirely possible to recover latent fingerprint evidence from the firearm, magazine, and ammunition. Firearms, even those heavily corroded, can have their markings readily restored by using a nylon or steel brush to remove the outer layer of corrosion; oil or a cleaning agent may be helpful as well. If the examiner has access to an ultrasonic cleaning machine of sufficient size, this works just as well.

To open a firearm that has seized shut from exposure to the elements, the author has immersed the firearm in a bucket filled with power-steering fluid. The primary concern with such firearms is the possibility that the firearm is loaded. The first effort should be made to render the firearm safe by ascertaining if cartridges are loaded, then removing them if they are. As a first preventive measure, the firing train should be disabled.

In the case of a revolver, the grips should be removed, which would reveal the mainspring, which then can also be removed. This removal relieves spring tension from the hammer, essentially rendering it inert. If the revolver has removable side plates, the side plates should be removed and the mechanical linkage (firing train) between the hammer and trigger removed, which isolates the hammer from the trigger. The examiner should ascertain if, for some reason, the firing pin has fixed in the extended position. If this is the case, care must be exercised that the firing pin not be allowed to contact the loaded cartridge; the possibility of a discharge does exist if the firearm is subjected to blows or rough handling. The firing pin can be removed, even if forcibly, from the frame, or it could be blocked using a thin piece of plastic or other material. If the revolver has a swing-out cylinder, the cylinder release latch may work free if allowed to immerse in oil for a period of time. Once the release latch works, the cylinder may be removed with hand pressure or with the assistance of a mallet by tapping on the cylinder to get it removed. Break-open revolvers would require that the action release be worked free if it is frozen, or that the mechanical parts be disassembled to allow the action to open. Solid-frame revolvers are generally the easiest to open if frozen, because the entire cylinder can be removed if the cylinder axis pin can be removed by relieving the release button.

Opening a frozen semiautomatic pistol is approached the same way as a revolver; the first concern is isolation of the ammunition source and rendering it safe. The principal difference is that it is not clear whether there is a live cartridge in the chamber, barring the presence of a loaded-chamber indicator that would tell the examiner otherwise. The examiner should first ascertain if the pistol is breech loaded or magazine fed. In the case of a breech-loaded example, the action-release latch is located and worked to open the action. If it is frozen, sufficient oil should be used to break the mechanism free. If the pistol is magazine fed, is it fed from an internal magazine or an external magazine? In the case of an external magazine, firearms generally have a release latch that will open the magazine bottom and allow the rounds to be removed. Once the ammunition source is isolated and removed, then attention is turned to the action. Semiautomatic pistols generally have reciprocating slides, and the slide will have to be sufficiently soaked or lubricated to allow it to work back and forth to clear the action.

Model Designations

The model is another piece of data that must be conspicuously displayed on the firearm, if a model is assigned. The model can take the form of a name, number, or both. Prior to 1968 in the United States, the model number and caliber were optionally marked, but it was not required by law. The majority of modern firearms have a model designation. In the case of many pistols, as well as rifles and shotguns, there is simply more real estate available on the slide or barrel to place markings than on the frame; thus information such as the manufacturer name, city and state of manufacture, model, and caliber may appear on the slide. However, a serial number must be affixed to the receiver, regardless of where the other information is placed.

Most manufacturers like to maintain a level of continuity within the product lines. Model designations can take the form of letters, numbers, names, or any combination thereof. Models may be marketed under one model designation but conceivably could carry a separate model number for factory use, so it is not uncommon to have a firearm identified by a particular model name or number but also referenced by yet another model by the manufacturer. The Taurus Judge revolver, for example, represents a series of revolvers of numerous model numbers, so designated based upon the build characteristics such as finish, accessories, and other particulars to a specific model within a broader product model. Subvariants of models are common, typically used to delineate a particular product that is somehow unique from the broader model from which it is derived. The Mossberg 500 is another such example. The model 500 is a slide-action shotgun that can be obtained in a multitude of configurations, as submodels of the basic 500 series include the 500A, 500C, 500CG, 500CL, and so forth. Manufacturers guard their trademarks carefully; however, certain names are in the public domain and are not protected, such as the 1911 or 1911A1 designations, which are commonly used by any number of the manufacturers producing a 1911-pattern pistol. Model designations can often incorporate the caliber of the weapon as part of the designation, such as the Masterpiece Arms MPA-380, a MAC-11 clone, MPA meaning "Masterpiece Arms" and 380 referencing the caliber of that firearm, .380 ACP.

As is the case with ammunition, firearms that have been taken into military service will have a military model designation and will not be identified by the commercial model. The Beretta 92, the standard U.S. military sidearm, is designated the M9. The SIG Sauer P225 is given the designation M11 in U.S. military terminology. This is equally true with arms in use by foreign governments. The Walther P.38 was taken into postwar Western German service as the P.1; the P.4 was a P.1 with a slightly shorter barrel. The P.5 was a modernized version of the P.1 and externally did not resemble its predecessors; there were compact and long-barrel versions produced as well. The P.6 was manufactured by SIG Sauer, sold commercially as the P225. The P.7, manufactured by H&K, was only ever designated the P.7, presumably since it was designed expressly for West German police and military forces, and although the P.7 was sold on the commercial market, H&K never saw the need to rebrand the firearm. This was not the case with the P.8, the German Bundeswehr designation for the H&K USP (universal self-loading pistol) chambered in 9×19mm. The Glock 17 is designated the Pistole 80 by the Austrian military, replacing the P.38 that had been in use since the end of the Second World War. The Colt M16 has been manufactured across the globe by various firms under license from Colt. The Canadian firm Diemaco's copy of the M16a2 is identified in Canadian military circles as the C7, itself having variants such as the C7A1 and C7A2. The Diemaco trademark is a capital D in a stylized script. The South Korean Pusan State Arsenal (later Daewoo Precision Industries) produced a copy of the M16a1, distinguished by its Korean ideographs on the receiver, but also marked in English as being manufactured under license from Colt. The Filipino version of the M16 was produced by Elisco Tool Company in Manila and later by the state-controlled defense arsenal. The Elisco variant is distinguished by the receiver markings indicating "Made in Philippines" and under license from Colt.

Discovery of a firearm marked with a "nonstandard commercial designation" should not be construed as a crime at face value. Surplus military and police firearms are frequently disposed of through commercial channels. Beretta USA markets the M9 commercially for consumers wishing to purchase one. Although military model numbers are within the public domain and not considered to be protected intellectual property, there are few examples where such terms are used to market nonmilitary arms, with the exceptions noted. The most likely explanation is to avoid confusion in identification to avoid mistaking a military weapon with a nonmilitary one.

Import Markings

Prior to 1968, firearms imported into the United States were required to have either the original manufacturer's name or the importing company's name affixed, but not necessarily both. Since the enactment of the Gun Control Act in 1968, all firearms imported into the United States must bear the name, city, and state of the importing firm. In addition, the manufacturer, nation of origin, model (if assigned), and the caliber must be identified. These markings must be made onto the receiver/frame, barrel, or slide. In lieu of full names, recognized abbreviations may be used. The model may be an arbitrary name or number that is assigned that firearm, and may or may not represent the actual model as recognized by the nation of origin or other areas, such as the AK-pattern rifles imported into the United States from Romania, which have used numerous model numbers, but are never identified as an AK. There is no legal prohibition as such, as a model is an arbitrary name or number; however, it seems logical that manufacturers and importers would refrain from using such terminology in the interest of avoiding confusion.

Offshore manufacturers that have a domestic branch that handles the importing of arms will ordinarily incorporate import markings into the other information affixed to the firearm. The same model firearm may be imported by different companies, so an individual firearm must be carefully examined to ensure that the importer of record for that arm is properly identified. Foreign-sourced firearms that do not bear import markings were likely imported into the country prior to 1968, but other possibilities cannot be excluded from consideration, such as a firearm that has been smuggled into the country via surreptitious means. Since imported firearms must indicate the nation of origin, it may be possible to encounter nations whose names are different, such as pre-1989 German-sourced firearms marked "West Germany." Import markings may be small, and all possible mark-bearing surfaces must be carefully examined. For identification purposes, the exact manufacturer may not be readily identified; however, the nation of origin can be used in lieu of this information. This is especially true when the arm in question was manufactured by a state arsenal that may not have been identified. Firearms that have significant collector interest may be inconspicuously marked so as not to detract from the collectible desirability of the firearm, while at the same time staying within the letter of the law.

Figures 5.6 and 5.7 show examples of importer information affixed to firearms.

Confusion often arises when a firearm thought to have originated from a foreign source lacks importer information, or when a domestically manufactured firearm bears importation markings. Aside from the reasons previously stated, it is likely that the firearm receiver

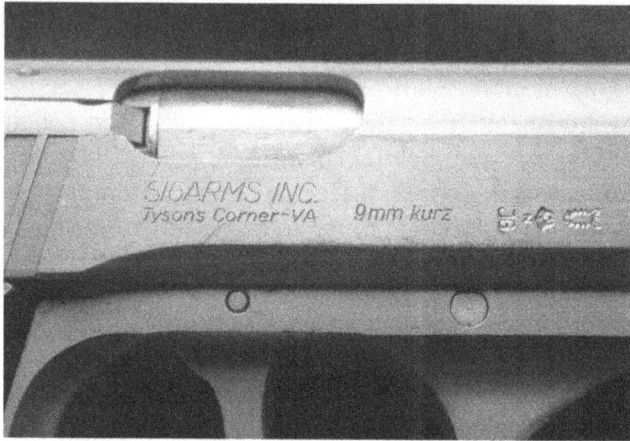

Figure 5.6 Close-up image of the right side of a Sig Sauer P230 slide showing importer information; note also the German proof marks. Compare these proofs to those shown in Figures 5.15 and 5.16. The Eagle over N indicates Nitro proof after 1973; the leaf shows proofing by the Kiel/Eckernforde proof house. The pistol was imported by Sigarms, Inc., and the legend is finely applied to the firearm. (Image from author's collection.)

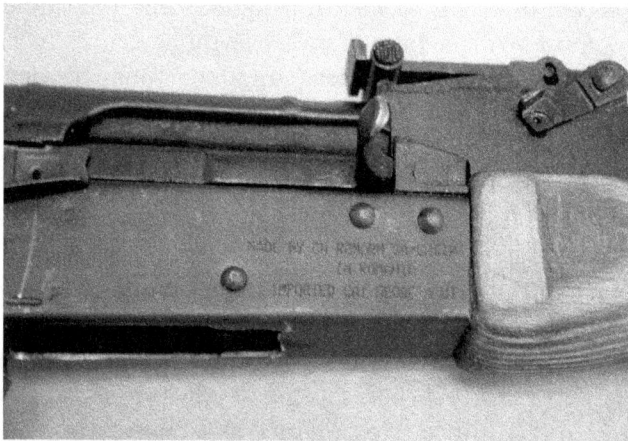

Figure 5.7 Importer information crudely pin stamped to the right side of a Romanian WASR-10 rifle, a semiautomatic copy of an AK rifle commercially imported into the United States. The stamping sources the rifle to have been built by CN Romarm CUGIR in Romania and imported by Century Arms International in Georgia, Vermont. (Image from author's collection.)

was domestically produced, although there may be foreign parts content. Foreign-sourced parts kits, which would include barrels, stocks, and other nonreceiver components, are used to assemble the complete firearm using an American-manufactured receiver. Once again, the golden rule applies: The receiver markings are those used for the legal identification, to the exclusion of any other markings that appear on other parts. Domestically produced firearms appear that have import stamps affixed. In such an instance, the firearm was exported to another country and later reimported into the United States and therefore must have been affixed with the importer stamp. Another clue that can be looked for in such a situation is the presence of proof marks on the firearm that are not particular to the manufacturer but indicate that that weapon underwent an inspection in another country. Firearms manufactured in the United States may bear an importer's marking. If the

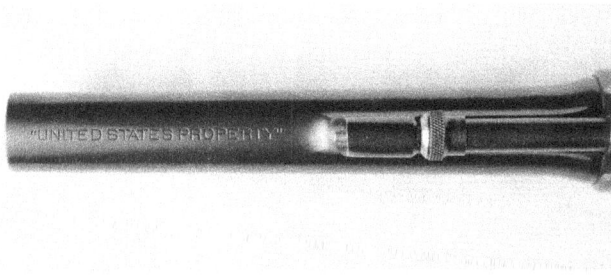

Figure 5.8 Smith & Wesson U.S. Army Model 1917 revolver. "United States Property" is inscribed on the underside of the barrel. (Image from author's collection.)

firearm were commercially exported from the United States to a foreign nation and then reintroduced into the United States, then the firearm must have come through a licensed importer, with the appropriate information affixed.

Other Markings

In addition to the required markings on firearms, secondary markings may be applied by the end user. These markings may be professionally done or be rather crude in appearance. Typically, these markings include police department names in the form of badges, crests, and initials. Other markings that have been observed include inventory or rack numbers, an individual's name, driver's license numbers, asset tag numbers, and even social security numbers have been applied for identification in event of theft or loss. One of the most ironic markings that the author has encountered is a public school system numbered asset inventory control label affixed to a rifle. Certain companies that hold firearms as corporate assets may affix their company name or crest. Manufacturer-affixed markings are encountered on ban-era firearms marked as restricted to law enforcement/government or export only.* Pistol versions of AR firearms may bear an inscription to indicate that the firearm is a "registered pistol" or similar language. Since these markings could be applied postmanufacture by a third party, such markings should be investigated if there is reason to suspect otherwise.

Historically, U.S. martial firearms have borne the inscription "United States Property," indicating that the firearm was military issued (see Figure 5.8). The U.S. Property mark may appear on the receiver, slide, or barrel. Handguns, shotguns, and certain rifles taken into U.S. military service were marked with the property stamp. If there is uncertainty about the actual manufacturer of the firearm, the inspector stamp could be used to create a list of candidate manufacturers. Certain military arms such as the M1 carbine and M1 Garand will not bear the U.S. Property inscription, but martial markings were affixed to the stock, whereas Enfield-pattern rifles manufactured in the United States will bear the U.S. Property marking.

U.S. military arms that passed through U.S. military arsenals to be refurbished may also have one or more arsenal markings present to indicate that they were rebuilt. In either case, as to inspector and proof stamps or arsenal-rebuild markings, this information cannot always be considered reliable, as the stamp may not appear on the firearm itself, instead

---

* The restricted markings are now archaic in the United States. The markings were only relevant 1994–2004, when the legislation expired.

appearing on the stock, which could easily have been replaced and attached to another firearm post-rebuild. By virtue of the fact that U.S. military firearms could have been arsenal refurbished a multitude of times, the receiver must serve as the basis of any definitive firearm identification. This is especially true with Model 1911 and 1911A1 handguns, where slides and frames were routinely swapped by government arsenals, private owners, and gunsmiths over the years. It is not uncommon to see frames made by one company but bearing the slide markings of another company. The natural tendency is to use the slide legend as the identifying information; however the slide markings cannot be relied upon for a definitive identification. Model 1911 and 1911A1 frame manufacturers can be truly identified by proof mark, serial number (different contractors were assigned different serial number blocks, and these were not always sequential), and inspector stamp. This presumption is based upon the premise that the 1911 or 1911A1 in question is an authentic period piece, as modern copies that replicate period examples have been manufactured by Colt, Auto Ordnance, and others.

National Crests

Historically, many nations who took arms into service had their national crest or icon affixed to the firearm. This practice was used to show ownership, instill a sense of pride and patriotism, and even for political purposes. The practice was especially prevalent in the late nineteenth century and up until the middle of the twentieth century. The Mauser 98–pattern rifle was widely exported and used by over 50 nations over its service life that exceeded half a century. Practically every nation that purchased a 98-pattern rifle, whether it was manufactured in Germany or another country, had its national crest affixed to the receiver (see Figure 5.9). It was not uncommon for any European nation to affix its national symbol to a weapon (see Figure 5.10), respective to the era when those crests or emblems were used, especially if the nation in question was a monarchy.

The Soviet Union stamped many, but not all, firearms with a hammer and sickle on the receiver. The presence of such a marking would indicate that firearm was made after the October 1917 revolution that deposed the tsarist rule over the country. The practice continued until the early 1950s, when the hammer and sickle was replaced by a star on the receiver. The practice appears to have ended with the transition to the AK-47, when only arsenal

Figure 5.9 The receiver markings of a Model 1937 Short Rifle, the Portuguese designation for the Mauser 98 pattern taken into service in 1937. The 1937 mark indicates the year of production, a standard marking that appears on Model 98 rifles. The emblem is the Portuguese crest and appeared on Model 98 rifles exported to Portugal from Mauser in Germany. (Image from author's collection.)

Figure 5.10 The Swiss cross affixed to the receiver ring of a Schmitt Rubin rifle. The Schmitt Rubin was a bolt-action rifle, using a straight-pull bolt. These rifles were used by Switzerland for some 80 years in various configurations. The cross is found on other arms used by the Swiss, including Luger pistols and Vetterli rifles. (Image from author's collection.)

markings were affixed to the left side of the trunnion, since there was no room on top of the receiver like there was on a Mosin-Nagant rifle or an SKS (Simonov Carbine Self-Loading).

The Israeli Defense Forces (IDF) emblem—a Star of David and an olive leaf—appears on Galil rifles, Mauser 98 rifles, and others used in IDF service. In addition, nonstandard, arsenal, or field-applied markings can also be observed, particularly on surplus weapons from around the world and may include symbols of faith, breakaway or revolutionary movements, and so forth.

Arsenal Markings

Absent trademarks, nations with state-operated arsenals will use the emblem or icon of the producing arsenal to identify the manufacturer; this is especially the case with Eastern Bloc weapons and other nations where arms production is a state enterprise (see Figure 5.11). Generally the same information can be used to identify both the firearm and ammunition. World War II Japanese arms were marked with the symbol of the producing arsenal, with the balance of the information in kanji (see Figure 5.12). Arsenal markings may also correspond to proof marks, as in the case of Russian firearms.

World War II-era Nazi firearms were stamped using a system of codes that were developed and evolved from the prewar era. Manufacturers involved in war industries were assigned a secret ordnance code. The code system changed from letters, numbers, and a mixed combination of letters and numbers to a series of letters from one to three characters in length. There were literally thousands of codes that covered manufacturers of small arms, ammunition, instruments, and even holsters and field equipment. Nazi-era firearms bearing commercial manufacturer information were typically not taken into military service but were intended for export or use by civic, political, or paramilitary organizations. Outside of collector interest, there is probably little need for an investigator to ascertain the precise manufacturer of such a firearm; often simply identifying the model of the firearm is sufficient. Table 5.1 shows select World War II German ordnance codes.

Figure 5.11 A Polish P64 pistol, based on the Walther PP. The year of manufacture is 1970; the arsenal is indicated by the circle 11 icon, indicating production by the Radom Arsenal. (Image from author's collection.)

Figure 5.12 A wartime Japanese Type 14 pistol, but simply known as the "Nambu," a generic term for the series of pistols of this design. The basic design mimics the Luger or Lahti pistol. (Image from author's collection.)

Table 5.1   Select World War II German Ordnance Codes

Manufacturer	Ordnance Code
Berlin Luebecker Maschinenfabrik	237, duv
Boehmische Waffenfabrik, Prague	fnh
Carl Walther Waffenfabrik	480, ac
Gustloff Werke, Weimar	bcd
J.P. Sauer & Son, Suhl	147, ce
Mauser Werke, Borsigwalde	S243, ar
Mauser Werke, Oberndorf	S/42, 42, byf, svw
Spreewerke	cyq, cvq
Steyr-Daimler-Pugh, Steyr, Austria	660, bnz

Germany was not the only nation concerned with industrial security and its industrial base coming under attack or sabotage. During the Second World War, the United Kingdom also devised a series of factory codes to identify firms engaged in war work. The United Kingdom was divided into three geographic zones or regions: North, Midlands, and the South. Each producer was assigned a letter prefix and a one- to three-digit numerical code, such as M/78, which stood for Elkington & Co. located in the British Midlands; S33 for Essex Engineering Works on the southeast coast of England; or M/47C for Birmingham Small Arms Factory in Shirley. The slash to separate the prefix with the number code was inconsistently used, so it may or may not appear on any marked article (Jones 2010). There are likely hundreds, if not thousands, of these codes that were used, and the exact identity of many of these firms has been lost to history.

Patent Dates, Legends, and Other Information

Firearms typically contain some design feature or other matter that bears protection by the granting of a patent or at least an indication of patent pending. If a patent for said firearm has been issued, the patent number and sometimes a date of issue will appear on the article. If the firearm has been manufactured under license from the patent holder or is using some patented feature licensed by another, then that information will appear on the firearm as well, typically on the receiver. Patent information can be applied for the purposes of identification in several ways:

- The firearm must have been manufactured after the last patent date stamped, giving the examiner a general idea of when the firearm would have been manufactured, at least within a time period of approximately 20 years of the last shown date. Obviously, firearms bearing "Patent Pending" are the earlier examples of such.
- If the firearm was not manufactured by the originating company or if the markings are absent that preclude manufacturer identification, the patent information can be used to trace back to the original firm or individual holding the patent.
- When faced with a legal challenge that the article was designed, redesigned, or manufactured as a firearm, the patent information can readily establish that the article or exhibit in question was in fact defined as such.

Proof Marks

The use of proof marks to identify firearms is well established. No less than Calvin Goddard himself, the father of the science of firearm identification, researched and published extensively on the subject. A proof mark is an indication that a particular firearm or firearm component—such as a barrel, frame, slide, or other piece—was inspected and test fired to ensure the quality of the firearm construction. Various grades of proof were given for various firearms, the proof stamp literally providing prospective buyers with the "proof" or evidence that the purchase they were contemplating had been independently tested and certified to the standard that the proof established. As such, various proof marks were created for arms that used black powder, arms that used smokeless powder, arms that were smooth bore, and so forth. Proofs and proof marking tend to be underwritten by government entities and are subject to formal legislative enforcement where proofing has a long and detailed history for both civilian and military arms, especially in European nations.

U.S. Commercial Proof Marks

The United States has never adopted formal proofing in the European tradition; U.S.-based arms makers did not engage in proofing to the degree of European arms makers, and the U.S. government never enacted laws requiring such markings or inspections. This is not to say that American arms were not tested before leaving the factory, but definitive, standardized proofing in the sense of the European tradition never came into being. Rather than relying on government bureaucracy or third parties, some American manufacturers opted to voluntarily proof mark their products based upon internal quality inspection. The American manufacturers who entered into proofing appear to have followed the standards set by the Birmingham, England, proof house. Remington Arms products are found with a myriad of markings and symbols on the barrel indicating in-house quality assurance inspections, and will be found bearing the REP marking, meaning "Remington English Proof." Colt handguns carry a small triangle bearing the letters VP, meaning "Verified Proof," which are usually affixed on the left-hand side of the frame on the trigger guard. Iver Johnson used as *P* within a circle as a provisional proof, and IPJ as a definitive proof. O.F. Mossberg has used a similar circle *P* for their shotguns. Figure 5.13 shows some of these proofs.

Proof Marks Used around the World

A proof mark can place the manufacture of a firearm within a particular era, or indicate a particular manufacturer or at least geographically where the firearm was made. Collectors, especially those whose interest is focused on military arms, look very closely for the evidence of ordnance stamps and proof marks to ensure that the piece they are contemplating for purchase is "right." Modern-era firearms may bear a plurality of proof marks. The weapon may exhibit proof marks from the factory where it was originally manufactured, but then may bear additional markings indicating it was taken into service in another nation that had additional inspection requirements above and beyond the initial factory testing. This may be further complicated by arms that changed hands again from one user to another, when the subsequent user required his or her own proofing requirements be satisfied. Figures 5.14 through 5.27 show proof marks from Austria (Figure 5.14), Belgium (Figure 5.15), England (Figure 5.16), Czechoslovakia (Figure 5.17), Finland (Figure 5.18), France (Figure 5.19), Germany (Figures 5.20 and 5.21), Hungary (Figure 5.22), Ireland (Figure 5.23), Italy (Figure 5.24), Russia (Figure 5.25), Spain (Figure 5.26), and Yugoslavia (Figure 5.27).

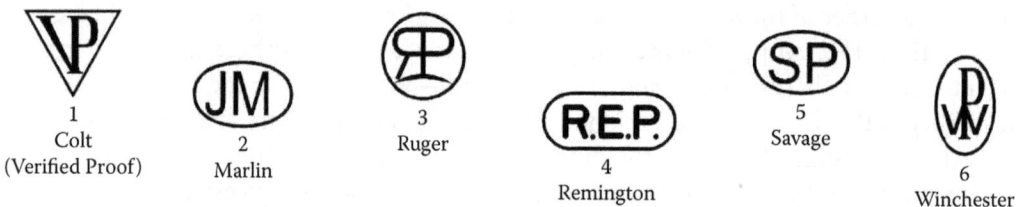

Figure 5.13 Proof marks of select U.S. manufacturers. (Source: BATFE Firearms Identification Guide.)

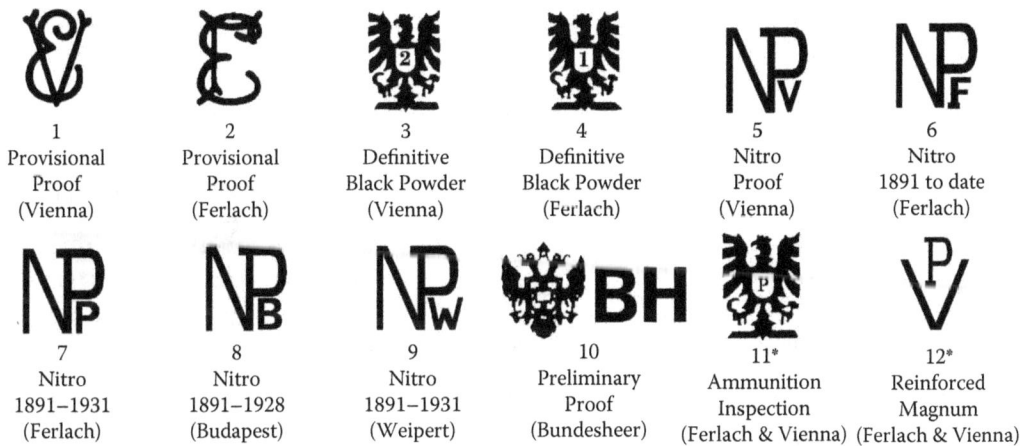

Figure 5.14 Austrian proof marks. * indicates that the proof mark is not unique to a single proof house. (Source: BATFE Firearms Identification Guide.)

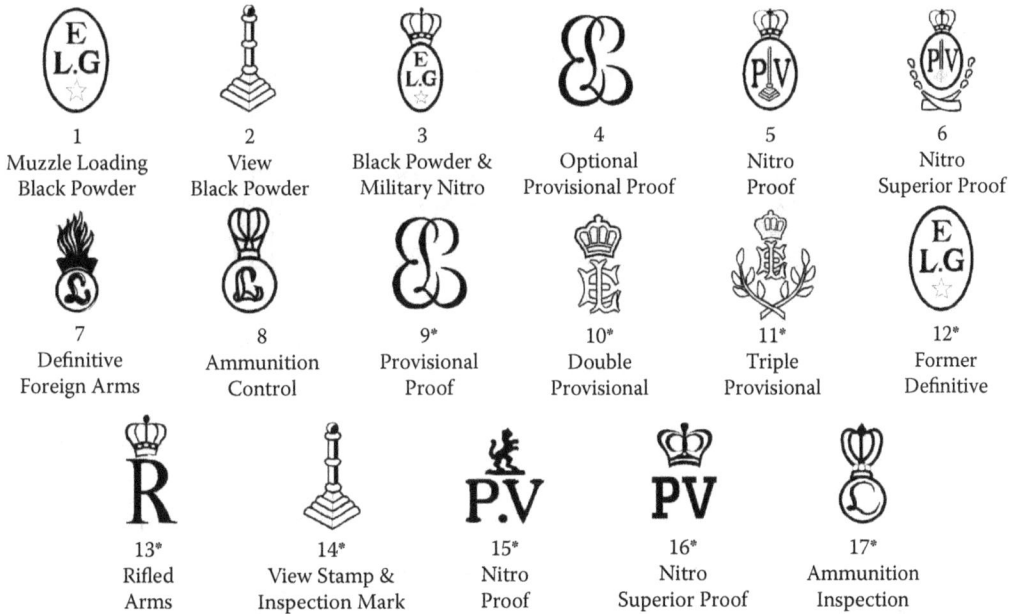

Figure 5.15 Belgian proof marks. * indicates that this proof mark was used prior to 1968, and may still be in use. (Source: BATFE Firearms Identification Guide.)

United States Military

The U.S. military used a system of inspectors to ensure quality standards on martial firearms, and their inspector mark, usually in the form of their initials, graced frames and stocks of firearms that passed through their hands. Two- and three-letter initials were used by the army inspectors, and many of them had long careers. The first proofed martial firearm is thought to have been manufactured around 1800. U.S. military inspectors in that sense phased out as the government left the making of

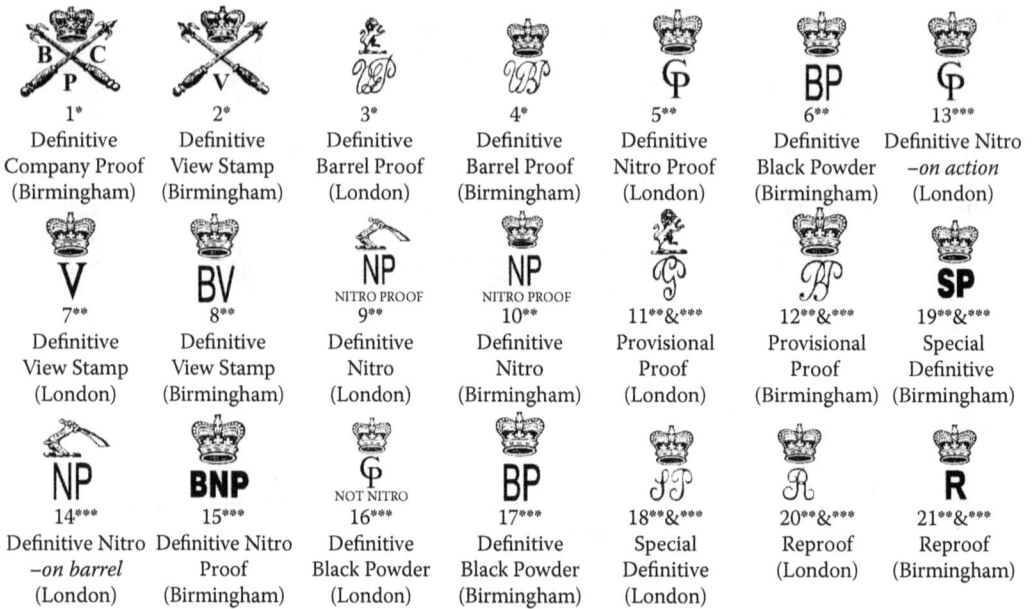

Figure 5.16 British proof marks. * indicates 1904 rules of proof. ** indicates 1925 rules of proof. *** indicates 1954, 1986, and 1989 rules of proof. (Source: BATFE Firearms Identification Guide.)

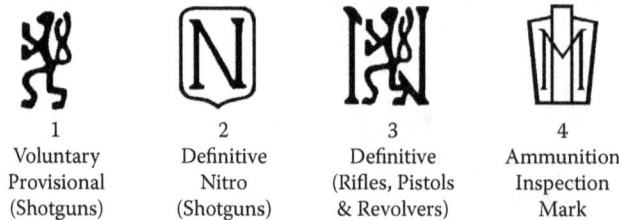

Figure 5.17 Czechoslovakian proof marks. (Source: BATFE Firearms Identification Guide.)

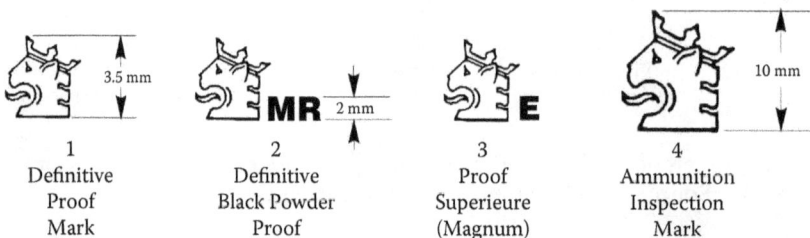

Figure 5.18 Finnish proof marks. (Source: BATFE Firearms Identification Guide.)

military arms to commercial manufacturers. The U.S. martial proof mark was the flaming bomb, the emblem of the U.S. Army Ordnance Corp, and this can be found on frames, receivers, and even smaller parts on certain firearms. Various other forms of the letter *P*, often coupled with a *V* and even an eagle's head have been used by U.S. arsenals going back to the eighteenth century. In more modern times, the U.S.

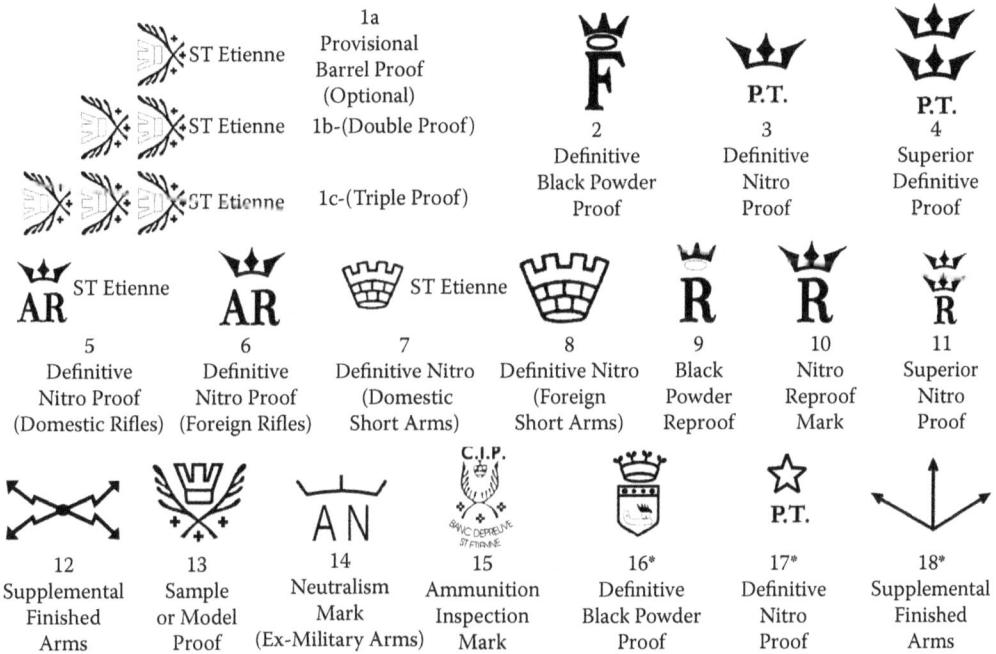

Figure 5.19 French proof marks. * denotes marks used in the proof house at St. Etienne and Paris. The Parisian proof house marks are seldom seen. (Source: BATFE Firearms Identification Guide.)

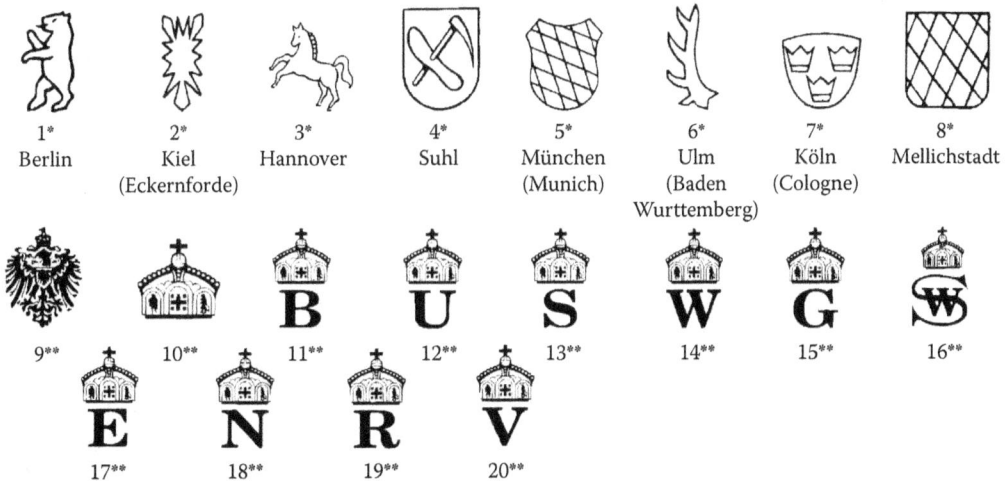

Figure 5.20 German proof marks. * indicates the German proof houses post-1955. ** indicates prewar proof marks. (Source: BATFE Firearms Identification Guide.)

ordnance marking was a *P* surrounded by a circle, which underwent several minor style revisions until the practice was discontinued in 1968. Other forms of U.S. martial proof marking included the use of an eagle head and a number to indicate inspection at the Springfield Arsenal, as well as two- and three-digit inspector name abbreviations. Additional, nonstandard markings would include the names of a ship or base where the firearm was in arsenal inventory.

N	V	J	SP	L		M	FB
21****	22****	23****	24****	25****	21–25	26*****	27*****
Definitive Nitro (post 1973)	Superior Magnum (post 1973)	Reproof Mark (post 1973)	Definitive Black Powder (post 1973)	Definitive Blank Guns (post 1973)	Eagle may appear in stylized form	Provisional Proof (pre 1973)	Voluntary Handgun Proof (pre 1973)

		N	L	V	R		
28*****	29*****	30******	31******	32******	33******	34******	35
Flobert Rifle Proof (pre 1973)	Kiel Oakleaf -Eckernforde- (pre 1973)	Definitive Proof (Suhl)	Definitive Proof (Blank Guns)	Superior Proof (Magnum)	Proof after Repair	Ammunition Inspection (Ulm)	Ammunition Inspection (Berlin)

G	N	R	S	U	W
36******	37******	38******	39******	40******	41******
Black Powder (Rifled Bore)	Nitro Proof Mark	Reproof after Repair	Black Powder (Smoothbore)	Inspection Proof Mark	Definitive Proof (Choked Bore)

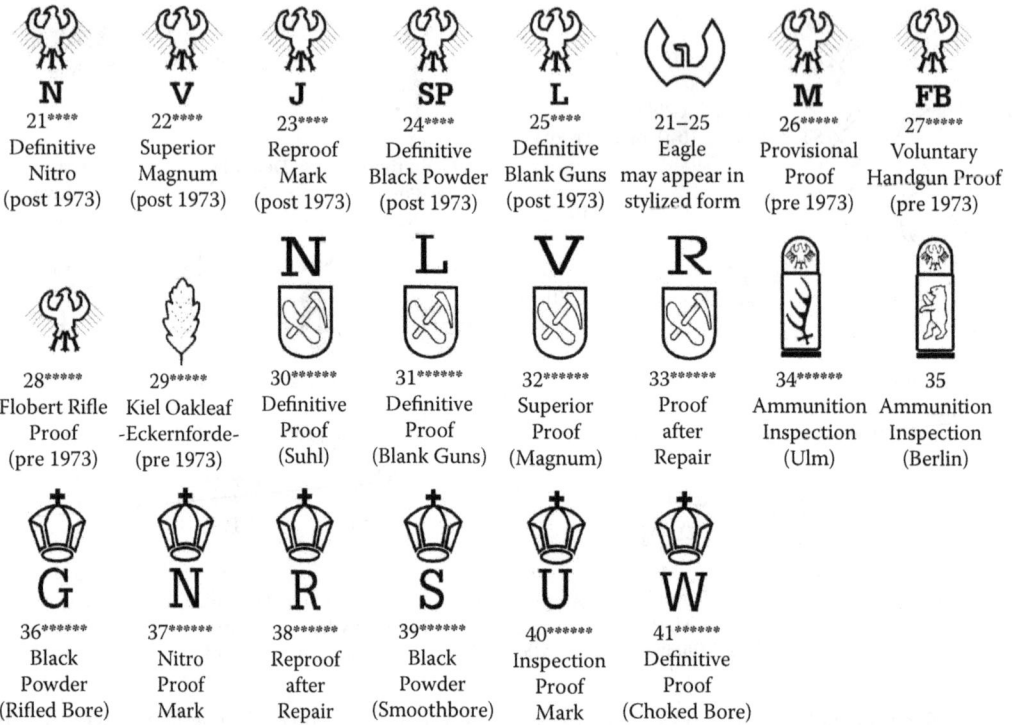

Figure 5.21 Additional German proof marks. *** indicates post-1973 proof for West Germany. **** indicates a pre-1973 proof for West Germany. ***** indicates proof mark used solely by the Suhl proof house. (Source: BATFE Firearms Identification Guide.)

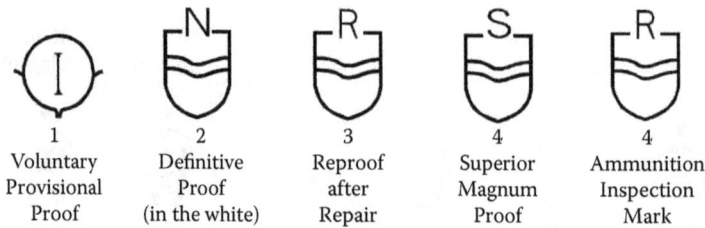

1	2	3	4	4
Voluntary Provisional Proof	Definitive Proof (in the white)	Reproof after Repair	Superior Magnum Proof	Ammunition Inspection Mark

Figure 5.22 Hungarian proof marks. (Source: BATFE Firearms Identification Guide.)

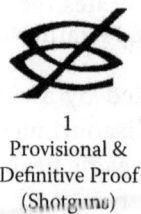

1
Provisional &
Definitive Proof
(Shotguns)

Figure 5.23 Irish proof mark. (Source: BATFE Firearms Identification Guide.)

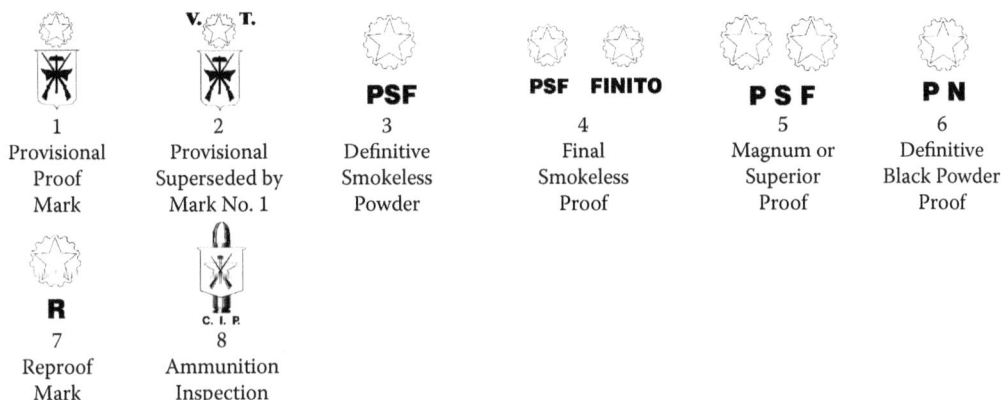

Figure 5.24 Italian proof marks. Note that the provisional mark represented by the crest of arms of Gardone Val Trompia. PSF means *polvere sensa fumo* (smokeless powder). The bore diameter in gauge, caliber, or millimeters may be stamped. The barrel weight (expressed in kilograms) may also be present. A note about Italian proofing: If the firearm was proofed before 1954, Arabic numerals were used to indicate the date. From 1954, a code of Roman numerals was used. From 1975, the year is indicated by a two-digit date in a rectangle. (Source: BATFE Firearms Identification Guide.)

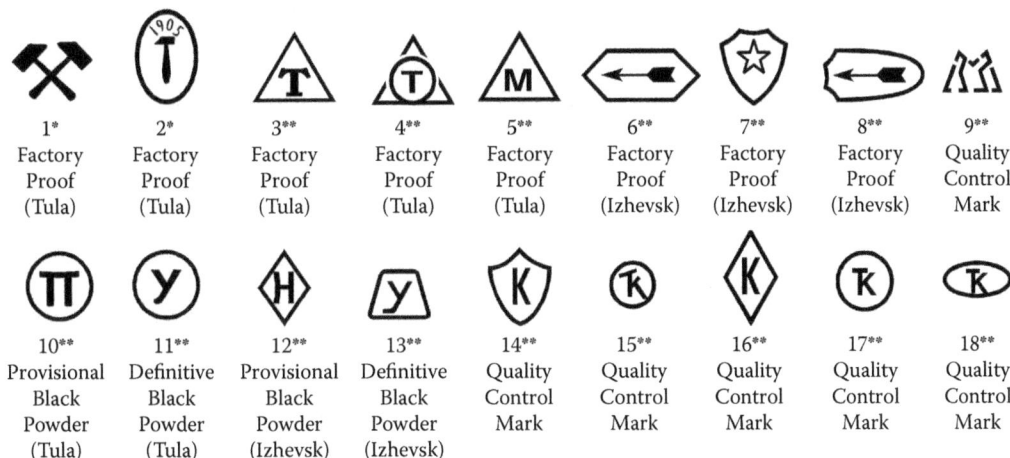

Figure 5.25 Russian proof marks for Tula and Izhevsk. * indicates markings used prior to 1917. ** indicates markings that were used after 1950. (Source: BATFE Firearms Identification Guide.)

Wartime firearms that fell under the wartime Lend-Lease program to the British Commonwealth carry myriad markings. Such arms may be marked "United States Property" and carry U.S. ordnance inspector stamps, but will also appear with markings such as NOT ENGLISH MAKE, foreign proof marks, and other service marks such as RNZAF (Royal New Zealand Air Force), RAAF (Royal Austrian Air Force), and RFC (Royal Flying Corps). The standard British proof mark seen on firearms they used is BNP, meaning British Nitro Proof; typically there is also a stamp that indicates what the pressure test load was, expressed in pounds per square inch. Collectors who specialize in the field can glean additional information from the various markings on such firearms. Surplus U.S. military arms can be found with any variation of markings present, indicating that any number of organizations took the weapons into service after they were released by the military.

1	2	3
Definitive Proof (Unique to Eibar)	Definitive Black Powder (Muzzle-loading Smoothbore)	Provisional Black Powder (Breech-loading Smoothbore)
4	5	6
Obligatory Nitro Proof (Shotgun)	Supplementary Magnum Nitro (B-Shotguns)	Definitive Proof (Small Bore)
7	8	9
Definitive Proof (Rifles)	Definitive Proof (Foreign Arms)	Ammunition Inspection Mark

Figure 5.26 Spanish proof marks for the proof house of Eibar. (Source: BATFE Firearms Identification Guide.)

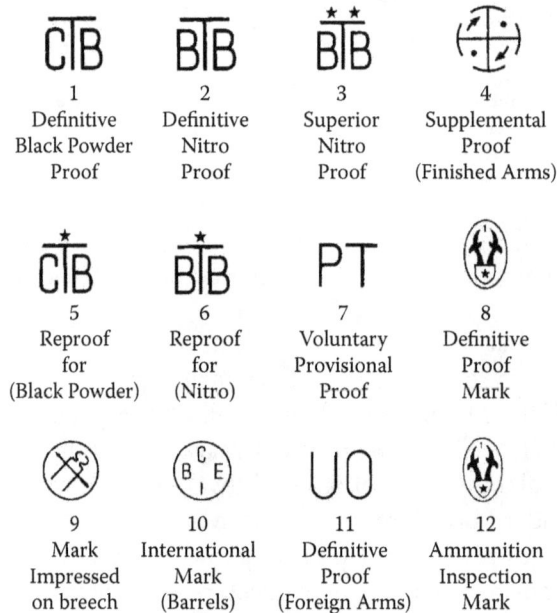

1	2	3	4
Definitive Black Powder Proof	Definitive Nitro Proof	Superior Nitro Proof	Supplemental Proof (Finished Arms)
5	6	7	8
Reproof for (Black Powder)	Reproof for (Nitro)	Voluntary Provisional Proof	Definitive Proof Mark
9	10	11	12
Mark Impressed on breech	International Mark (Barrels)	Definitive Proof (Foreign Arms)	Ammunition Inspection Mark

Figure 5.27 Yugoslavian proof marks. Kragujevec is the established proof house, established in 1969. (Source: BATFE Firearms Identification Guide.)

References

AB Giraites Ginkluotes Gamykla. n.d. About GGG. http://www.ggg-ammo.lt

Alliant Techsystems Inc. 2009. ATK awarded four-year contract to operate Lake City Army Ammunition Plant. http://atk.mediaroom.com/index.php?s=25280&item=58002.

———. 2010a. ATK Receives more than $200 million in orders for small-caliber ammunition. http://atk.mediaroom.com/index.php?s=25280&item=58132.

———. 2010b. ATK begins full rate production of the U.S. Army's new enhanced performance round. http://atk.mediaroom.com/index.php?s=25280&item=58128.

Anders, H. 2006. Ammunition: The fuel of conflict. Oxfam International, Oxford, UK. http://controlarms.org/wordpress/wp-content/uploads/2011/02/Ammunition-the-fuel-of-conflict.pdf.

Anders, H., and R. Weidacher. 2006. *Targeting ammunition: A primer.* Ed. S. Pézard and H. Anders. Geneva, Switzerland: Small Arms Survey. http://www.smallarmssurvey.org/de/publications/by-type/book-series/targeting-ammunition.html?0=.

Arex d.o.o. Šentjernej. 2011. Defence industry solutions. http://www.arex.si/eng/resitve_za_vojaska.php.

Arms Corporation of the Philippines. (2005). Armscor profile. http://www.armscor.com.ph/profile.htm.

Armstrong, D. 2007. M14 to MK 14: Evolution of a battle rifle. Defense Technical Information Center: http://www.dtic.mil/ndia/2007smallarms/5_9_07/Armstrong_12pm.pdf.

Arsenal JSCo. n.d. History of "Arsenal." http://arsenal-bg.com/history.htm.

Assensys B.V. n.d. Overview of technology, process and products. http://www.assensys.nl/products.html.

ATF. 2002. Department of the Treasury. Bureau of Alcohol, Tobacco and Firearms. 27 CFR pt. 178.11.

Bangladesh Ordnance Factories. 2008. About BOF. http://www.bof.gov.bd/about.php.

Barnaul Cartridge Plant. n.d. History. http://eng.barnaulpatron.ru/information/.

BATFE. 1997. ATF ruling and 97-2. Bureau of Alcohol, Tobacco, Firearms, and Explosives. http://www.atf.gov/regulations-rulings/rulings/atf-rulings/atf-ruling-97-2.html.

———. 2006a. Adding a vertical fore grip to a handgun. Bureau of Alcohol, Tobacco, Firearms and Explosives. http://www.atf.gov/press/releases/2006/04/041006-openletter-nfa-adding-vertical-fore-grip.html.

———. 2006b. Flare insert—Any other weapon. Bureau of Alcohol, Tobacco, Firearms, and Explosives. http://www.atf.gov/press/releases/2006/05/050406-openletter-nfa-flare-inserts.html.

———. 2006c. Various AK pattern rifles selector and rear sight markings. Washington DC: BATFE Firearms Technology Branch.

———. 2011. Annual Firearms Manufacturing and Export Report: 2010 Interim. Office of FESD, Washington DC. http://www.atf.gov/statistics/download/afmer/2010-interim-firearms-manufacturing-export-report.pdf.

Berden, H. 1866. March 20. Improvement in priming metallic cartridges. US Patent 53,388, issued March 20, 1866.

Blair, C. 1962. *European and American Arms, c.* 1100–1850. London: B.T. Batsford.

Bowling, M. N., and R. J. Frandsen. 2010. Federal firearms cases, FY 2008. Bureau of Justice Statistics. http://bjs.ojp.usdoj.gov/index.cfm?ty=pbdetail&iid=2175.

Brown, K., and V. Battaglia. 2009. Lightweight Ammunition Design. Presented at the NDIA International Infantry & Joint Services Small Arms Systems Symposium, Exhibition & Firing Demonstration. http://www.dtic.mil/ndia/2009infantrysmallarms/tuesdaysessioniii8550.pdf.

Bumar Amunicja S.A. 2011. History. http://www.mesko.com.pl/en/o-nas/kalendarium.html.

Bumar s.p. z o.o. 2011. About us. http://www.bumar.com/en/about-us/.

Bushmaster Firearms International. n.d. The world's most adaptive modular rifle. http://www.bush-master.com/acr/default.htm#/intro.

Buxton, C., and C. Marsh. 2003. MK 262 MOD 0. Defense Technical Information Center. www.dtic.mil/ndia/2003smallarms/bux.ppt.

Caracal Light Ammunition. 2011. http://www.caracalammo.ae/.

Carr, J. C. 1992. Enforcer conversion: "An obvious switch." *Association of Firearm and Tool Mark Examiners J.* 24 (1): 11–13.

Coal, Ltd. 1996. Home page. http://coal.en.ecplaza.net/.

Colt Defense, LLC. 2003. Colt Infantry Automatic Rifle. http://www.colt.com/mil/CAR.asp.

———. 2011. Colt defense weaponry. http://www.colt.com/ColtMilitary/Products.aspx.

Colt's Manufacturing Company LLC. 2011. Colt classic remake: 1877 Bulldog gatling gun. http://www.coltsmfg.com/About/NewsEvents/tabid/71/articleType/ArticleView/articleId/29/Colt-Classic-Remake—1877-Bulldog-Gatling-Gun.aspx.

Consumer Products Safety Commission. 2010. BB guns can kill. http://www.cpsc.gov/CPSCPUB/PUBS/5089.pdf.

Cook, C. W., and D. R. Rhoden. 1978. A training manual for the restoration of obliterated stamped markings. National Criminal Justice Reference Service: https://www.ncjrs.gov/pdffiles1/Digitization/51869NCJRS.pdf.

Cunningham, J. 1975. Hit gun. *Association of Firearm and Tool Mark Examiners J.* 7 (1): 40–42.

Dakota Ammo. 2012. DPX. Corbon Web site: http://www.shopcorbon.com/DPX-Handgun/200/200/dept (retrieved March 17, 2012)

Dakota Ammo Inc./Glaser, LLC. 2012. Multi-purpose green. Corbon website. http://www.corbon.com/MPG.html.

Defensa Nacional. 2003. Vision general. Industria Militar. http://www.cubagob.cu/otras_info/min-far/industria/industria_militar.htm.

Defense Industries Corporation of Nigeria. 2010. Our mission. http://www.dicon.gov.ng/index.html.

Defense Industries Group. 2006. Ammunition and Metallurgy Industries Group. http://www.diomil.ir/en/amig.aspx.

Department of the Navy Crane Division. 2009. Naval Sea Systems Command Naval Surface Warfare Center crane contract awards list. U.S. Navy Naval Sea Systems Command. http://www.nav-sea.navy.mil/nswc/crane/working/contracting/Lists/Contract%20List/Attachments/144/N00164_09116_NECO.pdf.

Departments of the Army and the Air Force. 1953. *Japanese Explosive Ordnance (Army Ammunition Navy Ammunition)*. Washington DC: United States Government Printing Office.

Doyle, A. C. 1992. *The annotated Sherlock Holmes: The four novels and fifty-six short stories complete.* Vols. 1 and 2, ed. W. S. Baring-Gould. Avenel, NJ: Wings Books.

DRT, LLC. 2011. Technology. Dynamic Research Technologies. http://drtammo.com/technology/.

DuPont. 2011. Heritage First Product: 1802. http://www2.dupont.com/Phoenix_Heritage/en_US/1802_a_detail.html.

Emanuel, L. 1982. A homemade silencer. *Association of Firearm and Tool Mark Examiners J.* 14 (4): 50.

EMPORDEFF. 2007. About us. http://www.empordef.pt/uk/main.html.

Fabrique Nationale Herstal. 2011. Meet the new addition to the FN SCAR family at MILIPOL Paris. http://www.fnherstal.com/index.php?id=750.

FAME S.A.C. 2011. FAME S.A.C. History. http://www.famesac.com/index.php?option=com_content&view=article&id=113&Itemid=53&lang=en.

FBI. 2001. *Forensic science communications*, ed. S. A. Schehl and C. J. Rosati. Federal Bureau of Investigation, Laboratory Services. http://www.fbi.gov/about-us/lab/forensic-science-communications/fsc/jan2001/schehl.htm/.

Federal Premium Ammunition. 2011. Handgun ammunition. http://www.federalpremium.com/products/handgun.aspx.

Felton, F. B. 1895. US Patent 535,097, issued March 5, 1895.

FNH USA. n.d. PS90 Standard Black. http://www.fnhusa.com/le/products/firearms/model. asp?fid=FNG006&mid=FNM0178.

———. n.d. FN 303 projectiles. http://www.fnhusa.com/le/products/firearms/family.asp?fid=FNF0 21&gid=FNG003&cid=FNC01.

Ford, J. 2010. Firearms examiner. Interview by R. Walker. December 15.

F.S.D.I.P. & VBR-Belgium. n.d. Home page: VBR-Belgium. Van Bruaene Rik. http://www.fsdip.com/ and http://www.vbr-belgium.be

General Dynamics. 2001. General Dynamics completes Santa Barbara acquisition. http://www.generaldynamics.com/news/press-releases/detail.cfm?customel_dataPageID_1811=13523,

General Dynamics European Land Systems. 2011. Index: GDELS. http://www.gdels.com/principal/ index.asp.

General Dynamics Ordnance and Tactical Systems. 2011. General Dynamics' small caliber ammunition program manufactures one billion rounds for U.S. Army. http://www.gd-ots.com/ News%20Release/2011/GDOTS%20SCA%201B%20rounds%20produced_03_15_2011.pdf.

Gerns, J. E. 1984. Wounding effects of unconventional ammunition. Association of Firearm and Tool Mark Examiners J. 16 (2): 104.

Glaser Safety Slug, Inc. n.d. Glaser safety slug. Foster City, CA: Glaser Safety Slug, Inc.

Golden Tiger Ammunition. n.d. Home Page. http://www.goldentigerammo.com/.

Government Arsenal of the Philippines. n.d. GA's manufacturing profile. http://www.arsenal.mil.ph/ index.html.

Griffard, P.B., and P.J. Troxell. 2009. Enhancing professional military education in the Horn of Africa. http:// www.csl.army.mil/usacsl/publications/IP_13_09_EnhancingProfessionalMilEdintheHOA.pdf.

H&R 1871. n.d. About us. http://www.hr1871.com/about/default.asp.

Haag, L. C. 1973. Home made suppressor that works. Association of Firearm and Toolmark Examiners Journal 5 (1): 29.

Hamilton, D. T. 1916. Cartridge manufacture, 1st ed. New York: The Industrial Press.

Harrington, G. H. 1871. US Patent 111,534, issued February 7, 1871.

Hast, R. H. 1999. Weaponry: Availability of military .50 caliber ammunition. Washington DC: U.S. General Accounting Office, Office of Special Investigations. http://archive.gao.gov/paprpdf2/162395.pdf.

Heckler & Koch USA. 2011a. Worth the wait…HK's MR556 rifle. http://www.hk-usa.com/civilian_ products/civ_newsroom_01192011.asp.

———. 2011b. USMC exercises Infantry Automatic Rifle full-rate production delivery order to HK. http://www.hk-usa.com/civilian_products/civ_newsroom_10042011.asp.

———. n.d. a. HK416. http://www.hk-usa.com/military_products/hk416_general.asp.

———. n.d. b. HK417. http://www.hk-usa.com/military_products/hk417_general.asp.

Heizer Defense. 2011. Introducing DoubleTap. http://heizerfirearms.com.

Hellenic Defence Systems S.A. 2007. Home page. http://www.eas.gr/index.php?lang=en.

High Standard Manufacturing Company, Inc. 2010. About us. http://www.highstandard.com/about-us.html.

HK Tactical Defense Systems, Inc. 2010. Products: Special purpose: High reliability steel magazine. http://hk-tac.com/item.asp?prodID=180.

Industria Militar. 2011. Products. (ENIGMIND). http://www.indumil.gov.co/.

International Bureau of Weights and Measures. n.d. The Metre Convention. http://www.bipm.org/ en/convention/.

Iver Johnson Arms, Inc. n.d. Home page. http://www.iverjohnsonarms.com/3001.html.

Izhmash OJSC. n.d. Saiga-12 self-loading smooth bored shotgun. http://www.izhmash.ru/eng/ product/saiga12.shtml.

Joint-Stock Co. 2009. Novosibirsk Cartridge Plant: History of the plant. http://www.lveplant.ru/ pages_en.php?id=01.

Jones, R. D. 2010. Email correspondence, March 15.

JSC Sumy Frunze NPO. n.d. A bit of chronology: Frunze Machine Tool Plant. http://frunze.com.ua/ index.php?option=com_content&view=article&id=299&Itemid=57&lang=en.

Kahlbaum, G. W., and F. V. Darbishire, eds. 1899. *The Letters of Faraday and Schöenbein 1836–1862*. Bale: Benno Schwabe; and London: Williams & Norgate.

Kahr Arms. 2011. Kahr Arms pursues patent infringement lawsuit. Kahr News. http://www.kahr.com/kahr-news2011.asp.

Kenya Ordnance Factories Corporation. n.d. About us. http://www.kofc.co.ke/about_us.htm.

KSPZ. 2011. About us: Klimovsk Specialized Ammunition Plant. http://eng.kspz.ru/kspz/gunfactory/about/.

Lamothe, D. 2010a. Corps continues testing so-called green round. *Marine Corp Times*. http://www.marinecorpstimes.com/news/2010/12/marine-corps-tests-new-environmentally-friendly-round-122710w/.

———. 2010b. Corps to use more lethal ammo in Afghanistan. *Marine Corps Times*. http://www.marinecorpstimes.com/news/2010/02/marine_SOST_ammo_021510w/.

LaVoy, T. A. 1979. Unpublished notes.

Leon, J. A. 1998. US Patent 5,705,763, issued January 6, 1998.

Lewis Machine & Tool. 2009. Lewis Machine & Tool Company to introduce new dual operating performance system. http://www.lewismachine.net/media/LMTDualSystem.pdf.

Lewis Machine & Tool Company, Inc. 2011. Fast change barrel integrates the Monolithic Rail Platform. http://www.lewismachine.net/media/PR_Fast_Change_Barrel.pdf.

Lewis Research Center. 1979. Restoration Process. http://www.sti.nasa.gov/spinoff/spinitem?title=Restoration+Process.

Lowden, R. A., and N. Vaughn. 2009. A powder metallurgy approach to non-lead bullets. http://www.ornl.gov/sci/physical_sciences_directorate/mst/SurfacePM/pbullets.shtml

Lutz, M. C. 1978. Velet exploding bullets. *Association of Firearm and Tool Mark Examiners J.* 10 (1): 28.

L.W. Seecamp Co., Inc. n.d. Care & maintenance. http://seecamp.com/caremaintenance.htm.

Marlin Firearms. n.d. The history of Marlin Firearms. http://www.marlinfirearms.com/about/history.asp.

Marsh, C., J. Stoll, and D. Leis. 2009. U.S. Navy small arms ammunition advancement. Presented at NDIA 2009 International Infantry & Joint Services Small Arms Systems Symposium, Exhibition & Firing Demonstration, Tuesday Session. www.dtic.mil/ndia/2009infantrysmallarms/tuesdaysessioniii8524.pdf.

Mauser, P. 1897. US Patent 584,479, issued June 15, 1897.

Memorandum decision. 2005. Memorandum decision on defendant's motion to exclude and recommended decision on defendant's motion for summary judgement, 04-240-P-S. United States District Court of Maine. September 20, 2005.

MIC Sudan. 2007. About Military Industry Corporation (MIC). http://mic.sd/english/about.htm.

Milbank, I. 1872. US Patent 123,352, issued February 6, 1872.

Ministry of National Defense R.O.C. 2011. Defense News: Ordnance Industry of R.O.C. was initiated by the arduous establishment of the 205th Arsenal. http://www.mnd.gov.tw/english/Publish.aspx?cnid=436&p=46371.

Morris, W. H. 1973. A revolver silencer. *Association of Firearm and Tool Mark Examiners J.* 5 (5): 11.

Nammo AS. 2011. NAMMO Group. http://www.nammo.com/Nammo-Group/.

———. 2012. Plastic training ammunition. http://www.nammo.com/Technologies/Plastic-training-ammunition/.

National Rifle Association. n.d. Girandoni Air Rifle as used by Lewis and Clark. http://www.nramuseum.com/the-museum/the-galleries/the-prospering-new-republic/case-28-romance-of-the-long-rifle/girandoni-air-rifle-as-used-by-lewis-and-clark.aspx.

Ordnance Developments Ltd. 2009. Home page. http://www.ordnance.co.nz/index.html.

Ordnance Factory Board, Ministry of Defense, Government of India. n.d. History. http://ofb.gov.in/index.php?wh=history&lang=en.

Pakistan Ordnance Factories. 2010. About us. http://www.pof.gov.pk/Introduction.aspx.

Parks, W. H. 1990. Memorandum for commander, United States Army Special Operations Command: Sniper use of open-tip ammunition. http://www.thegunzone.com/opentip-ammo.html.

PD Igman d.d. 2005. About us. http://www.igman.com.ba/JoomlaInst/index.php?option=com_content&view=article&id=68&Itemid=57.

Prvi Partizan. 2009a. Pistol ammunition with sinter bullets. http://www.prvipartizan.com/frangible.php.

———. 2009b. Company: Prvi Partizan. http://www.prvipartizan.com/company.php.

PT. PINDAD (Persero). 2011. Company profile: Brief history. http://www.pindad.com/showpro1.php?i=1&m=6&u=7&b=2.

Rock River Arms. 2011. RRA LAR-PDS pistol. http://www.rockriverarms.com/index.cfm?fuseaction=category.display&category_id=415

Rodman, T. J., and S. Crispin. 1863. US Patent 40,988, issued December 15, 1863.

Rosenberg, S. F. 1972. Compatible model numbers on Sears-Wards Guns. *Association of Firearm and Tool Mark Examiners J.* 4 (18): 9.

Royal Ammunition. 2009. Home page. http://royalammunition.com/index.asp.

Royal Canadian Mounted Police. 2011. Firearms Reference Table, 4.2. Canada: RCMP.

RUAG Holding. 2011. RUAG Ammotec history. http://www.ruag.com/Ammotec/Ammotec_Home/History.

Sauer, T. W. 2005. US Patent 6,836,991 B1, issued January 4, 2005.

Scanlon, J. J., J. B. Quilan, and E. F. Vanartsdalen. 1965. *Combustible ammunition for small arms. IV: Development of improved obturator devices for caseless ammunition.* Philadelphia: Frankford Arsenal Research and Development Directorate, U.S. Army.

Shipley, P. 2010. Caseless and case telescoped ammunition. Interview by R. Walker.

Skochko, L. W., and H. A. Greveris. 1968. *Silencers principles & evaluations.* Philadelphia: Frankford Arsenal, Department of the Army.

SME Ordnance Sdn Bhd. n.d. About us. http://www.smeordnance.com/front.htm.

Smith, H., and D. B. Wesson. 1854. US Patent 11,496, issued August 8, 1854.

Smith, L. L. 1971. Jokers in the field of firearms examination. *Association of Firearm and Tool Mark Examiners J.* 3 (12): 12–21.

ST Engineering. 2006. Milestones. http://www.stengg.com/about-us/milestones.

Sturm, Ruger, and Co., Inc. 2011. The Ruger LCR double-action revolver. http://ruger.com/products/lcr/.

Sullivan, M. J. 2006. Classifications of devices that are exclusively designed to increase the rate of fire of a semiautomatic firearm. ATF regulations and rulings. ATF Rul. 2006-2. http://www.atf.gov/regulations-rulings/rulings/atf-rulings/atf-ruling-2006-2.pdf.

———. 2008. ATF regulations and rulings. ATF Rul. 2008-1. http://www.atf.gov/regulations-rulings/rulings/atf-rulings/atf-ruling-2008-1.pdf.

Swartz, W. L. 1937. US Patent 2,140,946, issued April 13, 1937.

Thales Group. n.d. Australia>What We Do>Defence. http://www.thalesgroup.com/Countries/Australia/What_we_do/Defence/.

Thompson, R. C. 1978. Sears firearms source and stock numbers. *Association of Firearm and Tool Mark Examiners J.* 10 (2): 54–56.

Thompson, R., and A. Amble. 1979. "Exploder" ammunition. *Association of Firearm and Tool Mark Examiners J.* 11 (4): 19–20.

Tita, G., G. L. Pierce, and A. Braga. 2006. RAND study finds substantial amounts of ammunition bought by felons, others prohibited from buying bullets. http://www.rand.org/congress/|newsletters/safety_justice/1106/ammunition.html.

Truscott, C. J. 2005. ATF regulations and rulings. ATF Rul. 2005-4. http://www.atf.gov/regulations-rulings/rulings/atf-rulings/atf-ruling-2005-4.pdf.

TulAmmoUSA. 2010. Home page. http://tulammousa.com/.

United Nations Comtrade. 2009. United Nations Commodity Trade Statistics Database. http://comtrade.un.org/db/dqBasicQueryResults.aspx?px=HS&cc=93&r=842&p=0&rg=1&y=2010,2009,2008,2007,2006&so=8.

U.S. Army Environmental Command. 2008. Environmental update. http://aec.army.mil/usaec/newsroom/update/fall08/fall0805.html.

U.S. Army 1953. Technical Manual TM 9-1985-5.

U.S. Census Bureau. 2007. Industry quick report. http://smpbff1.dsd.census.gov/TheDataWeb_HotReport/servlet/HotReportEngineServlet?emailname=ec@boc&filename=mfg3.hrml&20120202092457.Var.NAICS2002=332992&forward=20120202092457.Var.NAICS2002

Utrata, D., and M. J. Johnson. 2003. Magnetic particle recovery of serial numbers. Ames Laboratory, Midwest Forensics Resource Center, Des Moines, IA. http://www.osti.gov/bridge/servlets/purl/832893-zGpBRD/webviewable/832893.pdf.

War Department. 1947. *Carbines, cal. .30 M1, M1A, M2, and M3.* Washington DC: U.S. Government Printing Office.

Watson III, A. 1984. *Small caliber ammunition identification guide.* U.S. Army Intelligence Agency, Defense Intelligence Agency. Charlottesville, VA: Foreign Science and Technology Center.

Westenberger, P. H. 1972. Wards Western field manufacturer identification code. *Association of Firearm and Tool Mark Examiners J.* 4 (4): 16.

Wilson, M. 2007. War stretches nation's ammo supply. http://www.usatoday.com/money/economy/2007-05-21-4211226901_x.htm.

Winchester Ammunition. 2010. Winchester Ammunition awarded U.S. Army contract. http://www.winchester.com/library/news/Pages/News-US-Army-Contract.aspx.

Windham, J. 1996. Fire to destruction test of 5.56mm M4A1 and M16A2 rifle barrels. U.S. Army Armament Research, Development, and Engineering Center, Department of the Army. Rock Island, IL: Rock Island Arsenal.

Wolf Harlow, C. 2001. Firearm use by offenders. http://bjs.ojp.usdoj.gov/index.cfm?ty=pbdetail&iid=940.

Wolf Performance Ammunition. 2011. Packaging alert. http://www.wolfammo.com/pdf/WOLF_Packaging_Alert.pdf.

Woods, J. K. 2010. News: The evolution of the M855A1 5.56mm enhanced performance round, 1960–2010. http://w4.pica.army.mil/PicatinnyPublic/news/images/highlights/2011/M855A1/32_The_Evolution_of_the_M855A1_5.56mm_Enhanced_Performance_Round,%201960-2010.pdf.

Zahn, M. S. 1978. Browning Hi-Power automatic pistols: Full auto modification. *Association of Firearm and Tool Mark Examiners J.* 10 (1): 13–14.

———. 1981. Browning Hi-Power automatic pistol: Another type of full auto modification. *Association of Firearm and Tool Mark Examiners J.* 13 (2): 24–26.

Index

026 head stamp, 145
027 head stamp, 145
04 head stamp, 158
05 head stamp, 158
0 8 head stamp, 143
10 head stamp, 150, 156
11 (circled) head stamp, 160
11 head stamp, 161
12 head stamp, 161
13 head stamp, 143
14 head stamp, 161
15 head stamp, 160
179 head stamp, 151, 154
17 head stamp, 150–151, 152
184 head stamp, 151
187 head stamp, 151
188 head stamp, 151, 154
21 head stamp, 159, 160
.22 ammunition for training cartridges, 96
.22 ammunition plating, 99
.22 cartridge interchangeability, 107, 109
22 head stamp, 158, 160
234 head stamp, 160
23 head stamp, 159
26 head stamp, 145
270 head stamp, 151, 161
27 head stamp, 145
304 head stamp, 151
321 head stamp, 160
322 head stamp, 160
323 head stamp, 160
324 head stamp, 160
325 head stamp, 160
.32 cartridge interchangeability, 108
33 head stamp, 140
343 head stamp, 160
37 head stamp, 140
38 head stamp, 151, 153
3B head stamp, 150
3-D Police Ammunition, 100
3 head stamp, 125
3 head stamp, 150, 155
.44 cartridge interchangeability, 107
44 head stamp, 151
.45 cartridge interchangeability, 107–108
46 head stamp, 151
50 head stamp, 151
528 head stamp, 151
529 head stamp, 151
531 head stamp, 151
539 head stamp, 150, 151, 153, 154

540 head stamp, 151
541 head stamp, 151
543 head stamp, 151
544 head stamp, 151
545 head stamp, 151
547 head stamp, 151
54 head stamp, 151, 160
58 head stamp, 151
5N7 cartridge, 71–72
60A head stamp, 146
60 head stamp, 151, 160
611 head stamp, 151
65 head stamp, 129
710 head stamp, 151
711 head stamp, 151, 154
71 head stamp, 158
7.62 rifle marking, 191
7 head stamp, 150, 152
93 head stamp, 141

A

A&D log, 309
A80 head stamp, 143
A A-C. head stamp, 138
AA head stamp, 132, 138
A AI head stamp, 140
AAI/Textron, 57
Abbreviations for ammunition cartridges, 66–67
A.C.B head stamp, 138
Accelerator cartridge, 89
Accidental discharges of revolvers, 171–172
ACE head stamp, 147
AC head stamp, 132
ACP, see Automatic Colt Pistol (ACP)
ACP and Makarov cartridge interchanges, 107
ACP head stamp, 142
Acquisitions and Disposal log, 309
ACR (Adaptive Combat Rifle), 22
Action Express handgun, 23
Action time, 65
Action types, 16
ad head stamp, 134
AD head stamp, 147
ADI head stamp, 126
AE (intertwined) head stamp, 143
AEG, see Automatic Electric Air Gun (AEG)
A.E.P. head stamp, 128
AEP head stamp, 128
AFF head stamp, 126
AF head stamp, 126

AFN head stamp, 141
Aguila subsonic cartridge, 88
A head stamp, 140, 141, 143, 151
AI C head stamp, 140
AIF head stamp, 126
AI head stamp, 140
Air guns, 224–227
    misrepresented as firearms in commission of
        crimes, 226–227
Airsoft guns, 225
    dangers of, 227
AK-pattern firearms; see also Kalashnikov-pattern
        firearms
    assault rifle description, 192
    can't be short-barreled, 196
    magazines, 229
    modification for fully automatic firing,
        253–254
    nation of origin, 289, 290
    selector switch markings, 289
AK series rifles; see also Kalashnikov rifles
    mechanism, 21–22
Albanian ammunition, 125
Alliant Techsystems (ATK) cartridge producers,
    34
Alloy jacketing, 99
Alteration of serial numbers, 311–316
Aluminum cartridge casings, 53–54
AMA head stamp, 132
am head stamp, 134
AM head stamp, 145
Ammunition; see also Cartridge
    American vs. European, 39
    design goals, 37
    failure to fire in revolvers, 172–173
    legal prohibition of possession, 46
    manufacturer match to weapon, 109
    proprietary sources, 50–51
    terminology, 38
    terminology for, 38
    terms for, 38
    types of by purpose, 37–38
Ammunition belts, 233
Ammunition offense categories, 6
Ammunition production, 34–36
Amtech, 57
AMT head stamp, 143
Amursk Cartridge Plant, 152
Anciens Establissements Pieper Hertsal, 128
an head stamp, 134
Anti-hijacking ammunition, 82–83
Antimony in lead projectiles, 102
Antique
    classification, 197
    firearms, 223–224
    replicas, 169
"Any Other Weapons" category, 222–223

AOA head stamp, 139
AOC head stamp, 138
AP, see Armor-piercing ammunition
AP head stamp, 134
ap head stamp, 134
AP head stamp, 142
API, see Armor-piercing incendiary (API)
    ammunition
APM head stamp, 134
APTI, see Armor-piercing tracer incendiary (APTI)
    ammunition
A.P.X. head stamp, 134
A.Px. head stamp, 134
APX head stamp, 134
AR-10 gas piston-driven system, 20–21
AR-15 gas impingement system, 20
Arabic head stamp, 137
    Egyptian ammunition, 132–133
Archaic calibers, 101
ARE head stamp, 132
A.R.E head stamp, 133
Argentinian ammunition, 125–126
Arisaka rifles, 139
Armalite Rifle (AR) manufacturers, 280–285
Arminius, 268–269
Armorers vs. gunsmiths, 12–14
Armor-piercing
    ammunition, 78–79
    incendiary (API) ammunition, 84
    projectile composition, 102
    tracer incendiary (APTI) ammunition, 84
Armscor, 142
Arms Export and Control Act, 167
AR-pattern firearms
    assault rifle description, 192
    can be short-barreled, 196–197
    commonly converted to fully automatic firing,
        244–249
    designation from Armalite Rifle,
        281–282
    indication of conversion, 248
    manufacturers of, 280–285
    two receivers, 282
    variations of, 280–285
Arrow ammunition, 130
Arrow head stamp, 136
Arrows head stamp, 155
Arsenal markings, 327–329
Arsenic in lead projectiles, 102
A R T head stamp, 134
asb head stamp, 134
asr head stamp, 134
Assault pistol, definition, 193
Assault rifle
    colloquial term, 192
    definition, 193
Assault shotgun, definition, 193

Assault weapon term, 192
ATF, see Bureau of Alcohol, Tobacco, Firearms and
      Explosives (BATFE)
ATF Curios and Relics List, 90, 197
ATF serial number prefix, 309
A T head stamp, 144
ATK, see Alliant Techsystems (ATK)
ATS head stamp, 134
ATZ head stamp, 161
Australian ammunition, 126
Austrian ammunition, 127
Austrian firearm imports into the U.S., 166
Austrian proof marks, 331
Austrian Steyr AUG gas-piston system, 21
Auto-loading handguns, 178
Automatic Colt Pistol (ACP) popularity, 37
Automatic Electric Air Gun (AEG), 226
Automatic weapons
      field testing, 239–240
      operation of, 20
      simulation of automatic operation, 259–260
Auto Rim cartridges, 107
auu head stamp, 134
aux head stamp, 134
auy head stamp, 134
auz head stamp, 134
A.V.E. head stamp, 134
AVE. head stamp, 134
av head stamp, 134
AVIS head stamp, 134
A.V.I.S. head stamp, 134
avt head stamp, 134
avu head stamp, 134
avy head stamp, 134
AW head stamp, 147
awt head stamp, 134
Axis pin, 173
axq head stamp, 134
aym head stamp, 134, 157
AYR head stamp, 141
A.Y.R head stamp, 141
AZF head stamp, 127

B

Bacon, Roger, 63
Baffled suppressor, 210
Bag-type projectiles, 94
BAI head stamp, 140
Bakelite firearm components, 164
Ballistics, 7–8
Ballistite, 64
Ball projectiles, 67–72
      shapes of, 70
Banana clip, 227
Bangladeshi ammunition, 127
Bar coded information on firearms, 311

Bare lead projectiles cause fouling, 98
Barnaul
      ammunition producer, 34
      shot shells, 55
Barnaul Machine Tool Plant, 152–156
Barnes Bullets lathe-turned projectiles,
      102–103
Barnes TSX projectile, 73
Barnes VOR-TX projectile, 73
Barnes XLC coated projectiles, 100
Barnes X projectiles, 100
Barrel attachments as suppressors, 213–214
Barrett Firearms Manufacturing, 43
Barricade-breaching operations using frangible
      projectiles, 81
"Barrier blind" projectile, 76–78
BATFE, see Bureau of Alcohol, Tobacco, Firearms
      and Explosives (BATFE)
Baton-type kinetic-impact projectiles, 95
Bazookas as destructive devices, 205
BB guns, 225
      dangers of, 227
BB head stamp, 126
B.B head stamp, 135
B blasting powder, 65
bd head stamp, 134
BD head stamp, 148
Bean bag projectiles, 94
Bear brands of ammunition, 153
BEAUX head stamp, 138
Bee Miller, 28
Behaviorisms, 8
be head stamp, 134
BE head stamp, 148
Belgian ammunition, 127–128
Belgian proof marks, 331
Belt buckle guns, 222
Belted casings, 48–49
Belts, 233
      plastic links for, 162
Benelli Renaissance Classic, 31
Benelli Super Nova, 32
Berden priming, 61–62
Beretta 96, 18
Beretta Holding Group, 133
Beretta Model 92
      blowback operation, 25
      delayed blowback, 25
      multiple manufacturers, 278
      restrike capability, 18
Bertram Bullet Works, 126
Beryllium armor-piercing projectiles, 78
Beschuss head stamp, 135
Beverage bottle suppressors, 210–211
BFA, see Blank fire adapter (BFA)
bf head stamp, 134
bg head stamp, 134

B head stamp, 127, 140, 147, 151, 157
Big bore revolvers, 178
Bimetal casings, 54
    Chinese, 130
Bismuth
    in lead projectiles, 102
    powder in chemical munitions, 95
    shot pellets, 103
bj head stamp, 134
bk head stamp, 134
BKIW head stamp, 135
Black powder, 62–63
    blasting powder, 65
Blank cartridges, 86–88
Blank fire adapter (BFA), 56
Blasting powder, 65
Blaze-orange muzzle cap on AEGs, 226
Blazer aluminum casings, 53–54
Blazer jacketed cartridges, 98
Blended-metal projectiles, 82
Blowback method of operation, 23–25
    additional forms of, 25
    delayed, 24–25
    retarded, 25
    simplicity of, 24
    vs. recoil operation, 25–27
Blowback pistols, see Recoil operation
Blow-forward action, 27
Bluing, 163
BMG (Browning machine gun), see Machine guns
BM head stamp, 146
BML Tool & Manufacturing, 58
BMRC head stamp, 147
bne head stamp, 134
Boat-tail projectile, 71–72
BOF head stamp, 127
BOLIVIA head stamp, 128
Bolivian ammunition, 128
Bolo rounds, 93
Bolt-action
    firearms, 31–32
    rifles, 190–191
    shotguns, 201
Bolt system roller-lock, 24–25
Bonderizing, 163
Bore, 44
Boxer priming, 62
bp3 head stamp, 152
B P D head stamp, 138
B.P.D. head stamp, 138
B P head stamp, 138
Brass
    armor-piercing projectiles, 78
    casing material, 52–53
    term for spent casings, 52
Brass casings
    military applications alternatives, 58

rolled brass, 51
    shotgun, 55
    supplanted by other materials, 52
Brazilian ammunition, 128
Brazilian firearm imports into the U.S., 166
Break action, 30–31
Break-top revolvers, 173–175
Break tops, 30–31
Breech face markings on cartridge casings, 7–8
Breech-loaded drillings, 203
Breech-loaded firearms, 30–31
    handguns, 178, 187
    revolvers, 168
    rifles, 191
    shotguns, 200
BREN light machine gun, 49
Brenneke-style slug, 91–92
Briefcase guns, 222
British ammunition, 147–148
British proof marks
    on firearms, 332
    in head stamps, 142
Broad-arrow proof mark, 136
Bronze armor-piercing projectiles, 78
"Broom handle" cartridge, 48
Browning, John, 179
Browning action, 25
Browning Buck Mark, 23
Browning development, 179
Browning Hi Power
    mechanism of, 25
    modification for fully automatic firing, 257–258
    P35, HP, Model 1935, 179–180
Browning Model 1911, 25
Brown powder, 63
Bryco Arms, 27–28
B.S.A head stamp, 147
B T head stamp, 145
Buckle guns, 222
Buckshot, 90–91
Buffer systems, 284
Bulgarian ammunition, 156–157
"Bull barrel," 214
Bullet as ambiguous term, 38, 102
Bull pup rifle, 194
Bump-fire principle, 260
Bureau of Alcohol, Tobacco, Firearms and
        Explosives (BATFE)
    ATF Curios and Relics List, 90, 197
    crime data, 36
    tracing of firearms, 309–310
Burgo, 268–269
Burkina Faso ammunition, 128–129
Burmese ammunition, 129
Burst firing of machine guns, 216
    components for, 245–246
Bushmaster ACR, 22, 297

Butt location for revolver serial numbers, 306–307
bxn head stamp, 157
byf head stamp, 133

C

C.AA head stamp, 138
CAC head stamp, 140
C A C head stamp, 140
C.A.C head stamp, 140
C A head stamp, 138
Caliber
    ambiguity of term, 38–40
    dual labeling with metric, 43
    interchangeability, 106–110
    length not a factor, 40
    for submachine guns, 216–217
CalWestCo., 27–28
Cambodian ammunition, 129
Camera guns, 222
Cameroonian ammunition, 129
Camouflage patterns on firearms, 234
Canadian ammunition, 129
Canadian Industries, 129
Cane guns, 222
Cannelures, 67
Cannons as destructive devices, 205
Cans, 206
Canuck ammunition, 129
Capacity limits imposed on magazines in the U.S.,
        230
CAR-9, 24
Carbine, 193–194
    definition, 17
Carcano bolt-action rifle, 31–32
Cartoucherie Voltaique (CARVOLT), 129
Cartridge capacity of revolvers, 168
Cartridges; see also Ammunition; specific
        designations
    abbreviations on, 66–67
    bottleneck casing, 47–48
    caliber and metric designations, 43
    caseless-telescoped, 59
    casing markings, 7–8
    casings, 46–58
    center-fire, 61
    center-fire introduction, 51
    chambering for interchangeability, 107–108
    components of, 46
    composite casings, 51
    electrically primed, 66
    fixed ammunition, 46
    functions of, 46–47
    historical development, 49–50
    hyphenated labeling, 40–41
    identification markings, 110
    identification of, 118, 123–125

interchangeability, 109
intermediate, 109
M855 cartridges, 68
markings, 66–67
metric and caliber designations, 43
metric designations, 42–43
naming, 41–42
Parabellum, 42–43
primer hardness, 172
purple ring, 149
recycling, 61–62
red ring, 149
rim designs, 48–49
rim-fire, 62
rims for revolvers, 177–178
shelf life, 61
size relative to chamber, 106
small sizes for Personal Defense Weapons
        (PDWs), 109–110
subsonic, 88
suffixes in labels, 43–44
tallow in, 51
technology development, 33–34
term ambiguity, 38
CARVOLT, 129
Cascade Cartridge International (CCI), 53–54
Cased-telescoped cartridge, 57–58
Case hardening, 163
Caseless ammunition, 58–59
Caseless-telescoped cartridge, 59
Casings
    Chinese-produced, 130
    coloring of Chinese-produced, 130
    modular, 58
    shifts in materials, 52
    use by other manufacturers, 118
Categories of crimes, 5–6
CAVIM head stamp, 148
CB&A head stamp, 157
CBC, see Companhia Brasileira de Cartuchos (CBC)
CCI, see Cascade Cartridge International (CCI)
CCMCK, see Close Combat Mission Capability Kit
        (CCMCK)
CD. head stamp, 138
cdp head stamp, 134
Cellular phone guns, 222
Center-fire
    cartridges, 61
    priming, 61–62
Centurion shot shells, 93
C F head stamp, 134
C.F. head stamp, 134
cg head stamp, 134
CG head stamp, 145
Chambered suppressor, 210
Chambering for cartridge interchangeability,
        107–108

Chamber variations and modifications, 51–52
Charge holes, 168
Charge vessel, 58
Charles Daley, 127
Charles Fusnot, 128
Chassis conversions, 234–235
C head stamp, 129, 150
Chemical agent delivery system, 95
Chemical munitions, 93, 95
Chemical obliteration of serial numbers, 315
Chemical restoration of serial numbers, 316–317
ch head stamp, 128, 134
Chilean ammunition, 130
Chilled shot, 103
China Sport, 130
Chinese
    paper-hulled shot shells, 55
    penetrator projectiles, 69
    use of counterfeit head stamps, 116–117
Chinese ammunition, 130–131
    banned by BATFE, 130
    red casing, 147
    Soviet head stamp practice used, 130
Chinese language head stamps, 130
Chokes, 204–205
Cigarette lighter guns, 222
CIM FNT head stamp, 144
CIM head stamp, 144
CIM-PS head stamp, 144
CIP, see Permanent International Commission for
        Firearms Testing (CIP)
Circle 11 head stamp, 160
Circle head stamps, 145, 156
Circle P proof mark, 330
cJ head stamp, 129
Classification of firearms, 167; see also specific types
    handguns vs. long guns, 198–199
    incomplete receivers, 262
    machine gun modifications, 245–246
Claw projectiles, 89
Clean Fire ammunition, 81
Clips, 231–232
    definition, 227
Close Combat Mission Capability Kit (CCMCK), 98
Closed-bolt weapons, 182
    field testing, 239–240
    fully automatic firing conversion, 238–239
    slam fires, 238–239
Close Quarter Training (CQT) ammunition, 96–97
CMC head stamp, 160
CNC machining, 164
CN tear gas, 95
Cobra Enterprises, 28
Cobra marking, 274
Cobray, 274
Cocked weapon safety, 175
CO head stamp, 144

Collapsible stocks, 281
Colloid in smokeless powder formulation, 64
COLOMBIA cartridge casing, 131
Colombian ammunition, 131
Color changes on firearms, 234
Color-coding of ammunition not standardized, 110
Colored handguns, 234
Colt Army revolvers, 107
Colt-Browning M1919 cartridge, 49
Colt-Browning Model 1911
    persistence of design, 15–16
Colt Defense, 58
Colt firearms licensed copies, 277
Colt Huntsman, 23
Colt Model 1905, 27
Colt Model 1908 Vest Pocket, 27
Colt Model 1911
    development, 179
    mechanism of, 25
    modification for fully automatic firing, 257
    multiple manufacturers, 276–277
Colt Model 1911A1 multiple manufacturers,
        276–277
Colt modular weapons systems, 283
Colt pistols popular, 37
Colt revolver serial number, 306
Colt submachine gun, 24
Colt Walker single-action-only revolver, 17
Colt Woodsman, 23
Commonly converted firearms, 240–255
Companhia Brasileira de Cartuchos (CBC), 128
    composite jacketing, 100
    hollow-point solid copper projectiles, 102
    shot shells, 55
Comparative ballistics, 7–8
Composite jacketing, 100
Compressed gas guns, 224–227
Computer numerical control (CNC) machining, 164
Concorde brand ammunition, 126, 142
Contraband firearms
    categories of crimes, 5–6
    modifications leading to, 6–7
Controlled-expansion projectiles, 72–75
    copper projectiles, 102
    designed to appear otherwise, 75
    Hague Convention prohibition, 71, 73
    semijacketed cartridges, 99
Conversion of semiautomatic firearms to automatic
        firing, 216, 218
Conversions commonly performed, 240–255
Conversions to fully automatic machine guns,
        237–239
"Cook off," 57–58
    avoiding in machine guns, 238
Copies of military weapons, 195–196
Copper
    armor-piercing projectiles, 78

jacketing, 99, 149
in lead projectiles, 102
projectiles, 102–103
Corbon
DPX hollow-point copper projectiles, 102
Glaser Safety Slug, 83
MPG cartridges, 81
subsonic cartridge, 88
Cordite, 65
Corrosive priming compound, 150
Corrugated casing, 86
Cottage industry in improvised firearms, 221
Counterfeit head stamps, 116–117
Country-specific cartridge head stamps and
manufacturing practices, 125ff
Country-specific military cartridge labeling, 44
Cowboy calibers, 101, 108
CP 99 head stamp, 138
CP head stamp, 134
C-P head stamp, 147
CQT, see Close Quarter Training (CQT)
ammunition
Crane, 173
Crescent moon head stamp, 146
C.R head stamp, 138
Crime categories, 5–6
"Crime conducive" firearms, 36
Criminal acquisition of ammunition and firearms,
36
Criminal ammunition preferences, 36
Criminal instrument and categories of crimes, 6
Cross-bar safety
in long guns, 189
pistols, 181
Crossed swords head stamp, 143
Crown head stamp, 132, 137, 141
CS tear gas, 95
Cuba head stamp, 132
Cuban ammunition, 131–132
Curio classification, 197
Curios and Relics List, see ATF Curios and Relics
List
Cut shot, 93
Cutting Edge Bullets, 102–103
CV head stamp, 128–129
Cylinder bolt, 170
Cylinder catch, 170
Cylinder indexing, 177
Cylinder lock up, 170
Cylinder rotation direction in revolvers,
169–170
Cyrillic characters on Soviet ammunition, 116,
150–151, 157
Cyrillic head stamp, 161
Czechoslovakian proof marks, 332
Czech Republic ammunition, 157–158
czo head stamp, 157

D

D&C head stamp, 135
D 02 head stamp, 143
DAG head stamp, 135
DA head stamp, 129
D.A. head stamp, 129
D A head stamp, 145
DAI head stamp, 140
Daisy V/L caseless ammunition rifle, 59
D.A.L. head stamp, 129
Danish ammunition, 132
DAO, see Double-action-only
DAQ head stamp, 129
Daubert Standard, 11–12
dbg head stamp, 134
DCCo head stamp, 129
D.C. Co head stamp, 129
D.C.E. head stamp, 138
D C head stamp, 129
DC head stamp, 129
DEactivated WAr Trophy (DEWAT), 262
Decocking lever, 181
Defacing of serial numbers, 311–316
Deflagration, 20
Department store firearms, 269–271
Depleted uranium armor-piercing projectiles,
78
Derringers, 168, 173, 187
Desert Eagle handguns, 22–23
Designers of firearms, 165
Destructive device classification, 203
Destructive devices defined, 205–206
Devastator cartridge, 84–85
DEWAT, see DEactivated WAr Trophy
(DEWAT)
DF head stamp, 136
D head stamp, 129, 135, 140
Diamond head stamp, 143
DIAS, see Drop-in auto sear (DIAS)
Digital imagery for reading obliterated serial
numbers, 317–318
DI head stamp, 129, 136
Direct gas operation, 20–21
DISA head stamp, 132
Disconnector, 238
Disintegrator ammunition, 81
DI Z head stamp, 129
dma head stamp, 134
dnf head stamp, 134
DNG head stamp, 135
dnh head stamp, 134
DNL head stamp, 143
DO head stamp, 140
dom head stamp, 134
Dominican Republic ammunition, 132
DOMINION head stamp, 129

Door-breaching operations with frangible
      projectiles, 81
Dot-matrix engraving of serial numbers,
      310–311
Dot peening of serial numbers, 310–311
Double-action firearms mechanism, 18
Double-action-only, 29–30
Double-action revolvers, 177–178
Double-action safety systems, 185–186
Double-action trigger types, 18
Double-based gunpowder compounds, 64
Double feed malfunction, 3
DoubleTap, 187
dph head stamp, 134
DPX hollow-point copper projectiles, 102
Dragunov sniper rifle, 49
Dram equivalent in shot shells, 45–46
Dreyse needle system, 50
Drilled cartridges, 86
Drilling firearm, 203
Drilling out serial numbers, 314
Drop-in auto sear (DIAS), 246–247
      for AK-pattern firearms, 254
      for AR-pattern firearms, 246–247
DRS head stamp, 132
DRT, see Dynamic Research Technologies (DRT)
Drum magazines, 229–230
D T head stamp, 145
dtp head stamp, 157
Dud cartridge, 3
Dum dum, 73
Dummy cartridges, 86
Duplex cartridges, 92–93
Duplex shot shells, 92–93
DuPont Company, 65
Durofol firearm components, 164
Dutch ammunition, 140
Duty cartridges, 81
DWA head stamp, 135
D W A head stamp, 135
D.W.A. head stamp, 135
DW head stamp, 135
DWMB head stamp, 135
DWM head stamp, 135
DWMK head stamp, 135
Dye components of bag-type projectiles, 94
dye head stamp, 134
Dynamic Research Technologies (DRT) blended-
      metal projectiles, 82
Dynamit Nobel; see also RUAG Ammotec Group;
      RWS
      caseless ammunition design, 59
      partner in Malaysian ammunition manufacture,
         139
      polymer-cased cartridges, 56
      subsonic cartridge, 88
   dza head stamp, 134

E

Eagle ammunition, 138
Easter European ammunition, 148–150
Eastern Arms, 269
East German ammunition, 158
eba head stamp, 134
EB head stamp, 147
ecc head stamp, 134
ecd head stamp, 134
EC Eley head stamp, 147
Ecology Line cartridge, 98
ECP head stamp, 134
edg head stamp, 134
E.D.P. head stamp, 136
edq head stamp, 134
eeg head stamp, 134
eej head stamp, 134
eem head stamp, 134
eeo head stamp, 134
eey head stamp, 134
EG. head stamp, 135
Egyptian ammunition, 132–133
E head stamp, 137, 139, 147, 151
EIGN head stamp, 161
Ejector marks on cartridge casings, 7–8
Ejector star, 173
EK head stamp, 136, 145
Electric priming, 65–66
Electroscribing of serial numbers, 310–311
ELEY BROs. head stamp, 147
ELEY head stamp, 147
Eley subsonic cartridge, 88
emp head stamp, 134
EMZ head stamp, 140
Enameling, 163
En bloc clip, 231–232
Enfield cartridges, 49
Enfield revolvers, 173
Enfield/Snider rifle, 51
Enforcer, 274
Enhanced Performance Round (EPR), 69–70
ENK head stamp, 136
eom head stamp, 134
EPR, see Enhanced Performance Round (EPR)
ESCUDO head stamp, 128
E S head stamp, 138
Ethiopian ammunition, 133
e-Trace program, 309–310
Etron X, 66
Evidence
      affected by handling, 4–5
      exculpatory, 10–11
      rules of, 8–9
Evidentiary concerns vary by type of crime, 6
Evidentiary inspections, 2–6
Exculpatory evidence, 10–11

Expert opinion, 9–10
Expert testimony standards, 11–12
Exploder cartridge, 84
Exploding bullets, 73
Explosive projectiles, 84–86
Extraction marks on cartridge casings, 7–8
ExtremeShock
    frangible projectiles, 82
    subsonic cartridge, 88

F

faa head stamp, 134
Fabico, 268–269
Fabrica Nacional de Cartuchos y Municoes, 128
Fabrica Realengo, 128
Fabricas y Maestranzas del Ejército (FAMAE), 130
Fabrique Nationale (FN)
    5.7 cartridges, 109
    Belgian manufacturer, 127
    chemical agent delivery system, 95
    Five-seveN pistol, 24
    pistol development, 179
Factory remanufactured ammunition, 105
fa head stamp, 134
Failure to fire in revolvers, 172–173
FAL-pattern rifle, 293–294
FAMAE head stamp, 130
FAME head stamp, 142
Farsi head stamp, 137
F.C.A.G. head stamp, 128
F C head stamp, 143
fde head stamp, 134
fd head stamp, 134
Federal Cartridge, 55
    Guard Dog cartridges, 74–75
    Nyclad projectiles, 100
    subsonic cartridge, 88
Federal Hyrda-Shok projectile, 73
Federal Rules of Evidence definition of expert, 8–10
Federal State Enterprise Production, 152
Feed ramp design and projectile shape, 72
Feed-ramp marks on cartridge casings, 7–8
Feed-ramp misfeeds, 28
fee head stamp, 134
FEG head stamp, 159
Felton Device, 175
fer head stamp, 134
FFV head stamp, 145
F-GY head stamp, 159
F head stamp, 130
Fiberlite stock, 281
Field-testing automatic weapons, 239–240
FI head stamp, 146
Filing serial numbers, 314
Filling serial numbers, 314–315
"Finger stack," 260

Finish modifications, 234
Finnish ammunition, 133
Finnish proof marks, 332
Fiocchi, 55
FIOCCHI head stamp, 138
Firearm designers, 165
Firearm examiner; see also Subject-matter expert
    expertise of, 14
    role in investigation, 12, 14
Firearm expert defined, 8–9
Firearm manufacturing history, 163–164
Firearm mechanical operation, 164–165
Firearm receiver, 261–262
Firearms
    air gun not legally a firearm, 226
    chassis conversions, 234–235
    definition, 15
    extreme environmental exposure, 320–321
    functional modifications, 236–237
    improvised, 221–222
    inert, 262
    for less-lethal munitions, 94
    manufactured by multiple manufacturers, 274–294
    manufacturers, 262–269
    mechanical modifications, 236–237
    purpose of, 16
    safe handling, 1–2
    subject-matter experts, 8
    unserviceable, 262
    U.S. production statistics, 165–166
    without serial numbers, 269, 307–308
Firearms examiner, see Subject-matter expert
Firearms identification methodology, 300–302
Firearms Imports and Exports (F.I.E.), 267–268
Firing conditions, 16
Firing-pin issues, 3
    breakage, 28
    impression on cartridge casings, 7–8
    slam fires, 188–189
Firing-pin safety, 177
    pistols, 181–182
Firing train, 165
Five-seveN pistol, 24
Fixed ammunition, 33
    functional unit, 46
F.J. head stamp, 147
Flare guns converted to firearms, 223
"Flash bang" devices, 95
Flash hiders, 215
Flash holes, 20, 61
Flashlight guns, 222
Flash powder, 63
Flat tops, 283
FLEA, 143
Flechettes, 93
Flintlock as archaic ignition system, 19

Fluted chamber, 54
FME head stamp, 130
FM EP head stamp, 130
FMJ/LEINAD nameplate, 274
FN 303 chemical agent delivery system, 95
FNB head stamp, 128, 133
F.N.C.M. head stamp, 128
FN FAL manufacturers, 293–294
FN FNC manufacturers, 293–294
FN head stamp, 128
FNH head stamp, 128
FNM head stamp, 143
FNP head stamp, 144
FN SCAR rifles, 22
Force-on-Force, 97
Forcing cone, 177
Foremost, 271
Forensics aspect of comparative ballistics, 7–8
Forging firearm components, 163
Format of serial numbers, 304
Foster-style slug, 91–92
Fragmentation of projectiles, 73, 74
Frame and slide combinations, 277
Francotte, May et Cie, 128
Frangible projectiles, 80–82
    breaching applications, 81
    low-energy, 143
FREMEL head stamp, 129
Fremel Manufacturing, 129
French ammunition, 133–134
French proof marks, 333
F R head stamp, 128
Frozen firearms, 320–321
Frye Standard, 11–12
FS head stamp, 145, 146
FTCI head stamp, 134
Full-metal-jacketed projectiles, see Total metal
    jacket (TMJ)
Fulminate, 60
Functional modifications of firearms, 236–237
Functional *vs.* operable, 15
Furniture, 163
fva head stamp, 134
FX cartridge markings, 96–97

G

G+A head stamp, 147
G.3, see under Heckler & Koch
G.43 gas-piston system, 21
ga head stamp, 134
GA head stamp, 147
GALAND head stamp, 134
Galil rifle
    AK-pattern firearm, 288
    based on Kalashnikov, 25
Gallery gun, 191

Garand's M1 rifle, see M1 rifle
Gas charging of air guns, 225
Gas impingement system, 20–21
    in AR variants, 284
Gas operation, 20
Gas-piston system, 20–21
    in AR variants, 284
    G.43, 21
Gas trap noise suppression, 209–210
Gatling guns, 218–220
Gauge in shotgun shell specification, 44–46
GAUPILLAT head stamp, 134
GBF head stamp, 147
GB head stamp, 147
GCA, see Gun Control Act (GCA)
G.C.D. head stamp, 135
GD head stamp, 135
GECADO-G.C. head stamp, 135
GECO head stamp, 135
Geco head stamp, 135
Ge head stamp, 135
GE head stamp, 135
GEL head stamp, 135
General Dynamics
    frangible projectiles, 81
    ordnance producer, 34
General Dynamics and Tactical Systems of Canada,
    129
    Simunition, 96–97
German ammunition, 134–135
German ordnance codes, 328
German proof marks, 333–334
GEVELOT head stamp, 134
GFL head stamp, 138
G.F.L. head stamp, 138
G.F.L head stamp, 138
G F L head stamp, 138
GG (intertwined) head stamp, 134
G.G & Cie head stamp, 135
G.G. & Co. head stamp, 135
G.G.C. head stamp, 135
GGG head stamp, 139
G head stamp, 134, 135, 136, 140, 147
Gilded jacket, 71–72
Gilding metal washes, 99
Glaser Safety Shot, 83
Glaser Safety Slug, 82–83
Glock 17
    FX, 96–97
    polymer framed, 28–29
Glock Auto Pistol, 107–108
Glock handguns emulated, 185–186
Glock pistols modification for fully automatic firing,
    251–252
Glock "Safe Action," 185
Glove guns, 222
GMx head stamp, 134

Goddard, Calvin, 330
Gold-colored primer, 149
Gouging of serial numbers, 313
GPC head stamp, 136
G.P. head stamp, 134
Grease Gun, 264
Greek ammunition, 135–136
Green Line ammunition, 81
Greenshield ammunition, 81
Grenade launchers, 94
    as destructive devices, 205
Grenade-launching cartridges, 88
Grenades
    as destructive devices, 205
    rifle-launched, 88
G.R. head stamp, 127
GR head stamp, 127
Grinding serial numbers, 314
Grip safety, 183–184
Groove function, 49
Grooveless cartridges, 49
G.S.F head stamp, 134
gtb head stamp, 134
G T head stamp, 134
Guard Dog cartridges, 74–75
Guatemalan ammunition, 136
Gun Control Act (GCA), 167
    import markings, 323–325
    serial number requirement, 303
Gun cotton, 64
Gunpowder; see also smokeless powder
    formulations, 37
Gunsmiths
    fine adjustments by, 236–237
    vs. armorers, 12–14
Gyro Jet cartridge, 90

H

H&K, see Heckler & Koch
H&R, see Harrington & Richardson
HAEC, see Homicho Ammunition Engineering
        Complex (HAEC)
HAEC head stamp, 133
Hague Convention prohibition of hollow-point
        projectile, 71
    Open-Tip Match (OTM) projectile, 75
HA head stamp, 132
ha head stamp, 134
Half-moon clip, 173
ham head stamp, 134
Hammer, spurred, 18
Hammer action, 17
Hammer and sickle firearm markings, 326–327
Hammer as indication of revolver action, 177
Hammer-block safety, 176
Hammer cock positions, 175

Hammerless designs, 17
Hammer "push off," 176
Hand-cranked rapid-fire mechanisms, 218
Handguns
    caliber interchangeability, 107
    calibers used in rifles, 108
    cartridges vs. rifle cartridges, 108
    classification as, 167–168
    in criminal activity, 36
    persistence of designs, 15–16
    stock attachments, 197
    vs. long gun classification, 198–199
Handler error with revolvers, 172
Handling in investigative context, 2–6
Hand loads, 104–106
Hang fires, 65, 188
Hardened steel projectiles, 102
Harrington & Richardson, 266
    automatic shell ejection system, 30
    revolvers, 173
HASAG head stamp, 135
has head stamp, 134
Haskell, 28
Hathcock, Carlos, 31
HB head stamp, 143
HCC head stamp, 161
Head stamps, 115ff
    alphabets used, 116
    cartridge markings, 110
    counterfeit, 116–117
    elements of, 115–116
    German ordnance codes, 328
    information content, 115
    manufacturer abbreviation and caliber, 109
    producer vs. inspector, 138
    U.S. ordnance plant marks, 121
    U.S. producer marks, 121–123
"Heavy ball" projectile, 68
Hebrew characters in head stamp, 137
Heckler & Koch
    caseless ammunition design, 59
    commonly converted to fully automatic firing,
        242–244
    G.3 manufacturers, 291–293
    importation of products, 242
    modular weapons system, 298
    rifles, 22, 24–25
    submachine gun, 109
    USP series, 18
    USP Tactical 45 pistol, 25
    V P-70 series pistols, 28
Heizer Defense, 187
Helwan handgun, 132
Hermann Weihrauch, 268–269
Herstal Group, 127
HERTER'S head stamp, 145
HG head stamp, 143

HG - NTN head stamp, 148
hgs head stamp, 134
H head stamp, 127, 128, 135
High-rate-of-fire weapons use of electric priming,
    65
High Standard, 266–267, 271
High velocity projectiles, 79
Hijacking defensive ammunition, 82–83
Hinged action, 30–31
Hi Point pistols, 28
Hi Power, 179
    mechanism of, 25
Hirtenberger, 127
    cartridges, 43
History of firearm manufacturing, 163–164
H K & C head stamp, 127
hla head stamp, 134
hlb head stamp, 134
hlc head stamp, 134
hld head stamp, 134
hle head stamp, 134
HL head stamp, 132
HN head stamp, 148
HO. & HO. head stamp, 147
Hollow-cavity nose projectile, 71–72
Hollow point, see Controlled-expansion projectiles
Homicho Ammunition Engineering Complex
    (HAEC), 133
Hornady Magnum Rimfire, 54
Hornady Manufacturing
    colored polymer inserts in ammunition, 110
    steel casings, 54
Hornady TAP projectile, 73
Hornady V-Max, 109
Hornet cartridge, 62
Howitzers as destructive devices, 205
HP head stamp, 127
HR head stamp, 127
htg head stamp, 134
HU head stamp, 135
Hulls, 44
Hungarian ammunition, 159
Hungarian proof marks, 333–334
Hunting
    influence on firearm popularity, 36
    semijacketed cartridges, 98
Hunting cartridges, 151
HXP head stamp, 136
Hy-Score ammunition, 140

I

Iberia Firearms, 28
ICI head stamp, 148
Identification of cartridges, 118, 123–125
Identification of firearms, 300–302
Identifying markings on cartridges, 110

IDF, see Israeli Defense Forces (IDF) markings
Ignition systems, 60–66
    archaic, 19
I head stamp, 127, 140
II head stamp, 150
III head stamp, 127
IK head stamp, 161
IMBEL head stamp, 128
IMG head stamp, 136
IM head stamp, 131
IMI, see Israeli Military Industries (IMI)
IMI head stamp, 126, 137
Imperial ammunition, 129
IMPERIAL head stamp, 129
Import markings on firearms, 323–325
Imports of firearms into the U.S., 166
Improvised explosive devices, 63
Improvised firearms, 221–222
Inadvertent discharges of revolvers, 171–172
Indian ammunition, 136
Indian revolvers, 173
Indirect gas-action system, 21–23
Indonesian ammunition, 136–137
Indoor gun range ammunition, 61, 80
INDUMIL marking, 131
Industria de War Material (IMBEL), 128
Industrial Society Burkina Arms and Ammunition,
    128–129
Industria Militar Colombia, 131–132
Inert cartridges, 86
Inert firearms, 262
Inert suppressors, 213
Inglis marking, 258
Ingrams, 241
Inked shot shell markings, 143
Integrated cartridge, 33
Internal magazines, 233
International Cartridge Corporation frangible
    projectiles, 81
Intratec pistols modified for fully automatic firing,
    250–251
Investigative procedures
    accidental revolver discharges, 171–172
    firearms discarded in harsh environments,
        320–321
    handling appropriately, 2–6
    photographic procedures, 320
    projectile masses, 101
    serial number restoration, 319–320
    suppressor suspected, 214
Investment casting, 163
IPJ proof mark, 330
Iranian ammunition, 137
Irish ammunition, 137
Iron armor piercing projectiles, 78
IRS serial number prefix, 309
Israeli ammunition, 137–138

Israeli Defense Forces (IDF) markings, 327
Israeli Galil rifle, see Galil rifle
Israeli Military Industries (IMI)
    ammunition, 138
    cartridges, 43
    M855 cartridges, 68
    subsonic cartridge, 88
Israeli UZI submachine gun, see UZI submachine
    gun
Israeli Weapons Industries Desert Eagle handguns,
    22–23
Italian ammunition, 138
Italian firearms manufacturers, 264
Italian proof marks, 335
Iver Johnson, 265–266
IVI head stamp, 129
IWI ammunition, 138
IWK head stamp, 135

J

Jacketing, see Projectile jacketing
Jamison International lathe-turned projectiles,
    102–103
Jap ammunition suffix, 139
Japanese ammunition, 138–139
Japanese use of wooden projectiles, 103–104
J.C. Higgins, 269
J.C. Penney firearms, 271, 273
Jennings, Bruce, 27–28
Jennings Firearms, 27–28
Jersey Arms Works, 274
J head stamp, 147
Jimenez Arms, 28
Jing An, 130
John Garand, 21
Joint Stock Company, 152–154
jtb head stamp, 134
Judge as gatekeeper of expert testimony, 11–12
Jungle carbine, 193

K

K-33 head stamp, 147
K34 head stamp, 147
K.50 head stamp, 147
K53 head stamp, 147
K55 head stamp, 147
K56 head stamp, 147
K.57 head stamp, 147
Kalashnikov, Mikhail, 21
Kalashnikov-pattern firearms, 285–291; see also
    AK-pattern firearms
Kalashnikov rifles, 22; see also AK series rifles
kam head stamp, 134
Kanji firearm markings, 327
KARCHEN head stamp, 134

Karl Gustav, 65
Kenyan ammunition, 139
kfg head stamp, 134
KF head stamp, 136
K-frame revolver, 178
K head stamp, 151
Kilgore Ammunition Products frangible projectiles,
    81
Kiwi ammunition, 140
Klimovsk ammunition plant, 154
Kmart firearms, 271
K.M. Polican, 125
"Knee knockers," 95
Knife gun, 223
KOF head stamp, 139
Korean head stamp, 141
Krag cartridge, 41–42
krb head stamp, 134
Krinkov, 289
KTW Co. Teflon-jacketed projectiles, 100
Kurz cartridges, 43
kye head stamp, 134
KY head stamp, 148
kyn head stamp, 134
KYNOCH head stamp, 147
kyp head stamp, 134

L

Lake City Ammunition Plant, 34
    head stamps, 118
Lakeside, 271
Lansco, 27–28
Lapua, 133
Lapua ammunition, 141
LAPUA head stamp, 133
LAR, see Light Automatic Rifle (LAR)
Large bore handguns, 23
Laser engraving of serial numbers, 310–311
Lathe-turned projectiles, 102–103
L. BEAUX & Co. head stamp, 138
Lead
    banned in shot, 103
    extra-hardened shot, 103
    projectiles, 102
Lead-free
    priming compounds, 54, 61
    projectiles, 102–103
Lee Enfield bolt-action rifle, 31–32
Legal standards, 11–12
    subcaliber insert, 89–90
Lesmoke, 65
Less-lethal ammunition, 38, 93–96
    in investigation efforts, 95–96
    purpose of, 95
Lethal ammunition, 37–38
Lever-action firearms, 31

Lever-action rifles, 190
Lever-action shotguns, 201
L head stamp, 141
Light Automatic Rifle (LAR), 293
Lightning Link, 247–248
Lightweight Polymer Cased Ammunition, 58
Lightweight Small Arms Technology (LSAT)
        Program, 57
LI head stamp, 146–147
"Limp wristing," 2
Linear coupling device, 209
Lion head stamp, 157
Lithuanian ammunition, 139
Live or spent ammunition terminology, 38
Ljungmann AG42 gas impingement system, 20
lkm head stamp, 134
Loaded-chamber indicator, 182–183
Loading data gives range of projectile diameter, 40
Loading gate, 173
Loading of a firearm, 16
Loading position, 175
"Lock up," 170
Long guns
        definition, 189
        vs. handgun classification, 198–199
Long-recoil operation, 26–27
LSAT Program, 57
Luballoy jacketing, 99
Luger, George, 179
Luger development, 179
Luger pistol operation, 24–25
Lunch-box guns, 307
LVE head stamp, 155

M

M&P series pistols, 29
M&S head stamp, 148
M-134 "mini-gun," 65
M14 and variants manufacturers, 278–280
M16 barrel failures, 282
M1A and variants manufacturers, 278–280
M1 carbine
        modification for fully automatic firing, 254–255
        multiple manufacturers, 274
M1/M1A rifles, 25
M1 rifle, 21
M4 designation, 194
M74 cartridge, 71–72
M74 projectile, 72
M80 cartridge, 139
M855A1 cartridge, 69–70
M855 cartridges, 68
        MK 318 replacement of, 76–78
M856A1 tracer cartridge, 70
M903 cartridge, 89
M962 cartridge, 89

M995 armor piercing cartridge, 70
Maadi handgun, 132
Machine guns, 215–221
        avoiding "cook off," 238
        classification as, 245, 246–248, 254
        conversions, 237–239
        definition, 215
        factory markings, 248–249
        field testing, 239–240
        operating principle, 179
Machine pistol, 217
Machine Pistole, 217
MAC-pattern firearms
        commonly converted to fully automatic firing,
                240–241
        multiple manufacturers, 274
MAC pattern firearms based on UZI, 24
Magazine release in handguns, 186
Magazines, 227–231
        for bolt-action firearms, 31
        capacity limits imposed in the U.S., 230
        components of, 230–231
        definition, 227
        internal, 233
        serial numbers on, 304
        shared architecture, 229
        tubular, 233
Magazine safety, 181
Magnaflux for reading obliterated serial numbers,
                318–319
Magna grips, 171
Magnetic Particle Inspection (MPI) fore reading
                obliterated serial numbers, 318–319
Magnum Research subsidiary of Israeli Weapons
                Industries, 22–23
Magnums in shot shell designations, 44–45
Magnum terminology, 39
Magtech ammunition, 128
        projectile masses, 101
Magtech Recreational Products (MRP), 128
Mainspring in revolver mechanism, 18
Makarov and ACP cartridge interchanges, 107
Malaysian ammunition, 139–140
Malfunction caused by riding the slide, 186
Malfunction types, 2–3
MAL head stamp, 139
Mannlicher bolt-action rifle, 31–32
Manually operated firearms, 30–32
MANUCAM head stamp, 129
Manufacture Camerounnaise de Munitions, 129
Manufacture de Stung Chral, 129
Manufacturers of firearms, 262–267
        Colt copies, 277
        firearms manufactured by multiple, 274–294
Mark 12 Special Purpose Rifle (SPR), 78
Marking cartridges, 96–97
Mark VII projectile, 70–71

Marlin Firearms, 266, 271
Marlin/Glenfield Model 60 modification for fully
    automatic firing, 258
Marlin Model 1895 lever-action shotgun, 31
Masterpiece Arms, 274
MatchKing ammunition, 76
Matchlock as archaic ignition system, 19
Mauser, Paul, 179
Mauser 98
    bolt-action rifle, 31–32
    persistence of design, 15–16
Mauser big-bore concept, 52
Mauser cartridge, 48
Mauser development, 179
Mauser HSc operation, 24
"The Maxim," 179
M. B. Associates Gyro Jet, 90
McMillan Group International solid brass
    projectiles, 102
MCS, see Modular Combat Shotgun (MCS)
Mechanical modifications of firearms, 236–237
Mechanical restoration of serial numbers, 315–316
MEN
    armor-piercing projectiles, 79
    owned by Brazilian manufacturer, 128
MEN head stamp, 135
Mercury fulminate, 60
Metal detectors
    erroneous claims of undetectable firearms, 185
    searching for casings, 53
Metal jacketing, 99
Metal projectiles, 102–103
Metal tins of Russian ammunition, 149–150
Metric ammunition labeling, 42–43
MF2 head stamp, 126
MF head stamp, 126
MFS head stamp, 159
MFT head stamp, 145
MG head stamp, 126
M head stamp, 144
MH head stamp, 126
Micro pistol, 22–23
Microsoft Windows (similar) head stamp, 152
MI head stamp, 134
Military cartridge labeling is country-specific, 44
Military model designations, 322–323
Military rifle copies, 192
Military weapon copies, 195–196
Mines as destructive devices, 205
Miniature ballistic rockets fired from handgun, 90
Misfeed malfunction, 3
Misfire malfunction, 3
Mismatched components, 277
MISR head stamp, 132
Missiles as destructive devices, 205
MJB head stamp, 126
MJ head stamp, 126

MK ammunition, 76
MKE head stamp, 146
MK.I. marking, 258
MM. head stamp, 129
M M head stamp, 143
MMM head stamp, 144
MNAM head stamp, 140
Model designations, 321–323
Model numbers confused with serial numbers, 304,
    306
Moderators, 206
Modifications of firearms, 233–240
    into contraband, 6–7
Modular casings, 58
Modular Combat Shotgun (MCS), 299
Modular weapons systems concept, 294–295
Monolithic Rail Platform (MRP), 294
Montgomery Ward firearms, 271, 272
Moon clip, 107, 173
Moroccan ammunition, 140
Mosin Nagant
    bolt-action rifle, 31–32
    cartridges for, 49
Mossberg, 194, 200, 271
MP-25, 27
MP designation, 217
MPI, see Magnetic Particle Inspection (MPI) fore
        reading obliterated serial numbers
mpr head stamp, 134
MQ head stamp, 126
mrb head stamp, 134
MR head stamp, 134
MRP, see Monolithic Rail Platform (MRP)
MRP head stamp, 128
MS head stamp, 126
M T head stamp, 145
Muffler defined, 206
Multiple manufacturers of firearms, 265
Multiple serial numbers, 308
mus head stamp, 143
Muzzle brake, 88, 215
Muzzle discipline
    hang fire handling, 65
    hang fires, 188
MW head stamp, 126
MZINGA head stamp, 146

N

Nammo AS cartridges, 57
Nammo Group, 133
Nammo Lapua, 141
Natec, 57
National ammunition preferences, 51–52
National crests on firearms, 326–327
National Firearms Act (NFA), 167
    categories, 222

National Tracing Center (NTC), 309–310
Nation-specific cartridge head stamps and
        manufacturing practices, 125ff; see also
        specific countries
NATO
    ammunition standards, 34
    cartridge markings, 110
    cartridges, 43
    cross head stamp, 134, 138, 139, 146
    head stamps, 117–118
Naval guns as destructive devices, 205
Nazi firearm markings, 327
Nazi proof mark, 258
ndn head stamp, 134
Needle gun, 50
Netherlands ammunition, 140
New Service revolver caliber interchangeability, 107
"New York" trigger, 237
New Zealand ammunition, 140
Nexter/GIAT, 128
NFA, see National Firearms Act (NFA)
N-frame revolver, 107, 178
nfx head stamp, 134
Nickel-plated casings, 54
Nielson device, 209
Nigerian ammunition, 141
Nitrocellulose, 64
Nobel, Alfred, 64
NOBEL head stamp, 148
Noise sources, 207
Noise suppressors, 206ff; see also Suppressors
Nonexpansive projectiles, 68
Nonfixed ammunition, 46
Nonfunctional suppressors, 213
Nonjacketed projectiles, 101
Non-lead projectiles, 102–103
NORINCO, see North China Industries
        Corporation (NORINCO)
NORINCO head stamp, 130
norma head stamp, 145
NORMA head stamp, 145
North American Ordnance Corporation Teflon-
        jacketed projectiles, 100
North American Treaty Organization, see NATO
North China Industries Corporation (NORINCO),
        32, 130
North Korean ammunition, 141
Norwegian ammunition, 141
Norwich Arms, 269
Nosler, 110
Nosler Accubond projectile, 73
Nosler Partition projectile, 74
Novosibirsk Low Voltage Equipment Plant,
        154–155
NPA head stamp, 136
NTC, see National Tracing Center (NTC)
NWM head stamp, 140

Nyclad projectiles, 100
Nylon 66 rifle, 164
Nylon/tungsten lead alternative, 103

O

OA head stamp, 126
Obliteration of serial numbers, 7, 311–316
Observation cartridge, 84
OC, see Oleoresin capsicum (OC)
Oerlikon Machine Tools, 139
OFN head stamp, 141
OFV head stamp, 136
Ogive, 73
O head stamp, 140
Oleoresin capsicum (OC), 95
Olin Mathieson Chemical Corporation caseless
        ammunition design, 58
Olive leaf firearm markings, 327
OLYMP head stamp, 136
Open-bolt designs
    field testing, 239–240
    fully automatic firing conversion, 238
    modification for fully automatic firing, 248
Open-Tip Match (OTM) projectile, 75–78
    colored inserts, 110
    legality of, 75
Operable vs. functional, 15
Operational sequence, 19–20
Opinion in rules of evidence, 8–9
Orange labeling for less-lethal platforms, 94
Orange muzzle cap on AEGs, 226
Orange tips on firearm analogs, 234
OTM, see Open-Tip Match (OTM) projectile
O/U (Over/Under), 200
Outsourced ammunition production, 105
oxo head stamp, 134
oyj head stamp, 134

P

P (circled) proof mark, 330
PA head stamp, 160
Pain compliance of target, 95
Paintball guns, 225–226
Paintball gun suppressors, 215
Paint munitions, 95
Pakistani ammunition, 141–142
Palm trees head stamp, 143
Paper-hulled shot shells, 55–56
Parabellum cartridge, 42–43
Paramilitary configuration, 192
Parkerizing, 163
Patent information for firearm identification, 329
Patent infringement claims on polymer-framed
        handguns, 185–186
Patenting of ammunition, 50–51

PAVA/OC chemical munition, 95
P cartridges, 101
PCA Spectrum, 57
P. code prefix, 134
PCP, 58
PDW, see Personal Defense Weapons (PDWs)
Pedaled projectiles, 89
Pederson hesitation lock system, 25
Peening to obliterate serial numbers, 314
Pelargonic acid vanillylamide/oleoresin capsicum
        chemical agent, 95
Pellet guns, 225
    dangers of, 227
Penetrator ammunition, 130
Penetrator projectile construction, 68, 69–70
Pen pistol, 222
Pentaerythritol tetranitrate (PETN) explosive
        ordnance, 84
Percussion cap, 49–50
Percussion detonation, 19
Percussion-fired weapons obsolete, 51
Permanent International Commission for Firearms
        Testing (CIP), 101
    standards for pressure loads, 106
Persistence of handgun designs, 15–16
Personal Defense Weapons (PDWs), 109–110, 217
Peruvian ammunition, 142
PETN, see Pentaerythritol tetranitrate (PETN)
        explosive ordnance
PG head stamp, 161
P head stamp, 140, 143, 144, 145
Philippine ammunition, 142
Phosphate, 163
Photographic images for firearm identification,
        300–302
Picatinny rail, 295
PINDAD head stamp, 137
Pin stamping of serial numbers, 310–311
Pipe (smoking) guns, 222
Pipe bombs using black powder, 63
Pistol calibers for submachine guns, 216–217
Pistolized versions of rifles, 23
Pistol receivers, 15
Pistols
    definition, 178
    machine pistols, 217
    noise suppression of, 212–213
    polymer framed, 27–30
    recoil operated, 23
    safety systems, 180–186
PJJ head stamp, 132
pjj head stamp, 134
Plastic hulls, 56
Plastic Man (PM) oeuvre cartridge, 56
Plastic short-range training ammunition (PSRTA),
        57
Plastic Training (PT) cartridge, 56

Plating of cartridges, 99
Plug-and-play reconfiguration, 237
PM, see Plastic Man (PM) oeuvre cartridge
PMC head stamp, 144
PM head stamp, 135
PMP head stamp, 143
POF head stamp, 142
Polish ammunition, 159–160
Polymer-cased cartridges, 56–58
Polymer firearm components, 163–164
Polymer framed pistols, 185–186
Polymer framed systems, 27–30
Polytech Ammunition Company, 57
Portable firearms, 261–262
Portuguese ammunition, 142–143
Possession of a firearm
    categories of crimes, 5–6
    possession of ammunition concurrent, 8
Powder metal projectiles, see Frangible projectiles
Powder ring indicating firing, 4–5
PP head stamp, 161
PPU Green Line ammunition, 81
PPU head stamp, 161
PP-YU head stamp, 161
Pre-fragmented projectile, 83
Premier shot shells, 93
Pressure load safety considerations, 106–107
Primer issues, 3
Priming, electric, 65–66
Priming compounds, 60, 61
    hardness issues, 172
Production statistics for firearms, 165–166
Prohibition of possession of ammunition, 46
    paper shells, 56
Projectile jacketing, 98–101
    composite materials, 100
Projectile mass, 101
Projectiles, 66–98; see also specific types
    fragmentation of, 74
    impact effects, 66
    materials for, 101–104
    round-nose, 70–72
    spitzer, 70–72
Proof cartridges, 101
Proof marks, 327, 329–336
Propellants, 60–66
Proprietary cartridges, 41
Prvi Partizan, 81, 98
PS head stamp, 144, 158
PSRTA, see Plastic short-range training
        ammunition (PSRTA)
PT, see Plastic Training (PT) cartridge
Pump action, 191
Pump-action firearms, 32
Pump action shotguns, 200–201
Punching to obliterate serial numbers, 314
Purple ring on cartridge casings, 149

Push-button safety
    in long guns, 189
    in pistols, 164
Pyrodex, 85

Q

Quick Response (QR) coded information on
    firearms, 311

R

RAI head stamp, 146
Ranger, 269
Rate of fire of machine guns, 216
Raven Arms, 27–28
Rebated rim casings, 49
Rebounding-hammer safety, 175–176
Receiver firearm frame, 13, 261–262
    classification as firearm, 293–294
    two-piece, 261
    two weapons from single receiver, 262
Reciprocating slide, 179
Recoil booster, 209
Recoilless rifles as destructive devices, 205
Recoil operation, 23
    short and long, 25
    spring mechanisms, 23
    vs. blowback system, 25–27
Red casings, 147
Red ring on cartridge casings, 149
Reduced-power training cartridges, 96
Refinishing firearms, 234
Regulation of firearms by type, 167
Release latch, 173
"Release trigger," 237
Relic classification, 197
Reloads, 104–106
Remanufactured ammunition, 105
Remington 700 sniper rifle, 31–32
Remington Arms
    Accelerator cartridge, 89
    duplex shot shells, 93
    Etron X, 66
    frangible projectiles, 81
    plastic shot shells, 56
    solid copper projectiles, 102
    subsonic cartridge, 88
Remington/Bushmaster ACR, 22
Remington cartridge interchangeability, 106–107
Remington Jet cartridge, 62
Remington Model 10 shotgun, 32
Remington Modular Combat Shotgun (MCS), 299
Remington Peters duplex shot shells, 93
Remington shotguns chamber variations, 110
Renumbered firearms, 308–309
REP proof mark, 330

Restricted cartridge sales, 109
Restrike capability, 18
Retarded blowback system, 25
Revelation, 271
Reversed trigger behavior, 237
Reverse engineering of firearms, 165
Revolvers, 168ff; see also specific models
    action type, 177–178
    caliber interchangeability, 107
    cartridge capacity, 168
    characteristics of, 168
    cylinder rotation direction, 169–170
    definition, 168
    frame, 261
    long guns, 172
    manufacturers, 169
    mechanical action, 170–171
    naming of models, 169
    noise suppression of, 212–213
RG head stamp, 148
R.G. head stamp, 148
RG Industries, 268
R head stamp, 148
"Ride the slide," 186
Rifle cartridges vs. handgun cartridges, 108
Rifled barrel ammunition matching, 39
Rifled slugs, 91–92
Rifle grenades, 88
Rifles, 189ff
    caseless ammunition design, 59
    copies of military rifles, 192
    definition, 189–190
    drillings, 203
    with handgun calibers, 108
    modified weapons, 199
    receivers, 261
    short-barreled, 196–199
Rifling twist rate modifications, 282
RIGB head stamp, 147
Rim designs, 48–49
Rim-fire head stamps, 117
Rim function, 49
Rimless casings, 49
Ring head stamp, 141
Rio Ammunition, 128
RL head stamp, 148
R. L. head stamp, 148
RMC head stamp, 143
R.M head stamp, 135
R.M.S. head stamp, 135
RNAD head stamp, 148
Rocket launchers as destructive devices, 205
Rock salt load in shot shells, 93
ROFB head stamp, 148
Rohm, 268
Roller-lock bolt system, 24–25
Roller marking of serial numbers, 310–311

Romanian ammunition, 160
RORG head stamp, 148
Rossi Circuit Judge, 172
Rotating bolt systems, 22–23
Round-nose projectiles, 70–72
Round term ambiguous, 38
Royal Ordnance Factory, 68
RPA head stamp, 142
RPB nameplate, 274
R P head stamp, 141
RPR head stamp, 160
R.R.Co. CAN head stamp, 129
R.R.Co. head stamp, 129
RS head stamp, 134
R T head stamp, 145
R T P head stamp, 141
RUAG Ammotec Group; see also Dynamit Nobel
    Austrian manufacturer, 127
    frangible projectiles, 81
    polymer-cased cartridges, 56
Rubber-ball shells, 94–95
Rubber bullets, 94
Rubber-finned projectiles, 95
Rubber projectiles, 94
Ruger firearms modification for fully automatic
    firing, 256, 258
Rules of Evidence, see Federal Rules of Evidence
Runaway gun, 258
Russian Federation ammunition, 148–150,
    151–152
Russian firearms manufacturers, 264–265
Russian proof marks, 335
RWS; see also Dynamit Nobel; RUAG Ammotec
    Group
    frangible projectiles, 81
    subsonic cartridge, 88
RWS/GECO head stamp, 135
R.W.S. head stamp, 135
RWS head stamp, 135
R.WS head stamp, 135
RY head stamp, 134

S

S&B head stamp, 157
SA, see Single-action
SAAF head stamp, 143
SAAMI, see Sporting Arms and Ammunition
    Manufacturers Institute (SAAMI)
Saboted Light Armor Piercing (SLAP) cartridge,
    34, 89
Sabots, 89–90
Sabot slug, 92
Sabot-type projectile, 79
SA/DA firing modes, 18
Safari-style arms have smooth bore, 44
"Safe Action," 28–29

Safe-action systems, 185–186
Safe handling, 1–2
Safety as primary consideration, 1–2
Safety designs, 19
    Colt 1908 Vest Pocket, 27
    Glock methods, 28–29
Safety devices, 165, 175–177
    pistols, 180–186
    reliance upon, 2–6
    for revolvers, 175–176
Safety notch, 175
Safety Slug, 82–83
Safety systems for long-guns, 189
Safety with firearms with extreme environmental
    exposure, 320–321
SA head stamp, 144
Saiga guns, 202, 291
Sako Finland, 133
SAKO head stamp, 133
SAM head stamp, 143
Samson ammunition, 138
SAO (Single-action-only), 17
S. A. Paul Schraff Bruxelles, 128
S.A.T. head stamp, 133
S A T head stamp, 133
SAT head stamp, 133
Saturday night specials, 163, 268
Saudi Arabian ammunition, 143
Savage Arms, 267, 271
Savage cartridge, 41
SAW, see Squad Automatic Weapon (SAW)
"Sawed off shotguns," 203–204
SB193 cartridge, 109
SB head stamp, 144, 157
SBP head stamp, 157
SB-T head stamp, 144
Schöenbein, Christian Friedrich, 64
Schwarz, Berthold, 63
Scottish ammunition, 143
Scratching serial numbers, 313
Screw-delayed blowback system, 25
Sear component, 165
Sears & Roebuck firearms, 269–270
Second Model Hand Ejector revolver caliber
    interchangeability, 107
Second strike capability, 18
Seecamp pistols, 25
Seeker ammunition, 142
Self-contained cartridge, 33
Self-inflicted wounds and stovepipe malfunction,
    2–3
Self-loading firearms operation, 20
Self-loading rifles, 191–193
Sellier & Bellot, 55
    coated projectiles, 100
    owned by Brazilian manufacturer, 128
    projectile masses, 101

Semiautomatic firearms, 17
   belt-fed, 233
   conversions to fully automatic firing, 216, 239
   conversion to automatic, 218
   double-action, 18
   field testing, 239–240
   gas-operated pistols, 22–23
   handguns, 23
   modified copies of classic weapons, 195–196
   pistols, 178ff
   rifles, 191–193
   shotguns, 201–203
Semijacketed hunting cartridges, 98
Semijacketed projectiles, 98–99
Semi-rimmed casings, 49
Semi-wad cutter projectile, 78
SEN head stamp, 134
Separation of projectile jackets, 99
Sequence of operations, 19–20
Serial numbers, 269
   application of, 310–311
   BATFE as repository, 309
   on components, 307
   confused with model numbers, 304, 306
   department store guns lack, 269
   on firearm parts, 303–304, 307
   firearms without, 307–308
   format, 304
   imported firearms, 311
   locations for, 306–307
   multiple, 308
   new number assigned by BATFE, 308–309
   obliteration of, 311–316
   re-assigned when remanufactured, 308–309
   requirement enacted, 7
   requirement for, 303, 310
   restoration of, 315–320
Set trigger, 17
S F head stamp, 134
SFM, see Société Française Munitions (SFM)
S.F.M. head stamp, 134
SFM head stamp, 134
SF S head stamp, 135
S. G. head stamp, 144
"Shaving lead," 177
S head stamp, 133, 135, 144
Shell casings as main evidence, xv
Shield head stamp, 141
Shooting-scene reconstruction using expert advice,
   8
Short-barreled rifles, 196–199
Short-barreled shotguns, 203–204
Short-range training ammunition (SRTA), 97–98
Short Stop ammunition, 81
Short stroke mechanism, 32
Shot charge, 45–46
Shotgun chokes, 204–205

Shotguns, 199ff
   definition, 199–200
   gauges manufactured, 44
   receivers, 261
   short-barreled, 203–204
Shotgun slugs, 91–92
Shot pellets, 90–91
   materials for, 103
Shot shells
   case head length, 56
   components of, 90–91
   duplex, 92–93
   frangible projectiles, 81
   gauge specification, 44–46
   hull materials, 54–56
   labelling information, 45–46
   length designations, 44–45
   load variations, 93
   in revolvers, 178
   Scottish, 143
   tracers, 93
   triplex, 92–93
Shot size, 45–46
Shoulder stocks in defining a weapon, 189
SIBAM head stamp, 128–129
Side plate as receiver, 261
Sierra MatchKing OTM projectile, 78
Sig 556, 22
Sigma series handguns, 29, 185
SIG Sauer decocking lever, 181
SIG Sauer P230, 24
SIG Sauer pistols, 256
Silencer defined, 206
Silver in lead projectiles, 102
SIM head stamp, 138
Simonov Carbine Self-Loading (SKS) modification
      for fully automatic firing, 252–253
Simulation of automatic firing, 259–260
Simunition, 96–97
Singaporean ammunition, 143
Single-action firearms mechanism, 17
Single-action revolvers, 177–178
Single-based gunpowder compounds, 64
Single-shot handguns, 187
Single-shot rifles, 191
Single-use suppressors, 210
SKd head stamp, 135
S.K.D. head stamp, 135
SKS, see Simonov Carbine Self-Loading (SKS)
SKS semiautomatic rifle, 88
Slack trigger action, 28–29
"Slam fire," 32, 188–189
   in closed-bolt weapons, 238–239
SLAP, see Saboted Light Armor Piercing (SLAP)
      cartridge
SLAP cartridges, 34
Slide-action firearms, 32

Slide-action rifles, 191
Slide-action shotguns, 200–201
Slide and frame combinations, 277
Slide-lock lever on pistols, 186
Slovak Republic ammunition, 157–158
Slugs, 91–92
    cut shot, 93
    sabot configuration, 92
SM head stamp, 145
SMI head stamp, 138
Smith & Wesson
    22 series, 23
    Army revolvers, 107
    cartridge designation, 41
    cartridge patent, 51
    M&P series pistols, 29
    magazine cutoff safety, 181
    Model 76 submachine gun, 65–66
    revolver serial numbers, 304, 306
    Sigma restrike capability, 18
    Sigma series handguns, 29
    SW40E, 29
Smokeless powder, 63–65
Smoking pipe guns, 222
Snail drum, 230
Snake Charmer, 187
SNC Technologies, 129
Sobrero, Ascanio, 64
Société Anonyme, 128
Société des Cartoucheries, 128
Société Française Munitions (SFM), 79
SO head stamp, 133
Solid-frame revolvers, 173–175
SOS head stamp, 127
South African AK-pattern firearms, 288
South African ammunition, 143–144
South Korean ammunition, 144
Soviet factory codes, 150–151
Soviet head stamp practice used by Chinese, 130
Soviet SVT 38
    gas-piston system, 21
Soviet weapons, see SVT
Spam cans, 130
Spanish ammunition, 144
Spanish firearms manufacturers, 264
Spanish proof marks, 336
SPAS, see Sporting-purpose automatic shotgun
        (SPAS)
Special Operations Forces Combat Assault Rifle
        (SCAR), 295–297
Speer cartridges, 54
    plastic shot shells, 56
Speer frangible projectiles, 81
Speer Gold Dot projectile, 73
Speer LE Force-on-Force, 97
Spencer rifle, 31
Spent or live ammunition terminology, 38

SP head stamp, 143
Spigot, 88
Spiral casing, 58
Spitfire cartridges, 57
Spitzer projectile, 70–72
Spitzer-type projectile, 43
Sporting Arms and Ammunition Manufacturers
        Institute (SAAMI), 101
    standards for pressure loads, 106
Sporting-purpose automatic shotgun (SPAS), 202
SPR, see Mark 12 Special Purpose Rifle (SPR)
Springfield Armory, 41
Springfield, U.S. Arsenal at, 41
Spurred hammer, 18
Squad Automatic Weapon (SAW), 98
Squib load, 188
Squires Bingham ammunition, 142
S R head stamp, 148
SRTA, see Short-range training ammunition (SRTA)
SS190 series cartridges, 109
SS 77/1 ammunition, 139
SSK subsonic cartridge, 88
Stamping, 163
Stamping of serial numbers, 310–311
Standardized chamber pressures, 106
Star head stamp, 141, 146, 151
Star head stamps, 145
Star of David firearm markings, 327
Star of David head stamp, 158
Starter guns converted to firearms, 223
STAR TRUST head stamp, 144
Steel armor-piercing projectiles, 78
Steel casings, 54
    introduction of, 52
    shotgun, 55
Steel shot pellets, 103
STEN gun modification for fully automatic firing,
        258–259
Steyr Mannlicher M95, 31
Stock attachments to handguns, 197
Stoner, Eugene, 294
Stovepipe malfunction, 2–3
Straight-pull bolt, 31–32
Straw purchasers, 310
Striker-fired mechanisms, 27–28
Stripper clips, 231–232
Subcaliber insert, 89–90
Subject-matter expert; see also Firearm examiner
    publication of unusual findings, 12
    role of, 14–15
Submachine guns, 216–217
Sub-sized ammunition, 20
Subsonic cartridges, 88
Sudanese ammunition, 145
SUD head stamp, 145
Suffix on cartridge designation, 43–44
SU head stamp, 145

suk head stamp, 134
Suppressor-like devices, 215
Suppressors, 206ff
    attachment of, 207–208
    definition, 206
    inert, 213
    integral, 88
    legality of possession, 212
    permanently attached, 208–209
Supreme Elite PDX1 projectile, 73
Survival guns, 203
SVT rifles gas-piston system, 21
svw head stamp, 133
svw MB head stamp, 133
Swartz safety system, 182
Swastika markings, 130
SWD nameplate, 274
Swedish ammunition, 145
Swedish K, 65
Swedish Ljungmann AG42 gas impingement
        system, 20
SW. GOV, head stamp, 145
Swinging link, 25
Swing-out cylinders, 173
Swiss ammunition, 145
Swiss Vetterli 1871 carbine, 17
Swords head stamp, 143
SxS, 200
SYI head stamp, 138

T

TAA head stamp, 146
Tactical Systems frangible projectiles, 81
ta head stamp, 134
TA head stamp, 137
T A head stamp, 144, 145
TA head stamp, 146
Taiwanese ammunition, 146
Talbot cartridge, 41
Tallow in cartridges, 51
Tanzanian ammunition, 146
Taurus International hammerless automatic
        handguns, 29
TC head stamp, 146
TCW head stamp, 153
Teardrop, circle and square head stamp, 154
Tear gas, 95
Teflon-jacketed projectiles, 100
TE head stamp, 134
T E head stamp, 139
Temperature variation method for reading
        obliterated serial numbers, 319
Terminology for ammunition, 38
Textron, 57
Thai ammunition, 146
T head stamp, 144, 145, 150, 153

"The Maxim," 179
TH head stamp, 148
Thompson Center, 108
Thompson Contenders, 108
Thompson submachine gun, 25
Threaded barrels, 214
Thumb piece, 173
THUN head stamp, 145
THV, see Tres Haute Vitesse (THV) projectile
Tin in lead projectiles, 102
Titan, 267
tko head stamp, 134
T.M head stamp, 138
TMJ, see Total metal jacket (TMJ)
Toggle action Luger pistol, 24
Tokarev cartridge, 48
Total metal jacket (TMJ), 98
TOYO head stamp, 139
TPZ head stamp, 153
Tracer ammunition, 83–84
    sabots, 89
    short-range training type, 98
    shot shells, 93
Tracing firearms by serial number, 309–310
Trademark protection, 322
Training ammunition, 38, 86, 96
    wood projectiles, 103–104
Transfer-bar-type safety, 19, 176
Tres Haute Vitesse (THV) projectile, 79
T.R. head stamp, 129
T.R head stamp, 138
Triangle head stamp, 141, 150
Trigger, second, 17
Trigger behavior reversed, 237
Trigger break defined, 16
Trigger disconnector, 189
Trigger overtravel defined, 16
Trigger press resistance, 16
Trigger reset defined, 16
Trigger resistance adjustment, 17
Trigger safety in pistols, 185
Trigger travel defined, 16
Triple-based gunpowder compounds, 64
Triplex shot shells, 92–93
Trophy Bonded Bear Claw ammunition, 76
T S head stamp, 134
T T head stamp, 145
"Tube gun," 259
Tubular magazines, 233
Tungsten
    powder in frangible projectiles, 82
    used in shot pellets, 103
Tungsten carbide
    armor-piercing projectiles, 78
    projectiles, 102
Tungsten-iron shot pellets, 103
Tungsten/nylon lead alternative, 103

Turkish ammunition, 146
Turning bolt, 31–32
T W head stamp, 145
TZ head stamp, 137
TZZ head stamp, 137

U

ua head stamp, 134
UAR head stamp, 132
Ugandan ammunition, 146–147
U head stamp, 140, 143
U intertwined with I head stamp, 144
Ukrainian ammunition, 160–161
Ultrasonic cavitation for reading obliterated serial
        numbers, 318
Ulyanovsk Machinery Plant, 155
Umbrella guns, 222
"Under the star," 171
Undetectable firearms, 185
UN head stamp, 135
Union of Military Industries, 131–132
United Arab Emirates ammunition, 147
United Kingdom ammunition, 147–148
United States Ammunition Company polymer-
        cased cartridges, 57
Unlawful firearm possession categories, 6
Unserviceable firearms, 262
Upper Volta, see Burkina Faso
Uranium armor-piercing projectiles, 78
U.S. Army's Lake City Ammunition Plant, 34
U.S. Army's Lightweight Small Arms Technology
        (LSAT) Program, 57
U.S. firearm production statistics, 165–166
U.S. government
        arsenals, 263–264
        firearms manufacturers, 278–280
        head stamps, 118
U.S. manufacturers proof marks, 330
U.S. military carbines, 194
        Model M1, 274
U.S. military magazines, 228
U.S. military proof marks, 331, 332, 333, 335
U.S. prosecution of firearms offenses, 166–167
uxa head stamp, 134
UZI brand ammunition cartridges, 43
UZI grip safety, 184
UZI modification for fully automatic firing, 258
UZI submachine gun, 24

V

va head stamp, 134
Valmet conglomerate, 133
VALMET head stamp, 133
Van Bruaene Rik (VBR), 128
        projectiles, 79–80

VBR-B head stamp, 128
V C head stamp, 129
VE head stamp, 134
Velet cartridge, 85–86
Velex cartridge, 85–86
VELO-DOG head stamp, 134
VELO-DOG PARIS head stamp, 134
Venezuelan ammunition, 148
VEN head stamp, 148
Very Pistols, 223
Vetterli 1871 carbine, 17
Vetterli military rifle, 62
VFM & CA V, 128
V head stamp, 155, 162
Video images for firearm identification,
        300–302
Viewpoints derived from applications, 1
Visual aids for witness identification of firearms,
        301–302
Voere
        caseless ammunition rifle, 59
        electrically primed cartridge, 66
Von Dreyse needle system, 50
VP-70 handgun, 164
VP proof mark, 330
V PT head stamp, 133
V.P.T. head stamp, 133
VS. head stamp, 134
VS head stamp, 138
Vympel cartridge plant, 152
VZK head stamp, 135

W

W&S head stamp, 148
Wad cutter projectile, 78
Wadding in shot shells, 90–91
wa head stamp, 134
Walking cane guns, 222
Walther P.38 modification for fully automatic firing,
        256–257
Walther PP and PPK
        blowback operation, 24
        multiple manufacturers, 274–275
Warsaw Pact ammunition, 148–150
Warsaw Pact cartridge markings, 110
wb head stamp, 134
WB head stamp, 148
WCC head stamp, 126
WCF, see Winchester Centerfire (WCF)
wc head stamp, 134
wd head stamp, 134
Weapon made from a rifle, 199
Weapon made from a shotgun defined, 204
Webley cartridge, 41
WEBLEY head stamp, 148
Webley revolvers, 173

we head stamp, 134
Western Auto Supply Company firearms, 271,
        273
Western Field, 271
wf head stamp, 134
wg head stamp, 134
W head stamp, 127
Wheel lock as archaic ignition system, 19
wh head stamp, 134
W H P head stamp, 127
Wildcat cartridges, 41
WILKINSON head stamp, 147
Winchester
    part of Herstal Group, 127
    slug/buckshot-loaded shot shell, 93
    subsonic cartridge, 88
Winchester Centerfire (WCF) cartridge designation,
        41
Winchester Model 1887, 31
Winchester Model 1894, 31
Winchester Model 70 sniper rifle, 31
Winchester Model 94, 31
Winchester Model 97 shotgun, 32
Winchester Silver Tip projectile, 73
Wipes in suppressors, 209
wj head stamp, 134
wk head stamp, 134
WOLF head stamp, 155
Wolf Performance Ammunition, 155–156
Wooden projectiles, 103–104
W over W head stamp, 127

WR head stamp, 148
W T head stamp, 145

X

xa head stamp, 134
X head stamp, 127, 132, 140, 158
X K&C head stamp, 127
XM-2010 sniper rifle, 31–32
X-ray analysis of obliterated serial numbers, 317–318

Y

ya head stamp, 134
"Yellow form," 309
y head stamp, 134, 139, 150
Yugoslavian ammunition, 161–162
Yugoslavian proof marks, 336

Z

zb head stamp, 134
ZG head stamp, 138
(z) head stamp, 157
Z head stamp, 157
ZI head stamp, 148
Zimbabwean ammunition, 148
Zip guns, 221
Zombie Max projectiles, 73
ZP over ZH head stamp, 130
zv head stamp, 157

For Product Safety Concerns and Information please contact our EU
representative GPSR@taylorandfrancis.com
Taylor & Francis Verlag GmbH, Kaufingerstraße 24, 80331 München, Germany

www.ingramcontent.com/pod-product-compliance
Lightning Source LLC
Chambersburg PA
CBHW080658220326
41598CB00033B/5250

* 9 7 8 0 3 6 7 7 7 8 3 0 9 *